America's Munitions
1917-1918

REPORT

OF

BENEDICT CROWELL

THE ASSISTANT SECRETARY OF WAR
DIRECTOR OF MUNITIONS

WASHINGTON
GOVERNMENT PRINTING OFFICE
1919

"THE END OF THE WAR.

A GRAPHIC RECORD.

One minute before the hour. All guns firing.

Nov. 11, 1918. 11 A. M.

One minute after the hour. All guns silent.

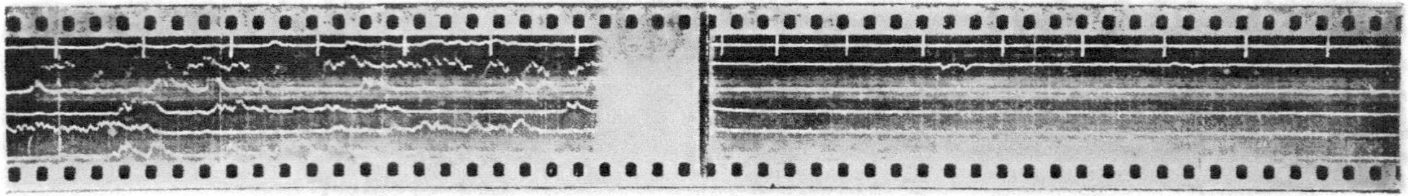

10.58-56 10.58-57 10.58-58 10.58-59 11.01-00 11.01-01 11.01-02 11.01-03

Hour – minutes - seconds

This is the last record by sound ranging of artillery activity on the American front near the River Moselle. It is the reproduction of a piece of recording tape as it issued from an American sound-ranging apparatus when the hour of 11 o'clock on the morning of November 11, 1918, brought the general order to cease firing, and the great war came to an end. Six seconds of sound recording are shown. The broken character of the records on the left indicates great artillery activity; the lack of irregularities on the right indicates almost complete cessation of firing, the two breaks in the second line probably being due to the exuberance of a doughboy firing his pistol twice close to one of the recording microphones on the front in celebration of the dawn of peace. The two minutes on either side of the exact armistice hour have been cut from the strip to emphasize the contrast. Sound ranging was an important means of locating the positions and calibers of enemy guns. A description of these wonderful devices, which were a secret with America and the Allies, is given in Book III, chapter 4.

WASHINGTON, D. C., *December 24, 1918.*

DEAR MR. CROWELL: American munitions production, which for some time has been in your charge, played an important part in the early decision of the war, yet the very immensity and complexity of the problem has made it difficult for this accomplishment to be adequately understood by the public or in fact by any except those who have had occasion to give the matter special study. As the whole people have been called upon to make sacrifices for the war, all the people should be given an opportunity to know what has been done in their behalf in munitions production, and I therefore ask that you have prepared a historical statement of munitions production, so brief that all may have time to read it, so nontechnical that all may be able readily to understand it, and so authoritative that all may rely upon its accuracy.

Cordially yours,

NEWTON D. BAKER,
Secretary of War.

HON. BENEDICT CROWELL,
The Assistant Secretary of War.

WASHINGTON, D. C., *May 10, 1919.*

DEAR MR. SECRETARY: Responding to your request, I transmit herewith a brief, nontechnical, authoritative history of munitions production during the recent war. The several chapters have been prepared in the first instance by the officers who have been directly responsible for production, and have been assembled and edited, under my direction, by Hon. Robert J. Bulkley, assisted by Capt. Robert Forrest Wilson and Capt. Benjamin E. Ling. Capt. Wilson has undertaken responsibility for the literary style of the report, and has rewritten the greater part of it, consulting at length with the officers who supplied the original material, and with officers of the statistics branch of the General Staff, in order to insure accuracy.

Maj. Gen. C. C. Williams, Chief of Ordnance; Brig. Gen. W. S. Peirce, Acting Chief of Ordnance; Maj. Gen. C. T. Menoher, Chief of Air Service; Maj. Gen. W. M. Black, Chief of Engineers; Maj. Gen. W. L. Sibert, Chief of Chemical Warfare Service; Maj. Gen. H. L. Rogers, Quartermaster General; Mr. R. J. Thorne, Acting Quartermaster General; Maj. Gen. G. O. Squier, Chief Signal Officer; Brig. Gen. Charles B. Drake, Chief of Motor Transport Corps; and Maj. Gen. W. M. Ireland, the Surgeon General, have cooperated in the preparation of the material transmitted herewith.

Special acknowledgment for the preparation and correction of the several chapters is due to the following officers:

The ordnance problem, Col. James L. Walsh.

Gun production, Col. William P. Barba.

Mobile field artillery, Col. J. B. Rose.

Railway artillery, Col. G. M. Barnes and Maj. E. D. Campbell.

Explosives, propellants, and artillery ammunition, Col. C. T. Harris and Maj. J. Herbert Hunter.

Sights and fire-control apparatus, Col. H. K. Rutherford and Maj. Fred E. Wright.

Motorized artillery, Col. L. B. Moody and Lieut. Col. H. W. Alden.

Tanks, Lieut. Col. H. W. Alden.

Machine guns, Col. Earl McFarland and Lieut. Col. Herbert O'Leary.

Service rifles, Maj. Lewis P. Johnson and Maj. Parker Dodge.

Pistols and revolvers, Lieut. Col. J. C. Beatty and Maj. Parker Dodge.

Small arms ammunition, Lieut. Col. J. C. Beatty, Maj. Lee O. Wright, Maj. A. E. Hunt, and Capt. C. J. Evans.

Trench warfare material, Lieut. Col. E. J. W. Ragsdale, Capt. J. R. Caldwell, Capt. R. D. Smith, and Lieut. J. T. Libbey.

Miscellaneous ordnance equipment, Lieut. Col. S. H. MacGregor, Maj. Bashford Dean, Capt. A. L. Fabens, and Capt. James S. Wiley.

The aircraft problem and airplane production, Lieut. Col. George W. Mixter.

The Liberty engine and other airplane engines, Lieut. H. H. Emmons, United States Navy.

Aviation equipment and armament, Lieut. Col. E. J. W. Ragsdale, Maj. E. Bradley, Capt. Robert D. Smith, Capt. H. E. Ives, and Lieut. John M. Hammond.

The airplane radio telephone, Col. C. C. Culver and Lieut. Col. Nugent H. Slaughter.

Balloons, Capt. H. W. Treat.

The Engineers in France, Lieut. Col. J. B. Cress and Capt. C. Beard.

Military railways, Col. J. M. Milliken and Mr. S. M. Felton.

Engineer activities at home, Lieut. Col. J. B. Cress and Lieut. Col. R. W. Crawford.

Sound and flash ranging and searchlights, Lieut. Col. J. B. Cress and Maj. W. D. Young.

Toxic gases, Col. M. T. Bogert, Col. W. A. Walker, Lieut. Col. E. M. Chance, and Lieut. Col. William McPherson.

Defensive gas equipment, Col. Bradley Dewey and Lieut. Col. A. L. Besse.

Subsistence, Lieut. Col. J. H. Adams and Capt. S. B. Johnson.

Clothing and equipage, Lieut. Col. F. A. Ellison and Capt. W. H. Porter.

Miscellaneous quartermaster undertakings: Music, Maj. George H. Richards; fuel, oil, and paints, Mr. J. Elliott Hall; brushes, Capt. T. W. S. Phillips; rolling kitchens, Capt. J. G. Williams and Mr. M. A. Dunning; tools and tool chests, Mr. W. F. Fusting and Mr. M. E. Moye; hardware, Lieut. Col. H. P. Hill and Mr. William A. Graham; factory enterprises, Lieut. Col. H. P. Hill; shoe fitting, Col. F. A. Ellison; meat cutting, Dr. W. O. Trone; packing, Capt. R. H. Moody; horses and mules, Maj. A. Cedarwald.

Motor and horse-drawn vehicles: Motor vehicles, Col. Fred Glover; horse-drawn vehicles, Maj. A. Volgeneau.

Medical and dental supplies, Lieut. Col. J. P. Fletcher and Capt. W. G. Guth.

Salvage, Col. J. S. Chambers and Capt. F. C. Simpson.

Mr. W. L. Pollard, Mr. Aaron Rachofsky, and Lieut. J. J. Cameron have rendered very valuable assistance in assembling data concerning quartermaster activities.

Cantonments and camps, and miscellaneous construction, Maj. W. G. Maupin.

Signal Corps material, Brig. Gen. C. McK. Saltzman and Capt. Donald MacGregor.

The accuracy of all statistics and direct statements of fact has been checked and approved by the statistics branch of the General Staff, under the direction of Maj. W. R. Burgess.

Respectfully submitted.

BENEDICT CROWELL,
The Assistant Secretary of War,
Director of Munitions.

Hon. NEWTON D. BAKER,
Secretary of War.

PREFACE.

Except in one or two instances, this account of the production of munitions in America for the war against Germany and her allies contains nothing about secret devices invented during the period under discussion. When the necessity for silence with respect to vital matters brought about a voluntary censorship in American publications, the land was filled with rumors of new and revolutionary developments in war matériel, particularly of new weapons of offense. It is fair to the American public to-day to state that such rumors were not without foundation. American inventiveness rose splendidly to the emergency. The expected American offensive in 1919 would have had its "surprises" in numbers, some of which might well have proved to be decisive. Certain of these inventions had been put in large production before the armistice was declared, others had been carried to an advanced experimental stage that insured their success. Since the value of these innovations as part of the Nation's permanent military assets depends largely upon their secret nature, it would be obviously unwise to mention or describe them at this time.

The Director of Munitions wishes to acknowledge the debt of America, so far as the production of munitions is concerned, to the Navy for its cooperation in industrial matters at home and its strong aid in the safe transport of munitions to France, and to all the other Government departments, each one of which contributed in numerous and important ways to the success of the munitions enterprise. The debt also extends heavily to the War Industries Board, its functions of creating facilities for manufacture, opening up new sources of raw materials, allocating materials, decreeing priorities, fixing prices, and acting as purchasing agent for the allies, making it the national industrial clearing house through which the War Department could work without waste effort. Acknowledgment is made to such essential agencies as the United States Railroad Administration, the United States Fuel Administration, the War Trade Board, and the United States Food Administration, and to all official or volunteer activities looking to the conservation and mobilization of our national resources. Without this entire cooperation the history set forth in these pages would not be what it is.

CONTENTS.

AMERICA'S MUNITIONS, 1917-18

INTRODUCTION.

As our war against Germany recedes into the past its temporal boundaries become more sharply defined, and it assumes the character of a complete entity—a rounded-out period of time in which the United States collected her men and resources, fought, and shared in the victory.

As such it offers to the critic the easy opportunity to discover that certain things were not done. American airplanes did not arrive at the front in sufficient numbers. American guns in certain essential calibers did not appear at all. American gas shells were not fired at the enemy. American troops fought with French and British machine guns to a large extent. The public is familiar with such statements.

It should be remembered that the war up to its last few weeks—up to its last few days, in fact—was a period of anxious suspense, during which America was straining her energies toward a goal, toward the realization of an ambition which, in the production of munitions, dropped the year 1918 almost out of consideration altogether, which indeed did not bring the full weight of American men and matériel into the struggle even in 1919, but which left it for 1920, if the enemy had not yet succumbed to the growing American power, to witness the maximum strength of the United States in the field.

Necessarily, therefore, the actual period of hostilities, between April 6, 1917, and November 11, 1918, was devoted in this country to laying down the foundations of a munitions industry that should bring about its overwhelming results at the appointed time. What munitions of the more difficult sort were actually produced in this period might almost be termed casual to the main enterprise—pilots of the quantities to come.

The decision to prepare heavily for 1919 and 1920 and thus sacrifice for 1917 and 1918 the munitions that might have been produced at the cost of any less adequate preparation for the more distant future, was based on sound strategical reasoning on the part of the Allies and ourselves.

On going back to the past we find that on April 6, 1917, the United States scarcely realized the gravity of the undertaking. There was a general impression, reaching even into Government, that the Allies alone were competent to defeat the Central Powers in time, and that America's part would be largely one of moral support, with expanding preparation in the background as insurance against any unforeseen disasters. In line with this attitude we sent the first division of American troops to France in the spring of 1917 to be our earnest to the governments and peoples of the Allies that we were with them in the great struggle. Not until after the departure of the various foreign missions that came to this country during that spring did America fully awake to the seriousness of the situation.

All through the summer of 1917 the emphasis upon American man power in France gradually grew, but no definite schedule upon which the United States could work was reached until autumn or early winter, until the mission headed by Col. Edward M. House visited Europe to give America place on the Supreme War Council and in the Interallied Conference. The purpose of the House mission was to assure the Allies that America was in the war for all she was worth and to determine the most effective method in which she could cooperate.

In the conferences in London and Paris the American representatives looked into the minds of the allied leaders and saw the situation as it was. Two dramatic factors colored all the discussions—the growing need for men and the gravity of the shipping situation. The German submarines were operating so effectively as to make exceedingly dark the outlook for the transport on a sufficient scale either of American troops or of American munitions.

As to man power, the Supreme War Council gave it as the judgment of the military leaders of the Allies that, if the day were to be saved, America must send 1,000,000 troops by the following July. There were in France then (on Dec. 1, 1917) parts of four divisions of American soldiers—129,000 men in all.

The program of American cooperation, as it crystallized in these conferences, may be summarized as follows:

1. To keep the Allies from starvation by shipping food.

2. To assist the Allied armies by keeping up the flow of matériel already in production for them in the United States.

3. To send as many men as could be transported with the shipping facilities then at America's command.

4. To bend energies toward a big American Army in 1919 equipped with American supplies.

In these conferences sat the chief military and political figures of the principal European powers at war with Germany. In the Supreme War Council were such strategists as Gen. Foch for the French and Gen. Robertson for the British, Gen. Bliss representing the United States. The president of the Interallied Conference was M. Clemenceau, the French prime minister. Mr. Winston Churchill, the minister of munitions, represented Great Britain, while Mr. Lloyd-George, the Prime Minister of England, also participated to some extent in the conferences.

Out of bodies of such character came the international ordnance agreement. It will be apparent to the reader that this agreement must have represented the best opinion of the leaders of the principal Allies, initiated out of their intimate knowledge of the needs of the situation and concurred in by the representatives of the United States. The substance of this agreement was outlined for Washington in a cabled message signed by Gen. Bliss, a document that had such an important bearing upon the production of munitions in this country that its more important passages are set down at this point:

The representatives of Great Britain and France state that their production of artillery (field, medium, and heavy) is now established on so large a scale that they are able to equip completely all American divisions as they arrive in France during the year 1918 with the best make of British and French guns and howitzers.

The British and French ammunition supply and reserves are sufficient to provide the requirements of the American Army thus equipped at least up to June, 1918, provided that the existing 6-inch shell plants in the United States and Dominion of Canada are maintained in full activity, and provided that the manufacture of 6-inch howitzer carriages in the United States is to some extent sufficiently developed.

On the other hand, the French, and to a lesser extent the British, require as soon as possible large supplies of propellants and high explosives: and the British require the largest possible production of 6-inch howitzers from now onward and of 8-inch and 9.2-inch shell from June onward.

In both of these matters they ask the assistance of the Americans.

With a view, therefore, first to expedite and facilitate the equipment of the American armies in France, and, second, to secure the maximum ultimate development of the ammunition supply with the minimum strain upon available tonnage, the representatives of Great Britain and France propose that the American field, medium, and heavy artillery be supplied during 1918, and as long after as may be found convenient, from British and French gun factories; and they ask: (A) That the American efforts shall be immediately directed to the production of propellants and high explosives on the largest possible scale; and (B) Great Britain also asks that the 6-inch, 8-inch, and 9.2-inch shell plants already created for the British service in the United States shall be maintained in the highest activity, and that large additional plants for the manufacture of these shells shall at once be laid down.

In this way alone can the tonnage difficulty be minimized and potential artillery development, both in guns and shells, of the combined French, British, and American armies be maintained in 1918 and still more in 1919.

This agreement had a profound effect upon American production of munitions. Most important of all, it gave us time; time to build manufacturing capacity on a grand scale without the hampering necessity for immediate production; time to secure the best in design; time to attain quality in the enormous output to come later as opposed to early quantity of indifferent class.

In the late autumn of 1917, shortly after Russia collapsed and withdrew from the war, it became evident that Germany would seize the opportunity to move her troops from the eastern front and concentrate her entire army against the French and British in 1918.

This intelligence at once resulted in fresh emphasis upon the man-power phase of American cooperation. As early as December, 1917, the War Department was anticipating the extraordinary need for men in the coming spring by considering plans for the transport of troops up to the supposed limit of the capacity of all available American ships, with what additional tonnage Great Britain and the other Allies could spare us. It is of record that the actual dispatch of troops to France far outstripped these early estimates.

Then came the long-expected German offensive, and the cry went up in Europe for men. England, "her back against the wall," offered additional ships in which to transport six divisions over and above the number of troops already scheduled for embarkation, agreeing further to feed and maintain these men for 10 weeks while they were brigaded with British units for final training. After the six additional divisions had embarked there was still need of men, and the British continued their transports in our service. The high mark of shipment was in July, when 306,000 American soldiers were transported across the Atlantic, more than three times the number contemplated for July in the schedule adopted six months earlier.

ACTUAL TROOP SAILINGS COMPARED WITH PROGRAMS.

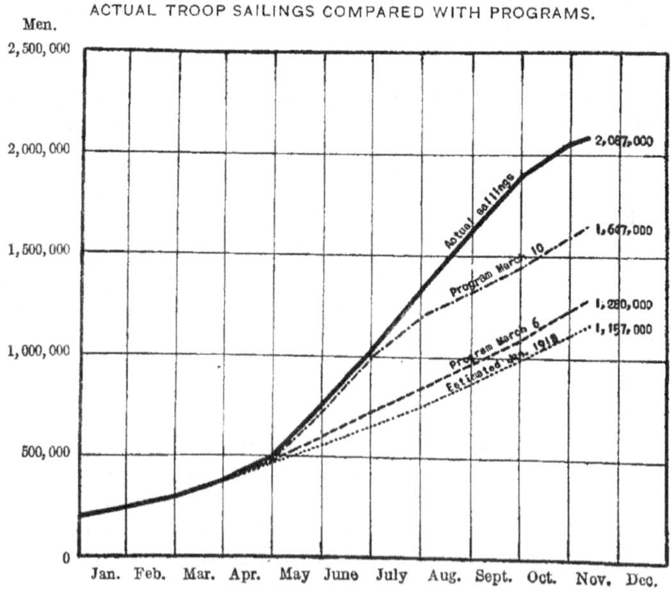

The effect of this stepping up of the man-power program upon the shipment of supplies was described by Lieut. Col. Repington, the British military critic, writing in the Morning Post (London) on December 9, 1918, in part as follows:

* * * they (the British war cabinet) also prayed America in aid, implored her to send in haste all available infantry and machine guns, and placed at her disposal, to her great surprise, a large amount of transports to hasten arrivals. * * *

The American Government acceded to this request in the most loyal and generous manner. Assured by their Allies in France that the latter could fit out the American infantry divisions on their arrival with guns, horses, and transport, the Americans packed their infantry tightly in the ships and left to a later occasion the dispatch to France of guns, horses, transport, labor units, flying service, rolling stock, and a score of other things originally destined for transport with the divisions. If subsequently—and indeed up to the day that the armistice was signed—Gen. Pershing found himself short of many indispensable things, and if his operations were thereby conducted under real difficulties of which he must have been only too sensible, the defects were not due to him and his staff, nor to the Washington administration, nor to the resolute Gen. March and his able fellow workers, but solely to the self-sacrificing manner in which America had responded to the call of her friends.

* * * * * * *

BRITISH AND AMERICAN EXPEDITIONARY FORCES ON WESTERN FRONT.

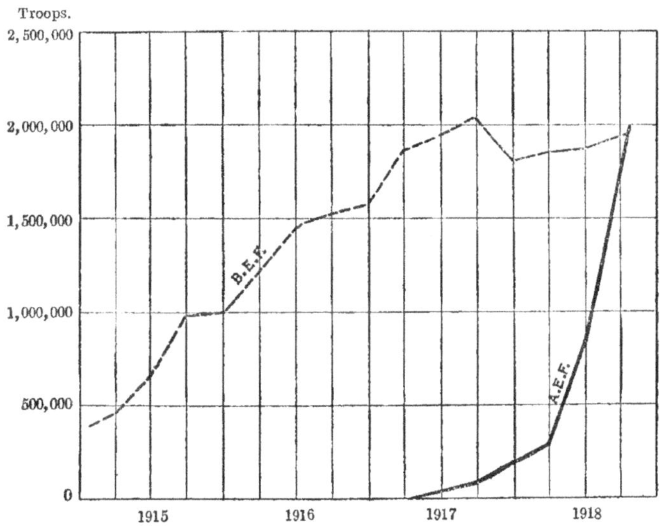

The really amazing thing which America did was to place in France in 19 months an army of the size and the ability of the American Expeditionary Force. The war taught us that America can organize, train, and transport troops of a superior sort at a rate which leaves far behind any program for the manufacture of munitions.

It upset the previous opinion that adequate military preparedness is largely a question of trained man power.

When the war touched us our strategical equipment included plans ready drawn for the mobilization of men. There were on file at the Army War College in Washington detailed plans for defending our harbors, our coasts, and our borders. There were also certain plans for the training of new troops.

It is worthy of note, however, that this equipment included no plan for the equally important and equally necessary mobilization of industry and production of munitions, which proved to be the most difficult phase of the actual preparation for war. The experience of 1917 and 1918 was a lesson in the time it takes to determine types, create designs, provide facilities, and establish manufacture. These years will forever stand as the monument to the American genius of workshop and factory, which in this period insured the victory by insuring the timely arrival of the overwhelming force of America's resources in the form of America's munitions.

B. C.

WASHINGTON, *May, 1919.*

BOOK I.

ORDNANCE.

CHAPTER I.

THE ORDNANCE PROBLEM.

To arm the manhood called to defend the Nation in 1917 and 1918, to make civilians into soldiers by giving them the tools of the martial profession—such was the task of the Ordnance Department in the late war.

The off-hand thought may identify ordnance as artillery alone. It may surprise many to know that in the American ordnance catalogue of supplies during the recent war there were over 100,000 separate and distinct items. Thousands of the items of ordnance were distinctly noncommercial, meaning that they had to be designed and produced specially for the uses of war.

While the principles of fighting essentially have changed not one whit since the age when projectiles were stones hurled by catapults, nearly every advance in mechanical science has had its reflection in warfare, until to-day the weapons which man has devised to destroy the military power of his enemy make up an intricate and an imposing list. When America accepted the challenge of Germany in 1917, part of the range of ordnance had already been produced in moderate quantities in the United States, part of it had been developed by the more militaristic nations of the world in the last decade or quarter century, and part of it was purely the offspring of two and one-half years of desperate fighting before America entered the great struggle. Yet all of it, both the strange and the familiar, had to be put in production here on a grand scale and in a minimum of time, that the American millions might go adequately equipped to meet the foe. Let us examine the range of this equipment, seeing in the major items something of the character of the problem which confronted the Ordnance Department at the outset of the great enterprise.

Starting with the artillery, there was first in order of size the baby two-man cannon of 37 millimeters (about an inch and a half) in the diameter of its bore—a European development new to our experience, so light that it could be handled by foot troops in the field, used for annihilating the enemy's machine-gun emplacements.

Then the mobile field guns—the famous 75's, the equivalent in size of our former 3-inch gun, the 155-millimeter howitzer, the

French 155 millimeter G. P. F. (Grand Puissance Filloux) gun of glorious record in the war, and its American prototypes, the 4.7-inch, 5-inch, and 6-inch guns—all of these employed to shell crossroads and harass the enemy's middle area.

Beyond these were the 8-inch and 9.2-inch howitzers and the terrific 240-millimeter howitzer, for throwing great weights of destruction high in air to descend with a plunge upon the enemy's strongest defenses.

Then there were the 8-inch, 10-inch, 12-inch, and 14-inch guns on railway mounts, for pounding the depots and dumps in the enemy's back areas. These weapons were so tremendous in weight when mounted as to require from 16 to 24 axles on the car to distribute the load and the recoil of firing within the limits of the strength of standard heavy railway track.

All of these guns had to be produced in great numbers, if the future requirements of the American forces were to be met, produced by the thousands in the cases of the smaller ones and by the hundreds and scores in the cases of the larger.

These weapons would be ineffective without adequate supplies of ammunition. In the case of the mobile field guns this meant a requirement of millions of shell or shrapnel for the incessant bombardments and the concentrated barrages which characterized the great war. The entire weight of projectiles fired in such an historic engagement as Gettysburg would supply the artillery only for a few minutes in such intensive bombardments as sowed the soil of Flanders with steel.

The artillery demanded an immense amount of heavy equipment—limbers, caissons, auto ammunition trucks, and tractors to drag the heavy and middle-heavy artillery. Some of them were fitted with self-propelled caterpillar mounts which could climb a 40° grade or make as high as 12 miles an hour on level ground. These, the adaptations to warfare of peaceful farm and construction machine traction, for the first time rendered the greater guns exceedingly mobile, enabling them to go into action instantly upon arrival and to depart to safety just as soon as their mission was accomplished.

Then, too, this artillery equipment must have adequate facilities for maintenance in the field, and this need brought into existence another enormous phase of the ordnance program. There must be mobile ordnance repair shops for each division, consisting of miniature machine shops completely fitted out with power and its transmission equipment and mounted directly on motor trucks. Then there must be semiheavy repair shops on 5-ton tractors, these to be for the corps what the truck machine shop was to the division. Each army headquarters called for its semipermanent repair shop for artillery and still larger repair shops for its railway artillery.

And in addition to all these were the base repair shops in France, which were erected on a scale to employ a force three times as large as the combined organizations of all the manufacturing arsenals of the United States in time of peace, having a capacity for relining 1,000 cannon and overhauling and repairing 2,000 motor vehicles, 7,000 machine guns, 50,000 rifles, and 2,000 pistols every month. This equipment of artillery and its maintenance organization implies the flow from American industry of enormous quantities of repair parts and spare parts to keep the artillery in good condition.

Coming next to the more personal equipment of the soldier, we find the necessity confronting the Ordnance Department to manufacture shoulder rifles by the million and cartridges for them by the billion. The great war brought the machine gun into its own, requiring in the United States the manufacture of these complicated and expensive weapons by the tens of thousands, including the one-man automatic rifle, itself an arm of a deadly and effective type.

Simultaneously with the mass employment of machine guns in the field came the development of the modern machine gun barrage, the indirect fire of which required sighting instruments of the most delicate and accurate sort, and tripods with finely calibrated elevating and traversing devices, so that the gunner might place the deadly hail safely over the heads of his own unseen but advancing lines and with maximum damage to the enemy. These thousands of machine guns required water jackets to keep their barrels cool and specially built carts to carry them.

The personal armament of the soldier also called for an automatic pistol or a revolver for use in the infighting, when squads came in actual contact with soldiers of the enemy. These had to be produced by the hundreds of thousands.

The requirements of the field demanded hundreds of thousands of trench knives, murderous blades backed by the momentum of heavily weighted handles, which in turn were protected by guards embodying the principle of the thug's brass "knucks" armed with sharp points.

Then there were the special weapons, largely born of modern trench warfare. These included mortars, ranging from the small 3-inch Stokes, light enough to go over the top and simple enough to be fired from between the steadying knees of a squatting soldier, to the great 240-millimeter trench mortar of fixed position. The mortars proved to be exceedingly effective against concentrations of troops, and so there was devised for them a great variety of bombs and shell, not only of the high explosive fragmentation type, but also containing poison gas or fuming chemicals. Great quantities both of mortars and their ammunition were required.

From the security of the trenches the soldiers first threw out grenades, which burst in the enemy's trenches opposite and created havoc. From the original device were developed grenades of various sorts—gas grenades for cleaning up dugouts, molten-metal grenades for fusing the firing mechanisms of captured enemy cannon and machine guns, paper grenades to kill by concussion. Then there were the rifle grenades, each to be fitted on the muzzle of a rifle and hurled by the lift of gases following the bullet, which passed neatly through the hole provided for it. The production of grenades was no small part of the American ordnance problem.

In addition to these trench weapons were the Livens projectors, which, fired in multiple by electricity, hurled a veritable cloud of gas containers into a selected area of enemy terrain, usually with great demoralization of his forces.

Bayonets for the rifles, bolos, helmets, periscopes for looking safely over the edges of the trenches, panoramic sights, range finders—these are only a few of the ordnance accessories of general application.

Then those innovations of the great war—the tanks—the 3-ton "whippet," built to escort the infantry waves, the 6-ton tanks, most used of all, and the powerful Anglo-American heavy tanks, each mounting a 37-millimeter cannon and four machine guns.

The war in the air put added demands upon ordnance. It required the stripped machine gun firing cartridges so rapidly that their explosions merged into a single continuous roar, yet each shot so nicely timed that it passed between the flying blades of the propeller. There had to be electric heaters for the gun mechanisms to prevent the oil which lubricated them from becoming congealed in the cold of high altitudes. The airplane guns required armor-piercing bullets for use against armored planes, incendiary bullets to ignite the hydrogen of the enemy's balloon or to fire the gasoline escaping through the wound in the hostile airplane's fuel tank, and tracer bullets to direct the aim of the aerial gunner. Other equipment for the airman included shot counters, to tell him instantly what quantity of ammunition he had on hand, and gun sights, ingeniously contrived to correct his aim automatically for the relative speed and direction of the opposing plane. These were all developments in ordnance brought about by the great war, and in each case they involved problems for the production organization to solve.

Then there were the drop bombs of aerial warfare, of many gradations in weight up to 500 pounds each, these latter experimental ones forecasting the day when bombs weighing 1,600 pounds would be dropped from the sky; then bomb sights to determine the moment when the missile must be dropped in order to hit its target, sights which corrected for the altitude, the wind resistance, and the rate of speed of the airplane; and then mechanisms to suspend the bombs from the plane and to release them at the will of the operator.

The list might be stretched out almost indefinitely—through pyrotechnics, developed by the exigencies in Europe into an elaborate system; through helmets and armor, revivals from medieval times to protect the modern soldier from injury; through the assortment of heavy textiles, which gave the troops their belts, their bandoleers, their haversacks, and their holsters; through canteens, cutlery for the mess in the fields, shotguns, and so on, until there might be set down thousands of items of the list which we know as modern ordnance.

It will be noted that the most important articles in this range are articles of a noncommercial type. In other words, they are not the sort of things that the industry of the country builds in time of peace, nor learns how to build. Many other war functions came naturally to a country skilled in handling food supplies for teeming populations, in solving housing problems for whole cities, and in managing transportation for a hundred million people; there was at hand the requisite ability to conduct war enterprises of such character smoothly and efficiently. Yet there was in the country at the outbreak of war little knowledge of the technique of ordnance production.

The declaration of war found an American Ordnance Department whose entire commissioned personnel consisted of 97 officers. Only 10 of this number were experienced in the design of artillery weapons. The projected army of 5,000,000 men required 11,000 trained officers to handle every phase of ordnance service. While a portion of this production would have to do with the manufacture of articles of a commercial type, such as automobiles, trucks, meat cans, mess equipment, and the like, yet the ratio of 97 to 11,000 gives an indication of the amount of ordnance knowledge possessed by the War Department at the outbreak of war as compared to what it would need to equip the first 5,000,000 men for battle.

The Government could obtain commissary officers from the food industry; it could turn bank tellers into paymasters, or convert builders into construction quartermasters; find transportation officers in the great railway systems, Signal Corps officers in the telegraph companies, or medical officers in professional life. But there was no broad field to which ordnance could turn to find specialized skill available. The best it could do was to go into the heavy manufacturing industry for expert engineers who could later be trained in the special problems of ordnance.

Prior to 1914 there were but six Government arsenals and two large private ordnance works which knew anything about the production of heavy weapons. After 1914, war industry sprang up in the United States, yet in 1917 there were only a score or so of firms engaged in the manufacture of artillery ammunition, big guns, rifles, machine guns, and other important ordnance supplies for the allies. When the armistice was signed nearly 8,000 manufacturing plants in the United States were working on ordnance contracts. While many of these contracts entailed production not much dissimilar to commercial output, yet here is another ratio—the 20 or more original factories compared with the ultimate 8,000—which serves as an indication of the expansion of the industrial knowledge of the special processes incident to ordnance manufacture.

When we found ourselves in the war the first step was to extend our ordnance knowledge as quickly as possible. The war in Europe had developed thousands of new items of ordnance, many of them carefully guarded as military secrets, with which our own officers were familiar only in a general way. As soon as we became a belligerent, however, we at once turned to the allies, and they freely and fully gave us of their store of knowledge—plans, specifications, working models, secret devices, and complete manufacturing processes.

With this knowledge at hand we adopted for our own program certain French types of field guns and howitzers and British types of heavy howitzers. The reproduction of the British types caused no unusual difficulties, but the adoption of French plans brought into the situation a factor the difficulties of which are apt not to be appreciated by the uninitiated.

This new element for consideration was the circumstance that the entire French system of manufacture in metals is radically different from our own in its practices and is not readily adapted to American methods.

The English and the American engineers and shops use inches and feet in their measurements, but the French use the metric system. This fact means that there was not a single standard American drill, reamer, tap, die, or other machine-shop tool that would accurately produce the result called for by a French ordnance drawing in the metric system. Moreover, the French standards for metal stocks, sheets, plates, angles, I-beams, rivet holes, and rivet spacing are far different from American standards.

It was discovered that complete French drawings were in numerous cases nonexistent, the French practice relying for small details upon the memory and skill of its artisans. But even when the complete drawings were obtained, then the American ordnance engineer was confronted with the choice of either revolutionizing the machining industry of the United States by changing over its entire equipment to conform to the metric system, or else of doing what was done—namely, translating the French designs into terms of standard American shop practice, a process which in numerous cases required weeks and even months of time on the part of whole staffs of experts working at high tension.

Nor do the French know the American quantity-production methods. The French artisan sees always the finished article, and he is given discretion in the final dimensions of parts and in the fitting and assembling of them. But the American mechanic sees only the part in which he is a specialist in machining, working with strict tolerances and producing pieces which require little or no fitting in the assembling room. Consequently, in the translating of French plans it was necessary to put into them what they never had before, namely, rigid tolerances and exact measurements.

FIGURE 1.

EXPENDITURE OF ARTILLERY AMMUNITION IN MODERN BATTLES.

Year.	Battle.	Days' duration.	Army.	Rounds of artillery ammunition expended.
1863..	Chickamauga..	2	Union........	7,325
1863..	Gettysburg....	3	Union.........	32,781
1870..	St. Privat......	1	German.......	39,000
1904..	Nan Shan......	1	Japanese......	34,047
1904..	Liao Yang.....	9	Russian.......	134,400
1904..	Sha Ho........	9	Russian.......	274,360
1915..	Neuve Chapelle	[1] 3	British........	197,000
1915..	Souchez........	[2] 1	French........	300,000
1916..	Somme........	[3] 7	British........	4,000,000
1917..	Messines Ridge.	[3] 7	British........	2,753,000
1918..	St. Mihiel......	[2] 4	United States.	1,093,217

[1] Artillery preparation lasted 35 minutes. [2] Artillery preparation lasted 4 hours.
[3] Artillery preparation intermittent 7 days.

One of the most striking developments of the present war has been the great increase in the use of artillery to precede infantry action in battle. This is illustrated by a comparison of the expenditure of artillery ammunition in characteristic battles of recent wars with that in important battles of the present war. The special features of the several battles should be kept in mind. Chickamauga was fought in a heavily wooded region; Gettysburg and St. Privat over open farm land. The latter battles, together with Nan Shan, and all the battles of the present war considered below, involved artillery preparation for assault upon armies in defensive position. The expenditures, therefore, are roughly comparable.

The high mark of the use of artillery in offensive battle was reached at the Somme and Messines Ridge, before the effective use of tanks was developed.

When an army of 100,000 men expands and becomes an army of 3,000,000, it becomes a job just 30 times bigger to feed the 3,000,000 than it was to feed the 100,000. A soldier of a campaigning army eats no more than a soldier of a quiet military post. The same is true approximately in the case of clothing an army. But the army's consumption of ammunition in time of war is far out of proportion to its numerical expansion to meet the war emergency.

For instance, an Army machine gun in time of peace might fire 6,000 rounds in practice during the year. This was the standard quantity of cartridges provided in peace. Yet it is necessary to provide for a single machine gun on the field in such a war as the recent one 288,875 rounds of ammunition during its first year of operation, this figure including the initial stock and the reserve

FIGURE 2.

RATES OF ARTILLERY FIRE PER GUN PER DAY IN RECENT WARS.

War.	Army.	Approximate rounds per gun per day.
1854–1856, Crimean	British and French	[1] 5
1859, Italian	Austrian	.3
1861–1865, Civil	Union	4
1866, Austro-Prussian	Austrian	2.2
	Prussian	.8
1870–71, Franco-Prussian	German	[2] 1.1
1904–5, Russo-Japanese	Russian	4
1912–13, Balkan	Bulgarian	7
PRESENT WAR.		
September, 1914	French	[2] 8
Jan. 1–Oct. 1, 1918	Italian	[2] 8
Jan. 1–Nov. 11, 1918	United States	[2] 30
Jan. 1–Nov. 11, 1918	French	[2] 34
Jan. 1–Nov. 11, 1918	British	[2] 35

[1] Siege of Sebastopol. [2] Field gun ammunition only.

The rates are based upon total expenditure and average number of guns in the hands of field armies for the period of the wars.

A large part of the heavy expenditure of artillery ammunition in the present as compared with other modern wars can be attributed to the increased rate of fire made possible by improved methods of supply in the field and by the rapid-fire guns now in use. In wars fought before the introduction of quick-firing field guns, four or five rounds per day was the greatest average rate. Even this was reached only in the siege of Sebastopol, where armies were stationary and supply by water was easy, and in the American Civil War, which was characterized by advanced tactical developments. The guns of the allied armies in France fired throughout the year 1918 at a rate about seven times greater than these previously high rates.

supply as well as the actual number of rounds fired. Thus the machine gun of war increases its appetite, so to speak, for ammunition 4,700 per cent in the first year of fighting.

In the case of larger weapons the increase in ammunition consumption is even more startling. Prior to 1917 the War Department allotted to each 3-inch field gun 125 rounds of ammunition per year for practice firing. Ammunition for the 75-millimeter guns (the 3-inch equivalent) was being produced to meet an estimated supply of 22,750 rounds for each gun in a single year, or an increased consumption of ammunition in war over peace of 18,100 per cent.

Thus when a peace army of 100,000 becomes a war army of 3,000,000 its ammunition consumption becomes not 30 times greater,

FIGURE 3.

EXPENDITURE OF ARTILLERY AMMUNITION IN RECENT WARS.

PAST WARS COMPARED WITH ONE MONTH OF PRESENT WAR.

Year.	War.	Army.	Rounds expended during war.
1859	Italian	Austrian	15,326
1861–1865	Civil	Union	5,000,000
1866	Austro-Prussian	Prussian	36,199
		Austrian	96,472
1870–71	Franco-Prussian	German	817,000
1904–5	Russo-Japanese	Russian	954,000
1912–13	Balkan	Bulgarian	700,000
1918	Present	British and French	In one month.[1] 12,710,000

EXPENDITURES FOR ONE YEAR, CIVIL AND PRESENT WARS.

1864[2]	Civil	Union	1,950,000
1918[3]	Present	United States	8,100,000
1918[3]	Present	British	71,445,000
1918[3]	Present	French	81,070,000

[1] Average, year ended Nov. 10, 1918. [2] Year ended June 30, 1864. [3] Year ended Nov. 10, 1918.

The industrial effort necessary to maintain modern armies in action may be measured to a certain extent by their expenditure of artillery ammunition. European wars of the past 100 years were for the most part decided before peace-time reserves had been exhausted. The American Civil War, however, required for its decision an industrial mobilization at that time unprecedented, which, like the use in that war of intrenchments by field armies, was more truly indicative of the trend of modern warfare than were the conditions of the more recent European wars.

but anywhere from 48 to 182 times 30 times greater—an increase far out of proportion to its increase in the consumption of food, clothing, or other standard supplies. Modern invention has made possible and modern practice has put into effect a greatly augmented use of ammunition. Figures 1, 2, and 3 show graphically how ammunition expenditure has increased in modern times.

Another circumstance that complicated the ordnance problem was the increasing tendency throughout the great war to use more and more the mechanical or machine methods of fighting as opposed to the older and simpler forms in which the human or animal factor entered to a greater extent.

At the time the United States entered the war the regulations prescribed 50 machine guns as the equipment for an infantry division. When the armistice was signed the standard equipment of a division called for 260 heavy machine guns and 768 light automatic rifles. Of the heavy machine guns with a division, only 168 were supposed to be in active service, the remainder being in reserve or in use for anti-aircraft work. However, the comparison in the two standards of equipment shows the tendency toward machine methods in the wholesale killing of modern warfare and indicates the fresh demands made upon the ordnance organization to procure this additional machinery of death. Moreover, when the fighting came to an end the A. E. F. was on the point of adding to its regimental and divisional equipment a further large number of automatic rifles.

The day of the horse was passing in the great war as far as his connection with the mobile artillery was concerned, and the gasoline motor was taking his place, this tendency being accelerated particularly by America, the greatest nation of all in automotivity. Trucks and tractors to pull the guns, motor ammunition trucks displacing the old horse-drawn caissons and limbers, even self-propelling platforms for the larger field guns, with track laying or caterpillar mounts supplying not only mobility for the gun but aiming facilities as well; these were the fresh developments. Some of these improvements were produced and put in the field, the others were under development at the signing of the armistice. The whole tendency toward motorization served to complicate ordnance production in this country, not only in the supply of the weapons and traction devices themselves, but in the production of increased supplies of ammunition, since these improvements also tended to increase the rapidity with which bullets and shell were consumed.

The total cost of the ordnance alone required to equip the first 5,000,000 Americans called to arms was estimated to be between \$12,000,000,000 and \$13,000,000,000. This was equal to about half of all the money appropriated by Congresses of the United States from the first Continental Congress down to our declaration of war against Germany, out of which appropriations had been paid the cost of every war we ever had, including the Civil War, and the whole enormous expenses of the Government in every official activity of 140 years. To equip with ordnance an army of this size in the period projected meant the expenditure of money at a rate which would build a Panama Canal complete every 30 days.

Above are sketched some of the difficulties of the situation. In our favor we had the greatest industrial organization in the world, engineering skill to rank with any, a race of people traditionally versatile in applying the forces of machinery to the needs of mankind, inventive genius which could match its accomplishments with those of the rest of the world added together, a capacity for organization that proved to be astonishingly effective in such an effort as the nation made in 1917 and 1918, enormous stores of raw materials, the country being more nearly self-sufficient in this respect than any other nation of the globe, magnificent facilities of inland transportation, a vast body of skilled mechanics, and a selective-service law designed to take for the Army men nonessential to the Nation's industrial efforts for war and to leave in the workshops the men whose skill could not be withdrawn without subtracting somewhat from the national store of industrial ability.

It only remains to sketch in swift outlines something of the accomplishments of the American ordnance effort. In general it may be said that those projects of the ordnance program to which were assigned the shorter time limits were most successful. There never was a time when the production of smokeless powder and high explosives was not sufficient for our own requirements, with large quantities left over for both France and England.

America in 19 months of development built over 2,500,000 shoulder rifles, a quantity greater than that produced either by England or by France in the same period, although both those countries in April, 1917, at the time when we started, had their rifle production already in a high stage of development. (See fig. 4.) However, the Franco-British production of rifles dropped in rate in 1918 because there was no longer need for original rifle equipment for new troops.

In the 19 months of war American factories produced over 2,879,000,000 rounds of rifle and machine-gun ammunition. This was somewhat less than the production in Great Britain during the same period and somewhat less than that of France; but America began the effort from a standing start, and in the latter part of the war was turning out ammunition at a monthly rate twice that of France and somewhat higher than that of Great Britain. (See fig. 4.)

Between April 6, 1917, and November 11, 1918, America produced as many machine guns and automatic rifles as Great Britain did in the same period and 81 per cent of the number produced by France; while at the end of the effort America was building machine guns and machine rifles nearly three times as rapidly as Great Britain and more than twice as fast as France. (Fig. 4.) When it is considered that a long time must elapse before machine-gun factories can be equipped with the necessary machine tools and fixtures, the effort of America in this respect may be fairly appreciated.

Prior to November 11, 1918, America produced in the 75-millimeter size alone about 4,250,000 high-explosive shell, over 500,000

FIGURE 4.

PRODUCTION OF RIFLES, MACHINE GUNS, AND AMMUNITION, FRANCE AND UNITED STATES COMPARED WITH GREAT BRITAIN.

AVERAGE MONTHLY RATE, JULY, AUGUST, AND SEPTEMBER, 1918.

		Per cent of rate for Great Britain.
Machine guns and machine rifles:		
Great Britain	10,947	100
France	12,126	111
United States	27,270	249
Rifles:		
Great Britain	112,821	100
France	40,522	36
United States	233,562	207
Rifle and machine-gun ammunition:		
Great Britain	259,769,000	100
France	139,845,000	54
United States	277,894,000	107

TOTAL PRODUCTION, APRIL 6, 1917, TO NOVEMBER 11, 1918.

		Per cent of rate for Great Britain.
Machine guns and machine rifles:		
Great Britain	181,404	100
France	229,238	126
United States	181,662	100
Rifles:		
Great Britain	1,971,764	100
France	1,416,056	72
United States	2,506,742	127
Rifle and machine-gun ammunition:		
Great Britain	3,486,127,000	100
France	2,983,675,000	86
United States	2,879,148,000	83

British and French production of rifles during 1918 was at a lower rate than had been attained because there was no longer need for original equipment of troops.

gas shell, and over 7,250,000 shrapnel. Of the high-explosive shell produced 2,735,000 were shipped to France up to November 15, 1918. In all 8,500,000 rounds of shell of this caliber were floated—nearly two-thirds of it being shrapnel. American troops on the line expended a total of 6,250,000 rounds of 75-millimeter ammunition,

THE MUNITIONS BUILDING, WASHINGTON, D. C.

The Lincoln Memorial and the Potomac River in the background.

A PARK OF AMERICAN-BUILT 155-MILLIMETER HOWITZERS AT ABERDEEN PROVING GROUND.

AMERICAN-BUILT G. P. F. 155-MILLIMETER GUNS STORED AT ABERDEEN PROVING GROUND.

3

GUNS OF VARIOUS SORTS AND SIZES RETURNED FROM FRANCE BY AMERICAN EXPEDITIONARY FORCES JUST AS THEY WERE UNLOADED FROM TRAIN AT ABERDEEN.

AMERICAN-BUILT ORDNANCE MATERIAL, PARKED ON ABERDEEN PROVING GROUND.

AMERICAN-BUILT ORDNANCE STORES AT ABERDEEN PROVING GROUND.

A PARK OF AMERICAN-BUILT CAISSONS, BACK FROM FRANCE, AT ABERDEEN PROVING GROUND.

CHARGING FLOOR OF AN OPEN-HEARTH FURNACE.

The charging floor of an "open-hearth" furnace building, showing two furnaces on the side into which the raw materials are "charged." Each of these furnaces is 75 feet long and 15 feet wide, and the melted steel lies in a shallow bath inside the three doors, into one of which the man is looking. The pool or "bath," as it is termed, is 33 feet long by 12 feet wide and approximately 2½ feet deep, weighs approximately 60 tons, and is composed of pig iron and well-selected scrap steel from previous operations, which are placed in the furnace through the three doors shown, the furnace being all the time at a temperature so high that the naked eye may not look within the furnace, but must be protected with blue glass or smoked glass, exactly as when looking at the noonday sun. The eye can see nothing in the atmosphere of the bath in which the steel is being melted and refined, due to the exceedingly high temperature, which gives a light as white as that of the sun.

largely high-explosive shell of French manufacture drawn from the Franco-American ammunition pool. American high-explosive shell were tested in France by the French ordnance experts and approved for use by the French artillery just before the armistice.

In artillery ammunition rounds of all calibers America at the end of the war was turning out unfilled shell faster than the French and nearly as fast as the British; but, due to the shortage in adapters and boosters, a shortage rapidly being overcome at the end of the war, the rate of production of completed rounds was only about

FIGURE 5.

PRODUCTION OF ARTILLERY AMMUNITION, FRANCE AND UNITED STATES COMPARED WITH GREAT BRITAIN.

[Types for use in A. E. F.]

MONTHLY RATE AT END OF WAR.

		Per cent of rate for Great Britain.
Unfilled rounds:		
Great Britain	7, 748, 000	100
France	6, 661, 000	86
United States	7, 044, 000	91
Complete rounds:		
Great Britain	7, 347, 000	100
France	7, 638, 000	104
United States	2, 712; 000	37

TOTAL PRODUCTION, APRIL 1, 1917, TO NOVEMBER 11, 1918.

		Per cent of rate for Great Britain.
Unfilled rounds:		
Great Britain	138, 357, 000	100
France	156, 170, 000	113
United States	38, 623, 000	28
Complete rounds:		
Great Britain	121, 739, 000	100
France	149, 827, 000	123
United States	17, 260, 000	14

one-third that of either Great Britain or France. In total production during her 19 months of belligerency America turned out more than one-quarter as many unfilled rounds as Great Britain did in the same time and about one-quarter as many as came from the French munition plants. In completed rounds alone did America lag far behind the records of the two principal allies during 1917 and 1918. (Fig. 5.)

The production of completed rounds of artillery ammunition was gaining rapidly, beginning with the early summer of 1918, and in the month of October was approaching half the rate of manufacture in

Great Britain or in France. Figure 6 shows graphically the rate at which the artillery ammunition deliveries were expanding.

In artillery proper the war ended too soon for American industry to arrive at a great production basis. The production of heavy ordnance units is necessarily a long and arduous effort even when plants are in existence and mechanical forces are trained in the work. America in large part had to build her ordnance industry from the

FIGURE 6.

COMPLETE ROUNDS OF ARTILLERY AMMUNITION PRODUCED FOR THE ARMY EACH MONTH DURING 1918 (FIGURES IN THOUSANDS OF ROUNDS).

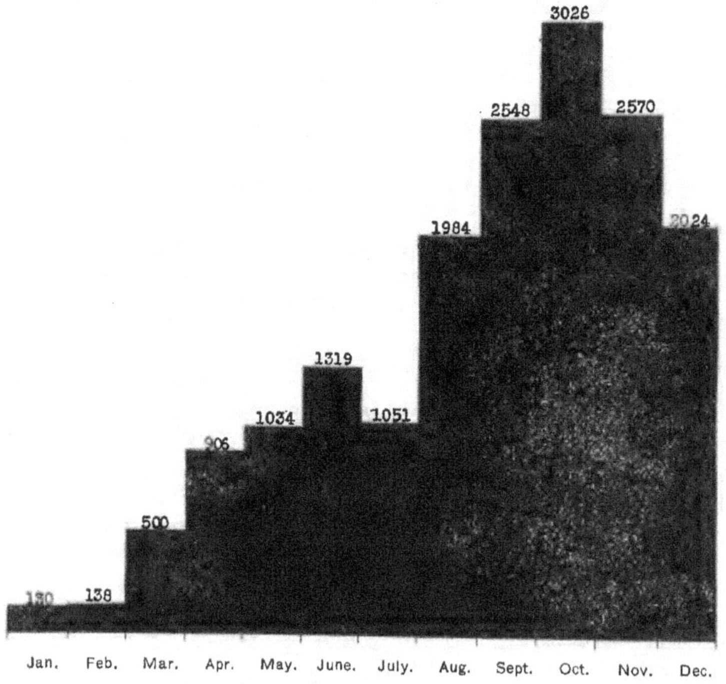

ground up—buildings, machinery, and all—and to recruit and train the working forces after that. The national experience in artillery production in the great war most like our own was that of Great Britain, who started in from scratch, even as we did. It is interesting, then, to know how Great Britain expanded her artillery industry, and the testimony of the British ministry of munitions may throw a new light on our own efforts in this respect. In discussing artillery in the war the British ministry of munitions issued a statement from which the following is an excerpt:

> It is very difficult to say how long it was before the British army was thoroughly equipped with artillery and ammunition. The ultimate size of the army aimed at was continually increased during the first three years of the war, so that the ordnance requirements were continually increasing. It is probably true to say that the equipment of the army as planned in the early summer of 1915 was completed by September, 1916. As a result, however, of the battle of Verdun and the early stages of the battle of the Somme, a great change was made in the standard of equipment per division of the army, followed by further increases in September, 1916. The army was not completely equipped on this new scale until spring, 1918.

FIGURE 7.

COMPLETE UNITS OF MOBILE ARTILLERY PRODUCED FOR THE ARMY EACH MONTH DURING 1918.

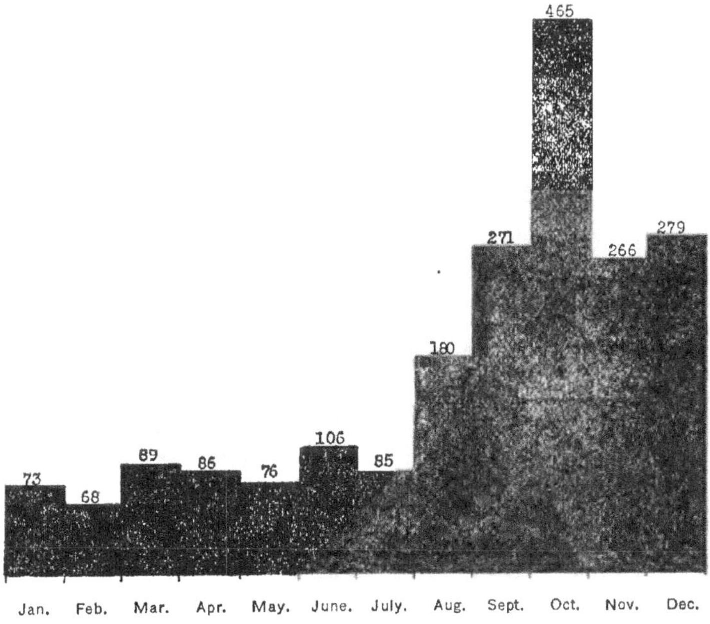

Thus it took England three and a half years to equip her army completely with artillery and ammunition on the scale called for at the end of the war. On this basis America, when the armistice came, had two years before her to equal the record of Great Britain in this respect.

As to the production of gun bodies ready for mounting, the attainments of American ordnance were more striking. At the end of the fighting America had passed the British rate of production and was approaching that of the French. In totals for the whole war period (Apr. 6, 1917, to Nov. 11, 1918) the American production of gun bodies could scarcely be compared with either that of the British or that of the French, this due to the fact that it required many months to build up the forging plants before production could go ahead.

In completed artillery units the American rate of production at the end of the war was rapidly approaching both that of the British and that of the French. In total production of complete units in the 19 months of war, American ordnance turned out about one-quarter as many as came from the British ordnance plants and less than one-fifth as many as the French produced in the same period. Figure 8 represents visually America's comparative performances in the production of gun bodies and complete artillery units.

FIGURE 8.

PRODUCTION OF ARTILLERY, FRANCE AND UNITED STATES COMPARED WITH GREAT BRITAIN.

AVERAGE MONTHLY RATE AT END OF WAR.

		Per cent of rate for Great Britain.
Gun bodies (new):		
Great Britain	802	100
France	1,138	142
United States	832	104
Complete units:		
Great Britain	486	100
France	659	136
United States	412	85

TOTAL PRODUCTION, APRIL 1, 1917, TO NOVEMBER 11, 1918.

		Per cent of rate for Great Britain.
Gun bodies (new):		
Great Britain	11,852	100
France	19,492	164
United States	4,275	36
Complete units:		
Great Britain	8,065	100
France	11,056	137
United States	2,008	25

Stress has sometimes been laid upon the fact that the American Army was required to purchase considerable artillery and other supplies abroad, the latter including airplanes, motor trucks, food and clothing, and numerous other materials. Yet, balanced against this fact is that every time we spent a dollar with the allied governments for ordnance, we sold ordnance, or materials for conversion into munitions to the allied governments to the value of five dollars.

The inter-allied ordnance agreement provided that certain munitions plants in the United States should continue to furnish supplies to the allies, and that additional plants for the allies should be built up and fostered by us. Thus, while we were purchasing artillery and ammunition from the allies we were shipping to them great quantities of raw materials, half-completed parts, and completely assembled units, and such war-time commodities as powder and explosives, forgings for cannon and other heavy devices, motors, and structural steel. The following table shows the ordnance balance sheet between America and the allied governments:

Purchase and sales from Apr. 6, 1917, to Nov. 11, 1918.

Purchases: By Army Ordnance Department from Allied governments	$450, 234, 256. 85
Sales:	
By Army Ordnance Department to Allied governments	200, 616, 402. 00
By United States manufacturers other than Army Ordnance Department to Allied governments	2, 094, 787, 984. 00
Total	2, 295, 404, 386. 00

The credit for the ordnance record can not go merely to those men who wore the uniform and were part of the ordnance organization. Rather it is due to American science, engineering, and industry, all of which combined their best talents to make the ordnance development worthy of America's greatness.

CHAPTER II.

GUN PRODUCTION.

The sole use of a gun is to throw a projectile. The earliest projectile was a stone thrown by the hand and arm of man—either in an attack upon an enemy or upon a beast that was being hunted for food. Both of these uses of thrown projectiles persist to this day, and during all time, from prehistoric days until now, every man who had a projectile to throw was steadily seeking for a longer range and a heavier projectile.

The man who could throw the heaviest stone the longest distance was the most powerfully armed. In the Biblical battle between David and Goliath, the arm of David was strengthened and lengthened by a leather sling of very simple construction. Much practice had given the young shepherd muscular strength and direction, and his longer arm and straighter aim gave him power to overcome his more heavily armed adversary.

Later, machines were developed after the fashion of a crossbow mounted upon a small wooden carriage which usually was a hollowed trough open on top and upon which a heavy stone was laid. The thong of the crossbow was drawn by a powerful screw operated by man power, and the crossbow arrangement when released would throw a stone weighing many pounds quite a distance over the walls of a besieged city or from such walls into the camps and ranks of the besiegers. This again was an attempt by mechanical means to develop and lengthen the stroke of the arm and the weight of the projectile.

With the development of explosives, which was much earlier than many people suppose, there came a still greater range and weight of projectile thrown, although the first guns were composed of staves of wood fitted together and hooped up like a long, slender barrel, wound with wet rawhide in many folds, which, when dried, exerted a compressive force upon the staves of the barrel exactly as do the steel hoops of barrels used in ordinary commercial life to-day.

This, the first gun, sufficed for a long while until the age of iron came. And then the same principle of gun construction was followed as is seen in that historical gun, the "Mons Meg," in the castle at Edinburgh. The barrel of that gun is made of square bars of iron, placed lengthwise, and similar bars of iron were wrapped hot around the staves to confine them in place and to give more resisting power than was possible with the wooden staves and the rawhide hooping.

Thus, all during the age of iron, gun development went steadily forward. Every military power was always striving by the aid of its best engineers, designers, and manufacturers to get a stronger gun, either with or without a heavier projectile, but in every case striving for greater power. As a special development we find in March, 1918, the now famous long-range gun of the Germans, which was at that time trained upon Paris, where it successfully delivered a shell approximately 9 inches in diameter, punctually every 20 minutes for a good part of each day until the gun was worn out. This occurred after a comparatively small number of shots, probably not more than 75 in all. The rapid wearing out was due to the immense demands of the long range upon the material of the gun. The Germans in the shelling of Paris used three of these long-range weapons and 183 shells are known to have fallen in the city.

The Germans evidently calculated with great care and experience upon the factors leading up to this famous long-range type of gun, which had an effective shooting distance of approximately 75 miles, which range, in the opinion of our experts, it is now quite easy for an experienced designer and manufacturer to equal and excel at will. In fact, one would hesitate to place a limit upon the length of range that could be achieved by a gun that it is now possible to design and build. In this connection it is interesting to note that the great French ordnance works at Le Creusot in 1892 produced the first known and well-authenticated long-range gun, which was constructed from the design of a 12-inch gun, but bored down to throw a 6-inch projectile. And instead of the usual 8 miles expected from the flight of a 6-inch shell this early Creusot long-range gun gave a range of approximately 21 miles with a 6-inch projectile, using a 12-inch gun's powder charge.

Closely connected with the development of the gun itself, and a necessary element of the gun's successful use, is the requirement that the weapon itself be easily transported from point to point, where its available range and capacity for throwing the projectile can be made of maximum use. This requires a gun carriage which has within itself various functions, the primary one being to establish the gun in the desired position where it can be made most effective against the enemy. Then, too, the gun carriage must have stability in order to withstand, absorb, and care for the enormous recoil energies let loose by the firing of the gun. It is obvious that the force which propels the projectile forward is equal to the reacting force to the rear, and in order to care for, absorb, and distribute to the earth this reacting force to the rear the carriage must have within itself some very peculiar and important properties. To this end there is provided what is known as a "brake," which permits the gun, upon the moment of firing, to slide backward bodily within the controlling apparatus mounted upon a fixed carriage.

The sliding of the whole gun to the rear by means of the mechanism of the brake is controlled, as to speed and time, by springs, by compressed air, by compressed oil, etc., either all together or in combinations of two or three of these agencies; so that the whole recoil energy is absorbed and the rearward action of the gun brought to rest in a fraction of a second and in but a very few inches of travel. The strains are distributed from the recoil mechanism to the fixed portion of the carriage that is necessarily anchored to the ground by means of spades, which the recoil force of each shot sets more firmly into the ground, so that the whole apparatus is thus steadily held in place for successive shots.

In mobile artillery, again, rapid firing is a prime essential. The 75-millimeter gun of modern manufacture is capable of being fired at a rate in excess of 20 shots a minute—that is, a shot every 3 seconds.

Rarely however, is a gun served as rapidly as this. The more usual rate of fire is 6 shots a minute or 1 about each 10 seconds, and this rate of fire can be maintained in the 75-millimeter gun with great accuracy over a comparatively long period.

The larger guns are served at proportionately slower rates, until as the calibers progress to the 14-inch rifles, which have been set up upon railway mounts as well as on fixed emplacements for seacoast defense, the rate of fire is reduced to one shot in three minutes for railway mounts, and to one shot a minute for seacoast mounts, although upon occasions a more rapid rate of fire can be reached.

Under rapid fire conditions, the gun becomes very hot, owing to the heat generated by the combustion of the powder within the gun at pressures as high as 35,000 pounds per square inch or more, which are generated at the moment of fire. This heat is communicated through the walls of the gun and taken off by the cooling properties of the air. Nevertheless, the wall of the gun becomes so hot that it would scorch or burn a hand laid upon it. The rapid fire and heating of the gun lessens the effective life of the weapon, due to the fact that the hot powder gases react more rapidly on hot metal than they do upon cold metal; hence a gun will last many rounds longer if fired at a slow rate than if fired at a rapid rate.

It may be helpful to keep in mind throughout that the sole purpose of a gun is to fire a projectile, as was stated at the very beginning of this chapter. All other operations connected with the life of a gun, its manufacture, its transportation to the place where it is to be used, its aiming, its loading and all its functions and operations are bound up in the single purpose of actually firing the shot.

Consider now for a moment, the life of, let us say, one of the 14-inch guns.

In the great steel mills it requires hundreds and perhaps thousands of workmen to constitute the force necessary to handle the enormous masses of steel through the various processes which finally result in the finished gun.

From the first operation in the steel mill it requires perhaps as long as 10 months to produce the gun ready for the first test. During the 10 months of manufacture of one of these 14-inch rifles there has been expended for the gun and its carriage approximately $200,000. Of course, while it requires 10 months to make a final delivery of one gun after its first operation is commenced, it should be remembered that yet other guns are following in series and that in a well-equipped ordnance factory two and perhaps three guns per month of this kind can be turned out continuously, if required.

Remembering now that it requires 10 months to produce one such 14-inch rifle and that its whole purpose is to fire a shot, consider now the time required to fire this shot. As the primer is fired and the powder charge ignited the projectile begins to move forward in the bore of the gun at an increasingly rapid rate, so that by the time it emerges from the muzzle and starts on its errand of death and destruction, it has taken from a thirtieth to a fiftieth of a second in time, depending upon certain conditions.

Assuming that a fiftieth of a second has been taken up and that the life of a large high-pressure gun at a normal rate of firing is 150 shots, it is obvious then that in the actual firing of these 150 shots only three seconds of time are consumed. Therefore, the active life of the gun, which it has taken 10 months to build, is but three seconds long in the actual performance of the function of throwing a shot.

However, after the gun has fired its life of 150 shots it is a comparatively simple and inexpensive matter to bore out the worn-out liner and insert a new liner, thus fitting the gun again for service, with an expenditure of time and money much less than would be required in the preparation of a new gun.

As the size of the powder charge decreases, a progressively longer life of the walls of the bore of a gun is attained, so that we have had the experience of a 75-millimeter gun firing 12,000 rounds without serious effect upon the accuracy of fire. Large-caliber guns, such as 12-inch howitzers, with the reduced powder charge required for the lower muzzle velocities employed in howitzer attack, have retained their accuracy of fire after 10,000 rounds.

From the fact that when in action guns are served with ammunition, aimed, fired, and cared for by a crew of men carefully trained to every motion involved in the successful use of the gun, it is most essential that the design and the material shall be such, both as to calculation in the design and as to manufacture in the material, as will insure the maintenance of the morale of the crew that serves the gun. Each man must be confident to the very last bit of fiber in his make-up that his gun is the best gun in the world, that it will behave properly, that it will protect him and his fellow soldiers who are caring for the welfare of their country, that it will respond accurately and well to every demand made upon it, that it will not yield or burst, that it will not shoot wild, but that it will in every respect give the result required in its operation.

To this end it has for generations been known that the requirements of manufacture of ordnance material, particularly for the body of the gun, are of the very highest order and call for the finest attainable quality in material, workmanship, and design.

It is well known and admitted that the steel employed in the manufacture of guns must be of the highest quality and of the finest grade for its purpose. It requires the most expert knowledge of the manufacture of steel to obtain this grade and quality. Until recently this knowledge in America was confined to the Ordnance officers of the Army and of the Navy and to a comparatively small number of manufacturers—not more than four in all—and only two of these manufacturers had provided the necessary equipment and appliances for the manufacture of complete guns.

Until 1914 the number of guns whose manufacture was provided for in this country as well as in the countries of Europe, excepting Germany, was very small. It might be stated that the sum total of guns purchased by the United States from the two factories mentioned did not exceed an average of 55 guns a year in calibers of from 3-inch to 14-inch, and that the stock of guns which by this low rate of increase of manufacture had been provided for us was pitifully small with which to enter a war of the magnitude of the one through which this country has just passed.

The two factories in question not having been encouraged by large purchases of ordnance material, as were similar industries in Germany, were not capable of volume production when we entered the war. But at the same time the gun bodies produced by these concerns at least equaled in quality those built in any other country on earth; so that while the big-gun-making art was in existence in this country and was maintained as to quality, it was most insufficient as to the quantity of the production available.

When the United States faced the war in April, 1917, arrangements were at once entered into to obtain in the shortest space of time an adequate supply of finished artillery of all calibers required by our troops and to get this supply in time to meet our men as they should set foot on the shores of France. Many thousands of forgings for guns, and finished guns too, had been ordered by the allies of the few gun makers in this country; and these makers were, at the time we got into the conflict, fully occupied for at least a year ahead with orders from the French and English ordnance departments. All of this production was immediately useful and available for the combined armies of the allies, and so it was allowed to go forward, the forgings preventing a gap in the output of the finished articles from the British and French arsenals which were then using the semifinished guns made in the old factories in existence in this country in April, 1917.

Some idea of the volume of this production in this country will be gained from the following table showing material supplied to the allies between April, 1917, and the date of the signing of the armistice, November 11, 1918.

Guns of calibers from 3-inch to 9.5-inch furnished to the allies.............	1,102
Additional gun forgings furnished to the allies......................tubes..	14,623
Shell and shell forgings furnished to the allies in this period ......pieces..	5,018,451

In supplying all of this material from our regular sources of manufacture in this country to the finishing arsenals of the allies we were but maintaining our position as a part of the general source of supply. The plan of the French and British ordnance engineers at the outbreak of the war in 1914 was to build their factories as quickly and as extensively as could possibly be done. By the time the United States entered the war all of these factories were in operation and clamoring for raw material at a rate which was far in excess of that which could be supplied by the home steel makers in Great Britain and France. Consequently their incursions into the semifinished ordnance material supplies in the United States were necessary. In sending these large quantities of our own materials abroad, when we needed them ourselves, we were distinctly adding to the rate and quantity of the supply of finished ordnance for the use of our own Army in the field as well as being at the same time of inestimable value to the allies. This was because the French and British had agreed to supply our first armies with finished fighting weapons while we were giving them the raw materials which they needed so badly.

The four gunmakers in America meanwhile were being expanded into a total of 19 makers. All of these 19 factories during the month of October, 1918, were practically in full operation. Many of them were producing big guns at a faster rate than that for which the plants had been designed. In the month of October, 1918, with 3 of the 19 factories yet to have their machine-tool equipment completed, there were produced 2,031 sets of gun forgings between the calibers of 3-inch and 9.5-inch, which is at the rate of upward of 24,000 guns a year. This figure, of course, does not indicate anything of the gun-finishing capacity of the country; yet this expansion may be contrasted to the fact that our supply of finished guns prior to 1917 amounted only to 55 weapons a year.

Monthly production of finished cannon, ranging in size from 75 millimeters to 240 millimeters, at the various machining and assembling plants.[1]

Caliber.	1917	1918					
	Dec.	Jan.	Feb.	Mar.	Apr.	May.	June.
75-millimeter	5	45	48	52	74	127	169
3-inch antiaircraft	3	16	24	16	2	....	11
4.7-inch	....	....	....	....	....	....	6
155-millimeter howitzer	....	....	3	10	16	28	75
155-millimeter gun	....	....	....	....	....	....	....
8-inch howitzer	....	....	....	34	38	8	....
240-millimeter howitzer	....	....	....	....	....	....	....
Total	8	61	75	112	130	163	261

Caliber.	1918						Total.
	July.	Aug.	Sept.	Oct.	Nov.	Dec.	
75-millimeter	142	204	199	214	320	214	1,813
3-inch antiaircraft	10	11	22	50	34	31	230
4.7-inch	8	15	29	71	50	39	218
155-millimeter howitzer	110	248	206	350	231	179	1,456
155-millimeter gun	2	....	14	51	22	40	129
8-inch howitzer	....	28	22	33	14	14	191
240-millimeter howitzer	....	1	....	....	1	....	2
Total	272	507	492	769	672	517	4,039

Monthly production of cannon forgings.

Caliber.	1917	1918					
	Dec.	Jan.	Feb.	Mar.	Apr.	May.	June.
75-millimeter	4	13	73	62	79	239	376
3-inch antiaircraft	....	....	6	7	5	4	12
4.7-inch gun	....	....	9	10	8	28	....
155-millimeter howitzer	2	13	26	61	44	146	133
155-millimeter gun	....	....	....	....	1	15	4
8-inch howitzer	....	....	....	34	38	8	....
240-millimeter howitzer	....	....	....	....	....	....	....
Total	6	26	114	174	175	440	525

Caliber.	1918						Total.
	July.	Aug.	Sept.	Oct.	Nov.	Dec.	
75-millimeter	574	678	754	1,385	674	310	5,221
3-inch antiaircraft	10	6	49	163	124	18	404
4.7-inch gun	70	100	84	35	25	53	422
155-millimeter howitzer	176	204	273	279	276	62	1,695
155-millimeter gun	42	28	56	105	79	24	354
8-inch howitzer	....	28	22	33	14	14	191
240-millimeter howitzer	....	30	21	31	22	49	153
Total	872	1,074	1,259	2,031	1,214	530	8,440

[1] Carriages, recuperators, and sights had to be added to these cannon to make them complete units ready for service.

Our chain of gun factories, that were making this remarkable production, were built as follows:

One at the Watertown Arsenal, Watertown, Mass., near Boston, for the manufacture of rough machined gun forgings of the larger mobile calibers. This factory was entirely built and equipped on Government land with Government money and is splendidly able to produce rough machined gun forgings of the highest quality at the rate of two sets a day for the 155-millimeter G. P. F. rifles, and one set a day of the 240-millimeter howitzers.

At Watervliet Arsenal, Watervliet, N. Y., large extensions were made to the existing plant that had always been the Army's prime reliance for the finishing and the assembly of guns of all calibers, including the very largest. This plant was extended to manufacture complete four of the 240-millimeter howitzers each day, and two a day of the 155-millimeter G. P. F. guns.

At Bridgeport, Conn., there was constructed a complete new factory by the Bullard Engineering Works for the United States to turn out four 155-millimeter G. P. F. guns a day.

At Philadelphia, the Tacony Ordnance Corporation, as agents for the Government, erected complete a new factory officered and manned by experts well-trained and experienced in the difficult art of the manufacture of steel and gun forgings. On October 11, 1917, the grounds for this great undertaking had been merely staked out for the outline of the buildings. Seven months later, on May 15, 1918, the entire group of buildings, comprising a complete steel works from making the steel to the final completion of 155-millimeter gun forgings, was entirely erected at a cost of about $3,000,000. This difficult and rapid building operation was carried through successfully during the extraordinarily severe winter of 1917-18. On June 29, 1918, the first carload of gun forgings was accepted and shipped from this plant, so we have the marvelous enterprise of building a complete steel works from the bare ground forward to the shipment of its first forgings in a total elapsed time of only eight and one-half months.

At another, the works of the Midvale Steel Co. in Philadelphia, large extensions were made to enable some of the larger guns to be produced, to be finished later at the Watervliet Arsenal.

At the Bethlehem Steel Co.'s plant, Bethlehem, Pa., as early as May, 1917, orders were placed and appropriations allotted for expansions to this enterprise to enable a rapid output of a larger number of gun forgings and finished guns.

Large extensions were made at the works of the Standard Steel Works Co., Burnham, Pa., to increase their existing forging and heat treating facilities, so that at this plant two sets of 155-millimeter howitzers and one set of 155-millimeter gun forgings were produced each day.

At Pittsburgh, Pa., the plants of the Heppenstall Forge & Knife Co. and the Edgewater Steel Co. were extended so as to provide for the daily production at the first plant of forgings for one 3-inch antiaircraft gun and one 4.7-inch gun, and at the second plant of forgings for one 155-millimeter G. P. F. gun and one 240-millimeter howitzer per day.

At Columbus, Ohio, the Buckeye Steel & Castings Co. in combination with the works of the Symington-Anderson Co. at Rochester, N. Y., had their facilities extended to provide for the manufacture each day of six sets of forgings for the 75-millimeter guns.

At the Symington-Anderson Co. in Rochester, N. Y., there was provided a finishing plant for the 75-millimeter gun with a capacity of 15 finished guns per day.

At Erie, Pa., one of the most remarkable achievements in rapid construction and successful mechanical operation was performed by the erection of a plant that was commenced in July, 1917, and out of which the first production was shipped to the Aberdeen Proving Grounds in February, 1918. The American Brake Shoe & Foundry Co. built and operated this plant as agents for the Ordnance Department, and much credit is due them for their energy and organizing capacity.

It is doubtful if history records any similar enterprise in which guns were turned out in a plant seven months from the date of beginning the erection of the factory. This plant was laid out to manufacture 10 of the 155-millimeter Schneider-type howitzers a day, and before the signing of the armistice it had more than fulfilled every expectation by regularly turning out up to 15 howitzers a day, or 90 a week.

At Detroit, Mich., the Chalkis Manufacturing Co. adapted an existing plant, and additional facilities were erected for the manufacture of three of the 3-inch antiaircraft guns each day.

At Madison, Wis., the Northwestern Ordnance Co. erected for the United States an entire new factory, beautifully equipped for the manufacture of four guns a day of the 4.7-inch model.

At Milwaukee, Wis., the Wisconsin Gun Co. put up for the Government an entirely new works capable of finishing six 75-millimeter guns each day. The plants at both Milwaukee and Madison acquitted themselves very well and gave us guns of the highest quality.

At Chicago, the Illinois Steel Co. expanded existing facilities to produce more of the necessary electric furnace steel, which was forged into guns at several works producing gun forgings, both for the Army and Navy.

At Indiana Harbor, Ind., the works of the Standard Forgings Co., whose sole business had been the volume production of forgings with steam hammers and hydraulic presses, were expanded to the enormous degree of producing each day 10 sets of gun forgings for the 155-millimeter howitzer and 25 sets a day for the 75-millimeter gun. It should be stated that this was a triumph of organizing ability and that this factory was one of our main reliances for these guns.

At Gary, Ind., the American Bridge Co. created what is perhaps the finest gun-forging plant in the world, comprising four presses from 1,000 tons to 3,000 tons forging capacity and all the other necessary apparatus for the production each day of two sets of 155-millimeter G. P. F. guns and the equivalent of one and one-half sets a day of the 240-millimeter howitzers.

At Baltimore, Md., the plant of the Hess Steel Corporation was enlarged from its peace-time capacity and caused to produce at three times its normal rate the special steels required for gun manufacture.

It will become evident that the collection of machinery, buildings, and equipment necessary to produce these guns in the short space of time required and at the rate of production stipulated, was an enormous task in itself. It required the production of vast quantities of raw materials and the congregating in one place of large numbers of men capable of undertaking the exceedingly intricate mechanical processes of manufacture. The success of this plan and its carrying out is due largely to the loyalty of the manufacturers who unselfishly came forward early in 1917 and agreed at the request of the Ordnance Department to turn over their plants, lock, stock, and barrel, to the requirements of the department; agreed also to undertake the manufacture of products totally unfamiliar to them; agreed likewise to lend all of their organizing ability and great material resources to the success of the plants which the United States found necessary to build in the creation of a new art, in new locations and in an extent theretofore undreamed of.

THE MAKING OF A BIG GUN.

Steel, of course, and steel in some of its finest forms is the basis of gun manufacture. The word "steel" for the purpose of producing guns means much more than is ordinarily carried by the word in its everyday and most commonly accepted use. Only steel of the very highest quality is suitable for gun manufacture, as was indicated previously when attention was directed to the complete reliance which the operating crews must place in their guns and the severity of the uses to which the big guns are put.

Let us take a hasty trip through a big gun plant, watching the processes through which is finally evolved from the raw materials one of our hardy and efficient big guns.

Entering an open-hearth furnace building at one of our big gun plants, we find two large furnaces in which the raw materials are charged. Each of these furnaces is 75 feet long and 15 feet wide, and in them in a shallow bath or pool lies the molten steel. The pool is about 33 feet long by 12 feet wide and approximately 2½ feet deep. This pool, or "bath" as it is termed, weighs approximately 60 tons and is composed of pig iron and well-selected scrap steel from previous operations.

The furnace is at all times during the operation of melting these raw materials in the bath kept at such a high temperature that the eye may not look within at the molten mass without being protected with blue glass or smoked glass, exactly as when looking at the noonday sun. The eye can see nothing in the atmosphere of the bath in which the steel is being melted and refined because the temperature is so exceedingly high that it gives a light as white as that of the sun.

After 12 or 15 hours of refining treatment in this furnace the metal is tested, analyzed in the chemical laboratory, and, if found to be refined to the proper degree, it is allowed to flow out of the furnace on the opposite side from that through which it entered. Flowing out of the furnace the entire charge of 60 tons finds its way into a huge ladle which is suspended from a traveling crane capable of safely carrying this great weight.

The ladle is then transferred by the crane to a heavy cast-iron mold which is built so as to contain as much of the 60 tons of molten metal as is required for the particular gun forging under manufacture.

The mold, which we have before us now on our imaginary trip through the gun plant, will provide an "ingot" from the molten metal that will be 40 inches in diameter and 100 inches high. On top of this ingot is a brick-lined so-called "sinkhead." This sinkhead is that portion of the molten metal that has been allowed to cool more slowly in the brick lining than the ingot does in the cast-iron mold proper. The ingot with the sinkhead will weigh approximately 60,000 pounds.

This sinkhead is to insure greater solidity to the portion of the ingot which is used for the gun forging. Only that part of the ingot below the sinkhead enters the forging. The sinkhead itself is cut off while hot under the press in a subsequent operation and afterwards remelted.

Next the ingot is placed under a 2,000-ton forging press which handles ingots up to 45 inches in diameter. There it is forged into a square shape after coming from the mold in an octagonal form. Previous to its being put under this press, however, a careful chemical analysis has been made of the ingot to determine that it is satisfactory for gun purposes, and then before being put under the press the whole ingot is heated in the charge chamber and fired either by a gas or oil flame.

VIEW OF LADLE, CONTAINING 60 TONS OF MOLTEN STEEL SUSPENDED FROM A TRAVELING CRANE.

The ladle is receiving metal from the furnace and the crane is conveying the ladle to the mold.

MOLTEN STEEL BEING POURED FROM LADLE INTO MOLD, WHICH IS OF HEAVY CAST-IRON CONSTRUCTION, AT THE TACONY ORDNANCE CORPORATION.

Arrow points from letter A to a completed ingot from a mold. The brick-lined sink head is a part of the mold and is to insure greater solidity to the portion of the ingot which is used for the gun forging; only the part below the sink head entering the forging, the sink head itself being cut off hot under the press in a subsequent operation.

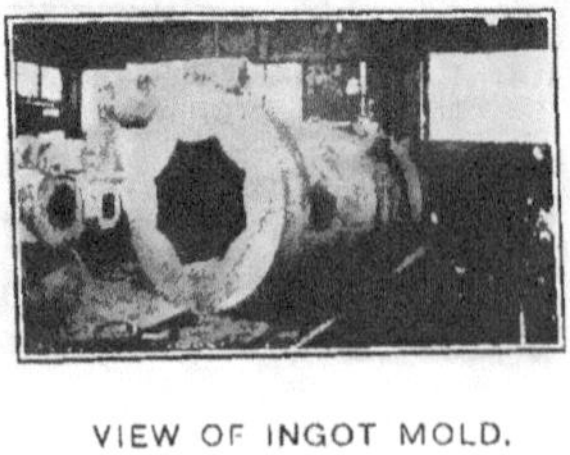

VIEW OF INGOT MOLD.

A 2,000-TON FORGING PRESS AT THE TACONY ORDNANCE CORPORATION PLANT.

This press can forge ingots up to 45 inches in diameter. The ingot under the press is shown in a partly forged state. Note that the original octagonal shape of the ingot as it came from the mold has been forged down to a square shape and later will be forged into a round shape. After coming from the mold, the ingot has been subjected to a careful chemical analysis to determine its fitness for use as a gun barrel.

A 9,000-TON HYDRAULIC FORGING PRESS IN THE PLANT OF THE MIDVALE STEEL CO.

This press is needed for such large caliber guns as the 14-inch and 16-inch guns. The piece of forging under the press is armor plate and not a gun forging.

THREE TUBES OF THE 155-MILLIMETER GUNS SUSPENDED AT FURNACE.

Three tubes of the 155-millimeter guns suspended at furnace ready for heating and quenching to give them the necessary combination of hardness and toughness. The door of the furnace is open. The tubes remain in this furnace for perhaps eight hours at a temperature of 1,500° Fahrenheit or until a bright yellow color, uniform in every part.

A TUBE FOR A 12-INCH GUN JUST OUT OF THE FURNACE, WHERE IT WAS TEMPERED AT WHITE HEAT AND IS NOW READY FOR QUENCHING IN THE PLANT OF THE MIDVALE STEEL CO.

The gun tube is 41 feet long.

TUBE OF A 155-MILLIMETER GUN BEING TURNED, PRIOR TO BORING, AT THE TACONY ORDNANCE CORPORATION PLANT.

TUBE OF A 155-MILLIMETER GUN IN A LATHE BEING BORED.

The ingot out of which this tube was made, came from the mold in an octagonal shape and later was forged into a square shape and finally made round. It now, too, has the hole bored partially into it. Through this hole, ultimately, will pass the projectile.

The forging press used for the larger caliber guns, such as 14-inch and 16-inch, is of a 9,000-ton weight capacity.

After the ingot forging has been reduced from squareness to a cylindrical shape under the press, it is allowed to cool, then taken to the machine shop, where it is turned and the hole through which the projectile ultimately will pass is bored into it. This hole is somewhat smaller than the diameter of the projectile, because in the finishing operation, when the gun is assembled finally and put together, the hole must be within one-one-thousandth of an inch of the diameter required, which is all the tolerance that is allowed from the accuracy to which the projectiles are brought. Otherwise the accuracy of the gun in firing would be injured and the reliability of its aim would not be satisfactory.

During all of these operations with the ingot, the steel is largely in the soft condition in which it left the forging press. As is well known, steel is capable of taking many degrees of "temper." Temper is an old term that no longer is quite descriptive of the condition desired or obtained, but it is sufficiently expressive of the condition desired for the purposes here. This condition is one of a certain degree of hardness—greater than that ordinarily carried by the soft steel—combined with the greatest obtainable degree of toughness. This combination of hardness and toughness produced to the proper degree resists the explosive power of the powder and also causes the wear on the gun in firing to be diminished and made as slight as possible.

To effect this combination of hardness and toughness it is necessary to take the bored and turned tubes of the guns and suspend them by means of a specially made apparatus in a furnace where they are heated for a period of perhaps eight hours to a temperature of approximately 1,500° F. or a bright-yellow color, uniform in every part of the piece.

After being subjected to this treatment for the time mentioned, the tube is then conducted by means of a traveling crane apparatus to a tank of warm water in which it is dipped and the heat rapidly taken from it down to a point of practically atmospheric temperature. This "quench" as it is called, produces the required degree of hardness called for by the ordnance officers' design; but the piece has not yet got the required degree of toughness. This toughness is now imparted to the hard piece by heating it once more in another furnace to a temperature of approximately 1,100° F., or a warm rosy red, for a period of perhaps 14 hours. From this temperature, the piece is allowed to cool naturally and slowly to the atmospheric temperature.

The ordnance inspectors at this point determine whether the piece has the required properties in a sufficient degree, by cutting from the tube a piece 5 inches long and ½ inch in diameter. The ends of this piece are threaded suitably for gripping in a machine. The piece is then pulled until the half-inch stem breaks. The machine registers the amount of force required to break this piece and this gives the ordnance engineer his test as to the degree of hardness and toughness to which the piece has been brought by the heat treatment processes just described.

A satisfactory physical condition having been determined by pulling and breaking the test pieces described, the whole forging is sent to the finishing shop where it is machined to a mirror polish on all its surfaces. The diameters are accurately measured and the forgings assembled into the shape of a finished gun.

In this process there is required a different kind of care and accuracy. Up until this time the care has been to provide a metal of proper consistency and quality. From this point forward the manufacture of a gun requires the machining and fitting of this metal into a shape and form so accurate that the full strength of the gun and the best accuracy of fire may be attained.

To explain how and why hoops are placed upon the gun tubing and how the various hoops are shrunk from the outside diameter of the gun will require a few lines.

Cannon are made of concentric cylinders shrunk one upon another. The object of this method of construction is twofold. The distinctly practical object is the attainment throughout the wall of each cylinder of the soundness and uniformity of metal which is more certainly to be had in thin pieces than in thick ones; the other object is more closely connected with the theory of gun construction.

When a hollow cylinder is subjected to an interior pressure the walls of the cylinder are not uniformly strained throughout their thickness, but the layer at the bore is much more severely strained than that at the outside. This can be readily seen if we consider a cylinder of rubber, for example, with a bore of 1 inch and an exterior diameter of 3 inches, which are about the proportions of many guns. If we put an interior air pressure on the cylinder until we expand the bore to 2 inches, the exterior diameter will not thereby be increased 1 inch. But supposing that it were increased as much as the bore, that is, 1 inch, we would have the diameter, and therefore the circumference, of the bore increased 100 per cent, and the circumference of the exterior increased 33⅓ per cent. That is, the layer at the bore would be strained three times as much as that at the exterior, and the interior layer would commence to tear before that at the exterior would reach anything like its limit of strength. The whole wall of the cylinder therefore would not be contributing its full strength toward resisting the interior pressure, and there would be a waste of material as well as a loss of strength.

Let us now consider, instead of our simple cylinder, a built-up cylinder composed of two concentric ones, the inner one of a bore originally a little greater than 1 inch, and the outer one of exterior diameter a little less than three inches, originally; so that when the outer one is pressed over the inner one (its inner diameter being originally too small for it to go over the inner one without stretching) the bore of the inner one is brought to 1 inch, and the exterior of the outer one to 3 inches. We now have a cylinder of the same dimensions as our simple one, but in a different state; the layers of the inner one being compressed and those of the outer one extended.

If now we commence to put air pressure on the bore, we can put on a certain amount before we wipe out the compression of the inner layer, and bring it to a neutral state, and thereafter can go on putting on more pressure until we stretch the inner layer 100 per cent beyond the neutral state, as before; which would take just as much additional pressure as the total pressure which we employed with our simple cylinder. We have therefore gained all that pressure which is necessary to bring the inner layer of our built-up cylinder from its state of compression to the neutral state. If we have so proportioned the diameter of junction of our inner and outer cylinders and so gauged the amount of stretching required to get the outer one over the inner one that we have not in the process caused any of the layers of the outer one to be overstrained, the gain has been a real one, attained by causing the layers of the outer cylinder to make a better contribution of strength toward resisting the interior pressure. This is the theory of the built-up gun.

The number of cylinders employed generally increases, up to a certain limit, with the size of the gun, practical considerations governing; and the "shrinkage," or amount by which the inner diameter of the outer cylinder is less than the outer diameter of the one which it is to be shrunk over, is a matter of nice calculation. Roughly speaking, it is about one and one half one-thousandths of an inch for each inch of diameter, varying with the position of the cylinder in the gun; and its accurate attainment, throughout the length of the cylinder of a large gun, is a delicate matter of the gun-maker's art and the machinist's skill.

The method of assembly is to have the cold tube set upright and prepared for a circulation of water within the bore of the tube to keep it cool. Then the hoop, whose inside diameter is smaller than the outside diameter of the tube on which it is to be shrunk, is measured and carefully heated to a temperature of approximately 450° F., or just about the temperature of a good oven for baking or roasting. This mild temperature so expands the material in the hoop that the difference of diameter is overcome and the hot hoop is expanded to a larger inside diameter than the outside diameter of the cold tube on which the hoop is to be placed. Next the hot, expanded hoop is placed in position around the breech end of the tube, and slowly and carefully cooled, so that in contracting from the high temperature to the low ordinary temperature, the hoop shrinks toward its original diameter and thus exerts an inclosing pressure or compressive strain upon the breech end of the tube.

Now when the gun is fired the tube tends to expand under the pressure and this expansion is resisted, first by the compressive force exerted by the shrunken hoop and later by the hoop itself, so that the built-up system is stronger and better able to resist the explosive charge of the burning powder than would be the case if the gun were made in one piece and of the same thickness of metal.

This brief explanation will show why so many pieces are provided for the manufacture of the finished gun and the reason for the large number of machine tools and machining operations necessary in order to carry forward the manufacture of the finished article. Sometimes one or more of the outer cylinders are replaced by layers of wire, wound under tension.

Both our 4.7-inch gun, model 1906, with which our troops have been equipped for a long time and which throws a projectile weighing 45 pounds a distance of about 6 miles, and the French 75-millimeter (2.95-inch) gun, successfully used by the French since 1897, were designed to be drawn by horses, and the guns are best used when drawn by teams of 6 or 8 horses. As the horse has a sustained pulling power of only 650 pounds, it is obvious that the weight to be drawn by the team of 6 horses must not be more than 3,900 pounds. So there is every incentive for making mobile artillery of this kind as light as possible, consistent with the strength required for the work to be done. Thus the pulling power of the horse coupled with his speed has been the limiting factor in the design and weight of mobile field artillery.

As one of our foremost United States ordnance engineers once said, "the limited power of the horse is what has governed the weight of our artillery," and that "if Divine Providence had given the horse the speed of the deer and the power of the elephant, we might have had a far wider and more effective range for our mobile artillery."

One of the answers of the United States ordnance engineers to this problem, as developed in the recent war, has been the production of a tractor to replace the horse, and this tractor has the speed of the deer and the power of the elephant. The most powerful tractors are mounted on track-laying devices and are colloquially known as caterpillars. One of these powerful caterpillars, on which is mounted an 8-inch howitzer with a range of 6 miles, which is

A SECTION OF THE MIDVALE STEEL CO. PLANT, SHOWING TYPES OF 6-INCH, 7-INCH, AND 8-INCH, NICKEL-STEEL, BREECH-LOADING RIFLES.

THREE FINISHED 8-INCH RIFLES, 45 CALIBERS, SET UP ON TURRET MOUNTS IN THE PLANT OF THE MIDVALE STEEL CO. WHERE THEY WERE MADE.

In the foreground are three large gun tubes partially completed. This company started the manufacture of the first piece of ordnance material in America in 1880.

manned and operated by only two men, and which can go up hill and down hill, over broken brushwood, trees, etc., was recently given a severe test at the Aberdeen Proving Grounds. Here it was sent through a dense wood in which it bumped square into a live locust tree that was 17 inches in diameter at the bottom. This tree, almost the tallest in the wood, was prostrated by the attack of the tractor, which rode over it and then emerged from the wood, took up its position, and fired its shot almost in as short a time as that which it takes to tell of the deed. Truly the power of the elephant and the speed of the deer has been brought to the aid of the ordnance engineer for any future warlike operations.

The number of workmen employed in gun production at once in this country totaled 21,329, and fully that many more are estimated to have been employed in the manufacture of gun carriages and fire-control instruments. Consequently in turning out the complete big guns there were fully 42,000 workmen engaged by the month of October, 1918. Furthermore, these men became so skilled in their work that it may be said that the difficult art of gun making has become firmly established in this country and that the United States may now and at any time in the near future rely on this trained body of artisans for the finest kind of gun-metal manufacture.

PRODUCTION OF CANNON FORGINGS.

Production of cannon forgings during the war at the various plants.

Caliber.	Contractor.	1917 Dec.	1918 Jan.	Feb.	Mar.	Apr.	May.	June.	July.	Aug.	Sept.	Oct.	Nov.	Dec.	Total.
75-mm. field gun, model 1916	Bethlehem Steel Co., Bethlehem, Pa	2	1	32	14	13	3	75	23	7	1	180	1		352
	Standard Forgings Co., Indiana Harbor, Ind.	1	5	11	10	19	5	67	29	2					149
	Buckeye Steel Co., Columbus, Ohio						154		10						164
75-mm. field gun, model 1897	do						44	162	419	325	322	658	181	224	2,335
	Standard Forgings Co							10	32	275	310	471	245	39	1,382
75-mm. field gun, model 1917	Bethlehem Steel Co	1	7	30	38	47	33	62	61	69	121	76	247	47	839
3-inch antiaircraft gun	do			6	7	5	4	12	10	6	46	112	109	5	322
	Heppenstall Forge & Knife Co., Pittsburgh, Pa.										3	51	15	13	82
4.7-inch gun	Bethlehem Steel Co			9	10	8	28		70	94	66	14		43	342
	Heppenstall Forge & Knife Co									6	18	21	25	10	80
155-mm. howitzer	Bethlehem Steel Co		10	26	51	9	37	25	5		11	52	19	32	277
	Standard Forgings Co	2	3		10	20	55	44	89	74	169	127	157	10	760
	Standard Steel Co					15	54	64	82	130	93	100	100	20	658
155-mm. gun	Bethlehem Steel Co					1	9		21	7	5	4	1		48
	Edgewater Steel Co., Pittsburgh, Pa									4	13	21	24	5	67
	Standard Steel Car Co., Burnham, Pa						6	4	21	14	23	41	21	9	139
	Tacony Ordnance Co., Philadelphia, Pa									3	15	31	26		75
	American Bridge Co., Gary, Ind											8	7	10	25
8-inch howitzer	Midvale Steel Co				34	38	8			28	22	33	14	14	191
240-mm. howitzer	Bethlehem Steel Co									30	16	16	19	16	97
	Edgewater Steel Co										1			14	15
	Tacony Ordnance Co										4	15	3	12	34
	Watertown Arsenal											(1)		7	7
Total		6	26	114	174	175	440	525	872	1,074	1,259	2,031	1,214	530	8,440

CANNON MACHINING AND ASSEMBLING.

Progress of the work of machining and assembling cannon at the various factories during the war.

Caliber.	Contractor.	1917 Dec.	1918 Jan.	Feb.	Mar.	Apr.	May.	June.	July.	Aug.	Sept.	Oct.	Nov.	Dec.	Total.
75-mm. field gun, model 1916	Symington-Anderson Co., Rochester, N. Y.					1	51	81	61	88	48	74	12		416
	Wisconsin Gun Co., Milwaukee, Wis.						8	18	20	38	27	5			116
	Watervliet Arsenal, Watervliet, N. Y.	4	38	18	14	26	35	8		8		5	5	5	166
	Bethlehem Steel Co.									1	1				2
75-mm. field gun, model 1897	Symington-Anderson Co.										1	52	50	136	239
	Wisconsin Gun Co.										1	2	6	26	35
75-mm. field gun, model 1917	Bethlehem Steel Co.	1	7	30	38	47	33	62	61	69	121	76	247	47	839
3-inch antiaircraft gun	Chalkis Manufacturing Co., Detroit, Mich.								1	7	19	48	29	30	134
3-inch antiaircraft gun, 15-pdr.	Watervliet Arsenal	3	16	24	16	2		11	9	4	3	2	5	1	96
4.7-inch, model 1906	Northwestern Ordnance Co., Madison, Wis.									5	7	31	23	32	98
	Watervliet Arsenal							6	8	10	22	40	27	7	120
155-mm. howitzer	American Brake Shoe & Foundry Co., Erie, Pa.			3	10	16	28	75	110	248	206	350	231	179	1,456
155-mm. gun	Bullard Engine Works Co., Bridgeport, Conn.								1		14	28	18	36	97
	Watervliet Arsenal								1			23	4	4	32
8-inch howitzer	Midvale Steel Co.				34	38	8			28	22	33	14	14	191
240-mm.	Watervliet Arsenal									1			1		2
Total		8	61	75	112	130	163	231	272	507	492	769	672	517	4,039

1 Figures in first table indicate delivery of completed sets of forgings only. Deliveries of finished and accepted gun forgings, not in complete sets, were made in carload lots and in other large quantities by various factories prior to the dates when their receipt of machine tools enabled them to produce completed sets. For instance, Watertown Arsenal made its first carload shipment of forgings on Oct. 28, 1918.

CHAPTER III.

MOBILE FIELD ARTILLERY.

The chance observer might assume that once the Ordnance Department had succeeded in putting in production the cannon of various sizes described in the preceding chapter the battle of providing artillery was as good as won. But such was not the case. Even after the ponderous tubes had come finished from the elaborate processes of the steel mills, the task of the ordnance officers had only just begun. Each one of these guns had to be rendered mobile in the field and it had to be equipped with a mechanism to take up the retrograde shock of firing (the "kick") and to prevent the weapon from leaping out of aim at each discharge.

Mobility to a gun is given by the carriage on which it rides. The device which absorbs the recoil and restores the gun to position is called the recuperator (in the case of the hydropneumatic French design) or the recoil mechanism. Carriage and recuperator, or recoil mechanism, together are known as the mount.

The forging, boring, reinforcing, machining, and finishing of the gun body is not half the battle of manufacturing a modern military weapon; it is scarcely one-third of it. No ordnance officer of 1917–18 will ever forget the heartbreaking experiences of manufacturing the mounts, a work which went along simultaneously with the production of the cannon themselves. The manufacture of carriages often presented engineering and production problems of the most baffling sort. As to the recuperators, a short analysis of the part they play in the operation of a gun will indicate something of the nature of the project of building them in quantities.

The old schoolbook axiom that action and reaction are equal has a peculiar emphasis when applied to the firing of a modern piece of high-power artillery. The force exerted to throw a heavy projectile 7 miles or more from the muzzle of a gun is equally exerted toward the breach of the weapon in its recoil. Some of these forces handled safely and easily by mechanical means are almost beyond the mind's grasp.

Not long ago a touring car, weighing 2 tons, traveled at the rate of 120 miles an hour along a Florida beach. Conceive of such a car going 337 miles an hour, which is much faster than any man ever traveled; then conceive of a mechanism which would stop this car, going nearly 6 miles a minute, stop it in 45 inches of space and half a second of time, without the slightest injury to the automobile. That is precisely the equivalent of the feat performed by the recuperator of a 240-millimeter howitzer after a shot.

Conceive of a 150,000-pound locomotive traveling at 53.3 miles an hour. The action of the 240-millimeter recuperator after a shot is equivalent to stopping that locomotive in less than 4 feet in half a second without damage.

The forging for the 155-millimeter howitzer's recuperator is a block of steel weighing nearly 2 tons—in exact figures, 3,875 pounds. This must be bored and machined out until it weighs, with the accessory parts of the complete recuperator placed on the scales with it, only 870 pounds. It is scarcely fair to a modern hydropneumatic recuperator to say that it must be finished with the precision of a watch. It must be finished with a mechanical nicety comparable only to the finish of such a delicate instrument as a navigator's sextant or the mechanism which adjusts the Lick telescope to the movement of the earth. No heavy articles ever before turned out in American workshops required in their finish the degree of microscopic perfection the recuperators called for.

We adopted from the French, the greatest of all artillery builders, four recuperators—one for the 75-millimeter gun, one for the 155-millimeter gun, another for the 155-millimeter howitzer, and the fourth for the 240-millimeter howitzer. These mechanisms had never been built before outside of France. Indeed, one could find pessimists ready to say that none but French mechanics could build them at all and that our attempt to duplicate them could end only in failure. Yet American mechanical genius "licked" every one of these problems, as the men in the greasy overalls say, and did it in little more than a year of time after the plans came to the workshops. There was not one of these beautiful mechanisms, in France the product of patient handiwork on the part of metal craftsmen of deep and inherited skill, that eventually did not become in American workshops a practical proposition of quantity production.

The problem of building French recuperators in the United States, in short, may be regarded as the crux of the whole American ordnance undertaking in the war against Germany, the index of its success. It presented the most formidable challenge of all to American industrial skill. There were men whose opinion had to be considered and who were convinced that it was impracticable to attempt to produce French recuperators here. Although the superiority of these recoil devices in their respective classes were universally conceded, Germany had never been able to make them, while England, with the cooperation of the French ordnance engineers freely offered, did not attempt them. The French built them one by one, as certain custom-built and highly expensive automobiles are produced. When American factories proposed to produce French recuperators not only but to manufacture them by making parts and assembling them according to the modern practice of quantity production, the ranks of the skeptics increased.

Yet, as we have said, the thing was done. The first of these recuperators ever produced outside of the French industry were produced in America and manufactured by typically American quantity methods.

The first of these recuperators to come into quantity production was that for the 155-millimeter howitzer. Rough forgings began to be turned out in heavy quantities by the Mesta Machine Co. in the spring of 1918, while the Watertown Arsenal, the other contractor, reached quantity production in rough forgings in September, 1918. At their special recuperator plant at Detroit the Dodge Bros. turned out the first finished 155-millimeter howitzer recuperator in July, 1918, and went into quantity production with them in September, producing 495 in the month of November alone, and turning out up to the end of April, 1919, the great number of 1,601 of them.

Next in order of time to be conquered as a factory problem was the 155-millimeter gun recuperator. The rough forgings at the Carnegie Steel Co., the sole contractor, were in quantity production in the spring of 1918. The first of these recuperators finished came from the Dodge plant in October, 1918; and although 30 issued from the plant and were accepted before the end of the year, quantity production may be said to have started on January 1, 1919, when the factory began producing them at the rate of more than four a day. In March the high mark of 361 recuperators was reached, and the total production up to the end of April was 880.

The heavy 240-millimeter howitzer recuperator was third to come into quantity production. The rough forgings were being turned out in quantity in the spring of 1918 by the Carnegie Steel Co., while the Watertown Arsenal, the other contractor, produced a number of these rough forgings in August, 1918. The two contractors for finishing and turning out the complete recuperators were the Otis Elevator Co., at its Chicago plant, and the Watertown Arsenal. The arsenal produced the pilot recuperator in October, 1918. In January the Otis Elevator Co. produced its first four, while quantity production began in February, 1919, both contractors that month sending out 19 recuperators, a number which may be regarded as good quantity when the size of this mechanism is taken into consideration. Both plants together in April turned out the large number of 89 recuperators for the 240.

Last to come through to quantity production was the hardest of the four to build, the one that promised to defy American industry to build it at all—the 75-millimeter gun recuperator. The two contractors for the rough forgings for this recuperator were the Carbon Steel Co. and the Bucyrus Co. The Carbon Steel Co. was in large series production of them in the spring of 1918, and the Bucyrus Co. reached the quantity basis of manufacture in October, 1918. In that month alone both contractors together turned out 1,305 sets of forgings.

The machining and finishing of the 75 recuperator was in the hands of the Rock Island Arsenal and the Singer Manufacturing Co., which built a costly plant especially for the purpose at Elizabethport, N. J. The first recuperator of this size to appear and be accepted under the severe tests came from the arsenal in October. Thereafter the production ceased for a while. The contractors indeed built recuperators in this period, but the recuperators could not pass the tests. The machining and production of parts seemed to be as perfect as human skill could accomplish, but still the devices would not function perfectly. Adjustments, seemingly of the most microscopical and trivial sort, had to be made—there was trouble with the leather of the valves and with oil for the cylinders. These matters, which could scarcely cause any delay at all in the production of less delicate machinery, indicate the infinite care which had to be employed in the manufacture of the recuperators. At length the producers smoothed out the obstacles and learned all the secrets and necessary processes, and then the 75-millimeter recuperators began to come—2 in January, 1919, and then 13 in February, 20 in March, and 23 in April.

It should be remembered that by quantity production in this particular is meant the production in quantity of recuperators of such perfect quality as to pass the inspection of the Government and to be accepted as part of our national ordnance equipment. In this inspection the Government was assisted by French engineers sent from the great artillery factories in France which had designed the recuperators and which until the successful outcome of the American attempt were their sole producers. Such inspection naturally required that the American recuperators should be the equals of their French prototypes in every respect.

Because the production of French recuperators stands at the summit of American ordnance achievement, here at this point, before there is given any account of the manufacture of field artillery, the theme of this chapter, a performance table is inserted to show the records written by the various concerns engaged in making these devices.

American acceptances of recuperators by firms on Army ordnance orders only.

Item, process, and firm.	1918							1919				Total, Nov. 11, 1918.	Total 1918.	Total, Apr. 30, 1919.
	To July 1.	July.	August.	September.	October.	November.	December.	January.	February.	March.	April.			
75-mm. gun recuperator:														
Forging—														
Carbon Steel Co.	259	259	254	750	1,005	300	552	407	49			2,600	3,379	3,835
Bucyrus Co.			29	78	300	173	111		68			435	691	759
Total	259	259	283	828	1,305	473	663	407	117			3,035	4,070	4,594
Finish machining and assembling—														
Singer Manufacturing Co.										3	8			11
Rock Island Arsenal					1			2	13	17	15	1	1	48
Total					1			2	13	20	23	1	1	59
155-mm. howitzer recuperator:														
Forging—														
Mesta Machine Co.	676	646	648	899	1,080	226		49		31		4,000	4,175	4,255
Watertown Arsenal				160	80	80		25	1			268	320	346
Total	676	646	648	1,059	1,160	306		74	1	31		4,268	4,495	4,601
Machining complete and assembling—														
Dodge Bros.		1	27	249	285	495	403	141				798	1,460	1,601
155-mm. gun recuperator:														
Forging—														
Carnegie Steel Co.	212	213	229	269	401	389	21					1,480	1,731	1,734
Finish machining and assembling—														
Dodge Bros.					1	10	19	116	270	361	103		30	880
240-mm. howitzer recuperator:														
Forging—														
Carnegie Steel Co.	286	99	115	61	70	79						678	710	710
Watertown Arsenal			21									21	21	21
Total	286	99	136	61	70	79						699	731	731
Finish machining and assembling—														
Otis Elevator Co.								4	14	41	62			121
Watertown Arsenal					1				5	19	27	1	1	52
Total					1			4	19	60	89	1	1	173

The process of manufacture of recuperators requires four steps—forging, rough machining, finish machining, and assembling. In the case of 155-millimeter howitzer recuperators all the machining was done by one firm; in the other cases rough machining was done by various firms, including, in the case of the 155-millimeter gun and 240-millimeter howitzer recuperators, the firms doing the forging. Complete records of rough machining are not available.

In discussing here, therefore, the production of field artillery in the war period, we are concerned chiefly with carriages and recuperators, for they offered the major difficulties. Since the production of gun bodies for these various units has been taken up in the preceding chapter, such reference to them as is necessary will be brief. For the sake of additional clearness in the mind of the reader inexpert in these things, the line should be sharply drawn between field artillery and the so-called railway artillery, which was also mobile to a limited degree. The mobile field artillery consisted of all rolling guns or caterpiller guns up to and including the 240-millimeter howitzer in size; and also included the antiaircraft guns of various sizes. All mobile guns of larger caliber than the 240-millimeter howitzer were mounted on railroad cars.

The list of the mobile field artillery weapons in manufacture here during the war period was as follows:

The little 37-millimeter gun, the so-called infantry cannon, one of which two husky men could lift from the ground—a French design;

The 75-millimeter guns—three types of them—the French 75, adopted bodily by the United States; our own 3-inch gun redesigned to the French caliber; and the British 3.3-inch gun, similarly redesigned;

The 4.7-inch gun of American design;

The 5-inch and 6-inch guns, taken from our coast defenses and naval stores and placed on mobile mounts;

The 155-millimeter gun, a French weapon with a barrel diameter of approximately 6 inches;

The 155-millimeter howitzer, also French;

The 8-inch and 9.2-inch howitzers, British designs, being manufactured in the United States when war was declared;

The 240-millimeter howitzer, French and American; and, finally,

The antiaircraft guns.

In modern times, but prior to 1917, the United States had designed types of field-artillery weapons and produced them in quantities shown by the following tabulation:

	Pieces.
2.95-inch mountain gun	113
3-inch gun	544
4.7-inch gun	60
5-inch gun	70
6-inch howitzer	40
7-inch howitzer	70
Total	897

A comparison of this list with the enumeration above of weapons put in production during the war against Germany indicates that we greatly expanded our artillery in types. That we were able to do this at the outset and go ahead immediately with the production of many weapons strange and unknown to our experience, without waiting to develop models and types of our own, is due solely to the generosity of the governments of France and Great Britain, with whom we became associated. We manufactured in all eight new weapons, taking the designs of three of them from the British and of five from the French.

It might seem to the uninitiated that the way of the United States to a great output of artillery would be made smooth by the action of the British and French Governments in agreeing to turn over to us without reservation the blue prints and specifications that were the product of years of development in their gun plants. Yet this was only relatively true. In numerous instances we were not able to secure complete drawings until months after we had entered the war, due to the practice of continental manufacturers that intrusts numerous exact measurements to the memories of the mechanics working in their shops. Consequently it required several months to complete drawings, and when we received them our troubles had only begun.

First there came the problem of translating the plans after we received them. All French dimensions are according to the metric system. A millimeter is one one-thousandth part of a meter, and a meter is 39.37 inches. An inch is approximately 0.0254 meter. Thus to translate French plans into American factory practice involves hundreds of mathematical computations, most of them carried out to decimals of four or five places. Moreover, the French shop drawings are put down on an angle of projection different from what is used in this country. This fact involved the recasting of drawings even when the metric system measurements were retained. When it is considered that such a mechanism as the recuperator on the 155-millimeter gun involves the translation of 416 drawings, the fact that the preparation of French plans for our own use never took more than two months is remarkable, particularly so since it was hard to find in the United States draftsmen and engineers familiar with such translation work.

Once our specifications were worked out from the French plans, it then became necessary to find American manufacturers willing to bid on the contracts. The average manufacturer would look at these specifications, realize what a highly specialized and involved sort of work would be required in the production of the gun carriages or recoil mechanisms, and shake his head. In numerous instances no such work had ever before been attempted in the United States.

However, as the result of efforts on the part of the Government an increased capacity for producing mobile field artillery was created as follows:

At Watertown, N. Y., the New York Air Brake Co., as agent for the United States, constructed a completely new factory to turn out

25 gun carriages a month for the 75-millimeter guns, model 1916—the American 3-inch type modified to the French dimensions.

At Toledo, Ohio, increased facilities were put up at the plant of the Willys-Overland Co. to manufacture a daily output of 17 French 75-millimeter gun carriages, model 1897.

At Elizabethport, N. J., the Singer Manufacturing Co. erected for the Government a complete new factory for finishing daily 17 French 75-millimeter recuperators.

At New Britain, Conn., the plant of the New Britain Machine Co. was adapted and increased facilities were created for the manufacture of two 3-inch antiaircraft gun carriages a day.

At Detroit, Mich., Dodge Bros., as agents for the Government, erected an entirely new factory, costing in the neighborhood of $11,000,000 to give the final machining to the rough-machine forgings for five recuperators daily for the 155-millimeter gun and to machine completely the parts for twelve recuperators daily for the 155-millimeter howitzer. Their huge new plant for this purpose established a record for rapidity of construction in one of the most severe winters of recent history.

At the plant of the Studebaker Corporation at Detroit, facilities were extended for turning out three carriages a day for the 4.7-inch guns.

At Plainfield, N. J., extended facilities were created at the factory of the Walter Scott Co. for manufacturing 20 carriages a month for the 4.7-inch guns.

At Worcester, Mass., at the plant of the Osgood Bradley Car Co. increased facilities were built for the daily manufacture of five carriages for the 155-millimeter howitzers.

At Hamilton, Ohio, at the works of the American Rolling Mill Co., extensions were made to provide for the manufacture each day of three carriages for the 155-millimeter howitzers.

The plant of the Mesta Machine Co., at West Homestead, Pa., near Pittsburgh, was extended to the enormous capacity of turning out the forgings for 40 recuperators a day for the 155-millimeter howitzers.

Extensively increased facilities were made at the shops of the Standard Steel Car Co., at Hammond, Ind., for the daily output of two carriages for the 240-millimeter howitzers.

Increased facilities were created in the plant of the Otis Elevator Co., Chicago, Ill., for the finish machining of the equivalent in parts of two and one-half recuperators a day for the 240-millimeter howitzers.

Large extensions were made to the plant of the Morgan Engineering Co., Alliance, Ohio, for the manufacture monthly of 20 improvised mounts for the 6-inch guns taken from the seacoast fortifications.

The facilities of the United States arsenals at Watertown, Mass., and at Rock Island, Ill., for the manufacture of field-gun carriages and recuperators were greatly increased.

This carriage construction for the big guns required the closest kind of fine machine work and fittings where the brake or recuperator construction entered the problem, and the great plants built for this purpose of turning out carriages and recuperators were marvels for the rapidity of their construction, the speed with which they were equipped with new and intricate tools, and the quality of their output.

Every mobile gun mount must be equipped with a shield of armor plate. The size of the artillery project may be read in the fact that our initial requirement for armor for the guns ran to a total of 15,000 tons to be produced as soon as it could be done. Now, we had no real source for getting armor in such large quantities, because the previous demands of our artillery construction had never called for it. The prewar manufacturers of artillery armor were three in number—the Simmons Manufacturing Co., of St. Louis; Thomas Disston & Sons, of Philadelphia; and the Crucible Steel Co. To meet the new demand two armor sources were developed—the Mosler Safe Co. plant of the Standard Ordnance Co. and the Universal Rolling Mill Co. The process of building this armor had been a closely guarded secret in the past, a fact entailing extended experiments in the new plants before satisfactory material could be obtained.

The new artillery program required the manufacture of 120,000 wheels of various types and sizes for the mobile carriages. The Rock Island Arsenal and two commercial concerns prior to the war had been building artillery wheels in limited quantities. One completely new plant had to be erected for the manufacture of wheels, while seven existing factories were specially equipped for this work. We had to develop new sources of supply of oak and hickory and to erect dry kilns especially for the wheel project.

The largest order for rubber tires in the history of the American rubber industry was placed as one relatively small phase of the artillery program, the order amounting to $4,250,000. Rubber tires on the wheels of all the heavier types of artillery carriages, so that the units might be drawn at good speed by motor vehicles, was essentially an American innovation. No tires of this size had ever been manufactured in this country. Consequently it was necessary for the firms who got the orders to build machinery especially designed for the purpose.

With practically all of the manufacturers of the American metal-working industries clamoring for machine tools, and with some branches of the Government commandeering the machine-tool shops in whole sections of the country, it is evident that the necessity for the heavier types of machine tools required by the manufacturers of artillery material offered a weighty problem at the outset. In fact, the machine-tool supply was never adequate at any time, and the shortage of this machinery hampered and impeded to a great degree the speed of our artillery production.

The Nation was raked with a fine-toothed comb for shop equipment. The Government went to almost any honorable length to procure this indispensable tooling. For instance, when the Dodge plant at Detroit was being equipped to manufacture the 155-millimeter recuperators, the Government agents discovered trainloads of machinery consigned to the Russian government and awaiting shipment. These tools were commandeered on the docks. One huge metal planer had dropped overboard while it was being lightered to the ocean tramp that was to carry it to a Russian port. Government divers fixed grappling hooks to this machine, and it was brought to the surface and shipped at once to the Dodge plant.

The 3-inch gun which we had been building for many years prior to the war was a serviceable and efficient weapon; but still we were unable to put it into production immediately as it was. Our earliest divisions in France, under the international arrangement, were to be equipped by the French with 75-millimeter guns; while we, on this side of the water, reaching out for all designs of guns of proven worth, expected to manufacture the 75's in large numbers in this country. The French 75 in its barrel diameter is a fraction of an inch smaller than our 3-inch gun, the exact equivalent of 75 millimeters being 2.95275 inches. Thus, if we built our own 3-inch gun (and the British 3.3-inch gun, as we intended) and also went ahead with the 75-millimeter project on a great scale, we should be confronted by the necessity of providing three sorts of ammunition of almost the same size, with all the delays and confusion which such a situation would imply. Consequently we decided to redesign the American and British guns to make their bores uniformly 75 millimeters, thus simplifying the ammunition problem and making available to us in case of shortage the supplies of shell of this size in France.

With all of the above considerations in mind, it is evident now, and it was then, that we could not hope to equip our Army with American-built artillery as rapidly as that Army could be collected, trained, and sent to France; and this was particularly true when in the spring of 1917 the Army policy was changed to give each 1,000,000 men almost twice as many field guns as our program had required prior to that date. Consequently, when on June 27, 1917, the Secretary of War directed the Chief of Ordnance to provide the necessary artillery for the 2,000,000 men who were to be mobilized in 1917 and the first half of 1918, the first thought of our officers was to find outside supplies of artillery which we could obtain for an emergency that would not be relieved until our new facilities had reached great production.

We found this source in France. The French had long been the leading people in Europe in the production of artillery, and even the great demands of the war had not succeeded in utilizing the full capacity of their old and new plants. Two days later, on June 29, 1917, the French high commissioner, by letter, offered us in behalf of France a daily supply of five 75-millimeter guns and carriages, beginning August 1, 1917. The French also offered at this time to furnish us with 155-millimeter howitzers; and on August 19, 1917, the French Government informed Gen. Pershing that each month, beginning with September, he could obtain twelve 155-millimeter Filloux guns and carriages from the French factories.

Before the signing of the armistice 75-millimeter guns to the number of 3,068 had been ordered from the French, and of this number 1,828 had been delivered. Of 155-millimeter howitzers, 1,361 had been ordered from the French and 772 delivered before November 11, 1918. Of 155-millimeter guns, 577 had been ordered from the French and 216 delivered previous to the granting of the armistice.

From British plants we ordered 212 Vickers-type 8-inch howitzers, and 123 had been delivered before the armistice had been signed; while of 9.2-inch howitzers, Vickers model, 40 of an order for 132 had been completed. In addition to this, 302 British 6-inch howitzers were in manufacture in England for delivery to us by April 1, 1919. These figures, with the exception of those relating to the order for British 6-inch howitzers, do not include the arrangements being made by this Government during the last few weeks of hostilities for additional deliveries of foreign artillery.

As to our own manufacture of artillery, when we had conquered all the difficulties—translated the drawings, built the new factories, equipped them with machine tools and dies, gages, and other fixtures needed by the metal workers, and had mobilized the skilled workers themselves—we forged ahead at an impressive rate. When the armistice was signed we were turning out 412 artillery units per month. Compare this with Great Britain's 486 units per month in the fall of 1918 and measure our progress, remembering that England had approximately three years' head start. Compare it with the French monthly production of 659 units per month, and remember that France was the greatest artillery builder in the world. When it came to the gun bodies themselves we obtained a monthly output of 832, as against Great Britain's 802 and France's 1,138. And our artillery capacity was then, in the autumn of 1918, only coming into production.

In the war period—April 6, 1917, to November 11, 1918—we produced 2,008 complete artillery units, as against 11,056 turned out by

France and 8,065 completed by Great Britain in the same period. In those 19 months we turned out 4,275 gun bodies, while in the same months France produced 19,492 and Great Britain 11,852.

THE 37-MILLIMETER INFANTRY FIELD GUNS.

The smallest weapon of all the field guns we built was the French 37-millimeter gun, the diameter of its bore being about 1½ inches in our measurements, the figure being 1.45669 inches. This was the so-called infantry field gun, to be dragged along by foot soldiers when they are making an advance. Its chief use in the war was in breaking up the German concrete pill boxes, machine gun nests, and other strong points of enemy resistance. In service it was manned by infantrymen instead of artillerists, a crew of eight men handling each weapon, the squad leader being the gunner. One of the men of the crew was the loader, and he was likewise able to fire the piece. The other six men served as assistants.

The 37-millimeter outfit as it exists to-day consists of the gun, with a split trail, mounted on axle and wheels. By means of a trailer attachment on the ammunition cart it can be drawn by one horse or one mule. The ammunition cart itself is merely a redesigned machine-gun ammunition vehicle. The wheels and axle can easily be removed and left a short distance in the rear of the place where it is desired to set up the gun. The whole outfit weighs only 340 pounds and is about 6 feet long.

The gun rests on its front leg which is dropped to form a tripod with the two legs of the split trail. The gun proper can be removed from the trail and the sponge staff can be inserted in the barrel through the opened breech. Two men can bear this part of the weapon in advancing action. Two other men are able to carry the trail, when its legs are locked together, while the four other members of the squad bring along the boxes of ammunition.

The ammunition cart holds 14 ammunition boxes, each containing 16 rounds. A spare-parts case, strapped to the trail, contains a miscellaneous assortment of such parts as can readily be handled in the field. A tool kit in a canvas roll is also transported on the cart, along with entrenching tools and other accessories.

Equipped with a telescopic sight for direct fire and a quadrant, or collimating sight, for indirect fire, great accuracy is obtained by this small piece of artillery. The length of the barrel of the gun proper is 20 calibers, which means that it is 20 times 37 millimeters in length, or about 29 inches. The length of the recoil when the gun is fired is 8 inches.

Two types of ammunition were provided for this gun at first; but, as the low-explosive type was not so effective as desired, it was abandoned entirely in favor of the high-explosive type contained in a projectile weighing 1¼ pounds. This projectile is loaded with 240 grains of T. N. T. and detonated by a base percussion fuse. The range of the gun is 3,500 meters, or considerably more than 2 miles. Only three to six shots from this gun were found to be necessary to demolish an enemy machine gun emplacement or other strongly held position.

In the great war the 37-millimeter gun found itself and proved its usefulness. The original model had been designed at the Puteaux Arsenal in France in 1885; but it was not until after 1914 that the weapon was produced in quantities.

In this country we took up the production of 37-millimeter guns in October, 1917. While our shops were tooling up for the effort, 620 of these weapons were purchased from the French and turned over to the American Expeditionary Forces. For the purposes of greater speed in manufacture our executives took the gun apart and divided it into three groups, known as the barrel group, the breech group, and the recoil group. Additional to these, as a manufacturing proposition, were the axle and wheels and the trail.

The barrel group went to the Poole Engineering & Machine Co., of Baltimore, Md., who subcontracted for some of the parts to the Maryland Pressed Steel Co., of Hagerstown, Md. The breech group was manufactured by the Krasberg Manufacturing Co., of Chicago. The C. H. Cowdrey Machine Works, of Fitchburg, Mass., turned out the recoil mechanisms. The axles and wheels were built by the International Harvester Co., of Chicago. The trails were turned out by the Universal Stamping & Manufacturing Co., also of Chicago.

When crated for overseas shipment, the gun, ammunition cart, and all accessories, weighed 1,550 pounds and occupied about 15 cubic feet of space.

The first delivery of completed 37-millimeter guns from our factories was made in June, 1918, and at the cessation of hostilities manufacturers were turning out the guns at the rate of 10 per day. Between June and November 122 American-built 37-millimeter guns were shipped abroad, and more were ready to be sent over when the armistice was signed. The gun had been so successful in use abroad that our original order of 1,200 had been increased to 3,217 before the signing of the armistice, including the 620 purchased from the French.

The various groups of this gun were shipped to the plant of the Maryland Pressed Steel Co., Hagerstown, Md., for assembly and were there tested at a specially built proving ground, 8 miles from the factory.

Three 37's were issued to each infantry regiment, making one for each battalion. The required equipment for a division was, therefore, 12 weapons.

THE 37-MILLIMETER INFANTRY CANNON.

TWO VIEWS OF 75-MILLIMETER FIELD GUN, MODEL 1918 (AMERICAN).

Figures on 37-millimeter gun production.

Guns procured from the French Government	620
Guns ordered manufactured in United States, October, 1917	1,200
Increase in order, September, 1918	1,397
Total number ordered in United States	2,597
Total number of guns completed prior to the signing of the armistice	884
Guns delivered for overseas shipment prior to the signing of the armistice	300
Guns shipped to various camps in this country	26
Guns shipped to other points in this country	4
On hand at Hagerstown Arsenal, proof fired	425
Completed and ready for proof firing	129

THE 75-MILLIMETER GUNS.

Next in order in the upward scale of sizes we come to the 75-millimeter gun, which was by far the most useful and most used piece of artillery in the great war. In fact the American artillery program might be divided in two classes, the 75's in one class, and all other sizes in the other, since it may be said practically that for every gun of another size produced we also turned out a 75. In number the 75's made up almost half of our field artillery. The 75-millimeter gun threw projectiles weighing between 12 and 16 pounds and it had an effective range of over 5½ miles.

We approached the war production of this weapon with three types available for us to produce—our own 3-inch gun; its British cousin the 3.3-inch gun or 18-pounder; and the French 75-millimeter gun, with its bore of 2.95275 inches. The decision to adopt the 75-millimeter size and modify the other two guns to this dimension, giving us interchangeability of ammunition with the French, was an historic episode in the American ordnance development of 1917.

While in 1917 the French with their excess manufacturing capacity began work on our first orders for 1,068 guns of this size to supply our troops during the interim until American factories could come into production, we were preparing our factories for the effort. Roughly speaking the 75 consists of a cannon mounted on a two-wheeled support for transportation purposes. This support also provides a means for aiming by suitable elevating and traverse mechanisms. As previously explained, a recoil mechanism is also provided to absorb the shock of firing, allowing a certain retrograde movement of the cannon and then returning it in position for the next shot—returning it "into battery," as the artillerists say. By its recuperator device the field gun of to-day is chiefly distinguished from its brother of the latter part of the nineteenth century. Without a recuperator the gun would leap out of aim at each shot and would have to be pointed anew; but one with a recuperator needs to be pointed only at the beginning of the action.

When we entered the war we found ourselves with an equipment of 544 field guns of the old 3-inch model of 1902. This gun had a carriage provided with the old-style single trail. By 1913, however, we had been experimenting with the split trail and it had been strongly recommended by our ordnance experts; and in 1916 we had placed orders for nearly 300 carriages of the split-trail type, which had come to be known as Model 1916. Of these orders 96 carriages were to come from the Bethlehem Steel Co., and the remainder from the Rock Island Arsenal.

Meanwhile for some time the Bethlehem Steel Co. had been engaged in turning out carriages for the British 3.3-inch guns. Here was capacity that might be utilized to the limit; and, accordingly, in May, 1917, we ordered from the Bethlehem Co. 268 of the British carriages. At the same time we ordered from the same company approximately 340 of our own Model 1916 carriages at a cost of $3,319,800. A few weeks later the decision had been made to make all our guns of this sort conform to the French 75-millimeter size, and these British and American carriages contracted for in May were ordered modified to take 75-millimeter guns. The carriages needed little modification and the guns not much. Subsequently, in rapid succession we placed orders with the Bethlehem Steel Co., calling for the construction of an additional 1,130 of the British carriages, all of them to be adapted to 75-millimeter guns.

Next it was the concern of the Ordnance Department to find other facilities for manufacturing carriages for these weapons. The artillery committee of the Council of National Defense located the New York Air Brake Co. as a concern willing to undertake this work; and in June, 1917, this company signed a contract to produce 400 American model 1916 carriages at a cost of $3,250,000.

By December we had the drawings for the French carriages of this size and made a contract with the Willys-Overland Motor Car Co. to produce 2,927 of them. The table at the end of this section shows the production attained at these various plants.

The manufacture of carriages for the 75's produced concrete results, as our factories here were turning them out for us at the rate of 393 per month when the fighting ceased, and our contract plants in France were making 171 per month. In all we received from American factories 1,221 carriages. At the rate of increase we would have been building 800 carriages per month by February, 1919.

It may be said we were thoroughly impressed with the difficulties attached to the transplanting to this country of the manufacture of French 75-millimeter recuperators. It was a question whether this device could possibly be built by any except the French mechanics trained by long years in its production. At first it seemed that we could secure no manufacturer at all who would be willing to assume such a burden. Not until February, 1918, were complete drawings and specifications of the recuperator received from France. At

TWO VIEWS OF THE FRENCH 75-MILLIMETER GUN.

This type of gun has been used by the French Army since 1897, and was the gun most used by the Allies in the Great War. This gun throws a shell weighing 12.3 pounds a distance of 8,400 meters or shrapnel weighing 16 pounds a distnce of 9,000 meters. The weight of the gun and carriage is 2,657 pounds. The service muzzle velocity of the shell is 1,805 feet per second, while for shrapnel it is 1,755 feet per second.

THE 75-MILLIMETER FIELD GUN, MODEL 1917 (BRITISH).

This gun throws a shell weighing 12.3 pounds a distance of 8,300 meters, and 16 pounds of shrapnel a distance of 8,900 meters. The weight of the gun and carriage is 2,887 pounds. Its muzzle velocity for shell is 1,750 feet a second and for shrapnel 1,680 feet a second.

length the Singer Manufacturing Co., builders of sewing machines, consented to take up this new work, and on March 29 the company contracted to produce 2,500 recoil systems for the 75-millimeter gun carriages. In April, 1918, the Rock Island Arsenal was instructed to turn out 1,000 of these recuperators.

The production of gun bodies for the 75-millimeter units was quite satisfactory. The Bethlehem Co., the Wisconsin Gun Co., the Symington-Anderson Co., and the Watervliet Arsenal were the contractors who built the gun bodies. Gun bodies of three types, but all of the same 75-millimeter bore, were ordered—the American type (the modified 3-inch gun), the British type (the modified 3.3-inch gun), and the French type.

Our ordnance preparation would have given us enough 75's for the projected army of 3,360,000 men on the front in the summer of 1919, together with appropriate provision for training in the United States. Of the 75's built in this country, 143 units were shipped to the American Expeditionary Forces before the armistice went into effect. Meanwhile the French had delivered to our troops 1,828 units of this size. The total equipment of 75's for our Army in France from all sources thus amounted to 1,971 guns with their complete accessories.

Unit.	Contractor.	Number ordered.	Number completed at signing of armistice.	Number floated overseas to Nov. 11, 1918.	Number completed up to April 17, 1919.
75-mm. gun carriage, model 1916.	Rock Island Arsenal	472	159	34	185
	Bethlehem Steel Co	455	14		25
	New York Air Brake Co	400	33		97
75-mm. gun carriage (French)	Willys-Overland Co	2,927	291	....	1,299
75-mm. gun carriage (British), complete.	Bethlehem Steel Co	2,868	724	124	921
75-mm. gun carriage limber (British), complete.	....do	968	489	....	1,010
75-mm. gun carriage limber, model 1918.	....do	436	436	....	441
	American Car & Foundry Co	3,661	3,661	980	3,661
75-mm. gun caisson, model 1918.	Bethlehem Steel Co	1,666	302	4,957	831
	American Car & Foundry Co	20,356	11,660	....	18,301
75-mm. caisson limber, model 1918.	Bethlehem Steel Co	1,916	1,210	....	1,916
	American Car & Foundry Co	20,675	15,526	4,126	20,675
75-mm. cannon, model 1916	Symington-Anderson Co	640	416	19	416
	Wisconsin Gun Co	160	116		116
	Watervliet Arsenal	264	161		192
	Bethlehem Steel Co	340	2		2
75-mm. cannon (French)	Symington-Anderson Co	4,300	103	....	860
	Wisconsin Gun Co	2,050	9	....	190
75-mm. cannon (British)	Bethlehem Steel Co	2,868	592	124	909

4.7-INCH GUNS.

In the 4.7-inch field gun, model of 1906, America took to France a weapon all her own. It was a proven gun, too, developed under searching experiments and tests. There were 60 of these in actual service when we got into the war. The 4.7-inch guns, with their greater range and power, promised to be particularly useful for destroying the enemy's 77-millimeter guns.

The carriage model of 1906 for the 4.7-inch gun is of the long recoil type, the recoil being 70 inches in length. The recoil is checked by a hydraulic cylinder, and a system of springs thereupon returns the gun to the firing position. The gun's maximum elevation is 15 degrees, at which elevation, with a 60-pound projectile, the gun has a range of 7,260 meters, or 4½ miles. With a 45-pound projectile a range of 8,750 meters, or nearly 5½ miles, can be obtained at 15 degrees elevation. It is possible to increase this range to about 10,000 meters, or well over 6 miles, by depressing the trail into a hole prepared for it, a practice often adopted on the field to obtain greater range. The total weight of the gun carriage with its limber is about 9,800 pounds.

An order for 250 of the 4.7-inch carriages was placed with the Walter Scott Co., at Plainfield, N. J., July 12, 1917, upon the recommendation of committees of the Council of National Defense, who were assisting the Ordnance Department in the selection of industrial firms willing to accept artillery contracts. Of the 250 ordered from this concern, 49 were delivered up to the signing of the armistice.

The Rock Island Arsenal had also been employed previously in turning out 4.7-inch carriages; and the capacity of that plant, although small, was utilized. Under the date of July 23, 1917, the arsenal was instructed to deliver 183 carriages. Late in December, 1917, the Studebaker Corporation was given an order for 500. On September 30, 1918, Rock Island Arsenal was given an additional order for 120 carriages, while the Studebaker order was reduced to 380. Additional plant facilities had to be provided at both the Walter Scott Co. and the Studebaker Corporation.

Up to December 12, 1918, a total of 381 carriages of the 4.7-inch type had been completed and delivered. These carriages included the recoil mechanism. In the month of October, 1918, alone, 113 were produced, and this rate would have been continued had the armistice not been signed.

Cannon for the 4.7-inch units were turned out at the Watervliet Arsenal and the Northwestern Ordnance Co., Madison, Wis. Deliveries from the Watervliet Arsenal began in June, 1918, totaling 120 up to December, while the Northwestern Ordnance Co., starting its deliveries in August, had completed 98 by December.

Up to the 15th of November, 64 complete 4.7-inch units had been floated for our forces overseas.

Forgings for the 4.7-inch gun cannon were made by the Bethlehem Steel Co. and the Heppenstall Forge & Knife Co., of Pittsburgh, Pa.

Owing to the great difference in cross section between muzzle and breech end of the jacket, great difficulty was experienced in the heat treatment of these forgings, particularly on the part of manu-

FRONT AND REAR VIEWS OF OUR 4.7-INCH GUN AND CARRIAGE, MODEL 1906, WITH WHICH OUR TROOPS HAVE BEEN EQUIPPED FOR A LONG TIME.

This gun throws a projectile weighing 45 pounds a distance of about 6 miles.

TWO VIEWS OF 155-MILLIMETER GUN, MODEL 1918, G. P. F.

The upper view shows the piece mounted on an auto truck for quick moving about.

facturers who had had no previous experience in the production of gun forgings.

In order to produce enough forgings to supply the finish-machining shops, an order for 50 jackets was later given to the Edgewater Steel Co., of Pittsburgh, Pa., where the jackets were forged. These were then sent to the Heppenstall Forge & Knife Co. for rough machining and finally returned to the Edgewater Steel Co. for heat treating. An order for 150 jackets was also given to the Tacony Ordnance Corporation.

Shortly before the signing of the armistice, the jacket was redesigned so that the heavy breech end was forged separately in the shape of a breech ring. This design, however, was not produced.

It was desired to develop a 4.7-inch gun carriage having the characteristics of the split-trail 75-millimeter gun carriage, model of 1916, so that greater elevation and wide traverse could be obtained. The Bethlehem Steel Co. was given a small order for 36 carriages of their own design prior to the war, and their pilot carriage had been undergoing tests at the proving ground. The design was, however, not sufficiently advanced to be used in the war.

Unit.	Contractor.	Number ordered.	Number completed at the signing of armistice.	Number completed up to April 17, 1919.
4.7-inch gun carriage, model of 1906	Rock Island Arsenal	303	183	183
	Studebaker Corporation	380	88	175
	Walter Scott Co	250	49	57
4.7-inch gun-carriage limber	American Car & Foundry Co	433	433	433
	Maxwell Motor Co	479	82	250
4.7-inch gun caisson	American Car & Foundry Co	1,848	320	848
	Ford Motor Co	1,001	106	400
4.7-inch cannon	Northwestern Gun Co	....	56	....
	Watervliet Arsenal	....	93	....

Sixteen of these units, also 48 which were previously on hand, were floated for overseas up to November 11, 1918.

THE 5-INCH AND 6-INCH GUN MOUNTS.

In the war emergency America sought to put on the front every pound of artillery she could acquire from any source whatsoever. Accordingly, before any of the manufacturing projects were even started, the Ordnance Department conducted a preparedness inventory of the United States to see what guns already in existence we might find that could be improvised for use as mobile artillery in France. The search discovered a number of heavy cannon that could serve the purpose—part of them belonging to the Army, these being the guns at our seacoast fortifications; part belonging to the Navy, in its stores of supplies for battleships; and part of them being the property of a private dealer, Francis Bannerman & Son, of New York.

The guns for this improvised use were obtained as follows:

From the Coast Artillery, a branch of the Army, we obtained ninety-five 6-inch guns, 50 calibers in length, and twenty-eight 5-inch guns, 44.6 calibers; from the Navy stores came forty-six 6-inch guns, ranging from 30 to 50 calibers in length; from Francis Bannerman & Son, thirty 6-inch guns, 30 calibers long. This was a total of 199 weapons of great destructive power, awaiting only suitable mobile mounts to make them of valiant service on the western front. It was the task of the Ordnance Department to take these guns and as swiftly as possible mount them on field artillery carriages of an improvised type that could be most quickly built.

Minor changes had to be made on many of the guns obtained in this manner in order to adapt them for use on field artillery carriages. The various seacoast guns were retained as they were in length, because it was planned to return them eventually to the fortifications from which they had been taken. The Navy guns, all of the 6-inch size, were shipped to the Watervliet Arsenal to be cut down to a uniform length of 30 calibers.

The need for speed in manufacture demanded that the carriages for these guns should be of the simplest design consistent with the ruggedness required for field operations and the accuracy necessary for effectiveness. When tests of the first carriages produced were made it was found that requirements had been more than met.

Orders were placed on September 24, 1917, with the Morgan Engineering Co., of Alliance, Ohio, for 70 mounts for the 6-inch units. A few days later this number was increased to 74, while on the 28th of September, 1917, the same company was given an order for 18 additional 6-inch gun mounts and 28 mounts for the 5-inch guns. Orders for limbers were placed with the same company on December 1.

It was soon discovered that big transport wagons would be required to carry the long 6-inch seacoast guns separately because of their great weight. On February 15, 1918, the Morgan Engineering Co. was ordered to build these necessary transport wagons.

Difficulties in securing skilled labor, necessary materials, and tools delayed production of these mounts, but the eighteen 6-inch gun mounts ordered September 28, 1917, were completed in March, 1918, while the twenty-eight 5-inch gun mounts ordered on the same date were finished in April. In August, 1918, the seventy-four 6-inch gun mounts were turned out. The production of an additional order for thirty-seven 6-inch gun mounts was just beginning when the armistice was signed.

The 6-inch gun carriage, bearing the gun, weighs about 41,000 pounds. A maximum range of over 10 miles can be obtained by this weapon. The complete 5-inch gun unit weighs about 23,500 pounds and has a maximum range of more than 9 miles. In understanding the difficulties that faced the Ordnance Department in building carriages for these guns, it should be recalled that these big weapons were originally built for fixed-emplacement duty and were therefore much heavier than mobile types. This fact complicated the problem of designing the wheeled mounts. They proved to be more difficult to maneuver than the lighter types of guns.

Model.	Size.	Number ordered.	Number completed prior to Nov. 11.	Number floated for overseas.
1897	5-inch	28	28	26
1917	6-inch	74	74	68
1917-A	6-inch	18	18	4
1917-B	6-inch	37	1	....

155-MILLIMETER HOWITZERS.

It is a testimonial to the adaptability and skill of American industry that we were able to duplicate successfully in this country the celebrated 155-millimeter howitzer, before 1917 built only in the factory of its original designer, the great firm of Schneider et Cie., in France. This powerful weapon is a fine example of the French gun builders' art, in a country where the art of gunmaking has been carried to a perfection unknown anywhere else.

The 155-millimeter howitzer's history dates back to the nineteenth century. In its development the French designers had so strengthened its structure, increased its range, and improved its general serviceability, that in 1914 it was ready to take its place as one of the two most-used and best-known weapons of the allies, the other being the 75-millimeter field gun.

As thus perfected the howitzer weighs less than 4 tons and is extremely mobile for a weapon of its size. It can hurl a 95-pound projectile well over 7 miles and fire several times a minute. The rapidity of fire is made possible by a hydropneumatic recoil system that supports the short barrel of the gun and stores up the energy of the recoil by the compression of air. With the gun pointing upward at an angle of 45 degrees, the recoil mechanism will restore it into battery in less than 13 seconds. The carriage of the gun is extremely light, being built of pressed steel parts that incorporate many ingenious features of design to reduce the weight. The shell and the propelling charge of powder are loaded separately.

The American-built 155-millimeter howitzer was practically identical with that built in France. Any of the important parts of the American weapon would interchange with those which had come from the Schneider factory. We equipped the wheels of our field carriage, however, with rubber tires, and gave the gun a straight shield of armor plate instead of a curved shield.

In the spring of 1917 we bought the plans of the howitzer from Schneider et Cie. and began at once the work of translating the specifications into American measurements. This work monopolized the efforts of an expert staff until October 8, 1917.

In order to facilitate the reproduction here, we divided the weapon, as a manufacturing proposition, into three groups—the cannon itself, the carriage, and the recuperator or recoil system—and placed each group in the hands of separate contractors. There was, of course, the usual difficulty in finding manufacturers willing to undertake production of such an intricate device and who also possessed machine shops that had the equipment and talent required for such work, and in procuring for these shops the highly specialized machinery that would be necessary.

The American Brake Shoe & Foundry Co., of Erie, Pa., whose magnificent work in building a special plant has been described in the preceding chapter, took an order in August, 1917, for 3,000 howitzer cannon and by October, 1918, was producing 12 of them every day. The company turned out its first cannon in February, 1918, approximately six months after receiving the contract, having in the interim built and equipped a most elaborate plant. It is doubtful if the annals of industry in any country can produce a feat to match this.

In fact, the production of cannon by the Erie concern so outstripped the manufacture of carriages and other important parts for the howitzer that it was possible by September, 1918, for us to sell 550 howitzer bodies to the French Government. When the armistice was signed on November 11, 1918, the company had completed 1,172 cannon.

In November, 1917, we placed orders for 2,469 carriages for this weapon, splitting the order between the Osgood-Bradley Car Co., of Worcester, Mass., and the Mosler Safe Co., of Hamilton, Ohio. Then followed a long battle to secure the tools and equipment, the skilled mechanical labor, and the necessary quantities of the best grades of steel and bronze, an effort in which the contracting companies were at all times aided by the engineers of the Ordnance Department. All obstacles were overcome and the first carriages were ready for testing in June, 1918. When the armistice was signed 154 carriages had been delivered, and production was moving so rapidly that one month later this number had been run up to 230.

The limbers were manufactured by the Maxwell Motor Car Co., which had orders to turn out 2,575 of them. The first deliveries of limbers came in September, 1918, and seven a day were being turned out in October, a total of 273 having been completed by the day of the armistice. A month later the number of completed limbers totaled 587.

It was in the making of the recuperator systems that the greatest problems were presented. No mechanism at all similar to this had ever been made in this country. No plant was in existence here capable of turning out such a highly complicated, precise, and delicate device.

Finally, after much Governmental search and long negotiation, the Dodge Bros., of Detroit, motor car builders, agreed to accept the responsibility. In this effort they built and equipped the splendid factory, costing $10,000,000, described elsewhere.

This howitzer recuperator is turned out from a solid forging, weighing 3,875 pounds, but the completed recuperator weighs only 870 pounds. Each cylinder must be bored, ground, and lapped to a degree of fineness and accuracy that requires the most painstaking care.

Difficulties of almost every sort were experienced with the forgings and other elements of the recuperators. The steel was analyzed and its metallurgical formulas were changed. The work of machining proceeded favorably until the very last operation—that of polishing the interior of the long bores to a mirrorlike glaze and still retaining the extreme accuracy necessary to prevent the leakage of oil past the pistons. Such precision had been theretofore unknown in American heavy manufacture. Until the many processes could be perfected, the deliveries were held back.

Even with the delivery of the first recuperator, difficulties did not vanish. This mechanism has no adjustments which can be made on the field, but depends for its wonderful operation upon the extreme nicety of the relation of its parts. It required the alteration of certain small parts before the first trial models could be made to function.

However, all obstacles and difficulties were finally overcome, and in the plant that had been erected during the bitter cold of one of our severest winters, and with practically entirely new machinery and workmen, production got under way, and the first recuperator was delivered early in July, 1918, nine months after the contract was signed. Production in quantity began to follow shortly after that month, and by November an average of 16 recuperators a day was being turned out. Of the 3,120 recuperators contracted for, 898 had been finished when the armistice was signed, and this quantity was increased to 1,238 one month later.

The steel required for the recuperators in these 155-millimeter howitzers, and also for those of the 155-millimeter guns, was of special composition; yet all the forge capacity in this country was being utilized in other war manufacture. New facilities for the manufacture of these forgings had to be developed by increasing the capacity of the Mesta Machine Co. of Pittsburgh, until it could meet our requirements. The Government itself contracted for these forgings and supplied them to Dodge Bros.

Each howitzer required some 200 items of miscellaneous equipment, such as air and liquid pumps and other tools. These were purchased from many sources, and many of these contractors had just as much difficulty with the small parts as the larger firms had with the more important sections of the howitzers.

Many of the problems involved in turning out the complete unit could not be known or understood until they were met with in actual manufacture. Mechanical experts representing Schneider et Cie. were on hand at all times to help solve difficulties as they arose.

The Government turned to France for an auxiliary supply of carriages for the American-built howitzers, placing orders for 1,361 with French concerns. Of this number 772 had been completed when the armistice was signed, and the French expected soon to turn out the carriages at the rate of 140 per month. It might also be noted here that we placed an order in England for 302 British 6-inch howitzers, a piece very like the French howitzer. The British contract was to be completed April 1, 1919.

The various parts of the 155-millimeter howitzer were assembled into complete units and tested at the Aberdeen Proving Grounds. After being assembled and tested, the whole unit was taken apart and packed into crates especially designed for overseas shipment. One crate held two howitzer carriages with recuperators in less space than would have been occupied by one carriage on its wheels.

It will be noted that the first gun body of the 155-millimeter howitzers made in this country was delivered in February and the first recuperator in July. Before the recuperators were ready, the other parts of the howitzer had been proof-tried by using a recuperator of French manufacture.

During the months of August and September, 1918, the first regiment equipped with 155-millimeter howitzers was made ready at Aberdeen. The big weapons were packed and on the dock for shipment overseas when the armistice was signed. These first ones were to be followed by a steady stream of howitzers. All arrangements had been made to assemble units and crate them for overseas at the Erie Proving Ground at Port Clinton, Ohio.

None of the 155-millimeter howitzers built here reached the American Expeditionary Forces, but French deliveries of the weapon up to the signing of the armistice totaled 747.

Unit.	Contractor.	Number ordered.	Number completed Nov. 11, 1918.	Number completed Apr. 17, 1919.
155-mm. howitzer carriage	Osgood Bradley Car Co	900	136	369
155-mm. carriage replacement parts.	do	49		0
155-mm. howitzer carriage	do	250		93
Do	American Rolling Mill Co. (old Mosler Safe contract).	1,270	18	26
Do	Rock Island Arsenal	172		0
155-mm. howitzer carriage limbers	Maxwell Motor Co	2,575	273	700
Do	Rock Island Arsenal	100		0
155-mm. howitzer caisson	Ford Motor Co	8,937	4,373	8,937
155-mm. howitzer cannon	American Brake Shoe & Foundry Co		1,172	1,789

THE 155-MILLIMETER G. P. F. GUNS.

The reproduction in the United States of the French 155-millimeter G. P. F. (the French designation) gun presents much the same story as that of the howitzer of equal size—a story of difficulties in translating plans, writing into them the precision of finishing measurements that the French factory usually leaves to the skill of the mechanic himself, difficulties in finding manufacturers willing to undertake the work, and then of providing them with suitable raw materials and machinery, and, above all, of locating the necessary skilled mechanics.

This strange, big monster of a weapon is of rugged design. The entire unit weighs 19,860 pounds. The gun has the extremely high muzzle velocity of 2,400 feet per second, a rate of propulsion that throws the 95-pound projectile 17.700 yards, or a little more than 10 miles.

The wheels of the carriage have a double tread of solid rubber tire. By an ingenious arrangement a caterpillar tread can be applied to the wheels in a few minutes whenever soft ground is encountered.

The center of gravity of the unit is low. The wheels are of small dimensions and the cradle is trunnioned behind in such a fashion as to reduce the height of the cannon. The carriage has a split trail, which allows for a large clearance for recoil at a high elevation and a large angle of traverse. The carriage when traveling is supported on semielliptical springs, as is also the carriage limber.

Two large steel castings make up the carriage of this unit. The bottom part of the carriage is supported by the axle, which carries the two sections of the split trail upon the hinge pins. The top part of the carriage is supported by and revolves upon the bottom carriage and carries in trunnioned bearings the recuperator. The principal difficulty in carriage manufacture was to obtain in this country the extremely large steel castings of light-section, high-grade steel.

The carriages, 1,388 in number, were ordered in November, 1917, from the Minneapolis Steel & Machinery Co. The first delivery of carriages was made in August, 1918, and in the last week of October they were being turned out at the rate of seven a day. Up to the armistice date 370 had been produced, of which 16 had been sent overseas.

We also placed orders in France for 577 of these carriages, of which 216 had been completed upon the signing of the armistice. The American monthly rate of production of carriages in October was 162.

The 155-millimeter gun itself is far from being simple to manufacture. It is of considerable length and is built of a number of jackets and hoops to give the required resistance to the heavy pressures exerted in firing, this being a high-velocity gun. Except for a slight change in the manner of locking the hoops to the jacket, our gun is identical with that of the French.

Orders for 2,160 cannon were given to the Watervliet Arsenal and the Bullard Engineering Works, at Bridgeport, Conn., in November, 1917. The Bullard Engineering Works had to construct new buildings and to purchase and install special equipment, and the Watervliet Arsenal had to extend its shops and also purchase and install much additional machinery—a job that took time at both places.

The first deliveries of cannon came from Watervliet Arsenal in July, 1918. During October 50 cannon were delivered, and it seemed certain that by early in 1919 the projected eight cannon per day would be the rate attained. We shipped 16 of the cannon overseas. By November 11 we had received 71 cannon, a number increased to 109 by December 12.

Limbers in the same quantity as carriages were ordered from the Minneapolis Steel & Machinery Co., which produced a limber to accompany each one of its delivered carriages. This limber has an extremely heavy axle, similar to the automobile front axle. Its size and weight caused difficulty in obtaining it as a drop forging.

To Dodge Bros. was assigned the task of producing the recuperators for this gun in their special plant. The 155-millimeter gun recuperators, however, were made secondary to the production of the recuperators for the 155-millimeter howitzers, which were the easier of the two sorts to build.

Forgings were available and work started on recuperators in April, 1918. No rapid completion of these intricate mechanisms was possible, however, as the first forgings encountered many delays in their machinings. In the cycle of operations, with everything speeded up to the limit, more than three months must elapse from the day the recuperator forging is received to the day when the completed mechanism can be turned over to the inspector as an assembled article.

155-MILLIMETER HOWITZER, MODEL 1918 (SCHNEIDER).

This weapon throws shell or shrapnel weighing 95 pounds. Muzzle velocity for shell is 1,420 feet per second. The weight of the howitzer and carriage is 7,600 pounds.

8-INCH HOWITZER, MODEL 1917.

It was in October, 1918, that the first 155-millimeter gun recuperator was delivered. The factory expected to reach a maximum capacity of 10 a day. The company built 12 more by December 1. After the armistice was signed the company's order was reduced to 880, which had all been completed by May 1, 1919.

In order to have recuperators available for use for the units shipped from the United States minus these mechanisms, 110 rough-machined recuperator forgings were shipped to France, where the work of machining and completing was done.

The translation of the French plans for this weapon furnished one of the most difficult pieces of work undertaken by the Ordnance Department. Without counting in the gun pieces, the carriage and limber is made up of 479 pieces, while the recoil mechanism itself has 372 pieces. A total of 150 mechanical tracings had to be made by our draftsmen for the carriage and test tools; 50 for the carriage limbers; 142 for the recoil mechanism; 74 for the tools and accessories; or a total of 416. It was extremely difficult to secure draftsmen who could do this work, and the translation, accomplished in a few weeks, is regarded as a remarkable achievement.

The cannon for this gun were tested at the Erie Proving Grounds and there packed for overseas shipment. We had many cannon and carriages awaiting shipment when the armistice was signed, the plan being to send them to France, where they would be equipped with recuperators.

Units.	Contractors.	Number ordered.	Number completed Nov. 11, 1918.	Number completed Apr. 17, 1919.	Number floated Nov. 11, 1918.
155-mm. gun carriage, model 1918 (Filloux).	Minneapolis Steel & Machinery Co.	1,446	370	800	16
155-mm. gun carriage limber, model 1918 (Filloux).	.....do	1,446	370	800	16
155-mm. gun cannon proper...	Bullard Engine Works	1,400	53	250	16
Do	Watervliet Arsenal	760	18	68	

8-INCH HOWITZERS.

In the early days of the war the British designed an 8-inch field howitzer that proved itself on battle fields in France. Great Britain loaded her own plants with orders for this weapon and then turned to the United States for additional facilities. The Midvale Steel & Ordnance Co. at Nicetown, Pa., was manufacturing this unit for the British at the time we entered the war.

On April 14, 1917, exactly eight days after we had formally announced our purpose of warring with Germany, an order for 80 of these 8-inch howitzers was placed with the Midvale Steel Co. It was understood that production on our order was to be begun upon the completion of the British contract on which the Midvale Co. was then engaged. The order included the complete units, with carriages, limbers, tools, and accessories, all to be built in accordance with British specifications.

Contracts for the trails were sublet by the Midvale Co. to the Cambria Steel Co; for the wheels, to the American Road & Machinery Co.; for the limbers and firing platforms, to the J. G. Brill Co.; and for the open sights, to the British-American Manufacturing Co. Panoramic sights for these guns were furnished by the Frankford Arsenal.

So satisfactory did the production proceed that on December 13, 1917, the first of the 8-inch howitzers was proof-tried with good results. Early in January, 1918, the complete units began to come through at the rate of three a week, increasing to four a week in April and to six a week in May.

A subsequent contract with Midvale brought the total number of howitzers ordered from that plant up to 195. These weapons, all of the model known as the Mark VI, were all produced and accepted before the signing of the armistice, 96 of them being shipped overseas, with their full complement of accessories. Each completed unit cost in the neighborhood of $55,000. These weapons throw a 200-pound projectile 11,750 yards.

The progress of the war moved so swiftly, however, that there soon was need for artillery units of this same size but with longer range. Accordingly, a new design, known as the Mark VIII½, was brought out, having a range of over 13,000 yards. On October 2, 1918, we placed with the Midvale Co. an order for 100 of these 8-inch howitzers, specifying carriages of the new, heavier type.

When we entered the war the Bethlehem Steel Co., at Bethlehem, Pa., was producing for the British Government a howitzer with a bore of 9.2 inches. The Bethlehem Co. expected to complete these British contracts in July, 1917. The 9.2-inch howitzer was approximately the same size as the 240-millimeter howitzer which we were getting ready to put into production. However, in our desire to utilize every bit of the production facilities of the country, we ordered 100 of the 9.2-inch howitzer units from the Bethlehem Steel Co. and placed additional orders for 132 of these units in England. The British concerns delivered 40 howitzers before the armistice was signed.

Mark.	Size.	Contractor.	Number ordered.	Number completed Nov. 11, 1918.	Number floated.	Number completed to Apr. 17, 1919.
VI	8-inch howitzer	Midvale Steel Co	195	167	96	195
VIII½	do	do	100			34
Model 1917	9.2-inch howitzer	Bethlehem Steel Co	100			1

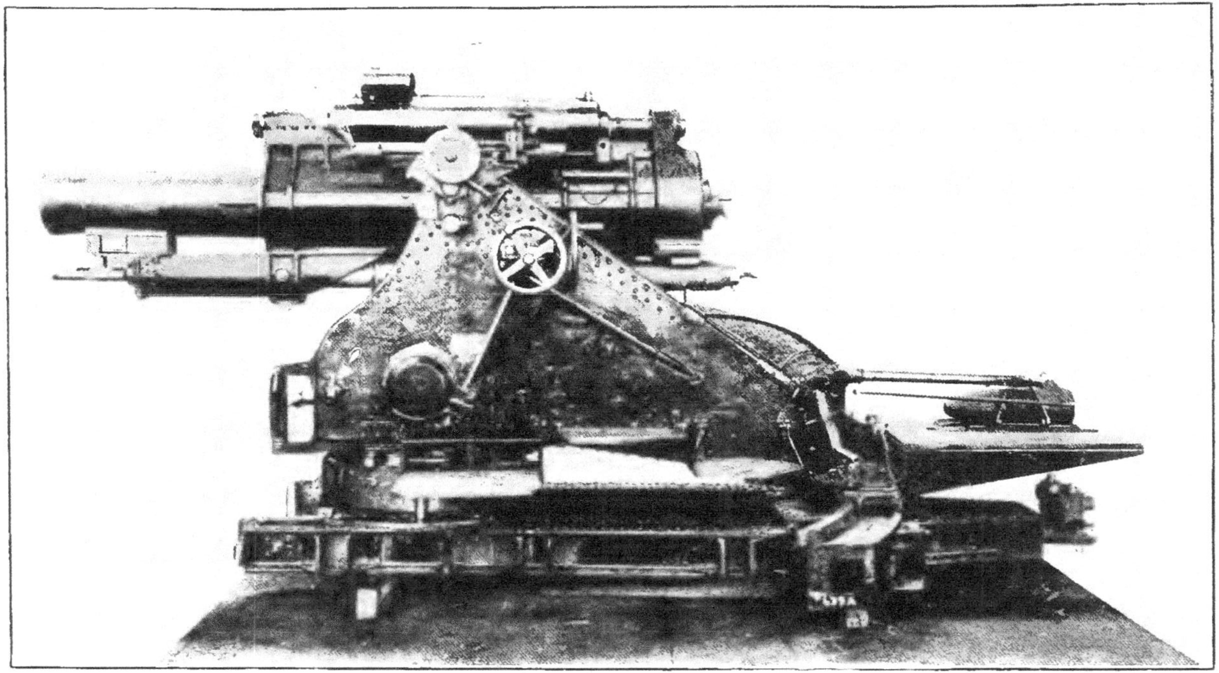

THE 9.2-INCH HOWITZER, MODEL 1917.

This gun shoots shell weighing 290 pounds 8,690 meters. The weight of the howitzer and carriage is 29,100 pounds.

TWO VIEWS OF THE 240-MILLIMETER HOWITZER, MODEL 1918.

THE 240-MILLIMETER HOWITZERS.

The scheme of production of the French 240-millimeter howitzers was entirely aimed at the year 1919; since even if American heavy manufacturing establishments had not been loaded with war orders, it would have been well-nigh impossible to turn out this mighty engine of destruction in quantities in any shorter period of time.

Although approximately the same size as the British 9.2-inch howitzer (the exact diameter of the bore of the 240 being 9.45 inches) and only a little larger than the 8-inch howitzer, the French gun was far more powerful than either. The 8-inch and the 9.2-inch howitzers had ranges in the neighborhood of 6 miles, while their shell weighed from 200 to 290 pounds. The 240, on the other hand, hurled a shell weighing 356 pounds and carrying a bursting charge of between 45 and 50 pounds of high explosive. Its range was almost 10 miles.

We produced the 8-inch and the 9.2-inch howitzers to fill the gap during the two years which must elapse before we could get into quantity production of the 240. The French and British governments in the fall of 1917 asserted their ability to equip our first 30 combat divisions in 1918 with heavy howitzers, so that if our production came along in the spring of 1919 it could meet the requirements of the war situation.

Consequently we planned to equip our first army of 30 divisions with 8-inch and 9.2-inch howitzers in equal numbers of each. Our second army of 30 divisions should be wholly equipped with 240-millimeter howitzers; and our expected production of these, being beyond our own contemplated needs, would serve to replace such 8-inch and 9.2-inch howitzers as had been lost in the meantime.

As we adapted it from the French Schneider model, the 240-millimeter howitzer consisted of four main parts—the howitzer barrel, the top carriage, the cradle with recoil and mechanism, and the firing platform. Each of these four parts had its own transportation wagon and limber drawn by a 10-ton tractor. The weapon was set up with the aid of an erecting frame and a small hand crane.

Each of the main sections is composed of numerous smaller assembled parts made up of various grades of iron and steel and raw materials, all requiring the greatest precision in their manufacture and all having to pass rigid and exacting tests for strength and dimensions.

The production of even one of these enormous weapons would have been a hard job for any American industrial plant, but to manufacture over 1,200 of them, and that within the comparatively limited time allowed and under the abnormal industrial and transportation conditions then prevailing, was a task of tremendous difficulty and complexity.

On September 1, 1917, an order was placed with the Watertown Arsenal for 250 carriages for the American 240's, to be turned out complete with the recoil mechanism, transportation vehicles, tools, and accessories. To show the size of the job, an allotment of $17,450,000 was set aside to cover the estimated expenses at the arsenal.

Well equipped as the Watertown Arsenal was said to be at the time for the production of heavy gun carriages, it was found necessary, in order to handle this job, to construct a new erecting shop that had a capacity practically as large as all the other buildings of the plant put together. The number of employees at the arsenal was increased from 1,200 to more than 3,000.

The greatest difficulty experienced was in obtaining the large number of heavy machine tools required, and experts were sent out to scour the country in an effort to locate these tools wherever they might be available. Raw materials could not be procured in sufficient quantities, while numerous transportation delays impeded the work.

Finally, in October, 1918, the pilot carriage was completed and sufficient progress had been made on the entire contract to assure production of the required number of units in the early part of 1919.

A second carriage contract (Nov. 16, 1917) went to the Standard Steel Car Co., of Hammond, Ind. This called for the delivery of 964 carriages complete with transportation vehicles, limbers, tools, etc., but not with recuperators. These the Otis Elevator Co., of New York, undertook to deliver.

The Standard Steel Car Co. is one of the most important builders of railway cars, freight and passenger, in the country, and it possessed a large and well-equipped plant. Nevertheless, the company was compelled to construct several additional buildings and practically to double the capacity of its huge erecting shop in order to prepare adequately for the tremendous task undertaken.

As a means to save time, subcontracts were immediately placed with more than 100 firms throughout the East and Middle West for the production and machining of as many as possible of the component parts needed by the Standard Steel Car Co. Wherever practicable, the subcontractors working on similar contracts for the Watertown Arsenal were retained by the Indiana company, so that better prices might be obtained, parts standardized, and the whole production greatly facilitated.

Once the work was well under way the ramifications of this one contract, with its subcontracts for parts, materials, tools, building construction, etc., extended throughout practically the entire industrial facilities of the eastern and central sections of the country.

As in the case of the contract given the Watertown Arsenal, there were many difficulties in obtaining tools and raw materials. In a large majority of cases allocations, partly of iron and steel products, had to be obtained through the War Industries Board. When allocations had been granted, priority orders had to be secured, as the producers of these materials were already overworked with Government orders of varying importance.

With the pilot carriage complete in the early part of October, production on all the main parts had progressed by November to such an extent that a large output of finished carriages was assured for December and thereafter, had not the signing of the armistice intervened and ended the necessity for further expedition of the work.

Orders for howitzer bodies were placed as follows:

	Sets.
Bethlehem Steel Co., Nov. 21, 1917	237
Edgewater Steel Co., Oct. 24, 1917	175
Tacony Ordnance Corporation, Nov. 14, 1917	175
Watertown Arsenal, Nov. 10, 1917	80
American Bridge Co., Mar. 31, 1918	800

The Watervliet Arsenal on November 20, 1917, was instructed to do the machining of forgings so as to turn out 250 gun bodies for the 240-millimeter howitzers, and three months later this order was doubled. On November 7, 1918, an additional 660 were ordered from Watervliet, making a grand total of 1,160 howitzer cannon of this caliber ordered machined and completed at the Watervliet Arsenal. The arsenal contracted to reach an output of 100 cannon a month and deliver the last of the 1,160 not later than September 30, 1919.

It was found necessary to erect an entirely new shop for the machining of these howitzers. This shop was completed in May, 1918. During the war period $13,164,706 was spent or allotted to the Watervliet Arsenal for increasing its facilities. Forgings were furnished to the arsenal by the Government, but the forging situation was never a delaying factor in the production of 240-millimeter howitzers.

In all, 158 sets, of 1,467 ordered, were delivered up to December 12, 1918. The pilot howitzer was delivered by the Watervliet Arsenal to the proving ground on August 24, 1918.

In the summer of 1918 the Watertown Arsenal contracted to build 252 additional recuperators for these howitzers. Work was started at once in the shops, and, though additional facilities had to be prepared and much new equipment added, the production of the first recuperator was begun without delay. It was found that the planing equipment at the arsenal was not sufficient to handle the work, and therefore a great deal of the rough planing was done by subcontractors.

The Watertown Arsenal was to furnish its own forgings, but it was quickly found that an additional source of supplies was required.

The Carnegie Steel Co. had been given an order on December 27, 1917, for 1,300 recuperator forgings, and some of these were sent to the Watertown Arsenal.

The first recuperator was completed October 28, 1918, and 16 had been finished up to December 31, 1918, when 280 forgings were in the process of machining.

To handle its order for 1,039 recuperators, the Otis Elevator Co., of New York, found it necessary to rebuild a plant which it owned in Chicago. Forgings were furnished by the Government.

On May 1, 1918, the Otis Elevator Co. started its rough machining. Hard spots were found in the metal, causing great trouble at first, but this difficulty was overcome by changes in the heat treatment. The Carnegie Steel Co. was then instructed to rough-machine the forgings before sending them to the Otis Elevator Co. An order was also given to the Midvale Steel Co. to rough-machine 24 forgings. Early in November, 1918, the Otis Elevator Co. finished its first recuperator.

One 240-millimeter howitzer unit was completed at the time of the signing of the armistice, out of a total of 1,214 contracted for; but had war conditions continued, the expectation was for a monthly capacity of 80 units by 1919. Actual deliveries are given below:

Units.	Contractors.	Number ordered.	Number completed Nov. 11, 1918.	Number completed Apr. 17, 1919.
240-mm. unit, complete, except howitzer.	Watertown Arsenal	250	[1] 1 [2] 4	[1] 41 [2] 25
240-mm. howitzer carriage units, except recuperators.	Standard Steel Car Co	964	5	67
Windlasses	Dodge Manufacturing Co	1,125	33	350
Rammer trucks	do	1,205	2	375
Shot trucks	do	3,214	2	1,000
240-mm. howitzer cannon	Watervliet Arsenal		2	19

[1] Carriage alone. [2] Carriages with recuperators.

FIGHTING THE AIRPLANE WITH ARTILLERY.

The American development of antiaircraft artillery had, previously to 1917, been confined almost exclusively to the task of designing and constructing stationary units of defense for our coast fortifications. It was naturally expected that it would be at those points that we would first, if ever, have to meet an attack from the air. Very little attention had been paid mobile artillery of this sort.

Before April, 1916, the Ordnance Department had designed a high-powered 3-inch antiaircraft mount for the fixed emplacement at coast fortifications. The gun on this mount fired a 15-pound projectile with a muzzle velocity of 2,600 feet a second. It is still to-day the most powerful antiaircraft weapon of its caliber. Between May, 1916,

TWO VIEWS OF ANTIAIRCRAFT GUNS ON TRUCK MOUNTS.

TWO OTHER VIEWS OF ANTIAIRCRAFT GUNS.

and June 18, 1917, orders for 160 of these mounts were placed with the Watertown Arsenal and the Bethlehem Steel Co. Up to April 10, 1919, a total of 116 of these had been completed and sent for emplacement at the points selected.

By the end of 1916, however, it was foreseen that it would be necessary to provide antiaircraft artillery of a mobile type as part of the equipment for any field forces that might be sent abroad. Since that contingency seemed entirely possible at that time, and as it appeared to be impossible to provide a suitable design that would have a sufficient period of time in which to get proper consideration and test, it was decided to improvise a simple structural steel design that would permit quick construction and on which a 75-millimeter field gun, that was already in production, could be mounted.

This design was completed May 1, 1917, and an order for 50 placed with the Builders Iron Foundry. Deliveries on these were made during the fall of 1917, and the carriages were at once shipped to France for equipment with French field guns and recuperators that had been already procured for the purpose.

In its mobility the improvised antiaircraft gun mount was far from perfect. It was necessary to disassemble it partly and mount it on trailers. The need for a mount that could be moved easily and speedily had been realized before our entrance in the war, and a design embodying these qualities was completed as early as December, 1916.

This truck was designed to be equipped with the American 75-millimeter field gun, model of 1916. Before the drawings were completed an order for the pilot mounts of this type was placed with the Rock Island Arsenal. The war came on, and it was decided not to wait for a test of the mounts before starting general manufacture. Accordingly the New Britain Machine Co., in July, 1917, was given an order for 51 carriages. No further orders were placed for carriages of this sort, as it was not thought best to go too heavily into production of an untried mount.

It may be noted here that our first 26 antiaircraft guns were mounted on White 1½-ton trucks.

It was also realized that the field guns with which these mounts were to be equipped did not have the power and range that the war experience was showing to be necessary. The only reasons that the field guns of the 75-millimeter caliber were used in this way was because they were the guns most quickly available and because the French were already using them for this purpose.

To meet the need of more powerful antiaircraft weapons, a need becoming more pressing each day, a 3-inch high-powered antiaircraft gun was designed and mounted on a four-wheel trailer of the automobile type. This mount permitted elevations of the gun from 10 degrees to 85 degrees and also allowed for "all around" firing. An order for 612 of these carriages was given to the New Britain Machine Co. in July, 1917, shortly after the contract for the 51 truck mounts had been placed with that concern.

Because of the urgency of the situation it was necessary to construct these carriages without the preliminary tests on a pilot carriage. This, of course, is a very undesirable practice, but under the existing conditions no other procedure would have been practicable. The French antiaircraft auto truck mount, which had the French 75-millimeter field gun with its recuperator placed upon a special antiaircraft mount, was not adopted at the time, because, in July, 1917, the whole question of the possibility of constructing French recuperators in this country was still entirely unsettled. It was imperative then that we develop our own designs.

All of the 51 truck mounts for the antiaircraft guns were delivered during the fall and early winter of 1918, and 22 of them were in France before December, 1918,

Delivery of the first carriage for the 3-inch high-powered gun mounted on the trailer carriage was made in August, 1917. It had been rushed ahead of general production in order to be given some sort of a test. No further deliveries were made, but manufacture reached a point where production in quantity could begin.

A representative of the Ordnance Department was sent to France and England in December, 1917, to gather all the information possible on antiaircraft artillery. As a result of his investigations it was determined that it would be best to procure the greater part of our fire-control equipment in France, since the instruments developed there were in some cases of a highly complicated nature and their manufacture entirely controlled by private parties. Orders were placed for enough of these instruments for the equipment of the first 125 batteries.

Meanwhile, fire-control instruments of various types were in the process of development in this country; but, as they were largely based upon theoretical construction derived from study of the French practices, it was deemed best not to manufacture any of these instruments in quantity, as better instruments of French design were available. Drawings of the French instruments were brought back by the Ordnance officer on his visit to France and were available in this country in the spring of 1918, when manufacture of some of them began in the United States.

At the signing of the armistice our forces in France were equipped almost wholly with antiaircraft artillery loaned to us and supplied by the French. This, of course, does not include the 101 improvised and truck mounts completed during 1917. Production here, however, had reached such a point that shipment of material would have begun in quantity in January, 1919.

The estimated requirements of antiaircraft artillery for 2,000,000 men in 48 divisions is only 120 guns. Other material, of course, would have been required previously for defense of depots, railheads, etc., dependent in a great measure upon the activities of German bombers. It is estimated that about 200 guns would have sufficed for this purpose.

To summarize, 50 of the so-called improvised 75-millimeter antiaircraft guns and mounts had been ordered and completed up to the time of the signing of the armistice; 51 of the 75-millimeter antiaircraft mounts, model of 1917, had been ordered and 46 completed; while 612 of the 3-inch antiaircraft trailer carriage mounts, model of 1917, had been ordered, of which 1 had been actually delivered at the signing of the armistice, the balance to come at the rate of 26 per month starting in December.

Artillery—Production of complete units, by months.

[Deliveries in the United States on U. S. Army orders only.]

	To Jan., 1918.	1918												Total.
		Jan.	Feb.	Mar.	Apr.	May.	June.	July.	Aug.	Sept.	Oct.	Nov.	Dec.	
75-mm. gun, model 1897	0	0	0	0	0	0	0	0	0	0	1	0	0	1
75-mm. gun, model 1916	0	0	0	9	4	6	21	2	60	42	51	11	45	251
75-mm. gun, model 1917	1	11	36	28	58	22	61	61	55	130	211	110	55	839
75-mm. antiaircraft gun	0	49	2	0	0	1	1	2	16	2	18	6	3	100
3-inch antiaircraft gun	0	0	0	0	0	0	0	0	1	0	0	0	0	1
4.7-inch gun	0	0	0	0	0	0	0	15	15	28	72	50	44	224
155-mm. howitzer	0	0	0	0	0	0	0	1	8	39	63	65	100	276
5-inch seacoast gun	0	0	1	27	(1)	(1)	(1)	(1)	(1)	(1)	(1)	(1)	(1)	28
6-inch seacoast gun	0	0	12	5	2	45	23	4	1	(1)	(1)	(1)	(1)	92
155-mm. gun	0	0	0	0	0	0	0	0	0	0	1	5	10	16
8-inch howitzer	7	12	17	20	22	2	0	0	23	27	33	13	15	191
9-2-inch howitzer [2]	0	0	0	0	0	0	0	0	0	0	0	0	0	[2] 0
240-mm. howitzer	0	0	0	0	0	0	0	0	0	0	0	1	0	1
8-inch seacoast gun	0	0	0	0	0	0	0	0	0	3	14	4	1	22
10-inch seacoast gun	0	0	0	0	0	0	0	0	0	0	0	0	0	0
12-inch gun	0	0	0	0	0	0	0	0	0	0	0	1	2	3
12-inch seacoast mortar	0	0	0	0	0	0	0	0	1	0	0	10	2	13
Total	8	72	68	89	86	76	106	85	180	271	464	276	277	2,058

[1] Project complete.
[2] No deliveries made by Bethlehem Steel Co. on U. S. Army orders until after signing of the armistice because of priority given to British orders placed before the American declaration of war.

By "complete units" is meant gun body complete, carriage, and recoil mechanism or recuperator. Units are given as complete when their component parts were complete, although the actual assembly of these parts at a common point, testing, and final delivery usually required from two weeks' to two months' additional time.

The 5-inch, 6-inch, 10-inch, and 12-inch seacoast guns and the 12-inch seacoast mortars were taken from the fortifications and modified for use with mobile carriages, all above 6 inches for railway mounts.

The 75-millimeter gun, model 1897, was the approved model for active service in France. Model 1916 and model 1917 were used for training purposes both in the United States and in France.

Production of mobile artillery (complete units), Apr. 1, 1917, to Nov. 11, 1918.

[Including all produced for France and Great Britain in United States.]

	Produced.	Shipped overseas.
75-mm. guns (or British 18-pounder)	970	181
3-inch and 75-mm. antiaircraft guns	97	[1] 26
4.5-inch howitzers	97	97
4.7-inch guns	157	64
155-mm. (5-inch and 6-inch seacoast guns)	121	[2] 114
155-mm. howitzers	144	0
7-inch guns on caterpillar mounts	[3] 10	0
Railway artillery	20	11
Heavy howitzers	[4] 418	322
Total	2,034	815

[1] Does not include 51 improvised mounts for which guns were furnished by French.
[2] Includes sixteen 155-mm. guns and carriages shipped without recuperators.
[3] Built for the Marine Corps.
[4] Includes sixteen 8-inch howitzers built for the Marine Corps.

1½-TON ANTIAIRCRAFT MACHINE GUN TRAILER.

TWO VIEWS OF THE 7-INCH NAVY RIFLE MOUNTED ON A PEDESTAL ON A RAILWAY CAR.

This rifle has a range of about 10 miles and throws a projectile weighing 165 pounds. Note the means of loading and the depression angle.

CHAPTER IV.

RAILWAY ARTILLERY.

As soon as war was declared against Germany the Ordnance Department, in its search for an immediate equipment of strong artillery, surveyed the ordnance supplies of the country and discovered some 464 heavy guns which might be spared from the seacoast defenses, obtained from the Navy, or commandeered at private ordnance plants where they were being manufactured for foreign Governments. There were six guns of this last-named class—powerful 12-inch weapons which had been produced for the Chilean Government. It was seen that if all, or if a large part, of these guns could be made available for service in France, America would quickly provide for herself a heavy artillery equipment of respectable proportions.

The guns thus available for mounting on railway cars ranged in size from the 7-inch guns of the Navy to the single enormous 16-inch howitzer which had been built experimentally by the Ordnance Department prior to 1917. The list of these guns according to number, size, length, and source whence obtained was as follows:

Number of guns.	Size.	Length.	Source whence obtained.
	Inches.	*Calibers.*	
12	7	45	Navy.
96	8	35	Seacoast defenses.
129	10	34	Do.
49	12	35	Do.
6	12	50	In manufacture for Chile.
150 (mortars)	12	10	Seacoast defenses.
21	14	50	Navy.

In addition to these there was the 16-inch howitzer, 20 calibers in length, which had been built by the Ordnance Department before 1917.

The expression 14-inch gun, 50 calibers, means that the gun has a barrel diameter of 14 inches and that the gun body is fifty times the caliber of 14 inches, or 700 inches (58 feet 4 inches) long.

The Ordnance Department conceived that the only way to make these guns available for use abroad would be to mount them on railway cars. These guns were not vital in the defense of our coast under the conditions of the war with Germany, but it was evident that they would make a valuable type of long-range artillery when placed on satisfactory railway mounts.

Mounting heavy artillery on railway cars, however, was not an idea born of the recent war. The idea was probably originally American. The Union forces at the siege of Richmond in 1863 mounted a 13-inch cast-iron mortar on a reinforced flat car, this being the first authenticated record of the use of heavy railway artillery.

In 1913 the commanding officer of the defenses of the Potomac, which comprise Forts Washington and Hunt, was called upon to report on the condition of these defenses. In reply, he advised that no further expenditure be made on any one of the fixed defenses, but recommended that a "strategic railroad" be built along the backbone of the peninsula from Point Lookout to Washington, with spurs leading to predetermined positions both on Chesapeake Bay and the Potomac River, so placed as to command approaches to Washington and Baltimore.

Further, he recommended that 4 major-caliber guns, 16 medium-caliber guns, and 24 mine-defense guns be mounted on railroad platforms, with ammunition, range finding, and repair cars making up complete units, so that this armament could be quickly transported at any time to the place where most needed. He suggested that this scheme be made applicable to any portion of the coast line of the United States. His argument was based upon the fact that guns in fixed positions, of whatever caliber, violate the cardinal military principle of mobility.

The nations engaged in the war now ending developed to a high stage the use of heavy artillery mounted on railway cars, bringing about a combination of the necessary rigidity with great mobility, considering the weight of this material.

Railway artillery came to be as varied in its design as field artillery. Each type of railway mount had certain tactical uses and it was not considered desirable to use the different types interchangeably. The three types of cannon used on railway mounts were mortars, howitzers, and guns. It was not practicable to use the same type of railway mounts for the different kinds of cannon. Moreover, these mounts differed radically from the mounts for such weapons at the seacoast defenses.

The three general types of railway mounts adopted were those which gave the gun all-around fire (360-degree traverse), those which provided limited traverse for the gun, and those which allowed no lateral movement for the gun on the carriage but were used on curved track, or epis, to give the weapons traverse aim.

The smaller weapons, such as the 7-inch and the 8-inch guns and the 12-inch mortars, were placed on mounts affording 360-degree traverse. The limited traverse mounts were used for the moderately long-range guns and howitzers. The fixed type of mount was used for long-range guns only, and included the sliding railway mounts, such as the American 12-inch and 14-inch sliding mounts and the French Schneider à glissement mounts.

The work of providing railway artillery—that is, taking the big, fixed-position guns already in existence within the United States and similar guns being produced and designing and manufacturing suitable mounts for them on railway cars—grew into such an important undertaking that it enlisted the exclusive attention of a large section within the Ordnance Department. This organization eventually found itself engaged in 10 major construction projects, which, in time, had the war continued, would have delivered more than 300 of these monster weapons to the field in France and, to a lesser extent, to the railway coast defenses of the United States.

As it was, so much of the construction—the machining of parts, and so on—was complete at the date of the armistice, that it was decided to go ahead with all of the projects except three, these involving the mounting of 16 guns of 14-inch size, 50 calibers long, the production of 25 long-range 8-inch guns, 50 calibers, and their mounting on railway cars, and the mounting of 18 coast-defense, 10-inch guns, 34 calibers long, on the French Batignolles type of railway mount.

Inasmuch as it will be necessary in this chapter to refer frequently to the barbette, Schneider, and Batignolles types of gun mounts for railway artillery, it should be made clear to the reader what these types are.

The barbette carriage revolves about a central pintle, or axis, and turns the gun around with it. When it was decided to put coast-defense guns on railway cars, the guns were taken from their emplacements, barbette carriages manufactured for them, and the whole mounted upon special cars. The barbette mount revolves on a support of rollers traveling upon a circular base ring. In the railway mount the base ring is attached to the dropped central portion of the railway car. The barbette railway mount is provided with struts and plates by which the car is braced against the ground.

The Schneider railway mount is named after the French ordnance concern Schneider et Cie, who designed it. In this mount the gun and its carriage are fastened rigidly parallel to the long axis of the railway car. Thus the gun itself, independently of any movement of the car, can be pointed only up and down in a vertical plane, having no traverse or swing from left to right, and vice versa. In order to give the weapon traverse for its aim, special railway curved tracks, called epis, are prepared at the position where it is to be fired. The car is then run along the curve until its traverse aim is correct, and the vertical aim is achieved by the movement of the gun itself. In the Schneider mount there is no recoil mechanism, but the recoil is absorbed by the retrograde movement of the car itself along the rails after the gun is fired. This movement, of course, puts the gun out of aim, and the entire unit must then be pushed by hand power back to the proper point.

In the Batignolles type, gun and cradle are mounted on a so-called top carriage that permits of small changes in horizontal pointing right and left. Thus with the railway artillery of the Batignolles type also, track curves, or epis, are necessary for the accurate aiming. The Batignolles mount partially cushions the recoil by the movement of the gun itself in the cradle. But, in addition, a special track is provided at the firing point and the entire gun car is run on this track and bolted to it with spades driven into the ground to resist what recoil is not taken up in the cradle. The unit is thus stationary in action, and the gun can be more readily returned to aim than can a gun on a Schneider mount.

THE 7-INCH RIFLES.

The conditions under which the war with Germany was fought virtually precluded any chance of a naval attack on our shores which would engage our fixed coast defenses. The British grand fleet, with the assistance of fleets of the other allies and America, had the German battle fleet securely bottled. On the other hand there was the prowling submarine able at all times to go to sea and even to cross the ocean, and some of the latest of these submarines were armed with long-range medium-caliber guns. It was not beyond possibility that some sort of an attack would be made on our shores by submarines of this character, yet it was safe to believe that these craft would keep well out of range of the guns at our stationary coast defenses.

To protect our coast from such attack the Ordnance Department conceived the plan of mounting heavy guns on railway cars. They might then be moved quickly to places on the seacoast needing defense. For this purpose the Navy turned 12 of its 7-inch rifles over to the Ordnance Department for mounting. Meanwhile our ordnance officers had designed certain standard railway artillery cars, known as models 1918, 1918 Mark I, and 1918 Mark II, for 7-inch and 8-inch guns and 12-inch mortars, respectively. These cars all had the same general features.

The model 1918 car was selected for the converted 7-inch Navy rifle. The rifle was mounted on a pedestal set on the gun car in such a manner as to give all-around fire, or 360-degree traverse. The pedestal mount permitted the gun to be depressed at an angle suitable for firing from high places along the coast down upon the low-lying submarines.

Contracts for the various parts for these cars and the pedestal gun mounts were let to concerns engaged in heavy steel manufacture, but the assembling was done by the American Car & Foundry Co., of Berwick, Pa. Twelve of the 7-inch rifles were so mounted. As this equipment was intended exclusively for use in this country, the gun cars were equipped with the American type of car couplings.

THE 8-INCH GUNS.

For the 8-inch guns taken from seacoast fortifications the Ordnance Department designed a barbette mount giving complete, 360-degree, traverse, thus providing for fire in any direction. There were 96 such guns available for railway mounts. Orders for 47 gun cars with carriages for mounting the weapons were placed with three concerns—the Morgan Engineering Co., of Alliance, Ohio, the Harrisburg Manufacturing & Boiler Co., of Harrisburg, Pa., and the American Car & Foundry Co., of Berwick. Two of the three contractors found it necessary to provide additional facilities and machine-tool equipment at their plants in order to handle this job.

The first railway mount for the 8-inch gun was completed and sent to the Aberdeen Proving Ground for test in May, 1918. In early June the test had shown that the weapon was efficient and entirely satisfactory. Before the end of the year 1918 a total of 24 complete units, with ammunition cars for standard-gauge track, shell cars for narrow-gauge track, transportation cars, tools, spare parts, and all the other necessary appurtenances of a unit of this character, had been completed. Three complete 8-inch units were shipped overseas before the armistice was signed.

When the armistice came the Harrisburg company had delivered 9 of these mounts and the Morgan Engineering Co. an equal number, making 18 in all. The former concern had reached an output of 5 mounts per month and the latter 10 per month.

An interesting feature of this mount is that it can be used either on standard-gauge or on narrow-gauge railroad track. The narrow gauge adopted was that in standard use in the fighting zones in France, the distance between the rails being 60 centimeters, or the approximate equivalent of 24 inches. Each gun car was provided with interchangeable trucks to fit either gauge. The artillery train necessary for the maneuvering of the weapon was also similarly equipped to travel on either sort of track.

As a rule the longer the barrel of a cannon, the greater its range. The 8-inch seacoast guns thus mounted were 35 calibers in length, that is, thirty-five times 8 inches, or 23 feet 4 inches. The requirements of our forces in the field in France called for guns of this same size but of longer range. Consequently an 8-inch gun of 50 calibers—that is, 10 feet longer than the seacoast 8-inch gun—was designed, and 25 of them were ordered. This project came as a later development in the war, the guns being intended for use abroad in 1920. The railway mounts for the weapons had not been placed in production when the armistice came. Because of the incomplete status of this project in the autumn of 1918, the whole undertaking was abandoned.

10-INCH AND 12-INCH GUNS.

There were at the seacoast defenses and in the stores of the Army a large number of 10-inch guns of 34 calibers. Of these 129 were available for mounting on railway cars. It was proposed to mount these weapons on two types of French railway mounts—the Schneider and the Batignolles.

The project to mount 36 of these weapons on Schneider mounts was taken up as a joint operation of the United States and French Governments, the heavy forging and rough machining to be done in this country and the finishing and assembling in the French shops. The American contractors were three. The Harrisburg Manufacturing & Boiler Co. undertook to furnish the major portion of the fabricated materials for the carriages and cars. The Pullman Car Co. contracted to produce the necessary trucks for the gun cars, while the American Car & Foundry Co. engaged to build the ammunition cars.

Eight sets of fabricated parts to be assembled in France had been produced before the armistice was signed. Gen. Pershing had requested the delivery in France of the 36 sets of parts by March 2, 1919. After the armistice was signed there was a natural letdown in speed in nearly all ordnance factories, but even without the spur of military necessity the contracting concerns were able by April 7, 1919, to deliver 22 of the 36 sets ordered. Had the war continued through the winter there is little question but that all 36 sets of parts would have been in France on the date specified.

The 10-inch seacoast gun, Batignolles mount project, was placed exclusively in the hands of the Marion Steam Shovel Co., of Marion, Ohio. It had been proposed also to mount 12-inch seacoast guns on this same type of equipment, and this work, too, went to the Marion concern. There were to be produced 18 of the 10-inch units and 12 of the larger ones.

The Marion Steam Shovel Co. had had a large experience in producing heavy construction and road-building equipment. The concern encountered numerous difficulties at the start in translating the French drawings and in substituting the American standard materials for those specified by the French. These difficulties, combined with struggles to obtain raw materials and the equipment

8-INCH RAILWAY ARTILLERY, BARBETTE TYPE.

This view shows gun in act of hurling projectile parallel to track.

8-INCH SEACOAST RIFLE, WITH RECOIL MECHANISMS, SET UP ON A SPECIALLY DESIGNED RAILWAY MOUNT.

This gun, thus mounted on a railway car, is capable of an all-around fire and can deliver a shot in any direction from its location on the car.

12-INCH RIFLE IN ACT OF FIRING.

The force of the recoil sends the entire car back on track about 5 feet.

12-INCH RIFLE ELEVATED TO ITS MAXIMUM POSITION.

It is capable of hurling a 700-pound shell 25 miles. This is a modified Schneider type of carriage.

TWO VIEWS OF 12-INCH RIFLE ON SLIDING TYPE RAILWAY MOUNT.

for the increased facilities which had to be provided at the factory, so delayed production that no mount for either the 10-inch or 12-inch guns had been delivered at the time of the armistice. The first mount of these classes—one with a 12-inch gun—reached the Aberdeen proving ground about April 1, 1919. The 10-inch project, calling for 18 mounts, was canceled soon after November 11, 1918. The work on the dozen mounts for 12-inch guns, however, had progressed so far that the Ordnance Department ordered the completion of the entire equipment.

As has been stated, the Government found in this country six 12-inch guns being made for the Republic of Chile. Their length of 50 calibers gave them a specially long range. It was decided to place the Chilean guns on a sliding mount. In a mount of this type the retrograde movement of the car along the track as and after the gun is fired takes up and absorbs the energy of fire.

The first sliding railway mount used on the allied side in the great war was of French design. But our manufacturers had so much trouble with French designs that when the project came up of mounting the Chilean guns in this fashion it was decided that it would be quicker to design our own mount. Consequently the French design was taken in hand by our ordnance engineers and redesigned to conform to American practice, with the inclusion in the design of all original ideas developed by the Ordnance Department in its creative work during the war period up to that time. The manufacturers who looked at the French design of the sliding railway mount estimated that it would take from 12 to 18 months before the unit could be duplicated in this country and first deliveries made. They looked at the American design and estimated that they could build it in 3 months.

It was decided to build three mounts of this character and thus have a reserve of one gun for each mount to serve as replacement when the original guns were worn out. Contracts were placed in the early summer of 1918, and all three mounts were delivered before the armistice was signed, the first mount being completed within 85 days after the order was placed. For these mounts the American Bridge Co. furnished the main girders or side pieces, the Baldwin Locomotive Co. built the railway trucks, and the Morgan Engineering Co. manufactured the many other parts and assembled the complete units. The speed in manufacture was made possible by the fact that the plant engineers of the three companies helped the ordnance officers in designing the details. With such intimate cooperation, the concerns were able to begin the manufacture of component parts while the drawings were being made.

All three weapons with their entire equipment, including supplies, spare parts, ammunition cars, and the whole trains that make up such units, were ready for shipment to France in November, 1918. Each mount as it stands to-day is 105 feet long and weighs 600,000 pounds. The load of the gun and the peak load put on the carriage when the gun is fired are so great that it requires four trucks of 8 wheels each, 32 car wheels in all, to distribute the load safely over ordinary standard-gauge track.

THE 12-INCH MORTARS.

In years past the Ordnance Department had procured a large number of 12-inch mortars for use at seacoast defenses. These great weapons are 10 calibers in length, or 10 feet in linear measurement, the diameter of the barrel being just an even foot. Of the number stationed at the coastal forts and in reserve it was decided that 150 could be safely withdrawn and prepared for use against Germany. When Gen. Pershing was informed of the proposal, he asked that 40 of these weapons mounted on railway cars should be delivered to the American Expeditionary Forces for use in the planned campaign of 1919. In order that there might be an adequate supply of them, the Ordnance Department let contracts for the mounting of 91 of these mortars on railway equipment, a project which would give the United States a formidable armament and still provide a reserve of 59 mortars to replace the service mortars on the carriages after repeated firing had worn them out.

This job proved to be one of the largest in the whole artillery program. The entire contract was let to the Morgan Engineering Co., of Alliance, Ohio. In order to handle the contract, a special ordnance plant, costing $1,700,000 for the building alone, had to be constructed at the company's works at Alliance. The work was so highly specialized that machine tools designed for the particular purpose had to be produced. The Government itself bought these tools at a cost of $1,800,000. Although work on this plant was not started until December 10, 1917, and although thereafter followed weeks and weeks of the severest winter weather known in recent years, with all the delays in the deliveries of materials which such weather conditions bring about, the plant was entirely complete on June 1, 1918, not only, but the work of producing the mounts had started in it long before that, some machines getting to work as early as April.

The gun car used for mounting the mortar carriage was of the same design as that for the 7-inch and 8-inch guns, except that each truck had six wheels. The carriage built upon this car was of the barbette type, and it allowed the gun to be pointed upward to an angle as high as 65° and provided complete traverse, so that the mortar could be fired in any direction from the car. A hydropneumatic system for absorbing the recoil of the mortar after firing was adopted. This recuperator in itself was a difficult problem for the manufacturer to

12-INCH SEACOAST GUN ON A CREUSOT RAILWAY MOUNT.

This huge weapon in this position is ready to fire half a ton of shot a distance of 25 miles. It requires only two men to operate the powerful elevating apparatus necessary to bring the gun into quick-firing position.

TWO VIEWS OF 12-INCH MORTAR ON RAILWAY MOUNT.

Lower view shows the mortar in its extreme position of recoil.

solve, being the first hydropneumatic recuperator of the size ever built in this country.

In spite of the weight and elaborate character of this unit it was put into production in an astonishingly short space of time. The pilot mount came through on August 22, 1918, less than nine months after the spade was first struck in the ground to begin the erection of the ordnance plant. By the end of August the pilot mortar had successfully passed its firing tests at Aberdeen, functioning properly at angles of elevation from 22 degrees to 65 degrees and in any direction from the mount. While this unit was put through hurriedly for these tests, the preparation for the rest of the deliveries was made on a grand scale, looking toward quantity production later on. When the armistice was signed, every casting, forging, and structural part for every one of the 91 railway mounts was on hand and completed at the works of the Morgan Engineering Co., and thereafter the process was merely one of assembling, although in a unit of such size the assembling job alone was one of great magnitude. Even at the reduced rate of production incident to the relaxation of tension after the armistice was signed, the company delivered 45 complete units to the Government up to April 7, 1919, or five more than Gen. Pershing said he would require during the whole campaign of 1919. Careful estimates show that if the war had continued the company would have delivered the mounts at the rate of 15 per month beginning on December 15, 1918, a rate which would have completed the entire project for 91 mounts by the middle of June, 1919.

As in the case of the 8-inch railway guns, the 12-inch mortars were provided with interchangeable wheel trucks allowing the unit to travel and work either on standard-gauge track or on the 60-centimeter, narrow-gauge track of the war zone in France.

14-INCH GUNS.

The War Department did not have any 14-inch guns which could be spared from the seacoast defenses for use abroad. The Ordnance Department, therefore, inaugurated the project for the construction of 60 guns of 14-inch caliber. For the construction of such guns complete new plants were required, as all available facilities were already taken over for other projects considered more important. This contract was to have been turned out by the Neville Island ordnance plant. The Navy Department in May, 1918, expressed willingness to turn over to the Army certain 14-inch guns, 50 calibers, then under construction and of which it was estimated that 30 would be completed by March, 1919.

It was decided to place some of these 14-inch guns on American sliding railway mounts, and 16 such mounts were ordered from the Baldwin Locomotive Works, deliveries to begin February 1, 1919. The 16 units were to be delivered prior to April, 1919, but due to the signing of the armistice work was suspended on the contracts, since the mounts were designed for use in France. The contract was canceled in March, 1919.

The Navy itself placed five of these guns on railway mounts of another design to be operated in France by naval forces on shore. Eleven such mounts were built by the Baldwin Locomotive Works under the supervision of the Navy Ordnance Bureau, and six of them were afterwards turned over to the Army.

THE 16-INCH HOWITZER.

Without discussing here the 12-inch howitzers, 20 feet long, which the Ordnance Department ordered produced and mounted on railway trucks, a development for use abroad in 1920, we come, finally, to the largest weapon of all in the railway artillery program, the 16-inch howitzer, the barrel of this mighty weapon being 26 feet 6 inches long. The American 16-inch howitzer had been forged out and finished prior to the date of America's entrance into the war. It was proposed to place this weapon on a railway mount and make it available for use on the western front.

The Ordnance Department completed the design for the mount on February 10, 1918. In order to turn out the unit in the shortest possible time, the project was placed with three manufacturers, each of whom was to produce different parts. The American Bridge Co. received the order to build the structural parts, the Baldwin Locomotive Works contracted for the trucks, while the Morgan Engineering Co. undertook to assemble the unit and also to build the top carriage and other mechanical parts. The contractors did a speedy job in producing the mount for this howitzer.

In nearly all railway artillery of this size it is necessary to provide bracing when the gun is set up in position for firing. The 16-inch howitzer mount was unique in that the weapon could be fired from the trucks without any track preparation whatsoever. An exhaustive test at the Aberdeen proving grounds demonstrated that this piece of artillery ranked with the highest types of ordnance in use by any country in the world.

In the meantime orders had been placed for 61 additional howitzers. The American Expeditionary Forces asked that 12 of these enormous weapons be sent overseas as soon as they could be produced, a job which would have extended over a period of months, if not years. Since none of the additional howitzers had been produced when the armistice was signed, the project of building mounts for them never got under way. The pilot howitzer and mount were not shipped abroad.

In the design of railway equipment for high-angle weapons such as howitzers, two loads must be considered by the builders in order to provide a gun car of sufficient strength to hold its freight. One

TWO VIEWS OF 14-INCH RAILWAY ARTILLERY.

This type was evolved entirely by the Ordnance Department. It is an excellent weapon for coast defense and hurls a 1,200-pound projectile more than 18 miles.

16-INCH HOWITZER ON RAILWAY MOUNT.

A 1,600-pound projectile being loaded into the 16-inch howitzer from which it will be sent on a journey of approximately 13 miles.

16-INCH HOWITZER ON RAILWAY MOUNT.

This view shows howitzer in the act of firing.

of these loads, the lighter one, consists merely of the ordinary weight of the gun and its carriage upon the car wheels. The other load, the so-called firing load, consists of the weight of the unit plus the additional weight of the downthrust of the howitzer when it recoils. In the case of the 16-inch howitzer the firing load is 748,231 pounds. The weight of 748,231 pounds must be distributed along the tracks by the numerous sets of wheels at the instant the gun is fired.

The mount for the howitzer is so constructed that this load is partly taken up by the slide of the gun car along the track. In addition, the howitzer is equipped with a hydraulic recoil cylinder. Thus the unit has a double recoil system. The car trucks in the tests comfortably transmitted, through a series of equalizer springs, this enormous load upon an ordinary rock-ballast track, without any distortion to the track or roadbed or impairment to the working parts of the unit. After each discharge the whole huge mount moves backward along the track for a distance of 20 or 30 feet.

Each railway artillery project called for the manufacture of a great equipment of ammunition cars, fire-control cars, spare-parts cars, supply cars, and the like, a complete unit being a heavy train in itself. Such armament-train cars, together with numerous other accessories and necessary equipment, were designed by the Ordnance Department and produced for each mount. In all, 530 ammunition cars were produced up to April, 1919. Most of them were shipped abroad, but 118 were retained for use in this country. Since the overseas cars were to be used with French railway equipment, it was necessary to fit them out with French standard screw couplers, air brakes, and other appliances for connecting up with French railway cars.

The matter of traction power for these gun and armament trains near the front set a problem for the Ordnance Department to solve. It was out of the question to use steam engines near the enemy's lines, since the steam and smoke would betray the location of artillery trains at great distances. The Ordnance Department adopted a gas-electric locomotive of 400 horsepower to be used to pull railway artillery trains at the front, and was on the point of letting a contract to the General Electric Co. for the manufacture of 50 of them when the armistice was signed.

NEVILLE ISLAND.

It seems fitting at this point to say something about the Neville Island ordnance plant, on an island in the Ohio River near Pittsburgh, which would have produced weapons of the character of those used with railway mounts and would have turned them out in large numbers had the armistice not come to put an end to this enormous project. The plant was being erected for the Government by the United States Steel Corporation without profit to itself. The estimated cost of this plant when finished was $150,000,000. Designed to supply the needs of the Army for artillery of the heaviest types, the Neville Island plant was being constructed on such a scale that it would surpass in size and capacity any of the famous gun works of Europe, including the Krupps.

It was being equipped to handle huge ordnance undertakings, such as the monthly completion of 15 great 14-inch guns and the production of 40,000 projectiles monthly for 14-inch and 16-inch guns. The plans of the Government contemplated the production of 14-inch guns to the number of 165 in all and their shipment to France in time to be in the field before May 1, 1920. An initial order for 90 of these weapons had been placed at the arsenal while it was being erected.

Besides 14-inch guns the plant was being equipped to turn out 16-inch and even 18-inch weapons. The immense size of the machinery necessary for such production can be understood when it is noted that an 18-inch gun weighs 510,000 pounds and a 14-inch gun 180,000 pounds. It requires from 12 to 18 months to produce guns of this size, yet Neville Island was being developed on a scale to build hundreds of them simultaneously. The entire plant was to cover 573 acres and was to employ 20,000 workmen when in full operation.

At the signing of the armistice work was suspended at Neville Island, and four months later the whole project was abandoned.

Type.	Total ordered.	Number produced Nov. 11, 1918.	Number produced to Apr. 7, 1919.	Number required by A. E. F. for campaign during 1919.	Guns available.	Remarks.
7-inch Navy gun, railway mount.	12	12	12	0	12	Produced for antisubmarine work along America's seacoast.
8-inch, 35-caliber seacoast gun, railway mount.	47	18	33	36	96	
10-inch, 34-caliber seacoast gun on French type railway mount.	36	8[1]	22[1]	36	111	Fabricated material and trucks, complete, produced within country, mount to be assembled in France.
Do	18	0	......	0	18	Project canceled on signing of the armistice, Batignolles.
12-inch, 35-caliber seacoast gun on French type railway mount.	12	0	1	12	49	French Batignolles type.
12-inch, 50-caliber gun on American sliding railway mount.	3	3	3	4	6	Guns obtained from Chilean Government manufactured in this country.
14-inch, 50-caliber naval gun on railway mount.	11	11	11	11	21	
12-inch, 10-caliber seacoast mortar on railway mount.	91	1	45	40	150	
16-inch howitzer, 20-caliber, on railway mount.	1	1	1	0	1	61 guns under construction.
14-inch, 50-caliber guns on American sliding railway mount.	16	0	....	....	....	Project canceled Mar. 11, 1919. Guns under construction.
12-inch, 20-caliber howitzer on railway mount.	1	0	....	....	....	If war had continued, 60 mounts contemplated.

[1] Sets, fabricated parts.

CHAPTER V.

EXPLOSIVES, PROPELLANTS, AND ARTILLERY AMMUNITION.

The Interallied Ordnance Agreement of the late fall of 1917, supplying to the United States as it did French and British artillery and other heavy ordnance supplies until the developing American ordnance industry could come into production, nevertheless called upon the United States to produce heavily the explosives and propellants that are of such major importance to a modern army. These commodities were needed by the armies of France and Great Britain more than any other sort of ordnance which America could supply.

The result was an enormous production of propellants and explosives in the United States during the period of American belligerency, no other prime phase of the ordnance program being carried to such a stage of development. The reader will clearly see the distinction between propellants and explosives. The propellant is the smokeless powder that sends the shell or bullet from the gun; the explosive is the bursting charge within the shell.

To realize the expansion of the American explosives industry during the war period, consider such figures as these: America in 19 months turned out 632,504,000 pounds of propellants—the powder loaded into small-arms cartridges or packed into the big guns behind the projectiles to send them against the enemy. In those same 19 months France produced 342,155,000 pounds of propellants and Great Britain 291,706,000 pounds. The American production was practically equal to that of England and France together.

In those 19 months we produced 375,656,000 pounds of high explosives for loading into shell. In the same 19 months England produced 765,110,000 pounds of high explosives and France 702,964,000 pounds. America was below both France and England in total output, but in monthly rate of output America had reached 47,888,000 pounds as against France's 22,802,000 pounds and England's 30,957,000 pounds. Our rate of manufacturing propellants at the end of the fighting was up to 42,775,000 pounds as against France's 17,311,000 and England's 12,055,000.

Figure 9 shows graphically the achievements of America in manufacturing propellants and explosives.

In the production of artillery ammunition a comparison with France and Great Britain shows that our monthly rate in turning out unfilled rounds of ammunition at the end of the war was 7,044,000 rounds, as against 7,748,000 rounds for Great Britain and 6,661,000 rounds for France. In producing complete rounds of artillery ammunition, our monthly rate at the signing of the armistice was 2,429,000 rounds while that of Great Britain was 7,347,000 rounds and that of France 7,638,000 rounds.

FIGURE 9.

PRODUCTION OF SMOKELESS POWDER AND HIGH EXPLOSIVES, FRANCE AND UNITED STATES COMPARED WITH GREAT BRITAIN.

AVERAGE MONTHLY RATE, AUGUST, SEPTEMBER, AND OCTOBER, 1918.

	Pounds.	Per cent of rate for Great Britain.
Smokeless powder:		
Great Britain......	12,055,000	100
France............	17,311,000	144
United States.....	42,775,000	355
High explosives:		
Great Britain......	30,957,000	100
France............	22,802,000	74
United States.....	43,888,000	142

TOTAL PRODUCTION, APRIL 6, 1917, TO NOVEMBER 11, 1918.

	Pounds.	Per cent of rate for Great Britain.
Smokeless powder:		
Great Britain......	291,706,000	100
France............	342,155,000	117
United States.....	632,504,000	217
High explosives:		
Great Britain......	765,110,000	100
France............	702,964,000	92
United States.....	375,656,000	49

In the 19 months of our participation in the war our production of unfilled rounds in ammunition was 38,623,000 rounds, while that of France was 156,170,000 rounds and that of Great Britain 138,357,000 rounds. In that time we had produced 17,260,000 complete rounds, while France had produced 149,827,000 rounds, and Great Britain 121,739,000 complete rounds.

The entrance of the United States into the war found the existing American explosives manufacturers operating to the very limit of their capacity in production for the allied governments and for general commercial purposes.

Since the outbreak of the war in 1914 the explosives business in this country had increased enormously and the trained men familiar with manufacturing operations and conditions in this highly specialized and extremely dangerous industry had fallen short of meeting demands.

When we entered the war, therefore, it became necessary at once to distribute this limited force of experts as equitably as possible and to put chemists, engineers and other specialists in the various plants under the supervision of this trained personnel so as to produce in as quick a time as possible a vastly enlarged force of competent operators and supervisors for the production of explosives.

Summed up, the problem that faced the Ordnance Department was, while maintaining the current great production of explosives, to expand enormously the facilities for further production, to provide personnel for operating these expanded facilities, to build up entirely new manufacturing plants for making both propellants and high explosives, and in addition to all of this, to bring into existence huge loading plants.

In all, 53 new plants for making explosives and propellants and for loading these were undertaken at a cost of approximately $360,000,000. When the armistice was signed a very large part of this construction work had been completed and was in an efficient state of operation.

How creditably this reflects upon America can be understood when it is made plain that in addition to the development of production there was also to be worked out the very intricate question of design, not only of the plants themselves but also of their products, which required an exceptional degree of technical skill and thorough control.

Prior to our entry into the war the Ordnance Department had depended upon ammonium picrate, known in the Army vernacular as explosive "D," as a bursting charge for our high-explosive shell.

During the progress of the European conflict the British had developed an explosive they called amatol, which is a mixture of trinitrotoluol—T. N. T.—and ammonium nitrate. As this had proved to be entirely satisfactory in actual service on European battle fields, and as ammonium nitrate could be produced here in large quantities, we adopted it.

The Ordnance Department eventually put into effect a standard policy for the use of high explosives. Every effort was being made to conserve the supply of T. N. T., and consequently this explosive was specified for the shell of smaller calibers only. The standard filling scheme was as follows: T. N. T. for shell between and including the calibers of 75-millimeter and 4.7-inch; amatol for shell of calibers between 4.7-inch and 9.2-inch, including the latter; ammonium picrate, or explosive D, for shell of 10-inch caliber and higher. While these were the standards the scheme was not always followed rigidly. As a matter of fact amatol was loaded into shell of all sizes and so was T. N. T., although explosive D was never used in shell smaller than those for the 10-inch guns. These departures from standard practice were due to the necessity for keeping certain plants in production and to other special causes and exceptional circumstances.

Production of large quantities of T. N. T. and ammonium nitrate was the first big problem to be solved by the high-explosives section of the Ordnance Department. All the work of the explosives section can be subdivided under four group heads—raw materials, propellants, high explosives, and loading.

RAW MATERIALS.

The first steps taken in the endeavor to meet the need for raw materials were to increase greatly the available means for obtaining toluol, phenol, caustic soda, sodium nitrate, sulphuric and nitric acids, ammonia liquor or aqua ammonia, and to attempt to provide a substitute for cellulose in case a shortage of cotton should render its use necessary.

How to increase the supply of toluol, the basic raw material from which T. N. T. is made, was the greatest and most pressing of all the problems in regard to the existing raw materials. Before the war the sole source of this ingredient was from by-product coke ovens. The monthly capacity of these ovens in 1914 was, approximately, 700,000 pounds. By April, 1917, when we stepped into the conflict, this capacity had been increased to 6,000,000 pounds a month.

By the time the armistice was signed our efforts for greater production had been carried on so successfully that the supply had been increased to 12,000,000 pounds a month, and the average cost of this was only 21 cents a pound. This tremendous increase of production not only took care of all demands for commercial purposes and permitted the shipment of about 11,000,000 pounds to the allied Governments, but was more than ample to take care of our own entire explosives program, leaving a stock on hand December 1, 1918, of 17,000,000 pounds.

A few details of how this tremendous increase in production was brought about through the energies of the officials charged with this task and the most efficient and whole-hearted cooperation of patriotic business concerns are interesting.

Three general sources existed from which toluol was obtained: first, from the by-product recovery coke ovens; second, by the stripping or absorbing of toluol from carbureted water and coal gas; and third, by the cracking or breaking down of oils.

In augmenting the supply of toluol through the first process, construction of additional by-product coke ovens by the following big steel companies was arranged:

Company.	Toluol capacity per year.
	Pounds.
Jones & Laughlin Steel Co., Pittsburgh, Pa	5,770,160
The Sloss-Sheffield Co., Birmingham, Ala	2,019,556
United States Steel Corporation, Clairton, Pa	2,308,064
International Harvester Co., Chicago, Ill	1,586,794
United States Steel Corporation, Birmingham, Ala	2,019,556
Rainey-Wood Co., Swedeland, Pa	2,163,810
The Seaboard By-Product Co., Jersey City, N. J	1,081,905
Pittsburgh Crucible Steel Co., Midland, Pa	2,019,556

The total cost of these additional ovens was about $30,000,000, which was met by private capital after contracts for the purchase of the product had been made, insuring a secure return on the investment. Production was to begin in 1919.

In addition to this there was arranged construction for 320 additional ovens at the following places:

Company.	Date of contract.	Estimated cost.	Estimated time of completion.
Donner Steel Co., Buffalo, N. Y	May, 1918	$6,000,000	Mar., 1920
Birmingham Coke Co., Birmingham, Ala	July, 1918	2,500,000	Oct., 1919
Domestic Coke Corporation, Fairmont, W. Va	Sept., 1918	2,700,000	Nov., 1919
Domestic Coke Corporation, Cleveland, Ohio	July, 1918	1,500,000	Feb., 1920
International Coal Products Corporation, Clinchfield, Va	May, 1918	2,000,000	Aug., 1919

From these sources the monthly production of toluol in 1920 would have been increased by 600,000 pounds a month.

While all these arrangements for vastly increasing the supply of this chemical in 1919 and 1920 were being made, technical experts of the Ordnance Department stimulated production by visiting existing by-product coke ovens and advising as to changes and alterations in the plants, both in regard to equipment and methods of operation.

Investigations were made early in the summer of 1917 on the possibility of recovering toluol by stripping illuminating gas, and a report was made on this subject in October, 1917. Construction of the necessary plants to carry out this plan was begun late in November, and the first plants were in operation in April, 1918. This was considered a remarkable record, in view of the fact that the operating personnel for the purpose had to be established and trained in this entirely new line of activity.

In this connection it is extremely interesting to note that the American people in 13 of the largest cities of the country played an unconscious part in contributing to the successful termination of the war by using artificial gas of considerably less heating power, as a result of the removal of the toluol for explosive purposes. For example, in New York City, due to the extraction of toluol, the artificial gas there was reduced in heating value approximately 6 per cent and the candlepower lowered from 22 to 16 because of this stripping process.

Contracts for taking the toluol from artificial gas were made with companies in the following cities: New York and Brooklyn, N. Y.; Boston, Mass.; New Haven, Conn.; Albany, N. Y.; Utica, N. Y.; Elizabeth, N. J.; Washington, D. C.; Detroit, Mich.; St. Louis, Mo.; New Orleans, La.; Denver, Colo.; and Seattle, Wash.

The total cost of the installations made for this purpose in these cities in connection with the gas plants was about $7,500,000.

For the production of toluol by cracking crude oils or petroleum distillates, three processes of the many submitted were officially approved and contracts awarded for operation.

The first and most important of these was that of the General Petroleum Co. of Los Angeles, Calif. Under their scheme a yield of 6 per cent toluol was obtained from a petroleum distillate, of which there was a large quantity available, by treatment under temperature and pressure. To facilitate production of toluol by this means, two large plants, one at Los Angeles and the other at San Francisco, were erected at a cost of approximately $5,000,000. These plants have a monthly capacity of 3,000,000 pounds of toluol and their construction destroyed all possibility of a shortage in this vital raw material.

Another process was that known as the Rittman process, evolved by a scientist of the Bureau of Mines. This scheme, which called for producing toluol from solvent naphtha or light oils by cracking under high pressure and temperature, was finally demonstrated to be capable of operation under war conditions, and production had just started at a plant on Neville Island, Pittsburgh, Pa., at the time of the signing of the armistice.

A third process was that known as the Hall process, by which toluol was also obtained by cracking solvent naphtha under high pressure and temperature by another, different, mechanical system. This scheme was in operation on a small scale during 1918 at the Standard Oil Plant, Bayonne, N. J.

Phenol, one of the essentials in the manufacture of picric acid, was another raw material, the production of which was greatly augmented. At the time of our entry into the war the monthly production amounted to 670,000 pounds, while in October, 1918, it had been increased to 13,000,000 pounds. In December, 1917, the price of phenol as fixed by the War Industries Board was 46 cents a pound, while Government contracts in force a year later had reduced this figure to 31 cents a pound.

The price of sulphuric acid jumped from $14 a ton to $60 a ton early in the war, while nitric acid advanced from 5¼ cents a pound to 10 cents. The shortage of sulphuric acid was met by the erection of both chamber and contact plants in all high-explosives factories built for or under direction of the Ordnance Department.

Both pyrites and sulphur were used at the beginning of the war, but the submarine warfare stopped the importation of the pyrites from Spain, and therefore sulphur deposits in Texas and Louisiana were depended upon. A destructive storm in the early part of 1918 temporarily curtailed the production from Louisiana deposits, but repairs were made in time to prevent its effect being felt by the acid manufacturers.

The submarine also had the effect of lessening the importations from Chile of sodium nitrate, which prior to the war were depended upon entirely in the production of nitric acid. It became necessary, therefore, to develop other methods of production. After investigations a plant for the fixation of nitrogen under what is known as a modified Haber process was erected at Sheffield, Ala., while a plant for the same purpose using the cyanamide process was erected at Muscle Shoals, Ala.

Both of these were equipped for the oxidation of ammonia to nitric acid, each using a different process. When the armistice was signed these plants were just coming into production. The existence of these two nitrate plants insures the independence of this country in its supply of commercial nitrogen, either for peace or for war.

There were also in course of erection, though not in operation on November 11, 1918, great plants for the extraction of nitrogen from the air, at Toledo and Cincinnati, Ohio, but construction on these two plants, each of which was to cost $25,000,000, was stopped when the armistice was signed.

PROPELLANTS.

In army usage the term "propellant" includes both smokeless powder and black powder.

At the outbreak of the European war, the producing capacity in this country for smokeless powder was approximately 1,500,000 pounds a month. By the time the United States got into the war this capacity had been increased from 25 to 30 times, and under the explosives program laid down by us it was indicated that even this capacity would have to be greatly increased.

The increase in the production of smokeless powder was helped by the construction of two of the largest smokeless-powder plants in the world—one known as the Old Hickory Plant, located almost on the site of Andrew Jackson's old home at Nashville, Tenn., and the other at Nitro, near Charleston, W. Va.

The Old Hickory Plant was the larger and more complete of the two. It is probably the biggest plant of its kind in the world and is entirely self-contained; in other words, the plant actually takes the crude, raw cotton and, producing both the acid and solvents used, puts it through every process until the final product is attained.

Nine powder lines were planned for this enterprise, each with a capacity of 100,000 pounds per day, although developments from the early operations indicated that the ultimate production of the plant would reach 1,000,000 pounds a day.

The estimated cost of this huge undertaking was in the neighborhood of $90,000,000. Negotiations were begun in October of 1917 and led to a contract with the du Pont Engineering Co., under which this concern was to construct the plant and operate it for a six months' period after its completion.

Operation of the first powder line in the plant was to start September 15, 1918, or seven and one-half months after the signing of the contract. Ground was broken March 8, 1918, and work was pushed so efficiently and successfully that on July 1, 1918, the first powder line was put in operation, 75 days ahead of the schedule called for in the contract.

Some idea of the magnitude of this enterprise can be realized in the statements that the plant covers an area of 5,000 acres and that in addition to the powder plant proper there was built a city, housing twenty odd thousand people, complete with schools, churches, and all other elements that go to make up a town. There was also built in connection with the plant a number of subprocess plants for the manufacture of purified cotton, sulphuric acid, nitric acid, diphenylamine, and other chemicals used in powder manufacture. Each of these was an undertaking of no little size in itself.

Operation of the plant during the four and one-half months preceding the signing of the armistice showed a production in excess of contract requirements. On November 11, 1918, the plant was over 90 per cent complete and about 50 per cent in operation. At that time 6,000,000 pounds of powder over and above contract expectations had been produced, the total capacity having reached 423,000 pounds a day.

The second powder plant, located at Nitro, is somewhat smaller than the Old Hickory Plant. It has a capacity of 625,000 pounds of smokeless powder a day. It was built under the direction of D. C. Jackling, director of United States Government explosive plants, by the Thompson-Starrett Co., of New York. The contract was dated January 18, 1918, and ground was broken February 1. A contract for the operation of the plant was signed with the Hercules Powder Co., and at the time of the armistice the output was running approxi-

NITRO, WEST VIRGINIA.

D. WEST VIRGINIA.

mately 109,000 pounds a day, with the expectation of early and speedy increase. As in the case of the Old Hickory Plant, a large village and many subprocess plants were constructed in connection with this enterprise.

When the war began smokeless powder was dried by the circulation of warm dried air for a long period of time over the damp powder as it came from the solvent recovery house. This process required from six weeks for small-caliber powder to nine months for large-caliber powder. This time-consuming method being obviously impracticable in war, the Ordnance Department authorized the so-called water-drying process. This consists in the immersion of the powder as it comes from the solvent recovery house in warm water for varying periods up to 72 hours, the water then being expelled by filtration or centrifugal force and the surplus external moisture dried off by hot air. By this method the time of drying was reduced to 4 days for the small-caliber powder and to 22 days for powder for the larger caliber guns.

Just prior to the signing of the armistice an entirely new drying process had been experimentally tried out. This was known as the Nash or alcohol-drying process. The preliminary tests indicated that this method was a great improvement both in safety and in the reduction of cost. The indications were that drying could be reduced from days to hours by this new method. The Nash process also insured apparently a more uniform and tougher grade of powder, both of which characteristics were greatly to be desired.

In spite of the rise in price of labor and of almost everything else, the cost of powder was being reduced. At the beginning of the war cost figures were 80 cents a pound for small-arms and 53 cents a pound for cannon powder. When the armistice was signed these costs had been reduced to 62 cents for small-arms powder and 41¼ cents for cannon powder.

At the time of the signing of the armistice there was on hand approximately 200,000,000 pounds of smokeless powder.

It early became evident that the supply of cellulose, even though all available sources of supply were utilized to the utmost, would nevertheless be insufficient to meet our vast production program. For years it had been rumored that the Germans in the manufacture of their smokeless powder had been using, with great success, cellulose produced from wood pulp. Following out this idea, experimental work was undertaken in an effort to develop cellulose that could be produced from wood pulp in suitable physical form for nitration and which would meet the chemical requirements.

In the southern and southwestern portions of the United States there are large tracts of land from which timber has been removed and there are also vast acreages of swamp lands. Processes developed by the Ordnance Department had in view the idea of taking as much of these lands as possible for farming and reforesting and utilizing the tree stumps thereon. These stumps contained quantities of turpentine and resin that could be recovered and the resultant pulp after proper treatment could be prepared in suitable form as cellulose for nitration purposes.

The question of black powder, while an important one, did not present many difficulties excepting one, the necessary supply of potassium nitrate. This was because Germany was the principal source of the potash. It was thought that sodium nitrate might possibly have to be used as a substitute. Experimental work along these lines indicated that by using certain precautions, this substitution, if necessary, could be made, although it was never adopted.

Black powder of all grades for military purposes was being produced at the rate of 840,000 pounds a month, at a cost of 25 cents a pound, at the time the armistice was signed. At that time there was on hand 6,850,000 pounds of black powder.

If the war had continued the United States could have produced during the year 1919 more than 1,000,000,000 pounds of smokeless powder. Two-thirds of this would have been available for our overseas forces and the balance would have gone to the allied governments. This rate of production would have amounted to about seven times the quantity of explosives normally manufactured in peace times.

LOADING THE PROPELLANTS.

In addition to solving the problem of producing a sufficient quantity of propellant powder there was also the problem, just as important, of assembling this powder into fixed ammunition, or loading it into bags. The Frankford Arsenal and commercial cartridge factories, after expansion, were enabled to take care of the expanded small-arms program. But it became necessary for the Government to erect and operate several great bag-loading plants. These were located at Woodbury, N. J., Tullytown, Pa., and Seven Pines, Va.

The ordinary cartridge fired from the rifle is familiar to most people. The projectile is fitted into the metal case in which the explosive force is contained. Projectiles for big guns are made along similar lines, until the 4.7-inch gun is reached. Up to and including guns of this caliber the projectile is fired with what is known as fixed ammunition—that is to say, the shell itself is fixed into a metal container which holds the powder.

Guns above the caliber of 4.7 inches, however, are fired with unfixed ammunition—that is, the powder is loaded in silk bags, the projectile placed in the gun, and a number of bags, depending upon the size of the charge necessary, put into the breach of the gun behind the projectile. The powder is then ignited and the big shell ejected by the gases generated.

From the mills the powder is shipped to the bag-loading plants in bulk. The silken bags are manufactured in huge quantities by industrial plants and forwarded to the bag-loading plants, where are also daily received large quantities of metal and fiber containers, into which are loaded bags packed for overseas shipment not to be unpacked again until they have reached the battle field.

Filling the bags is a precise and delicate operation. Chances can not be taken or averages struck. Errors may mean the possible loss of battles. A battery commander who has figured his range and who is about to drop a number of high-explosive shell on an enemy battery must know exactly how much powder he has behind his charge. If more powder is in the bag than he calculates on, he will overshoot his mark; if less, the shell instead of dropping upon an enemy battery may explode in midst of his own advancing troops.

The three bag-loading plants the Government constructed at Woodbury, Tullytown, and Seven Pines were built to load bags that were to be used in firing guns from 155-millimeter caliber up to a caliber of 10 inches. The estimated average capacity of each plant was 20,000 bags a day, but as a matter of fact a maximum capacity of 40,000 bags a day at each plant had been reached before the signing of the armistice. Two shifts a day were used at these plants most of the time. In each shift there were approximately 3,500 operatives, most of them women.

At each of these plants, which are located in comparatively isolated points, because of the dangerous work, special housing facilities had to be constructed. For example, at Tullytown there were 70 bungalows, 13 residences for officers and executive heads, and six 98-room dormitories, while at Woodbury 19 great dormitories were built to house workers.

The number of buildings at Tullytown is 215. They range from guardhouses to electrical generating stations for power and light. Besides this construction there are between 22 and 30 miles of railroad track laid at each of these points. The extremely dangerous nature of the work makes it necessary to store not more than 400,000 pounds of explosives in a single building, and where powder is stored the buildings are at least 350 feet apart.

Up to the time of the signing of the armistice there were loaded into small-arms ammunition 19,741,500 pounds of powder; there were assembled into fixed ammunition approximately 33,000,000 pounds of smokeless powder; and there were assembled into bags, properly packed for shipment, approximately 32,300,000 pounds of smokeless powder.

HIGH EXPLOSIVES.

When Europe was plunged into the great war in August, 1914, the American production of trinitrotoluol for commercial purposes amounted to approximately 600,000 pounds a month of varying grades of purity. This quantity was almost entirely consumed in the making of explosives for blasting purposes. When we entered the war this production had been increased to 1,000,000 pounds a month, exclusive of that which was being used here commercially. Under pressure of our own war-time needs the production of this highly important explosive chemical had been run up to 16,000,000 pounds a month at the termination of hostilities in November, 1918.

During the early stages of the war the average price of T. N. T. for military purposes was $1 a pound. Largely, however, because of the tremendous quantity production and enormous economies effected by reason of this, and despite the scarcity of raw materials, and notwithstanding the greatly increased labor cost, this price had been reduced at the time of the signing of the armistice to 26½ cents a pound. There were in the course of erection at the time of the armistice, two great Government T. N. T. plants—one at Racine, Wis., that was to have a capacity of 4,000,000 pounds a month, and one at Giant, Cal., with a capacity of 2,000,000 pounds a month.

During the war three grades of T. N. T. were produced. Grade I was used for booster charges—that is, those charges which initiated the explosive wave in the main shell charge. Grade II was used as a shell filler; while Grade III was utilized with ammonium nitrate in producing amatol.

In view of the fact that high explosives were produced in such enormous quantities and that it was necessary to carry on these tremendous manufacturing operations with an inexperienced force, the toll of life taken in the production was remarkably small. Only two explosions of any magnitude occurred in plants where explosives were manufactured and both of these took place in T. N. T. producing plants. One of these happened at Oakdale, Pa., in the plant of the Aetna Explosives Co. in May, 1918. This cost the lives of 100 persons. The other took place on July 2, 1918, at Split Rock, N. Y., in the plant of Semet-Solvay Co., where 60 people lost their lives. At the time of the explosions neither of these plants was operating on War Department contracts.

Before the great war about 58,000,000 pounds of ammonium nitrate used in the manufacture of commercial explosives were being produced annually in this country, at an average cost of about 12 cents a pound. By January, 1917, the commercial explosives manufacturers had extended their facilities so that they had increased their production by 1,700,000 pounds monthly. This expansion, however, was insufficient to meet our demands, and a Government ammonium nitrate plant was erected at Perryville, Md. This plant was operated under the supervision of the Atlas Powder Co., who also cooperated in its erection.

It did this manufacturing under the Brunner-Mond process that was developed in England under the patents of Capt. Freeth. Under this process ammonium nitrate is produced by the double decomposition of ammonium sulphate and sodium nitrate.

In December, 1917, the Atlas people detailed several technical men to go to England and study the Brunner-Mond process as carried on there. In 1918 these men returned to the United States and prepared designs as a result of the information they had gained abroad.

Ground was broken for the plant at Perryville March 8, 1918, and it was in production by July 15. This plant is a large one, of excellent construction, and absolutely fireproof, as is necessary because of the nature of the work conducted in it. Because of the type of the building the rapidity of its construction may well be classed as phenomenal. Even while the plant was being put up, experimental work of a highly technical nature was being carried on.

At the time of the signing of the armistice production of ammonium nitrate at the Perryville plant had reached 452,000 pounds a day, and this was greatly in excess of that being obtained at the English plant of a similar size that had been in operation for months before ground had been broken for our American plant.

Each of the Government-owned nitrogen fixation plants at Muscle Shoals, Ala., and Sheffield, Ala., was also equipped to produce ammonium nitrate by neutralization. Our total capacity from all sources at the time of the signing of the armistice was 20,000,000 pounds monthly. Ammonium nitrate is the one material in the field of explosives that shows an increase in price over that of normal times. The average cost of this substance used for military purposes was 17½ cents a pound. There were on hand 60,500,000 pounds of ammonium nitrate on November 11, 1918.

Picric acid as such is not used by this country directly for military purposes. But it is one of the raw materials used in producing ammonium picrate, or explosive D, and in the manufacture of the poisonous gas known as chlorpicrin.

Picric acid is, however, the main explosive used by the French, who had placed enormous contracts for this material with explosives manufacturers prior to the entry of the United States in the war. Because of our purchase of early large supplies of ammunition and guns from the French Government, to be largely paid for by picric acid, large contracts were entered into by our Government for this explosive, which was produced here in accordance with French specifications and subject to joint inspection by our officers and the French.

In November, 1917, we were turning out 600,000 pounds of picric acid monthly, and a year later this had been increased to a monthly production of 11,300,000 pounds; the average cost was 56 cents a pound.

To insure production quickly for the needs of the times, three Government picric-acid plants were authorized. One of these was located at Picron, near Little Rock, Ark., to be operated by the Davis Chemical Corporation; another was located at New Brunswick, Ga., to be operated by the Butterworth-Judson Corporation; and the third was located at Grand Rapids, Mich., for operation by the Semet-Solvey Co. All of these contracts were made on a cost-plus basis. Each of these plants was to have a capacity of 14,500,000 pounds of picric acid a month. The plant at Picron in Arkansas was the only one that had started production before the signing of the armistice.

Ammonium picrate, otherwise known as explosive D in our Army annals, is produced by the ammoniation of picric acid, and because it is more insensitive than picric acid and is less liable to form sensitive salts with metals it is used as the explosive charge for all armor-piercing projectiles.

Our average monthly production of ammonium picrate in May, 1917, was 53,000 pounds, and this had been increased without the erection of any Government plants to a monthly capacity in November, 1918, of 950,000 pounds. There was on hand at the time of the signing of the armistice 6,500,000 pounds of this explosive, the average cost of which was 64 cents a pound.

Tetryl, on account of its high cost and the lack of manufacturing facilities for its production, was not used except as a loading charge for boosters. It is more sensitive than T. N. T. and has a higher rate of detonation.

Only two companies, the du Pont Powder Co. and the Bethlehem Loading Co., manufactured tetryl. Expansion of these two plants increased the monthly capacity of 8,700 pounds in December, 1917, to 160,000 pounds in November, 1918, while its cost was reduced from $1.30 a pound to 90 cents a pound.

This increased capacity, however, was not in excess of our explosives requirements, and there was authorized by the Government the erection of a plant at Senter, Mich., that was to be operated by the Atlas Co., and which was to have a monthly capacity of 250,000 pounds. This plant had not reached production when the armistice was signed.

The Aetna Powder Co. at the time we entered the war was manufacturing for the Russian Government tetranitroaniline that was to be used in the loading of boosters and fuses. This company's plant at Nobleston, Pa., was destroyed by an explosion. Ordnance officers learned that this material was equal to tetryl as a military explosive. Consequently a contract was entered into with Dr. Bernhardt Jacques Flurschein, the holder of the patent rights, to have manufactured T. N. A. for our own uses. A Government plant was authorized for erection on the ground of the Calco Chemical Co., Bound Brook, N. J., to be operated by that concern. Production at this plant was to be on a cost-plus basis, the estimated cost of the material being 70 cents a pound. When the armistice was signed, about 8,000 pounds of T. N. A. had been produced, but none had been utilized.

Mercury fulminate, a very sensitive and powerful explosive, was used only in caps, primers, detonators, etc., as a means of initiating detonation, on account of its own high rate of detonation. The three plants operating in this country to produce this explosive for commercial purposes, the du Pont Co., Pompton Lake, N. J., the Atlas Powder Co., Tamaqua, Pa., and the Aetna Powder Co., Kingston, N. Y., expanded their facilities sufficiently to meet our program. Their average monthly production in 1918 was 50,000 pounds at a cost of $3.21 per pound, and there was on hand in November, 1918, 330,900 pounds of this explosive.

In the early stages of the war to meet the apparent shortage of T. N. T. and ammonium nitrate then existing because of our enormous explosives program, it was necessary to develop an explosive for trench warfare purposes that could be used for filling hand and rifle grenades, trench-mortar shell, and drop bombs. To meet this need, the Trojan Powder Co., of Allentown, Pa., submitted a nitrostarch explosive. After exhaustive investigations and complete tests, this explosive was authorized for use in loading the hand and rifle grenades and the 3-inch trench-mortar shell.

Development of a nitrostarch explosive for commercial purposes had been under consideration and investigation by two other large experienced manufacturers for a number of years, but the difficulties incident to the production and purification of nitrostarch were such that their efforts had met with little success.

The Trojan Powder Co., operating under secret process, solved this problem, and all nitrostarch explosives used were produced by this company, although another nitrostarch explosive known as "grenite," which was produced by the du Pont Co., was tested and authorized for use.

Our country was the only Government that used nitrostarch explosives during the war, and the development of this explosive made the loading problem easier and made possible the use of materials that were available and whose cost was low. The average cost of this explosive was 21.8 cents a pound. In July, 1918, the average monthly production of nitrostarch was 840,000 pounds and this had been increased by November, 1918, to 1,720,000 pounds a month.

There were loaded with nitrostarch explosive 7,244,569 defensive hand grenades; 1,526,000 offensive hand grenades; 9,921,533 rifle grenades and 813,073 three-inch trench-mortar shell. At the time of the signing of the armistice there was on hand of this explosive 1,650,500 pounds.

The du Pont Co. developed an explosive called lyconite, and this was authorized for use in the loading of drop bombs.

Anilite, a liquid explosive used by the French, was thoroughly investigated and improvements were made in it to render its use safer, but development had not progressed far enough to warrant authorization for its use prior to the signing of the armistice.

Chlorate and perchlorate explosives were also investigated and several types developed that were considered entirely satisfactory for use, but these never got into production before the end of the war.

AMMUNITION AND SHELL LOADING.

When we entered the war the quantity of field artillery ammunition on hand was considerably less than a single month's supply, basing our rate of expenditure on the estimated rate for November, 1918. There were no facilities of any degree of magnitude available to take care of our projected program for filling the high-explosive shell necessary for use by our overseas forces.

Consequently it became necessary at once to plan and to develop the resources of the country for the production of metallic parts, such as the shell proper, the fuse, boosters and adapters, as well as to design and build entirely new plants and to train completely new forces for the loading of the shell with the high explosives.

The explosion of an H. E. shell is really a series of explosions. The process of the burst is about as follows: The firing pin strikes the percussion primer, which explodes the detonator. The detonator is filled with some easily detonated substance, such as fulminate of

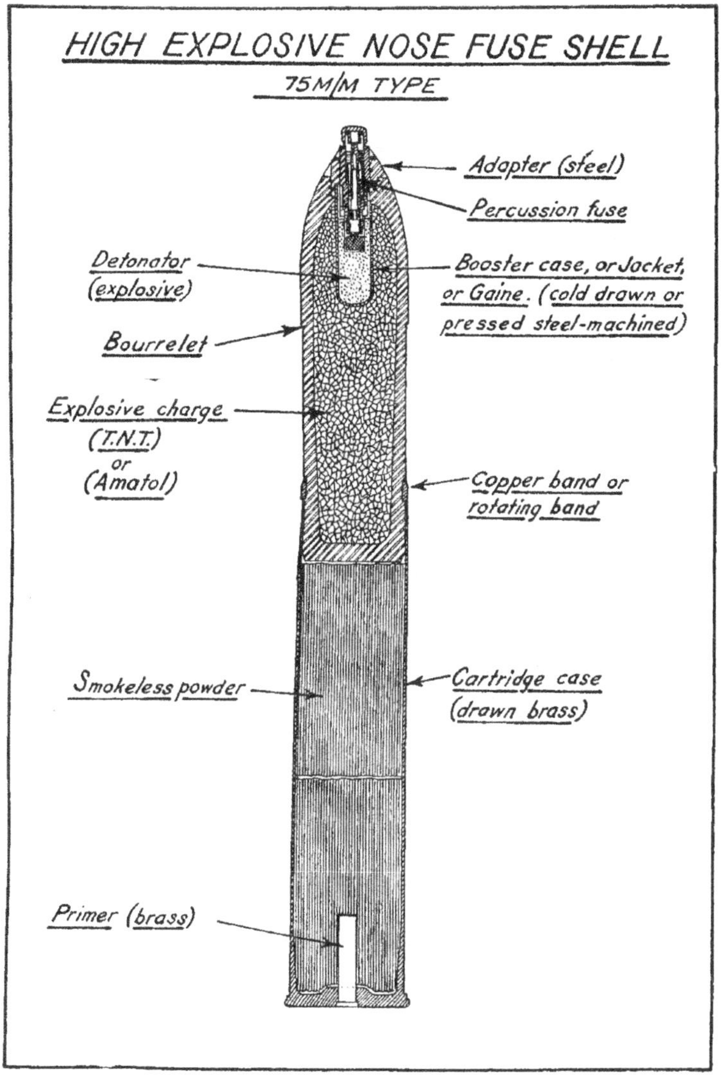

HIGH EXPLOSIVE NOSE FUSE SHELL
75M/M TYPE
Adapter (steel)
Percussion fuse
Detonator (explosive)
Booster case, or Jacket, or Gaine. (cold drawn or pressed steel-machined)
Bourrelet
Explosive charge (T.N.T.) or (Amatol)
Copper band or rotating band
Smokeless powder
Cartridge case (drawn brass)
Primer (brass)

GIRLS LOADING POINT-DETONATING FUSES FOR HIGH-EXPLOSIVE SHELL IN BODY-MACHINING DEPARTMENT OF A LOADING PLANT.

mercury. The concussion of this explosion sets off the charge held within the long tube which extends down the middle of the shell and which is known as the booster. The booster charge is a substance easily exploded, such as tetryl or trinitroaniline (T. N. A.). The explosion of the booster jars off the main charge of the shell, T. N. T. or amatol. This system of detonator, booster, and main charge gives control of the explosives within the shell, safety in handling the shell, and complete explosion when the shell bursts. Without the action of the booster charge on the main charge of the shell, the latter would be only partially burned when the shell exploded, and part of the main charge would thus waste itself in the open air.

The shell used by our Army before the war had been largely of the base-fuse type. Interchangeability of ammunition with the French required that we adopt shell of the nose-fuse type. The boosters and adapters that went with this type were unfamiliar to our industry.

The adapter is the metallic device that holds the booster and fuse and fastens them in the shell. The adapter, therefore, is a broad ring, screw-threaded both outside and inside. The inside diameter is uniform, so as to allow the same size of booster and fuse to be screwed into shell of different sizes. The outside diameters of the adapters vary with the sizes of the shell they are made to fit, the rings thus being thicker or thinner as the case may require. Fuses of several sorts are employed by the modern artillerist; and with shell equipped with adapters, any fuse may be inserted in the field right at the gun.

Unexpectedly the manufacture of boosters and adapters proved to be much more difficult than it appeared to be at the start, and the shortage of these devices was a limiting factor in the American production of shell.

On May 1, 1917, drawings and specifications were sent to the principal manufacturers of ammunition and ammunition components inviting bids on 3-inch ammunition. These bids were opened on May 15, 1917, and after full discussion with the Council of National Defense orders were placed for 9,000,000 rounds of 3-inch shell and shrapnel ammunition. The bids for shell and shrapnel ammunition for all the other calibers of guns and howitzers we had on hand then were about to be asked, when the French mission to this country arrived; and the sending out of proposals was deferred, while discussion ensued as to changing our 3-inch and 6-inch artillery to 75-millimeter and 155-millimeter calibers, so as to make our ammunition interchangeable with that of the French. This decision was made June 5, 1917.

There then took place much discussion and consideration of the French ammunition. The French had several distinct types of shell, ranging from the very thin walled high capacity kind to the thicker walled types. The French specifications were radically different from our own or those of the British. The steel shell in the French practice was subjected to a drastic heat treatment, which did not seem necessary to us for the thicker walled types of shell.

The French fusing system also was entirely different from that used by our service. French fuses were carried separately, and the adapter and the booster casing were screwed permanently into the shell.

Our decision to adopt French types of ammunition made it necessary to rearrange all our plans, and to obtain drawings of the shell, boosters, adapters, and fuses from France. This caused much negotiating, and a considerable amount of time was consumed in getting the necessary specifications and drawings here.

As a result of recommendations from French officials against production in this country during 1917 of the so-called "*obus allongé*" and the semisteel type of shell, no attempt was made to produce these for the 155-millimeter guns and howitzers during the first year of the war, but as a result of new recommendations and investigations of our officers in France in the spring of 1918 both of these types of shell were put into quantity production here. When the armistice was signed they were being turned out in such quantities that it appeared that there was sure to be an ample supply on hand in the early spring of 1919.

Radical differences of manufacture existed between the French and British in the matter of specifications and methods of production. Large quantities of British ammunition had been made in this country, and we had adopted the British 8-inch howitzer, so that it appeared we should use British practice in the manufacture of shell. Manufacturers claimed that great delay would result in the production of shell here if the heat treatment and hydraulic tests were insisted upon as the French specifications called for, and investigation proved this to be essentially true, as no facilities for heat treating and hydraulic testing existed.

The upshot of the entire matter was that it was decided to use French dimensions and shell for the 75-millimeter and 155-millimeter calibers so as to obtain uniformity of ballistics, but to permit American metallurgical practice to obtain in the manufacture. Shells made under these specifications were tested by the French commission in France. The verdict on these shell can be summarized in this quotation from their report:

> To sum up, from the test of 10,000 cartridges of 75 millimeter, it may be concluded that American ammunition is in every way comparable to French ammunition and that the two may be considered as interchangeable.

Our designs for shrapnel and time fuses had been proven to be entirely satisfactory, and they were continued as they were. In fact it was generally agreed that ours was the best time fuse used on the allied side during the war. That our decision in the matter of continuing production of shrapnel and time fuses was warranted, is borne out by the fact that we obtained early deliveries in sufficient quantities to meet requirements.

In the use of the adapters and boosters, which introduced an entirely new component to our service in shell making, we had had no experience, and subsequently met with great difficulties due to this lack of experience. Delays were encountered because in this part of shell manufacture it was generally necessary to await information from France whenever difficulties were encountered, or to conduct experiments before we could proceed.

When we began receiving our bids for 3-inch gun ammunition there were comparatively few factories in the United States that were able to turn out complete rounds of ammunition. There were many factories, however, capable of turning out one or more of the shell components. It was necessary to place orders for complete rounds of ammunition with those factories that could furnish them, and have the remaining components manufactured separately, and to provide assembling plants. To get as many factories as possible on a production basis in anticipation of the future large orders for ammunition that must necessarily follow with extension of operations by our field forces, orders for our initial quantities of ammunition were distributed as widely as possible.

To prevent confusion and loss of time because of the scramble for steel forgings and other raw materials it was decided that the Government would purchase all raw materials as well as furnish components for ammunition.

How successful we were in getting into quantity production on ammunition after the numerous and large obstacles in the early months of the war can be indicated best by the fact that of the 11,616,156 high-explosive shell for 75-millimeter guns machined up to November 1, no less than 2,893,367 passed inspection in October; while of the 7,345,366 adapters and boosters for 75-millimeter guns that had been machined up to the 1st of November, 2,758,397 passed inspection in October.

The figures for the 4.7-inch and 155-millimeter guns and howitzers follow:

Kind of ammunition.	Machined high-explosive shell accepted up to Nov. 1.	Machined adapters and boosters accepted up to Nov. 1.
4.7-inch	994,852	[1] 636,096
155-millimeter	2,083,782	2,516,216

[1] For use in 4.7-inch and other sizes.

Ammonium picrate or explosive D upon which this country had depended almost entirely up to the time of our entry into the war was forced into the shell under hydraulic pressure. The adoption of the point-fused shell and an explosive for shell filling new to this country, namely, amatol, made necessary the provision of new methods for shell loading and the expansion of plant facilities for these new methods capable of loading the vast and tremendous numbers of shell required in modern warfare. As a result of reports, following investigations by our officers of methods used abroad, various new shell-loading plants were built in the United States.

The names, location, and output of the shell-loading plants in our country are as follows:

Company.	Location.	Total capacity daily (shell).
T. A. Gillespie Loading Co	Morgan, N. J	47,000
Do	Parlin, N. J	25,000
Do	Runyon, N. Y	3,500
Poole Engineering & Machine Co	Texas, Md	15,000
United States Arsenal	Rock Island, Ill	1,000
Sterling Motor Car Co	Brockton, Mass	10,000
American Can Co	Kenilworth, N. J	20,000
Atlantic Loading Co	Amatol, N. J	53,500
Bethlehem Loading Co	Mays Landing, N. J	41,000
Do	New Castle, Del	27,400
Do	Redington, Pa	4,000
duPont Engineering Co	Penniman, Va., G plant	41,000
Do	Penniman, Va., D plant	13,330
J. D. Evans Engineer Corp	Old Bridge, N. J	30,000
Total		284,730

It was found necessary in the early stages of the war to fill all shell with T. N. T., regardless of cost, until there could be built the required and properly equipped plants for the mixing and loading of amatol.

Two methods for loading T. N. T. were adopted. The one most largely used, however, was the casting method by which the chemical was brought to a molten condition in a steam jacketed kettle and poured into the shell. To do this two operations were usual. First, the shell was filled approximately two-thirds full with the molten material, and then as soon as a crust was formed this was broken through and the second filling took place. This process was necessary to prevent the formation of cavities in the filling charge. Such cavities cause breakdowns, resulting almost invariably in incomplete or entire failure of detonation.

The ammonium nitrate first produced in this country during the war was of such a character that proper densities could not be obtained when mixed with T. N. T. to form amatol. This difficulty was overcome after much investigation, and proper methods were outlined for the ammonium nitrate manufacturers, with the result

EIGHT-INCH SHELL BEING LOADED WITH AMATOL.
View of extruding machine bulkhead in background.

MARK V FUSE ASSEMBLY.

This picture shows two complete units for this assembly work. The operation begins in the foreground with cap assembly and progresses toward background, the fulminate detonator being inserted midway down table. The protecting bulkhead for cap supply is shown in the foreground.

that Grade III ammonium nitrate was produced as a sharp, hard crystal at a setting point of not less than 290° F. This was found to be perfectly satisfactory.

The so-called 50–50 amatol, composed of 50 parts ammonium nitrate and 50 parts T. N. T., is loaded into shell by a casting method similar to that used in loading T. N. T. alone.

The so-called 80–20 amatol, composed of 80 parts ammonium nitrate and 20 parts T. N. T., was originally loaded cold, by hand, and then followed up with mechanical pressing. As a substitute for this method, which is accompanied by a certain element of danger, the use of hot 80–20 amatol, was resorted to in England. This was tamped by hand to the proper density, it being more compressible than cold amatol.

As this is an exceedingly tedious method of operation it was entirely done away with in England, except for large shell, by the use of what is known as the horizontal extruding machine. With this machine the British were able to load 80–20 amatol with great success into the 75-millimeter shell and higher calibers up to 8 inches.

This machine took a mixture of T. N. T. and ammonium nitrate in a jacketed hopper, so that the temperature might be maintained, and the hopper fed it down through a funnel upon a screw that was placed against the shell by counterweights to give the proper density. One of these machines was imported here from England, but, as it was unsatisfactory from a construction standpoint, new and satisfactory machines were built on the same principles of construction in our own amatol loading plants.

Experimental work with these machines was carried on at the Government testing station Picatinny Arsenal, Dover, N. J., and the du Pont Experimental Station, Gibbstown, N. J., as well as experimental plant operations at the Morgan plant of the T. A. Gillespie Co., Parlin, N. J., and the Penniman plant of the du Pont Co., Penniman, Va. All difficulties of the operations were overcome so satisfactorily that the greater portion of the loaded shell was produced by this method.

The metal parts as received at the shell-filling plant are inspected and cleaned to remove all traces of foreign matter such as grit or grease before being sent to the loading room. After being loaded the shell are again inspected. At intervals a split shell is loaded and then taken apart and examined, so that any loading defects may be found quickly and conditions remedied, before any large quantities of shell are produced.

The cavity left in the amatol by the tube of the extruding machine is filled with molten T. N. T., and a cavity is produced in this T. N. T.

into which the booster fits. This is necessary in order to provide for complete detonation. The booster cavity is produced either by the use of a former, which upon removal leaves a cavity of the proper size, or by plunging the booster into the shell filling before this is cooled, or by drilling out a cavity for the booster after the filling has been thoroughly cooled.

A large number of rounds of ammunition of all calibers had also to be loaded with a flashless compound that was inserted in the propelling charges, so that the discharge of the guns would not betray their positions to the enemy at night, while a smoke compound was inserted in a large quantity of shell so that each missile of this character might be located after firing to determine the accuracy of the shot.

Coordination of manufacture of metallic parts so as to cause the proper quantities of shell, fuse, and boosters to be produced without leaving any incomplete rounds that would have to be held awaiting other components caused the greatest difficulty.

The magnitude of the task of providing the necessary shell components in the tremendous quantities required can be better appreciated by a realization of the fact that the various parts of each component must be made to fit each other properly and perfectly. Gauging had to be resorted to frequently in the process of manufacture to make certain that there was perfect interchangeability of parts of each component to prevent any waste of time in selecting parts to fit each other.

The complete components, too, must themselves be made with equal care and scrupulous attention to make certain that they fit properly. Thus, the booster had to be made in such a fashion and with such precision and accuracy that it would fit perfectly into the shell as well as into the booster cavity in the shell filling into which it is screwed and also at the same time accommodate the fuse which screws into the booster.

This extreme accuracy made necessary a large number of gauges, which had to be designed at the same time as, and in coordination with, the design of the component. For example, in a complete round of artillery ammunition, 80 dimensions must be gauged. To standardize the gauges used for these 80 dimensions, 180 master gauges are required, while the actual number of different gauges used during the various stages of manufacture of a complete round is over 500.

Government inspectors required over 200 gauges in their work of inspecting and gauging the finished components for the shell, so in all about 800 gauges were used in the process of manufacturing a complete round of artillery ammunition, to insure interchangeability

LOADING SMOKELESS POWDER.

Notice safety door at the girl's elbow. A flash in this room will not communicate to an adjoining room. The room is heated by overhead hot-air heating system.

FULMINATE COMPOSITION CHARGING, STEEL SHIELD, WITH WINDOW OF HEAVY GLASS TO THE RIGHT.

Girl operating the same device on the left. The view shows the bulkhead between the operations.

SHELL PAINTING.

This view shows the exhaust hood open and turntable lowered. Operator raises turntable by foot lever and closes hood before spraying.

GENERAL VIEW OF SHELL-PAINTING ROOM.

Shell is received on the elevated platform and trucked to the edge on hand trucks, where the trolley hook just enters the eyebolt as shell is removed from truck, thus making it unnecessary to lift the shell during any operation in this room.

of parts, proper fit for the projectile in the gun, and perfect functioning of the various parts.

All fixed ammunition was assembled at the shell-filling plants, making it necessary to install at these points storage capacity and equipment to handle the propellant powder as well as to fill the high-explosive shell. Boosters and fuses were loaded at separate plants and shipped to the shell-filling assembly places to be packed for shipment with the shell for transportation overseas.

The cost of a loaded 75-millimeter shell with the fuse and propellant charge ready to be fired is about $11. Such a shell contains a little over 1½ pounds of high explosive, which costs $1. The loading and assembling of the complete round costs $4.

A loaded 155-millimeter shell complete with fuse costs about $30, exclusive of the propellant charge of powder, which is loaded separately. A shell of this caliber holds about 14¼ pounds of high explosive, which costs $10, while the loading and assembling costs $4.

The 75-millimeter and 155-millimeter shell were used in the greatest quantities on the European battle fields, and at the time of the signing of the armistice our American loading plants were concentrating on filling ammunition for guns of these two calibers.

The nature of the work carried on at these shell-loading plants, of course, made the danger of a disaster ever present. Prior to our entry into the war an explosion at the Canadian Car & Foundry Co.'s plant, Kingsland, N. J., resulted in the entire destruction of the plant with large loss of life.

In October, 1918, the Morgan plant of the T. A. Gillespie Co., South Amboy, N. J., was wiped out by an explosion in which about 100 employees lost their lives. Plans for rebuilding this plant, had progressed far when the armistice was signed. In the fall of 1917, 40 people were killed in an explosion at the Eddystone Loading Plant, Eddystone, Pa.

For the successful carrying out of our program for the production of vast quantities of explosives and propellants, as well as shell loading, the women of America must be given credit, on account of the highly important part they took in this phase of helping to win the war. Fully 50 per cent of the number of employees in our explosive plants were women, who braved the dangers connected with this line of work, to which they had been, of course, entirely unaccustomed, but whose perils were not unknown to them.

In connection with the production of shell themselves, the American Ordnance Department adopted certain changes of design which were not only radically different from what we had known before the war but were interesting for the way in which they were brought about and for the results they accomplished.

The modern shell as we knew it before the war was simply a metal cylinder cut off squarely at the base and roundly blunted at the nose. The shell is zoned with a so-called rotating ring, a circular band of copper which by engaging the rifling channels of the gun gives to the shell the whirl that keeps it from tumbling over and over and thus holds it accurately on its course in flight.

In the proof-firing of the 6-inch seacoast guns it was discovered that their fire was none too accurate; and the American ordnance engineers began studying the shell to see if the fault lay there. One of these experts was Maj. F. R. Moulton, who before accepting a commission in the Army had been professor of astronomy at the University of Chicago. Maj. Moulton began a study of the 6-inch shell; and soon it was discovered that the mathematics which could chart the orbits of comets could also deal with the flight of projectiles, calculate the influences of air resistance and gravitation, and eventually work out new, scientific contours for offsetting these influences as much as possible.

Maj. Moulton first dealt with the inaccuracy of our 6-inch shell. He discovered the cause in the rotating band. Although but a slight portion of this band was upraised above the surface of the shell's circumference, yet the enormous force exerted upon the projectile to start it from the gun actually caused the cold copper to "flow" backward. The result was that when the shell emerged from the muzzle of the gun it bore around its sides an entirely unsuspected and undesirable flange. This flange not only shortened the range of the shell by offering resistance to the air, but it was seldom uniform all the way around, a condition giving rise to the idiosyncracies of our 6-inch shell as they were fired at the target.

The remedy for this was a redesigned rotating band, making it somewhat thicker in front. The "flow" of the copper could thus be accommodated without causing any detrimental distortion to the projectile. When this improvement was made the 6-inch shell became as accurate as any.

But Maj. Moulton was to make an even greater contribution to the 6-inch shell. This shell, like those of our other types, was square ended at the base. Maj. Moulton in his new design tapered in the sides somewhat, making the shell "boat ended." He elongated the nose, bringing it out to a much sharper point. The result was the first American "streamline" design for a shell. Shell of this new model were built experimentally and tested. The 6-inch gun could fire its old shell 17,000 yards, while the streamline shell went 4,000 or 5,000 yards farther—2 or 3 miles added to the range of an already powerful weapon by the application of brains and mathematics.

FIGURE 10.

IMPROVEMENT OF FIELD GUNS SINCE THE NAPOLEONIC WARS.

MUZZLE VELOCITY.

Type.	Date.	Feet per second.
Early rifled guns	1863–1870	1,090
Later rifled guns	1870–1893	1,466
Early quick firers	About 1900	1,696
Modern quick firers	1914–1918	1,770

RANGE WITH SHRAPNEL.

Smooth bores	1815–1850	1,257
Early rifled guns	1863–1870	2,004
Later rifled guns	1870–1893	4,120
Early quick firers	About 1900	6,160
Modern quick firers	1914–1918	6,500

RANGE WITH SHELL.

Smooth bores	1815–1850	1,670
Early rifled guns	1863–1870	3,965
Later rifled guns	1870–1893	6,168
Early quick firers	About 1900	7,340
Modern quick firers	1914–1918	8,500
With streamline shell	1918–19	12,130

The limiting factor in the development of light field guns has always been the continuous hauling power of 6 horses, which is about 4,000 pounds. The gun has been as powerful as possible within the limits of this weight, which includes the carriage and limber as well as the cannon itself. Improved technique and materials have reduced the necessary weight of the cannon from 1,650 pounds in 1815 to about 800 pounds to-day, permitting the use of weight for recoil mechanism and shield of armor plate without exceeding the limit.

The 800-pound nickel-steel gun of 1918 fires as heavy a projectile (12–15 pounds) as the 1,650-pound bronze gun of the Napoleonic wars. The improved material permits a more powerful propellant charge, which results in greater muzzle velocity, a flatter trajectory, and longer maximum range. The latter is due in part also to improved shapes of projectiles and the introduction of rifling. The efficiency of artillery is further increased by the introduction of high-explosive bursting charge. The modern 75-millimeter shell contains about 1.76 pounds of high explosive as against about 0.5 pound of black powder in shell prior to 1893.

The French were experimenting with streamline shell. We adopted the French streamline 75-millimeter shell and put it into production, calling it our Mark IV shell. Our regular 75-millimeter shell, known as the Mark I 1900 shell, had a maximum range of 9,000 yards. The

Mark IV shell proved to have a maximum range of 12,130 yards, giving an increase in range of well over a mile. America up to April 3, 1919, turned out about 524,000 of these streamline shell.

The French also built shell of semisteel, steel to which iron was added. It was claimed that these shell, by bursting into fine fragments upon exploding, were more effective against troops than all-steel shell, because the fragments of the latter were larger. We adopted this shell also and produced it experimentally. In contour it was a compromise between the old cylindrical shell and the extreme streamline type and was easier to make than the latter.

Artillery ammunition, complete rounds—Acceptances in United States and Canada on U. S. Army orders only.

[Figures in thousands of rounds.]

	To Jan. 1.	1918												Total.
		Jan.	Feb.	Mar.	Apr.	May.	June.	July.	Aug.	Sept.	Oct.	Nov.	Dec.	
Calibers for American Expeditionary Force program.														
75-mm. gun H. E.							235	287	809	1,168	1,122	1,175	790	5,586
75-mm. gun shrapnel	20	121	124	483	888	1,011	1,049	730	732	802	1,057	812	738	8,567
75-mm. gun gas									188	164	213	15		580
75-mm. A. A. shrapnel									92	97	185	134	126	634
3-inch A. A. shrapnel											11	59	2	72
4.7-inch gun H. E.										32	45	43	46	166
4.7-inch gun, shrapnel	9	9	14	17	18	23	35	23	38	29	28	15	19	277
5-inch S. C. gun H. E.										7			5	12
6-inch S. C. gun H. E.							2			1	36	23		62
155-mm. gun H. E.[1]											9	33	51	93
155-mm. howitzer H. E.[1]								11	113	193	119	173	140	749
155-mm. shrapnel									12	22	66	41	93	234
8-inch howitzer H. E.											91	8		99
9.2-inch howitzer H. E.										13	8	24	3	48
240-mm. howitzer H. E.											2			2
8-inch S. C. gun H. E.											20	11		31
10-inch S. C. gun H. E.										20	50	4	11	85
Total	[2] 20	[2] 130	[2] 138	[2] 500	[2] 906	[2] 1,034	1,321	1,051	1,984	2,548	3,062	2,570	2,024	17,297
Calibers for use in United States only.														
2.95-inch mountain gun H. E.	22													22
2.95-inch mountain gun shrapnel	37							9	14	2				62
3-inch F. G. H. E.	333	73	212	142	128	95	3	1	84					1,071
3-inch F. G. shrapnel	957	164	231	174	55	60	15							1,656
3.8-inch howitzer H. E.	3	3	2	2	1									11
3.8-inch howitzer shrapnel	12	1												13
4.7-inch howitzer H. E.	14	4			5	1	1	1	1			12		39
4.7-inch howitzer shrapnel						4		23	5	8	10	10		60
6-inch howitzer H. E.	20	1	3	24	35									83
6-inch howitzer shrapnel							1		3					4
Total	1,398	246	448	342	224	160	20	34	107	10	10	22		3,021
Grand total	1,427	376	586	842	1,130	1,194	1,341	1,085	2,091	2,558	3,072	2,592	2,024	20,318

[1] All thick walled type; not all supplied with fuses.

[2] Shrapnel only.

The following table lists the name of each manufacturer of the various types and sizes of shell for big guns and states the quantity turned out by each:

Contractor.	Forgings.		Machinings.	
	Quantity ordered to Nov. 1, 1918.	Quantity accepted to Nov. 1, 1918.	Quantity ordered to Nov. 1, 1918.	Quantity accepted to Nov. 1, 1918.
3-inch antiaircraft high-explosive shell.				
Hydraulic Pressed Steel Co., Cleveland, Ohio	1,938,806	135,435		
John Inglis Co., Toronto, Ontario	500,000	131,542		
Saskatchewan Bridge & Iron Works, Moose Jaw, Saskatchewan			84,000	
West Shell & Box Co., North Edmonton, Alberta			83,000	
Manitoba B. & I. Co., Winnepeg, Manitoba			83,000	
Medicine Hat P. & B. Co., Medicine Hat, Alberta			83,000	
Dominion Bridge Co., Winnepeg, Manitoba			84,000	
Salisbury Wheel & Axle Co., Jamestown, N. Y			500,000	1,097
3-inch antiaircraft shrapnel.				
Symington Machine Corporation, Rochester, N. Y	1,052,099	1,013,199	1,000,000	1,000,000
75-millimeter antiaircraft high-explosive shell.				
Moline Forge Co., Moline, Ill	939,866	540,532		
Jackson Munitions, Jackson, Mich	225,000			
Spencer Engine Co., Toledo, Ohio	500,000	28,293		
Chamberlain Machine Works, Waterloo, Iowa	365,000	23,669		
75-millimeter antiaircraft shrapnel.				
Symington Manufacturing Co., Rochester, N. Y	672,625	672,625	672,625	672,625
75-millimeter gas and high-explosive shell.				
T. A. Gillespie, Parlin, N. J	1,400,000	1,400,000	4,400,000	1,977,149
American International Corporation, New York City	3,000,000	2,433,438		
American Can Co., New York City	7,000,000	2,563,151	4,000,000	399,728
Hydraulic Pressed Steel Co., Cleveland, Ohio	12,000,000	4,455,090		
Valley Forge Co., Verona, Pa	4,000,000	880,263		
New York Air Brake Co., New York City	2,000,000	192,774	1,300,000	17,652
Worthington Pump Machine Co., New York City	2,650,000	1,473,929	2,650,000	634,159
The Canadian Allis-Chalmers Co., Toronto, Ontario	2,267,062	1,802,117	435	140,647
Canada Car & Foundry Co., Montreal, Quebec	1,656,302	1,592,877		
A. P. Smith Co., Orange, N. J			125,000	
S. A. Wood Manufacturing Co., Boston, Mass			1,500,000	405,344
Vermont Farm Machine Co., Bellows Falls, Vt			750,000	188,300
American Machinery Corporation, Port Huron, Mich			200,000	
Consolidated Car Heating Co., Albany, N. Y			810,000	181,885
Wire Wheel Corporation, Springfield, Mass			300,000	71,239
The Canadian Crocker Wheeler, St. Catherines, Ontario			475,000	160,935
Lachine Manufacturing Co., Lachine, Quebec			660,000	255,264
The Electric Steel & Metal Co., Welland, Ontario			11,458	11,458
J. Bertram & Co., Dundas, Ontario			100,000	51,141
Canadian Fairbanks Morse, Toronto			1,584,548	1,377,800
W. H. Banfield & Sons, Toronto			1,620,000	670,000
Canadian Bridge Co., Walkerville, Ontario			1,450,000	456,993
Canadian Metal Co., Toronto	3,250,000	1,154,371		
Goldie & McCullough, Galt, Ontario	1,100,000	921,206	410,000	61,476
John Inglis Co., Toronto	1,700,000	775,033	75,000	42,400
Cluff Ammunition Co., Toronto	600,000	509,343		
G. W. McFarland Engineering Co., Paris, Ontario	1,500,000	285,335		
Dayton, Ohio, Products Co., New York City	3,500,000	732,842		
E. W. Bliss Co., Brooklyn, N. Y	1,300,000	701,804		
Lymburner (Ltd.) Co., Montreal, Quebec	800,000	630,978	2,474,000	1,126,556
Moline Forging & Machining Co., Moline, Ill	1,500,000	471,281		
Laconia Car Co., Laconia, N. H	550,000			
Symington Machine Co., Rochester, N. Y	4,025,000		6,025,000	1,200,686
Roberts Filter Co., Darby, Pa			600,000	151,975
Auto Transportation Co., Buffalo, N. Y			350,000	107,441
Dominion Bridge Co., Montreal, Quebec			795,000	301,144
Canadian Ingersoll Rand Co., Sherbrooke, Quebec			1,100,000	290,431
Steel Co. of Canada, Brantford, Ontario			515,000	162,399
Allis-Chalmers Co., Milwaukee, Wis			1,520,000	347,635
Jackson Munitions, Jackson, Mich			775,000	67,570
Maxwell Motor Co., Detroit, Mich			800,000	61,761

Contractor.	Forgings.		Machinings.	
	Quantity ordered to Nov. 1, 1918.	Quantity accepted to Nov. 1, 1918.	Quantity ordered to Nov. 1, 1918.	Quantity accepted to Nov. 1, 1918.
75-millimeter gas and high-explosive shell—Continued.				
Batavia Steel Products, Batavia, N. Y.	....	....	1,175,000	311,417
Wheeling Mold & Foundry, Wheeling, W. Va.	....	....	500,000	118,496
Eddystone Munitions, Eddystone, Pa.	....	....	1,000,000	190,100
Lachine Manufacturing Co., Lachine, Quebec	....	....	....	....
The International Clay & Machine, Dayton, Ohio	....	....	124,000	3,812
Smead & Co., Jersey City, N. J.	....	....	1,100,000	246,841
Manufacturing Production Co., Dayton, Ohio	....	....	1,600,000	340,885
Chicago Pneumatic Tool Co., Chicago, Ill.	....	....	250,000	132,321
Mueller Manufacturing Co., Port Huron, Mich.	....	....	500,000	78,300
The Westfield Manufacturing Co., Westfield, Mass.	....	....	1,740,000	413,578
The Platt Iron Works, Dayton, Ohio.	....	....	1,600,000	170,312
The Mueller Metal Co., Wayne, Mich.	....	....	750,000	....
75-millimeter field-gun shrapnel.				
American Can Co., New York City	969,039	969,039	904,067	904,067
Eddystone Munitions Co., Eddystone, Pa.	769,961	769,961	750,000	750,000
Bartlett-Hayward Co., Baltimore, Md.	6,565,519	4,272,900	6,200,000	3,492,863
Symington Machine Co., Rochester, N. Y.	5,459,378	4,868,942	8,375,000	3,329,025
Frankford Arsenal, Philadelphia, Pa.	650,000	4,713	750,000	4,713
Laconia Car Co., Laconia, N. H.	450,000	369,483	....	....
Bossert Corporation, Utica, N. Y.	200,000	....	....	....
Hydraulic Pressed Steel Co., Cleveland, Ohio	2,285,000	10,000	....	....
Canada Forge Co., Welland, Ontario	730,000	....	....	....
The Liberty Ordnance Co., Bridgeport, Conn	1,000,000	27,000	....	....
155-millimeter howitzer high-explosive shell, Mark I, type B.				
Whittaker Glessner, Portsmouth, Ohio.	130,000	137,406	....	....
American Rolling Mills, Middletown, Ohio.	100,000	49,785	....	....
Pressed Steel Car Co., Pittsburgh, Pa.	600,000	552,867	....	....
American Car & Foundry Co., New York City	2,800,000	1,110,964	....	....
New York Air Brake Co., New York City.	350,000	1,158	138,316	....
Wm. Wharton Manufacturing Co., Philadelphia, Pa.	280,000	61,224	....	....
Standard Steel Car Co., Pittsburgh, Pa.	450,000	....	....	....
Standard Forging Co., Chicago, Ill.	21,141	....	....	....
Curtis & Co., Manufacturing Co., St. Louis, Mo.	500,000	404,645	....	....
American Steel Foundry Co., Chicago, Ill.	412,042	412,042	....	....
Midvale Steel & Ordnance Co., Philadelphia, Pa.	130,000	130,000	....	....
Detroit Shell Co., Detroit, Mich.	....	....	500,000	45,563
J. J. Cavrick, Batavia, N. Y.	....	....	300,000	92,974
Standard Sanitary Co., Pittsburgh, Pa.	....	....	600,000	94,409
Potter & Johnson, Pawtucket, R. I.	....	....	175,000	....
North American Motor Co., Pottstown, Pa.	....	....	30,000	29,446
Minneapolis Steel & Machine Co., Minneapolis, Minn.	....	....	400,000	245,344
W. J. Oliver Manufacturing Co., Knoxville, Tenn.	....	....	130,000	88,662
Twin City Forge & Foundry Co., Stillwater, Minn.	....	....	600,000	54,483
Winslow Bros. Co., Chicago, Ill.	....	....	600,000	176,081
American Brake Shoe & Foundry Co., New York City.	....	....	750,000	184,697
American Clay & Machine Co., Bucyrus, Ohio	....	....	700,000	....
Elyria Machine Co., Elyria, Ohio.	....	....	100,000	32,139
American Machine & Manufacturing Co., Atlanta, Ga.	....	....	240,000	75,063
Haroun Motor Corporation, Wayne, Mich.	....	....	200,000	23,899
Wagner Electric Manufacturing Co., St. Louis, Mo.	....	....	300,000	12,569
155-millimeter howitzer high-explosive shell, Mark IV, type D.				
National Tube Co., Pittsburgh, Pa.	800,000	48,263	....	....
P. Lyall & Sons, Montreal, Quebec.	400,000	4,774	150,000	2,559
National Iron Works, Toronto, Ontario.	400,000	9,137	....	....
Dominion Steel Foundry, Hamilton, Ontario.	400,000	23,270	....	....
Studebaker Corporation, Detroit, Mich.	800,000	....	800,000	....
Fairfax Forge Co., Montreal, Quebec.	400,000	....	150,000	....
Pressed Steel Car Co., Pittsburgh, Pa.	1,000,000	15,122	....	....
Cleveland Crane Co., Wickliffe, Ohio.	500,000	....	....	....
Bethlehem Steel Co., South Bethlehem, Pa.	600,000	139,103	....	....
G. W. McFarland, Paris, Ontario.	370,000	521	....	....
LaClede Gas Light Co., St. Louis, Mo.	850,000	....	850,000	....
Standard Forging Co., Chicago, Ill.	500,000	....	....	....
Whittaker Glessner Co., Portsmouth, Ohio.	900,000	31,909	....	....
Curtis & Co., St. Louis, Mo.	130,000	....	....	....
Warden King & Co., Montreal, Quebec.	180,000	....	....	....
John Inglis Co., Toronto, Ontario.	400,000	....	....	....
Canada Iron Foundry Co., Montreal, Quebec.	100,000	....	100,000	....
Cluff Ammunition Co., Toronto, Ontario.	500,000	....	....	....
Taylor Forbes (Ltd.), Toronto.	90,000	....	....	....
Moon Motor Co., St. Louis, Mo.	....	....	200,000	....

Contractor.	Forgings.		Machinings.	
	Quantity ordered to Nov. 1, 1918.	Quantity accepted to Nov. 1, 1918.	Quantity ordered to Nov. 1, 1918.	Quantity accepted to Nov. 1, 1918.
155-millimeter howitzer high-explosive shell, Mark IV, type D—Continued.				
Standard Sanitary, Pittsburgh, Pa			150,000	
Holden Morgan Thread Co., Toronto, Ontario			100,000	
E. Leonard & Sons, London, Ontario			80,000	
Otis Fenson Elevator Co., Hamilton, Ontario			200,000	
Dominion Copper Products, Montreal, Quebec			150,000	2,056
Caron Bros., Montreal, Quebec			125,000	235
Potter & Johnson, Pawtucket, R. I			350,000	
Biscoe Motor, Jackson, Mich			325,000	
Hudson Motor, Detroit, Mich			400,000	
Munition & M. N. (Ltd.), Sorel			50,000	
John Bartram Sons, Dundas, Ontario			450,000	
155-millimeter howitzer gas shell.				
American Rolling Mills, Middletown, Ohio	500,000	492,399		
Midvale Steel & Ordnance Co., Philadelphia, Pa	120,000	96,799		
American Radiator Co., Washington, D. C	625,000	500	416,667	
Wilson Foundry & Machine Co., Pontiac, Mich	400,000		300,000	
Rathbone Sard & Co., Aurora, Ill	600,000		400,000	
155-millimeter gun high-explosive shell, Mark III, type B.				
Standard Steel Car Co., Pittsburgh, Pa	1,000,000	568,092	1,000,000	431,238
Whittaker Glessner Co., Portsmouth, Ohio	350,471	350,471		
Standard Forging Co., Indiana Harbor, Ind	800,000	730,950		
Mead Morris & Co., Gloucester, Mass	300,000	2,056		
Twin City Forge & Foundry Co., Stillwater, Minn	425,000	136,053		
Chicago Rlg. Equipment Co., Chicago, Ill	400,000	23,356		
Minneapolis Steel & Machine Co., Minneapolis, Minn			200,000	41,254
International Arms & Fuse, Bloomfield, N. J			500,000	310,130
North American Motors, Pottstown, Pa			70,000	
Potter & Johnson, Pawtucket, R. I			100,000	73,836
Templer Motor Co., Cleveland, Ohio			450,000	45,014
New York Air Brake Co., New York City			211,684	
Jackson Munitions, Jackson, Mich			177,500	25,981
Pullman Co., Pullman, Ill			300,000	
New Home Sewing Machine Co., Orange, Mass			200,000	
155-millimeter gun high-explosive shell, Mark V, type D.				
Symington Chicago Corporation, Chicago, Ill	1,000,000		805,000	
American Rolling Mills, Middletown, Ohio	755,000	36,161		
Milton Manufacturing Co., Milton, Pa	10,000			
Whittaker Glessner Co., Portsmouth, Ohio	750,000			
Dominion Foundry & Steel Co., Hamilton, Ontario	500,000			
Winslow Bros., Chicago, Ill			400,000	
Grant Motor Car Co., Cleveland, Ohio			260,000	
Cribbon Sexton Co., Chicago, Ill			200,000	
155-millimeter gun gas shell.				
Bethlehem Steel Co., South Bethlehem, Pa	100,000	92,430		
Kohler Co., Kohler, Wis	850,000	100	657,000	100
American Radiator Co., Washington, D. C	125,000		83,333	
Whittaker Glessner Co., Portsmouth, Ohio	5,000	5,000		
American Car & Foundry Co., New York City			1,350,000	63,914
155-millimeter gun and howitzer shrapnel.				
Dayton, Ohio, Production Co., Dayton, Ohio	850,000	131,329		
Wm. Wharton, jr., Philadelphia, Pa	540,947	345,457		
Bartlett-Hayward Co., Baltimore, Md	200,000		1,600,000	135,590
Frankford Arsenal, Philadelphia, Pa	100,000			
3.8-inch howitzer shell.				
Frankford Arsenal, Philadelphia, Pa	1,000	1,000	15,928	11,757
Hydraulic Pressed Steel Co., Cleveland, Ohio	105,000			
F. R. Wilford & Co	105,000			
3.8-inch howitzer shrapnel.				
Frankford Arsenal, Philadelphia, Pa	18,522	14,264	43,522	14,264
Hydraulic Pressed Steel Co., Cleveland, Ohio	35,000			

Contractor.	Forgings.		Machinings.	
	Quantity ordered to Nov. 1, 1918.	Quantity accepted to Nov. 1, 1918.	Quantity ordered to Nov. 1, 1918.	Quantity accepted to Nov. 1, 1918.
4.72-inch shell.				
National Tube Co., Christie Pks. Works	12,500	5,614		
United States Government	1,850	1,850	1,850	1,850
Buffalo Pitts Co., Buffalo, N. Y			12,705	
Twin City Forge, Stillwater, Minn			2,500	
4.7-inch antiaircraft shell.				
National Tube Co., Christie Pks. Works	230,000	188,495		
Maritime Manufacturing Co., Montreal, Quebec			100,000	
Spartan Manufacturing Co., Montreal, Quebec			45,000	45,159
Darling Bros., Montreal, Quebec			42,500	15,060
Alberta Foundry & Machinery Co., Alberta			42,500	6,170
4.7-inch antiaircraft shrapnel.				
The E. W. Bliss Co., Brooklyn, N. Y	10,000			
Frankford Arsenal, Philadelphia, Pa	60,000		60,000	
National Tube Co., Christie Pks. Works	100,000	42,840		
Alberta Foundry & Machinery Co., Alberta			42,500	
4.7-inch drill projectile.				
Grand Rapids Brass Co., Grand Rapids, Mich	2,975	404	2,975	405
4.7-inch gun gas shell.				
Milton Manufacturing Co., Milton, Pa	400,000	194,612	400,000	92,342
American Radiator Co., Buffalo, N. Y			189,360	
4.7-inch gun shrapnel.				
Frankford Arsenal, Philadelphia, Pa	22,897	22,440	22,897	22,440
Bartlett-Hayward Co., Baltimore, Md	312,005	327,183	701,500	306,635
National Tube Co., Christie Pks. Works	754,777	338,507		
Metal Production Co., Beaver, Pa			150,000	11,264
4.7-inch howitzer shell.				
Frankford Arsenal, Philadelphia, Pa	87,833	26,614	87,833	26,614
4.7-inch howitzer shrapnel.				
Bartlett-Hayward, Baltimore, Md	46,115	46,294	40,000	40,000
Frankford Arsenal, Philadelphia, Pa	79,865	19,999	79,865	20,379
4.7-inch gun high-explosive shell.				
National Tube Co., Christie Pks. Works	1,284,848	903,543		
Allegheny Steel Co., Pittsburgh, Pa	900,000	435,978		
The E. W. Bliss Co., Brooklyn, N. Y	10,000			
Frankford Arsenal, Philadelphia, Pa	40,286	12,047	40,286	12,047
Milton Manufacturing Co., Milton, Pa	700,000	351,731	700,000	285,000
Hydraulic Pressed Steel Co., Cleveland, Ohio	200,000			
Darling Bros., Montreal, Quebec			65,000	
Spartan Machine Co., Montreal, Quebec			165,000	
Robb Engineering Co., Amherst, N. J			95,000	3,720
Motor Trucks Co., Brantford, Ontario			205,000	11,083
P. Lyall & Sons, Montreal, Quebec			845,000	318,578
Steel Products Co., Huntington, W. Va			100,000	9,023
Armstrong Ck. Co., Lancaster, Pa			475,000	20,238
Campbell Howard Machine Co., Sherbrooke, Quebec			350,000	
Thurlow Steel Works, Chester, Pa			136,500	35,116
Bell Manuacturing Co., Fairmont, Ind			75,000	5,289
Buffalo Pitts Co., Buffalo, N. Y			350,000	70,975
Indiana Fiber Co., Marion, Ind			75,000	12,520
Canadian Westinghouse Co., Hamilton, Ontario			300,000	94,156
Ry. Ind. Engineering Co., Greensburg, Pa			100,000	34,347
Sherbrooke Ironworks, Sherbrooke			60,000	14,026
Bridgeport Project Co., Bridgeport, Conn			20,000	16,802
American & British Manufacturing Co., Bridgeport, Conn			87,319	57,932
Maritime Manufacturing Co., St. Johns, New Brunswick			100,000	
Alberta Foundry & Machinery Co., Alberta			50,000	

Contractor.	Forgings.		Machinings.	
	Quantity ordered to Nov. 1, 1918.	Quantity accepted to Nov. 1, 1918.	Quantity ordered to Nov. 1, 1918.	Quantity accepted to Nov. 1, 1918.
8-inch gun and howitzer high-explosive and gas shell.				
Carnegie Steel Co., Pittsburgh, Pa.	561,548	210,171		
Root & Vandervoort Engineering Co., East Moline, Ill.	40,000	40,928	190,000	144,815
Wagner Electrical & Manufacturing Co., St. Louis, Mo.	40,000	40,000	170,000	48,586
McMyler Interstate Co., Cleveland, Ohio	500,000	263,674	450,000	238,470
Pollak Steel Co., New York City	100,000			
Curtis & Co., St. Louis, Mo.	295,000	167,202		
Midvale Steel & Ordnance Co., Philadelphia, Pa.	140,000	135,176		
Standard Steel Car Co., Butler, Pa.	100,000	6,072		
Pressed Steel Car Co., Pittsburgh, Pa.	250,000			
Westinghouse Electric & Manufacturing Co., Pittsburgh, Pa.			360,000	166,803
Willys Overland Co., Toledo, Ohio			600,000	
Motor Products Corporation, Detroit, Mich.			100,000	
British War Mission, Munsey Building, Washington, D. C.	101,817	100,277		
Imperial Munitions Board, Ottawa	8,612	7,722		
Pollak Steel Co., New York City	75,000	22,681		
American Steel Foundry Co., Chicago, Ill.	570,000	247,649		
Dominion Steel Foundry Co., Hamilton, Ontario	100,000	91,191		
Canada Cement Co., Montreal, Quebec	150,000	22,304	650,000	4,700
British Forgings (Ltd.), Toronto, Ontario	275,000	24,933		
Dominion Bridge Co., Montreal, Quebec	150,000	55,324		
Standard Forging Co., Chicago, Ill.	300,000	38,659		
Pressed Steel Car Co., Pittsburgh, Pa.	250,000	85,750		
Wm. Wharton, jr., & Co., Philadelphia, Pa.	125,000			
Dominion Foundries & Co. (Ltd.), Hamilton, Ontario.	250,000	10,746		
American Brake Shoe & Foundry Co., New York City.			250,000	197,250
Maritime Manufacturing Corporation, St. John, New Brunswick.			460,000	26,000
9.2-inch howitzer high-explosive shell.				
Russell Motor Car Co., Toronto, Ontario			335,000	15,049
St. Lawrence Bridge Co., Montreal			335,000	31,880
United States Ammunition Corporation, Poughkeepsie, N. Y.			250,000	6,486
Fisher Motor Co., Orilla, Ontario			180,000	100
Canadian Bridge Co., Walkersville, Ontario			110,000	
240-millimeter high-explosive shell.				
Carnegie Steel Co., Pittsburgh, Pa.	190,000	92,316		
Curtis & Co. Manufacturing Co., St. Louis, Mo.	275,000	174,174		
American Car & Foundry Co., New York City	90,000		400,000	47,953
American Steel Foundries Co., Chicago, Ill.	80,000	3,277		
Scullin Steel Co., St. Louis, Mo.	350,000			
A. F. Smith Manufacturing Co., East Orange, N. J.			25,000	
Motors Truck (Ltd.), Brantford, Ontario			125,000	
Laclede Gas Light Co., St. Louis, Mo.			526,014	
5-inch seacoast gun shell.				
Cleveland Crane & Engineering Co., Wickliffe, Ohio.	244,812	122,324		
McMyler Interstate Co., Cleveland, Ohio	5,000	5,107		
Milton Manufacturing Co., Milton, Pa.	30,000	29,121		
Machine Products Co., Cleveland, Ohio			75,000	21,532
A. J. Vance & Co., Winston-Salem, N. C.			40,000	1,578
Twin City & Foundry Co., Stillwater, Minn.			400	
A. B. Ormsby Co. (Ltd.), Torouto, Ontario			50,000	10,029
P. Tyrall Construction Co., Montreal			105,000	38,385
6-inch seacoast gun shell.				
Frankford Arsenal, Philadelphia, Pa.	40,950	25,957	40,950	25,957
Bethlehem Steel Co., Bethlehem, Pa.	16,000	22,053	16,000	15,910
Columbian Iron Works, Chattanooga, Tenn.	40,000	40,346	132,542	149,281
The Pressed Steel Car Co., McKeesport, Pa.	385,000	370,677		
Standard Steel Car Co., Hammond, Ind.	400,000	376,827		
Anniston Steel Co., Anniston, Ala.	243,812			
Westinghouse Electric Manufacturing Co., Pittsburgh, Pa.	35,000	31,310	385,000	192,684
Wm. Wharton, jr., Easton, Pa.	24,000			
The Southern Machinery Co., Chattanooga, Tenn.			447,458	19,537

Contractor.	Forgings.		Machinings.	
	Quantity ordered to Nov. 1, 1918.	Quantity accepted to Nov. 1, 1918.	Quantity ordered to Nov. 1, 1918.	Quantity accepted to Nov. 1, 1918.
10-inch seacoast gun shell.				
American Car & Foundry Co., New York City	24,360	24,360	275,000	130,040
Carnegie Steel Co., Pittsburgh, Pa	60,000	61,770		
Carnegie Steel Co., Munhall, Pa	225,000	137,168		
12-inch seacoast gun shell.				
Carnegie Steel Co., McKees Rocks, Pa	165,000	7,627		
Watertown Arsenal, Watertown, Mass	15,000	1,449		
Washington Steel & Ordnance Co., Giesboro Manor, D. C	28,631	6,129	38,000	1,907
Leaside Munitions Corporation, Toronto, Ontario	105,000		105,000	
Standard Forging Co., Chicago, Ill	15,000			
Bethlehem Steel Co., Bethlehem, Pa	32,000			
American Clay Machine Co., Bucyrus, Ohio			15,000	
14-inch seacoast gun shell.				
Carnegie Steel Co., McKees Rocks, Pa	10,000	220		
Watertown Arsenal, Watertown, Mass	9,000			
Washington Steel & Ordnance Co., Washington, D. C			80	
16-inch seacoast howitzer shell.				
Washington Steel & Ordnance Co., Washington, D. C	140		140	

CHAPTER VI.

SIGHTS AND FIRE-CONTROL APPARATUS.

At the threshold of the war with Germany we were confronted with the problem of providing on a large scale those instruments of precision with which modern artillerists point their weapons. As mysterious to the average man as the sextant and other instruments which help the navigator to bring his ship unerringly to port over leagues of pathless water, or as those devices with which the surveyor strikes a level through a range of mountains, are the instruments which enable the gunner to drop a heavy projectile exactly on his target without seeing it at all.

The old days of sighting a cannon point-blank at the visible enemy over the open sights on the barrel of the weapon passed with the Civil War. As the power of guns increased and their ranges lengthened, the artillerists began firing at objects actually below the horizon or hidden by intervening obstacles. These conditions necessarily brought in the method of mathematical aim which is known as indirect fire.

In the great war indirect firing was so perfected that within a few seconds after an aviator or an observer in a captive balloon had definitely located an enemy battery, that battery was deluged with an avalanche of high-explosive shell and destroyed, even though the attacking gunners were located several miles away and hills and forests intervened to obscure the target from view. With the aid of correlated maps in the possession of the battery gunners and the aerial observer, a mere whisper of the wireless sufficed to turn a torrent of shell precisely upon the enemy position which had just been discovered. So accurate had indirect artillery fire become that a steel wall of missiles could be laid down a few yards ahead of a body of troops advancing on a broad front, and this wall could be kept moving steadily ahead of the soldiers at a walking pace with few accidents due to inaccurate control of the guns firing the barrage.

The chief difference between the old and the new methods of artillery practice is the degree of precision attained. At the time of the Civil War the artillery was fired relatively blindly, reliance being placed upon the weight of the fire regardless of its accuracy and its effectiveness; but modern artillery has recognized the importance of the well-placed shot and demands instruments that must be marvels of accuracy, since a slight error in the aiming at modern ranges means a miss and the total loss of the shot. Such uncanny accuracy is made possible by the use of those instruments of precision known as fire-control apparatus. The gunner who is not equipped with proper fire-control instruments can not aim correctly and is placed at a serious disadvantage in the presence of the enemy. These instruments must not only be as exact as a chronometer, but they must be sufficiently rugged to withstand the concussion of close artillery fire.

Equipment classified under "Sights and fire-control apparatus" comprises all devices to direct the fire of offensive weapons and to observe the effect of this fire in order to place it on the target. Included in this list are instruments of a surveying nature which serve to locate the relative position of the target on the field of battle and to determine its range. For this purpose the artillery officer uses aiming circles, azimuth instruments, battery commander telescopes, prismatic compasses, plotting boards, and other instruments. Telescopes and field glasses equipped with measuring scales in them are also employed in making observations.

Instruments of a second group are attached directly to the gun to train it both horizontally and vertically in the directions given by the battery commander. These devices include sights of different types, elevation quadrants, clinometers, and other instruments. The intricate panoramic sight which is used especially in firing at an unseen target is one of the most important instruments of this group.

Still another set of instruments comprises devices such as range deflection boards, deviation boards, and wind indicators which, together with range tables and other tables, assist the battery commander to ascertain the path of the projectile under any condition of range, altitude, air pressure, temperature, and other physical influences. When it is understood that the projectile fired by such a weapon as the German long-range gun which bombarded Paris at a distance of 70 miles mounts so high into the air that it passes into the highly rarified layers of the air envelope surrounding the earth and thus into entirely different conditions of air pressure, it can be realized how abstruse these range calculations are and how many factors must be taken into account. The fire-control equipment enables the artilleryman to make these computations quickly.

In addition to the above items many auxiliary devices are needed by the Artillery, notable among these being the self-luminous aiming posts and other arrangements which enable the gunners to maintain accuracy of fire at night. This whole elaborate set of instruments is supplied to the field and railway artillery—the big guns—and in part to trench-mortar batteries and even to machine guns, which in the latter months of the war were used in indirect firing.

Still another group of pointing instruments is used by antiaircraft guns against hostile aircraft to ascertain their altitude, their speed, and their future location in order that projectiles fired by the antiaircraft guns may hit these high and rapidly moving targets. Sights are also used on the airplanes themselves to aid the pilot and the observer in the dropping of bombs and in gunfire against enemy planes or targets. One of these sights corrects automatically for the speed and direction of the airplane.

Fuse setters, which enable the gunner to time the fuse in the shell so that the projectile moving with enormous speed explodes at precisely the desired point, were required in large numbers.

The responsibility for the design, procurement, production, inspection, and supply of the above equipment to the American Expeditionary Forces was lodged in the Ordnance Department. The effectiveness of the artillery on the field of battle depended directly on the fire-control equipment furnished by this bureau.

The optical industry in this country before the war was in the hands of a few firms. Several of these were under German influence, and one firm was directly affiliated with the Carl Zeiss Works, of Jena, Germany; the workmen were largely Germans or of German origin; the kinds and design of apparatus produced were for the most part essentially European in character; optical glass was procured entirely from abroad and chiefly from Germany.

It was easier and cheaper for manufacturers to order glass from abroad than to develop its manufacture in this country. Educational and research institutions obtained a large part of their equipment from Germany and offered no special inducement for American manufacturers to provide such apparatus. Duty-free importation favored and encouraged this dependence on Germany for scientific apparatus.

With our entrance in the war the European sources of supply for optical glass and optical instruments were cut off abruptly and we were brought face to face with the problem of furnishing these items to the Army and Navy for use in the field. Prior to 1917 only three private manufacturers in the United States had built fire-control apparatus in any quantity for the Government. The Bausch & Lomb Optical Co., Rochester, N. Y., had made range finders and field glasses for the Artillery and Infantry, and gun sights, range finders, and spy glasses and field glasses for the Navy; the Keuffel & Esser Co., Hoboken, N. J., had produced some fire-control equipment for the Navy; the Warner & Swasey Co., Cleveland, Ohio, with J. A. Brashear, Pittsburgh, Pa., had furnished depression-position finders, azimuth instruments, and telescopic musket sights to the Army. The only other source of supply in this country had been the Frankford Arsenal.

Prior to 1917 the largest order for fire-control equipment which our Army had ever placed in a single year amounted to $1,202,000. The total orders for such instruments placed by the Ordnance Department alone during the 19 months of war exceeded $50,000,000, while the total orders for fire-control apparatus placed by the Army and Navy exceeded $100,000,000.

To meet the situation, existing facilities had to be increased, new facilities developed, and other, allied, industries converted to the production of fire-control material.

Quantity production had to be secured through the assembling of standardized parts of instruments which heretofore had either never been built in this country or only in a small, experimental way. A large part of the work had of necessity to be done by machines operated by relatively unskilled labor. The manufacturing tolerances had to be nicely adjusted between the different parts of each instrument, so that wherever less precise work would answer the purpose the production methods were arranged accordingly. Only by a careful coordination of design, factory operations, and field performance could quantity production of the desired quality be obtained in a short time. Speed of production meant everything if our troops in the field were to be equipped with the necessary fire-control apparatus and thus enabled to meet the enemy on even approximately equal terms.

To accomplish this object a competent personnel within the Army had to be organized and developed; the Army requirements had to be carefully scrutinized and coordinated with reference to relative urgency; manufacturers had to be encouraged to undertake new tasks and to be impressed with the necessity for whole-hearted cooperation and with the importance of their part in the war; raw materials had to be secured and their transportation assured. These and other factors were faced and overcome.

Although American fire-control instruments did not reach the front in as large numbers as were wanted, great quantities were under way, and we had attained in the manufacturing program a basic stage of progress which would have cared for all of our needs in the spring and summer of 1919. Incidentally there has been developed in this country a manufacturing capacity for precision optical and instrument work, which, if desired, will render us independent of foreign markets. At the present time there exists in this country a trained personnel and adequate organization for the production of precision optical instruments greatly in excess of the needs of the country. One of the problems which we now have to consider is the conversion of this development brought about by war-time conditions into channels of peace-time activity.

At the present time American manufacturers are in a position to make instruments of precision equal to the best European product, and the industry will continue, provided there is an adequate market for its product. Such a market will exist if the universities and commercial laboratories of the country will obtain scientific apparatus from American manufacturers rather than import it from abroad as has heretofore been the custom.

In April, 1917, the most serious problem in the situation was the manufacture of optical glass. Prior to 1914 practically all of the optical glass used in the United States had been imported from abroad; manufacturers followed the line of least resistance and preferred to procure certain commodities, such as optical glass, chemical dyes, and other materials difficult to produce, direct from Europe rather than to undertake their manufacture here. The war stopped this source of supply abruptly, and in 1915 experiments on the making of optical glass were under way at five different plants—the Bausch & Lomb Optical Co. at Rochester N. Y.; the Bureau of Standards at Pittsburgh, Pa.; the Keuffel & Esser Co. at Hoboken, N. J.; the Pittsburgh Plate Glass Co. at Charleroi, Pa.; the Spencer Lens Co. at Hamburg, Buffalo, N. Y.

By April, 1917, the situation had become acute; some optical glass of fair quality had been produced, but nowhere had its manufacture been placed on an assured basis. The glass-making processes were not adequately known. Without optical glass fire-control instruments could not be produced; optical glass is a thing of high precision and in its manufacture accurate control is required throughout the factory processes. In this emergency the Government appealed to the Geophysical Laboratory of the Carnegie Institution of Washington for assistance.

This laboratory had been engaged for many years in the study of solutions, such as that of optical glass, at high temperatures and had a corps of scientists trained along the lines essential to the successful production of optical glass. It was the only organization in the country with a personnel adequate and competent to undertake a manufacturing problem of this character and magnitude. Accordingly, in April, 1917, a group of its scientists was placed at the Bausch & Lomb Optical Co. and given virtual charge of the plant; its men were assigned to the different factory operations and made responsible for them. By November, 1917, the manufacturing processes at this plant had been mastered and large quantities of optical glass of good quality were being produced. In December, 1917, the work was extended, men from the Geophysical Laboratory taking practical charge of the plants of the Spencer Lens Co. and of the Pittsburgh Plate Glass Co.

The cost to the Geophysical Laboratory of contributing to the Government the solution of the optical glass problem amounted to about $200,000, but the results attained surely more than justified these expenditures. These results could not have been attained, however, without the hearty cooperation of the manufacturers and of the Army and Navy, which assisted in the procurement and transportation of the raw materials. An ordnance officer was in charge of the Rochester party from the Geophysical Laboratory and was responsible for much of the pioneer development work accomplished there. It was at this plant, that of the Bausch & Lomb Optical Co. at Rochester, that the methods of manufacture were first developed and placed on a production basis. The Bureau of Standards aided in the development of a chemically and thermally resistant crucible in which to melt optical glass; also in the testing of optical glass, and especially in the testing of optical instruments. The Geological Survey aided in locating sources of raw materials, such as sand of adequate chemical purity.

By February, 1918, the supply of optical glass was assured; but the manufacture of optical instruments was so seriously behind schedule that a military optical glass and instrument section was formed in the War Industries Board and took charge of the entire optical instrument industry of the country. Through the efforts of its chief, Mr. George E. Chatillon, of New York, the entire industry was coordinated. By September, 1918, the production of fire-control instruments in sufficient quantities to meet the requirements of both the Army and Navy during 1919 was believed to be assured.

To the accomplishment of this result the Ordnance Department contributed most effectively. The information and long experience of Frankford Arsenal in instrument manufacture and in the work of precision optics were placed at the service of contractors; trained officers of the Ordnance Department were stationed at the different factories; in many factories these officers rendered valuable aid in devising and developing proper and adequate factory operations, in establishing production on a satisfactory basis, in securing the proper inflow of raw materials, in devising testing fixtures, in establishing proper manufacturing tolerances, and in testing the performance of the assembled instruments. Schools for operatives in precision optics were established at Frankford Arsenal, Philadelphia, Pa., at Rochester, N. Y., and at Mount Wilson Observatory, Pasadena, Cal. To many contractors financial aid had to be extended. The fire-control program required, in short, all the available talent and resources of the country to carry it to a successful finish.

The general procedure adopted by the Ordnance Department was to assign the more difficult instruments to manufacturers who had had experience along similar lines. To others, who had produced articles allied only in a distant way to fire-control instruments, less intricate types of instruments were awarded. In certain instances the optical elements were produced by one firm, the mechanical parts by another, the final assembly of the instrument being then accomplished by the latter.

Because our Army had adopted a number of French guns for reproduction here, it became necessary to build sights for these weapons according to the French designs. This gave us much trouble, not only because of the delay in securing samples and drawings from France, but because of the difficulties in producing articles from these French drawings by American methods and with American workmen.

The most intricate of these French sights was the Schneider quadrant sight. It was used with the French 155-millimeter gun, the 155-millimeter howitzer, and the 240-millimeter howitzer. The structure of this sight was highly complicated, and extreme accuracy was required at every stage of production. These sights were put into production by the Emerson Engineering Co. of Philadelphia, the Raymond Engineering Co. of New York, and by Slocum, Avram & Slocum of New York.

The design of this sight was received from France early in 1918, yet it was the 1st of November—10 days before the armistice was signed—when the first Schneider sight was delivered to the Army; but at all times the progress made was as rapid as could be expected. A total of 7,000 Schneider quadrant sights was ordered, which meant a year's work for 1,000 men. Of this order 3,500 sights were to be manufactured by the Schneider Co. in France and the rest by the three firms in this country. On November 11 the American factories had delivered 74 sights and since that time over 560 have been completed.

The amount of labor involved in the case of Schneider quadrant sights is shown by the fact that while the raw material for it cost about $25, the finished sight is worth about $600. In order to expedite production the Government extended financial assistance to some of the factories to aid in the procurement and installation of additional equipment. On November 11 the number of these sights completed was short of requirements for installation on completed carriages by about 400, but the rate of progress which had been attained in production would have overtaken the output of gun carriages by January 1, 1919.

Another difficult task was the construction of telescopic sights for the French 37-millimeter guns, the "Infantry cannon" which we adopted for reproduction in this country. Here again we encountered the same difficulty of adapting French plans to our methods. The original contract was placed with a firm which had had no experience with optical instruments of precision, but no other company was available for the work. When by May, 1918, this concern had produced only a few sights the contract was taken from it and placed with a subcontractor, the Central Scientific Co., of Chicago, who had been building mechanical parts for the sights. In this plant the complete force had to be educated in the art before any production could begin. When the armistice was signed the gun factories had produced 884 of the 37-millimeter guns, but only 142 telescopic sights had been completed. The rate of production of these sights by the Central Scientific Co. was such, however, that the shortage would have ceased to exist shortly after January 1, 1919.

The French design for the telescopic sight for the 37-millimeter gun used on the tanks was also adopted by the Army. Here again difficulty was experienced in manufacture, but excellent progress was made especially by one firm (Burke & James of Chicago, Ill.), and the output in adequate quantities was assured for 1919. The French collimator sight for the 75-millimeter gun presented difficulties to the manufacturer, especially in the optical parts. These were, however, overcome by the Globe Optical Co., who furnished the optics to the Electric Auto-Lite Corporation and to the Standard Thermometer Co. of Boston, with the result that at the time of the signing of the armistice the production of these sights was progressing well.

Periscopes from 20 inches to nearly 20 feet in length were produced in quantity. These periscopes enabled the men in the front-line trenches to look over the top with comparative safety. The long periscopes were used in deep-shelter trenches and bomb proofs. The production of the short-base periscopes and also of the battery commanders' periscopes by the Wollensak Optical Co., Rochester, N. Y., and of the 3-meter and 6-meter periscope by the Andrew J. Lloyd Co. of Boston, Mass., was progressing at such a rate that the needs of the Army for 1919 would be met on time.

At the outbreak of the war the policy followed by the Ordnance Department was to place orders for standard fire-control apparatus, such as range finders of different base lengths, battery commander telescopes, aiming circles, panoramic sights, musket sights, and prismatic compasses with firms of established reputation and experience. The result was that when requests from the Army in France came for instruments of new design, new sources of manufacture had to be sought out and these organizations educated in the methods of precision optics. Such a procedure necessarily caused delay, but it was the only course of action left. Wherever possible part of the total contract was awarded to an experienced manufacturer, so that some production was assured.

PANORAMIC SIGHT.

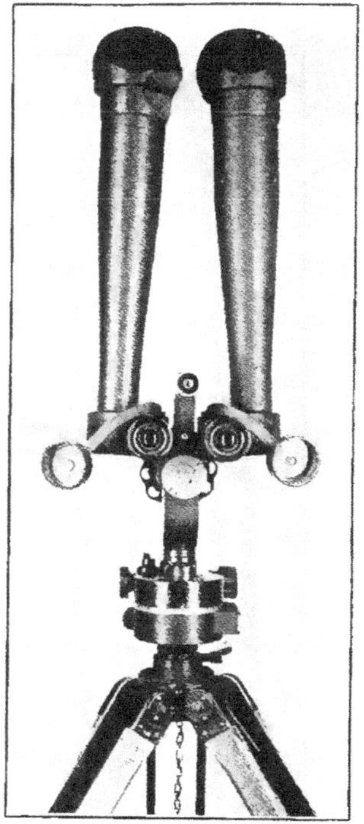

BATTERY COMMANDER'S TELESCOPE.

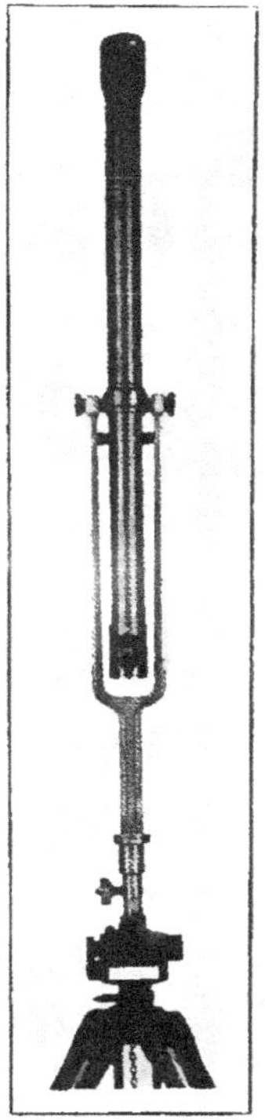

BATTERY COMMANDER'S PERISCOPE.

AIMING CIRCLE.

BRACKET FUSE SETTER, MODEL OF 1916.

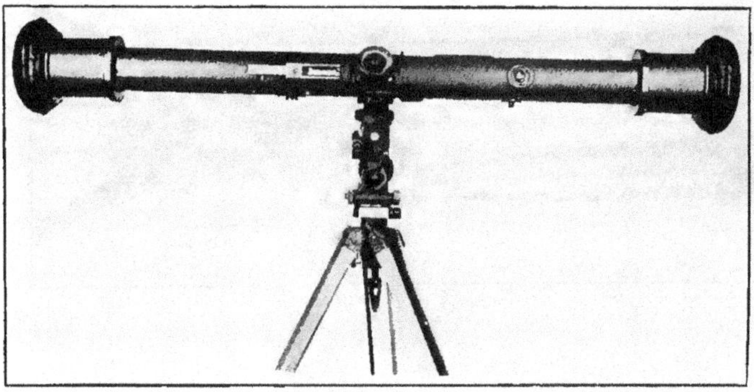

RANGE FINDER.

FRENCH QUADRANT SIGHT WITH AMERICAN PANORAMIC SIGHT.

SIGHT FOR 75-MILLIMETER FIELD GUN.

The records show that the experienced manufacturers overcame the difficulties encountered and had obtained in general a rate of output which was satisfactory at the time of the signing of the armistice. Thus the Bausch & Lomb Optical Co. had delivered large numbers of range finders of base lengths of 80 centimeters, 1 meter and 15 feet, and battery commanders' telescopes; Keuffel & Esser had made many prismatic compasses and a few range finders; the Spencer Lens Co. had produced aiming circles in quantity; the Warner & Swasey Co., with J. A. Brashear of Pittsburgh, had furnished large numbers of the valuable panoramic sights with which much of the artillery fire is directed. Much credit is due the above organizations for the efficient manner in which they placed the manufacture of these items on a high-speed production basis. Frankford Arsenal proved to be a most reliable source of supply for battery commander telescopes, panoramic sights, azimuth instruments for, 3-inch telescopes, plotting boards, and other ordnance fire-control instruments.

The manufacture of many other types of instruments was undertaken in this country. Among these the French sitogoniometer, a device which assists the battery commander in obtaining data for the direction of fire, was successfully produced by the Martin-Copeland Co. of Providence, R. I.; quadrant sights for the 37-millimeter gun by the Scientific Materials Co. of Pittsburgh; lensatic compasses and Brunton compasses were furnished by Wm. Ainsworth & Sons of Denver, Colo.; prismatic compasses by the Sperry Gyroscope Co. of Brooklyn, N. Y.; telescopes for sights on antiaircraft carriages by the Kollmorgen Optical Corporation of Brooklyn; altimeters, gunners' quadrants, elevation quadrants, and aiming stakes by the J. H. Deagan Co. of Chicago, Ill.; panoramic telescopes and fuse setters by the Recording & Computing Machines Co. of Dayton, Ohio; battery commander telescopes by Arthur Brock of Philadelphia; tripods for fire-control instruments by the National Cash Register Co. of Dayton, Ohio. Optics for different sights were furnished by the American Optical Co. of Southbridge, Mass., and by the Mount Wilson Observatory of Pasadena, Calif. These and other organizations entered into the task and devoted their energy to the production of equipment desired by the Government.

At no time during the fighting did our artillery units have a sufficient supply of fire-control instruments. This was due to the fact that we were not able to secure in Europe the amount of this equipment required to take care of our needs while our own industry was being developed.

With almost a total lack of optical glass in this country, with an equal lack of factories and workmen familiar with military optical instrument-making, we were suddenly called upon to produce about

200 different types of instruments in large quantities. These included many new designs of fire-control apparatus made necessary by new artillery developments both among the allies and in our own factories, by the adoption of trench warfare in place of open warfare, by the development of weapons for use against aircraft, by the extension of indirect fire-control methods to weapons which formerly had been fired by direct sighting, and by the use of railway and seacoast artillery.

While we did not solve all the difficulties in this development, we had met and conquered the worst of them, and we were making such great strides in production when the war ended that all the requirements of the Army would have been met early in 1919. It has been a source of inspiration to witness the high sense of patriotic duty and cooperation shown by the manufacturers which made possible the remarkable expansion of the optical glass and instrument industry in the United States during the period of the war.

The following table shows the principal items of sights and fire-control apparatus, the firms that did the work, the quantity of the various kinds of instruments ordered, and the deliveries made up to November 11, 1918, and to February 20, 1919:

Material.	Firm.	Total ordered.	Deliveries to—	
			Nov. 11, 1918.	Feb. 20, 1919.
Aiming circle, model 1916	Spencer Lens Co., Buffalo, N. Y	1,473	717	1,117
Do	Frankford Arsenal, Philadelphia	98	98	98
Aiming stakes for machine gun	J. C. Deagan Co., Chicago, Ill	16,618	....	1,320
Aiming posts, field artillery	Metropolitan Manufacturing Co., Detroit, Mich.	16,791	25	250
Do	Dahlstrom Metallic Door Co., Jamestown, N. Y.	10,791	....	....
Aiming devices	National Vitaphone Corporation, Plainfield, N. J.	5,700	....	150
Angle of site instruments	Atwater Kent Manufacturing Co., Philadelphia.	4,468	4,401	4,468
Do	Blair Tool Machine Co., New York City	1,090	1,090	1,090
Azimuth instruments, model 1910	Warner & Swasey Co., Cleveland, Ohio	129	126	129
Azimuth instruments, model 1918	Spencer Lens Co., Buffalo, N. Y	669	....	1
Boards, gun deflection	Premier Metal Etching Co., New York City.	13	....	....
Boards, Pirie deviation	Metallograph Corporation, New York City.	628	....	....
Boards, plotting	McFarlan Motor Co., Connersville, Ind.	4,811	4,811	4,811
Boards, Pratt range	F. F. Metzger, Philadelphia, Pa	134	....	65
Boards, range deflection	Gorham Manufacturing Co., Providence, R. I.	741	....	....
Boards, rocket	Liquid Carbon Co., Chicago, Ill	3,000	....	630
Chronographs	Precision Thermometer Co., Philadelphia, Pa.	19	9	19
Chronographs, Aberdeen	Leeds Northrup Co., Philadelphia, Pa.	20	18	20
Clinometers, machine-gun	Atwater Kent Manufacturing Co., Philadelphia, Pa.	26,972	8,270	21,972
Do	Central Scientific Co., Chicago, Ill	10,644	....	....
Clinometers, machine gun, model 1912.	F. F. Metzger, Philadelphia, Pa	25	25	25
Compass, lensatic	Wm. Ainsworth & Sons, Denver, Colo.	11,651	8,150	11,651
Compass, prismatic	Sperry Gyroscope Co., Brooklyn, N. Y.	9,575	600	3,000
Do	Keuffel & Esser Co., Hoboken, N. J	4,028	3,828	4,028
Compass, transit, pocket, Brunton.	Wm. Ainsworth & Sons, Denver, Colo.	1,500	1,500	1,500
Cylinders, cannon pressure	Wilton Tool Co., Boston, Mass	8,000	8,000	8,000
Depression position finders, Lewis.	Pratt-Whitney Co., Hartford, Conn.; J. A. Brashear, Pittsburgh, Pa.	90	81	90
Electrical equipment for aiming posts.	Line Material Corporation, Milwaukee, Wis.	26,888	....	11,765

Material.	Firm.	Total ordered.	Deliveries to—	
			Nov. 11, 1918.	Feb. 20, 1919.
Electric lighting devices	Guide Motor Lamp Co., Cleveland, Ohio.	5,352	....	....
Flash lights	Delta Electric Co., Marion, Ind	136,861	73,066	125,448
Do	Novo Manufacturing Co., New York City.	13,563	13,563	13,563
Do	American Ever-ready Works, Long Island City, N. Y.	341,373	194,878	257,258
Glass, optical, lbs	Pittsburgh Plate Glass Co., Charleroi, Pa.	45,000	23,761½	24,010
Do	Spencer Lens Co., Buffalo, N. Y	3,490	....	517½
Do	Bausch & Lomb Optical Co., Rochester, N. Y.	4,450	4,450	4,450
Goniometers, model 1917	Sloane & Chase Manufacturing Co., Newark, N. J.	90	....	....
Levels, longitudinal, 3-inch	Young & Sons, Philadelphia, Pa	1,310	934	1,201
Levels, longitudinal	Arthur Brock, jr., Philadelphia, Pa	1,474	864	864
Levels, sight	Electric Auto-Lite Corporation, Toledo, Ohio.	1,277	560	1,277
Levels, testing	Carlson-Wenstrom Co., Philadelphia, Pa.	1,620	196	590
Lighting devices for field carriages.	Globe Machine & Stamping Co., Cleveland, Ohio.	27,240	....	8,000
Night firing boxes for machine gun.	New Method Stove Co., Mansfield, Ohio.	16,618	....	3,196
Do	Delta Electric Co., Marion, Ind	16,818	....	4,417
Periscopes, battery commander's	....do	11,701	289	5,000
Periscopes, mirror	J. R. Young Co. (Penn Toy Co.), Pittsburgh, Pa.	60,000	60,000	60,000
Do	Seneca Camera Co., Rochester, N. Y	36,625	72	72
Periscopes, rifle, model 1917	Oneida Community, Oneida, N. Y	140,527	115,236	115,236
Do	John W. Browne Manufacturing Co., Detroit, Mich.	58,313	58,313	58,313
Periscopes, 3 m. deep, shelter	A. J. Lloyd Co., Boston, Mass	2,234	....	16
Periscopes, 6 m. deep, shelter	....do	2,234	276	700
Periscopes, trench, No. 10	Wollensak Optical Co., Rochester, N. Y.	32,512	2,948	9,252
Plane tables	Pfau Manufacturing Co., Norwood, Ohio.	4,928	3,200	4,928
Protractors, Alidade	Metallograph Corporation, New York City.	13,945	....	....
Do	Wm. Ainsworth & Co., Denver, Colo	1,000	....	108
Protractors and straightedges	Frankford Arsenal, Philadelphia, Pa	1,284	1,284	1,284
Do	Eugene Dietzgen Co., Chicago, Ill	35,112	35,112	35,112
Do	Whitehead Hoag Co., Newark, N. J	5,000	....	3,500
Do	Celluloid Co., New York, N. Y	12,422	12,422	12,422
Do	Keuffel & Esser Co., Hoboken, N. J	6,509	6,509	6,509
Quadrants, elevation	Recording & Computing Machine Co., Dayton, Ohio.	214	74	106
Do	J. C. Deagan Co., Chicago, Ill	120	45	120
Quadrants, gunner's	International Register Co., Chicago, Ill.	72	72	72
Do	Central Scientific Co., Chicago, Ill	6,245	....	2,852
Do	J. C. Deagan Co., Chicago, Ill	6,245	....	2,552
Do	Gorham Manufacturing Co., Providence, R. I.	491	137	329
Quadrants, range	Talbot Reel Manufacturing Co., Kansas City, Mo.	200	101	186
Do	Slocum, Avram & Slocum, Newark, N. J.	1,386	431	940
Range finders, 80-cm	Bausch & Lomb Optical Co., Rochester, N. Y.	5,470	2,167	2,600
Do	Keuffel & Esser Co., Hoboken, N. J	1,000	....	....
Range finders, 1-meter	Bausch & Lomb Optical Co., Rochester, N. Y.	7,131	1,508	1,665
Range finders, 15-foot	....do	65	55	55
Range finders, 9-foot	Keuffel & Esser Co., Hoboken, N. J	86	....	....
Recording thermometers	Bristol Co., Waterbury, Conn	439	439	439
Rules, battery commander's	Wescott Jewel Co., Seneca Falls, N. Y.	26,406	26,406	26,406
Do	Stanley Rule & Level Co., New Britain, Conn.	1,500	1,500	1,500
Rules, elevation, slide, model 1918.	J. E. Sjostrom Co., Detroit, Mich	200	200	200
Rules, Hitt-Browne, for machine gun.	U. S. Infantry Association, Washington, D. C.	24,058	....	24,058
Rules, musketry	Taft-Pierce Manufacturing Co., Woonsocket, R. I.	80,000	80,000	80,000
Do	Metallograph Corporation, New York.	55,067	....	55,067
Rules, slide	J. H. Weil Co., Philadelphia, Pa	4,852	4,852	4,852
Rules, slide, model E	Frankford Arsenal, Philadelphia, Pa	1,500	1,500	1,500

Material.	Firm.	Total ordered.	Deliveries to—	
			Nov. 11, 1918.	Feb. 20, 1919.
Rules, 2-foot	Stanley Rule & Level Co., New Britain, Conn.	52,519	52,519	52,519
Do	Lufkin Rule Co., Saginaw, Mich	38,540	38,540	38,540
Do	Upson Nut Co., Cleveland, Ohio	14,358	14,358	14,358
Do	Chapin-Stephens Co., Pine Meadow, Conn.	7,040	7,040	7,040
Rules, zinc, for machine guns	Clapp Eastman Co., Cambridge, Mass.	5,193		5,193
Rules, 3-foot	L. S. Starrett Co., Athol, Mass	343	343	343
Rules, boxwood	Stanley Rule & Level Co., New Britain, Conn.	2,000	2,000	2,000
Do	Lufkin Rule Co., Saginaw, Mich	15,630	3,000	7,509
Rule, zigzag	do	2,312	2,312	2,312
Sights, antiaircraft, model 1917	Recording & Computing Machines Co., Dayton, Ohio.	25	25	25
Do	New Britain Machine Co., New Britain, Conn.	60	1	60
Sights for antiaircraft carriages	do	519	27	63
Sights, telescopes, for antiaircraft carriages.	Kollmorgen Optical Co., Brooklyn, N. Y.	519	66	255
Sights, telescopes, for goniometers	do	90		16
Sights, optics, for altimeter telescope, model 1917.	Mount Wilson Observatory, Pasadena, Calif.	467		467
Sights, bomb	Globe Optical Co., Boston, Mass	100	100	100
Sights, bore	Benjamin Electric Manufacturing Co., Chicago, Ill.	2,191	1	2,191
Do	Poole Engineering & Machine Co., Hagerstown, Md.	1,500	524	1,357
Do	Buffalo Forge Co., Buffalo, N. Y	900	900	900
Sights, panoramic, for machine guns.	Atwater-Kent Manufacturing Co., Philadelphia, Pa.	6,000		525
Do	Scientific Materials Co., Pittsburgh, Pa.	4,510		
Sights for 1917 6-inch gun carriages.	Recording & Computing Machines Co., Dayton, Ohio.	123	123	123
Sights, luminous	Radium Luminous Material Corporation.	1,250	1,215	1,250
Sights, luminous, for machine gun.	Watson Luminous Gunsight Co., New York.	123,236	18,018	87,236
Sights, panoramic, model 1917	Warner & Swasey Co., Cleveland, Ohio	9,500	1,336	2,180
Do	Frankford Arsenal, Philadelphia, Pa	800	800	800
Do	Recording & Computing Machines Co., Dayton, Ohio.	6,000	100	230
Sights, panoramic, model 1915	Frankford Arsenal, Philadelphia, Pa	237	237	237
Sights, panoramic, for 8-inch gun.	Recording & Computing Machines Co., Dayton, Ohio.	30	30	30
Sights, quadrants, Schneider	Emerson Engineering Co., Philadelphia, Pa.	800		
Do	Raymond Engineering Co., New York City.	764		1
Do	Slocum, Avram & Slocum, New York City.	3,800	74	567
Sights, telescopic, rifle, style B	Winchester Repeating Arms Co., New Haven, Conn.	89		89
Sights, telescopic, rifle, 5A, mounted on rifle.	do	400	400	400
Sights, telescopic, rifle, model 1918.	do	32,000		
Sights, telescopic, rifle, model 1913.	Warner & Swasey Co., Cleveland, Ohio.	4,000	4,000	4,000
Sights, optics for telescopic, rifle, model 1918.	Eastman Kodak Co., Rochester, N. Y.	42,607		
Sights, telescopic, 37-mm. Infantry gun.	Central Scientific Co., Chicago, Ill	4,100	142	578
Sights, telescopic, for 37-mm. Infantry gun.	Universal Optical Co., Providence, R. I.	1,225		
Sights, telescopic, for 37-mm. gun.	Globe Optical Co., Boston, Mass	50	50	50
Sights, optics, clinometer, for 37-mm. gun.	American Optical Co., South Bridge, Mass.	1,692	910	1,692
Sights, telescopic, for 37-mm. tank gun.	Burke & James Co., Chicago, Ill	6,576	50	386
Sights, optics for telescopic, for 37-mm. gun.	American Optical Co., South Bridge, Mass.	784	784	784
Sights, quadrant, for 37-mm. gun.	Scientific Materials Co., Pittsburgh, Pa.	3,192	600	1,207
Sights for 75-mm. gun	Electric Auto-Lite Corporation, Toledo, Ohio.	2,632	221	1,100
Do	Standard Thermometer Co., Boston, Mass.	2,000		
Sights, master, for 75-mm. gun	Electric Auto-Lite Corporation, Toledo, Ohio.	820		7
Do	Standard Thermometer Co., Boston, Mass.	410		26

Material.	Firm.	Total ordered.	Deliveries to—	
			Nov. 11, 1918.	Feb. 20, 1919.
Sights, optics for model 1901, for 75-mm. gun.	Globe Optical Co., Rochester, N. Y.	2,632	385	1,500
Sights, model 1918, for 75-mm. gun.	Ansco Co., Binghamton, N. Y.	3,142		
Sights, shanks for telescopic, model 1918, for 75-mm. gun.	American Standard Motion-Picture Machine Co., New York.	2,178		
Sights, for 3-inch gun, model 1916.	Peerless Printing Press Co., Palmyra, N. Y.	1,456	455	591
Sights, peep, for 3-inch gun	Standard Thermometer Co., Boston, Mass.	2,000	900	1,600
Sights for 3-inch gun	Frankford Arsenal, Philadelphia, Pa.	366	366	366
Sights, model 1916, for 3.8-inch howitzer carriage.	do	40	40	40
Sights, peep, for Schneider quadrants.	Electro Auto-Lite Corporation, Toledo, Ohio.	2,632	96	960
Sights, peep, for 4.7-inch gun	do	720		24
Sights, for 4.7-inch gun	Carlson-Wenstrom Co., Philadelphia, Pa.	286	70	126
Do	Emerson Engineering Co., Philadelphia, Pa.	500		125
Sights for 5-inch improvised gun carriage.	Blair Tool & Machine Co., New York City.	26		
Sights for 6-inch improvised gun carriage.	do	143		
Sights, dial, 8-inch howitzer	Arthur Brock, jr., Philadelphia, Pa.	75		
Sights, clinometer, 8-inch howitzer.	do	75		
Rocking bar, 8-inch	do	75		
Sights, lens for master, for 75-mm. gun.	Central Scientific Co., Chicago, Ill.	615		615
Sitogoniometer	Martin Copeland Co., Providence, R. I.	5,100		5,100
Squares, zinc	Metallograph Corporation, New York.	13,551	13,551	13,551
Squares, zinc, for machine gun	Clapp Eastman Co., Cambridge, Mass.	12,752	456	12,752
Staffs, sighting	Colson Co., Elyria, Ohio	1,205	1,205	1,205
Staffs, Jacob's, for field glass supports.	McFarlan Motor Co., Connersville, Ind.	15,745		15,745
Tapes, steel, 5 feet	Justus Roe & Sons, Patchogue, N. Y.	50,000		50,000
Do	Lufkin Rule Co., Saginaw, Mich.	31,791	31,791	31,791
Tapes, steel, 60 feet	do	4,250	4,250	4,250
Tapes, steel	do	1,422		1,422
Tapes, metallic linen	do	10,441	5,608	8,988
Telescopes, azimuth instrument, model 1918.	Spencer Lens Co., Buffalo, N. Y.	1,579		
Telescopes, battery commander's.	Bausch & Lomb Optical Co., Rochester, N. Y.	6,428	2,820	3,698
Do	Arthur Brock, jr., Philadelphia, Pa.	2,029		
Do	Central Scientific Co., Chicago, Ill.	2,000		
Do	Frankford Arsenal, Philadelphia, Pa.	52	52	52
Telescopes, battery commander's, tripod.	Naitonal Cash Register Co., Dayton, Ohio.	15,730	9,858	15,730
Telescopes for panoramic 4 and 10 power.	Recording & Computing Machines Co., Dayton, Ohio.	217	41	50
Telescopes, periscopic	Keuffel & Esser Co., Hoboken, N. J.	1,579		
Tripods for machine-gun sights	Herschede Hall Clock Co., Cincinnati, Ohio.	7,854		

CHAPTER VII.

MOTORIZED ARTILLERY.

Complete motorization of field artillery and its ammunition supply is almost certain to be one of the far-reaching and highly important results of our country's experiences in its participation in the war.

Practically all field artillery was of the horse-drawn type previous to our entry into the war, but with the evolution and perfection of the heavier siege artillery, 5-ton, 10-ton, and even heavier, traction engines were brought into play as means of motive power for the big guns and howitzers, with such success that the horse in the field artillery operations was being supplanted to a large degree by mechanical power.

Strictly speaking, the foundation for this departure had been laid before 1917, in the Mexican campaign of 1916 and in experiments that had been conducted at the Rock Island Arsenal. Insufficiency of funds, however, had prevented the experiments from being either thorough or extensive.

A consideration of the difficulties that vehicles of all sorts had to contend with in the battle areas of Europe made it evident at the outset that two general types of motor carriers would be required by the Army so far as ordnance was concerned—one type for far-advanced work, for hauling artillery over the worst possible kind of shell-torn and water-soaked earth, and the other for bringing up ammunition, supplies, equipment for repairs and the like in less advanced zones and areas, but over roads and country that had been cut and hacked and made almost impassible by the activities of the contending forces.

TRUCKS.

The standard four-wheel-drive commercial trucks, modified to meet the special needs of the service, were adopted immediately after war began, while experimental work was put under way to develop a standard type that would set this country far in advance of all others in this line of activity.

A total of 30,072 of the four-wheel-drive trucks was ordered, and before the armistice 12,498 of this number had been completed, while 23,499 had been turned out by the 31st of January, 1919.

In round numbers, 25,000 of these trucks were to be equipped with bodies for the hauling of ammunition, and the balance with special

ARTILLERY SUPPLY TRUCK ON F. W. D. CHASSIS.

This truck was designed with special bodies and loads for varied classes of artillery supply, and the bodies could be mounted on either Nash or F. W. D. chassis.

ARTILLERY REPAIR TRUCK ON F. W. D. CHASSIS.

This is another special body equipped with suitable machinery and tools for minor repairs and capable of being mounted on either F. W. D. or Nash four-wheel-drive chassis.

AMMUNITION TRUCK MOUNTED ON STANDARD ORDNANCE FOUR WHEEL DRIVE TRACTOR.

This is the vehicle, designed by the Ordnance Department and civilian experts, that was intended to supersede both the Nash and the F. W. D. and become the standard Army wheeled tractor. It is shown here with standard ammunition body mounted thereon.

ORDNANCE EQUIPMENT REPAIR TRUCK ON F. W. D. CHASSIS.

Special body, carrying tools and machinery for doing repair work to harness, personal equipment, etc., and capable of being mounted on either F. W. D. or Nash four-wheel-drive chassis.

bodies and equipment suitable for artillery supply and repair, for repair of equipment, and for heavy mobile ordnance.

Special bodies were manufactured by these concerns:

American Car & Foundry Co., Berwick, Pa.
J. G. Brill Co., Philadelphia, Pa.
Hale & Kilburn Corporation, Philadelphia, Pa.
Dumbar Manufacturing Co., Chicago, Ill.
Pullman Co., Pullman, Ill.
Kuhlman Car Co., Cleveland, Ohio.
C. R. Wilson Body Co., Detroit, Mich.
Insley Manufacturing Co., Indianapolis, Ind.
Lang Body Co., Cleveland, Ohio.
Heil Co., Milwaukee, Wis.
Variety Manufacturing Co., Indianapolis, Ind.
J. E. Bolles Iron & Wire Co., Detroit, Mich.

The first contract for these trucks was placed on August 18, 1917, and 9,420 were shipped to the American Expeditionary Forces overseas by the date of the armistice.

It required considerable time to work out and perfect all the details of the special bodies and equipment, as most of these were exceedingly complicated, and in a number of cases there were as many as 700 items of equipment on a single truck.

Representatives of the allied governments were not hesitant in asserting that the line of artillery repair trucks developed for our Army was the most complete and well worked out in detail that any army ever received.

These manufacturers did the work of turning out the special trucks:

Nash Motors Co., Kenosha, Wis.
Four-Wheel-Drive Auto Co., Clintonville, Wis.
Mitchell Motor Car Co., Racine, Wis.
Premier Motor Corporation, Indianapolis, Ind.
Kissel Motor Car Co., Hartford, Wis.
Hudson Motor Car Co., Detroit, Mich.
National Motor Car Co., Indianapolis, Ind.
Paige Motor Car Co., Detroit, Mich.
Commerce Motor Car Corporation, Detroit, Mich.
White Co., Cleveland, Ohio.
Dodge Motor Car Co., Detroit, Mich.

About 4,000 of the 5,000 special body type of trucks were delivered before the middle of December, 1918.

TRAILERS.

There were developed five different types of four-wheeled trailers. Each type, being for a particular use, required a special study and individual design, with all the consequent specially prepared machines and specialized shop work.

For antiaircraft service, a 1½-ton and a 3-ton trailer were worked out; for the 75-millimeter field gun, a special 3-ton trailer; for the mobile repair shops, a 4-ton trailer; and for the small tank, a special 10-ton trailer.

By the middle of December, 2,157 of these trailers had been delivered of the 4,847 that had been ordered and put in production.

Concerns engaged in turning out trailers were:

Sechler & Co., Cincinnati, Ohio.
Trailmobile Co. of America, Cincinnati, Ohio.
Ohio Trailer Co., Cleveland, Ohio.
Grant Motor Car Corporation, Cleveland, Ohio.

It might also be stated at this point, too, that two special types of passenger motor vehicles were designed and built. One of these was for staff observation and the other for reconnaissance. Nearly all of the total of 2,250 that were ordered of these two types were completed by mid-December, 1918, delivery of them having started in the month of April, 1918.

CATERPILLAR TRACTORS.

It was found after a comprehensive study of the needs of the various branches of ordnance and the requirement of the big guns that five sizes of caterpillar tractors would be required—of capacities of 2½ tons, 5 tons, 10 tons, 15 tons, and 20 tons.

Commercial types of machines of the 15-ton and 20-ton sort, with only slight alterations, were found to be suitable, but special designs were made for those of 2½-ton, 5-ton, and 10-ton capacity. Our experience in Mexico and the experiments at the Rock Island Arsenal had taught us the need of the special designs of machines of those sizes.

In all, 24,791 of these five types of caterpillar tractors were ordered. The 5-ton machine reached production in the summer of 1918 and the 2½-ton machine in the fall. By the end of the following January, 5,940 of the tractors had been delivered. Manufacturers who had orders for the caterpillar tractors were:

Holt Manufacturing Co., Peoria, Ill.
Chandler Motor Car Co., Cleveland, Ohio.
Reo Motor Car Co., Lansing, Mich.
Maxwell Motor Car Co., Detroit, Mich.
Federal Motor Truck Co., Detroit, Mich.
Interstate Motor Co., Indianapolis, Ind.

Throughout the production of tractors during the war period there was continuous and persistent experimentation, and satisfactory solutions of many of the problems were being reached at the time of the signing of the armistice.

Self-propelled caterpillar gun mounts were the subject of the most important of these experiments. The self-propelled caterpillar gun mounts differ from the ordinary caterpillar tractors in that they have the guns mounted directly on them, the guns forming an integral part of the entire machine. Six types of these were being developed, and 270 had been ordered when the armistice came.

A 2½-ton tractor mounting a 75-millimeter gun and a 5-ton tractor containing a gun of the same size were far along the road to success in their first state of development.

Development of caterpillar cargo carriers or caissons for bearing supplies over any sort of terrain, no matter how rough the going might be and regardless of whether there were roads or not, was so far along the pathway of success that two sizes were about to go into production on November 11.

A 2½-ton ammunition trailer, a 2-ton 11-inch trench mortar trailer, and a 4.7-inch antiaircraft gun trailer were also in development, but not in production, at the time of the signing of the armistice.

So successful were the experiments with new types of four-wheel-drive trucks and tractors that orders for what would probably have proven the best type of four-wheel-drive truck and the best type of four-wheel-drive tractor ever produced had been placed, but the signing of the armistice forced cancellations of these orders. In the course of the experiments, all types of American four-wheel-drive vehicles were examined and two of the best French types.

The purchase of $365,000,000 worth of trucks, trailers, and tractors was obligated in about 3,000 separate orders.

SELF-PROPELLED CATERPILLAR GUN MOUNTS.

In Europe, the French had been the only people to experiment with caterpillar mounts for guns. They produced the St. Chamond type, but this had not gone far beyond the experimental stages.

Prior to the early months of 1918, our own efforts along this line consisted in the building of one caterpillar mount, self-propelled by a gasoline engine and carrying an antiaircraft gun. Around this nucleus an ambitious caterpillar program was built.

An 8-inch howitzer was placed on this antiaircraft caterpillar mount and fired at angles of elevation varying up to 45°. Maneuvered over difficult ground, the machine withstood the firing strains and road tests in a highly satisfactory manner.

As a result of the success of these tests, orders were placed for three more experimental caterpillars to mount 8-inch howitzers. Tests of two of these completed units were so gratifying that it was felt they warranted quantity production. Accordingly, orders were placed for 50 units of the 8-inch howitzer caterpillars to cost about $30,000 apiece, for 50 caterpillar units mounting 155-millimeter guns, and for 250 units mounting 240-millimeter howitzers.

The Standard Steel Car Co., Hammond, Ind., was to produce the 240-millimeter howitzer caterpillars, the Harrisburg Manufacturing & Boiler Co., Harrisburg, Pa., was to turn out the 8-inch howitzer caterpillars, and the Morgan Engineering Co., of Alliance, Ohio, was to produce the 155-millimeter gun caterpillars.

Mountings for the 8-inch howitzer and 155-millimeter gun were practically identical. Both utilized many of the standard Holt caterpillar parts. The only real change was in the carriage for the 155-millimeter gun. This was made sufficiently sturdy to carry higher-powered guns. A 194-millimeter gun is now being machined in France, and when finished it will be shipped to this country to be mounted upon the 155-millimeter caterpillar mount for experiment.

The 240-millimeter howitzer mounts were of two types—one following closely the St. Chamond type of the French and the other being a self-contained unit designed by Ordnance Department engineers. The self-contained type is a single unit that mounts both the power plant and the howitzer and for which it is necessary to provide additional cargo-carrying caterpillars to haul ammunition and fuel. Two units make up the St. Chamond type. One mounts the gun and electric motors; the other, a limber, mounts the power plant and carries ammunition.

In the battle area the St. Chamond type had the peculiar advantage that the power-plant unit could be run to shelter and be available for a rapid advance or change of location of the gun mount as the situation might demand. With the self-contained unit a direct hit by the enemy would put both gun and power plant out of commission.

Contracts for the caterpillar mounts called for the completion of the entire program not later than February, 1919. All the firms engaged on the work of production were putting forth every effort when the armistice was signed and there was every reason to believe deliveries would be as scheduled. The termination of hostilities caused all contracts to be reduced. Provisions have been made for only enough caterpillars of each type to provide for further experimental work.

Twenty mounts equipped with caterpillar treads and mounting 7-inch Navy rifles were built by the Baldwin Locomotive Co. for the Navy Department. These were so successfully operated that orders were placed for 36 similar units for the use of the Army, but since the signing of the armistice this order has been cut to 18.

The great gun on a caterpillar mount fires its death-dealing projectile, and almost before the shot has reached its destination the caterpillar mount has moved the gun to another point. With motor still running the gun is fired again and once more quickly moved

TEN-TON ARTILLERY TRACTOR.

FIFTEEN-TON ARTILLERY TRACTOR.

TWO AND ONE-HALF TON ARTILLERY TRACTOR.

FIVE-TON ARTILLERY TRACTOR.

20-TON ARTILLERY TRACTOR.

STAFF OBSERVATION CAR.

Special body for field touring on a White one-ton truck chassis.

RECONNAISSANCE CAR ON WHITE CHASSIS.

The machine-gun truck is similar except for the addition of gun racks under rear seat and on Commerce chassis.

CATERPILLAR TRACTOR WITH 3-INCH GUN MOUNTED ON IT.

ANOTHER VIEW OF CATERPILLAR TRACTOR WITH 3-INCH GUN.

CATERPILLAR TRACTOR MOUNTING AN 8-INCH HOWITZER WHICH HAS A RANGE OF 6 MILES WITH A PROJECTILE WEIGHING 200 POUNDS.

LIGHT REPAIR TRUCK ON DODGE CHASSIS.

Special body with tools for making minor motor repairs.

THREE-INCH FIELD GUN TRAILER.

A specially designed vehicle for carrying different loads, including a 3-inch field-gun carriage and limber in one load and two 3-inch field-gun caissons for another load.

4-TON SHOP TRAILER CHASSIS.

3-INCH ANTIAIRCRAFT GUN TRAILER.

240-MILLI METER TRENCH MORTAR TRUCK.

on to another location, so that the enemy's artillery is unable to get its range.

Ordnance motor production table.

TRACTORS.

Size.	Quantity ordered.	Quantity accepted Nov. 11, 1918.	Quantity accepted Jan. 31, 1919.	Floated to Nov. 11, 1918.
2½-ton	5,586	10	25	2
5-ton	11,150	1,543	3,480	459
10-ton	6,623	1,421	2,014	628
15-ton	267	267	267	232
20-ton	1,165	126	154	81
TRAILERS.				
1½-ton antiaircraft machine gun	2,289	150	562	126
3-inch field gun	830	235	472	15
4-ton shop bodies	576	101	384	12
4-ton shop chassis	576	260	555	....
10-ton	540	104	245	1
3-inch antiaircraft	612	543	611	109
TRUCKS.				
F. W. D. chassis	13,907	5,361	10,615	3,561
Nash chassis	16,165	7,137	12,884	5,859
Ammunition bodies	24,729	18,212	21,709	....
Ammunition mountings	24,729	9,615	11,024	6,955
Artillery repair	1,332	1,318	1,332	350
Artillery supply	5,474	813	1,838	444
Light repair	1,012	1,012	1,012	362
Dodge chassis	1,012	1,012	1,012	436
Commerce chassis	1,500	1,500	1,500	24
Machine-gun body, mounted on Commerce or White 1-ton chassis	1,500	486	1,306	241
1-ton supply	60	60	60	55
White chassis	2,695	1,929	2,695	575
Reconnaissance	1,081	712	1,003	320
Staff observation	1,175	1,164	1,175	189
Equipment repair	310	310	310	121
H. M. R. S. trucks	624	287	416	12

CHAPTER VIII.

TANKS.

The tank, more than any other weapon born of the great war, may be called the joint enterprise of the three principal powers arrayed against Germany—America, France, and Great Britain. An American produced the fundamental invention, the caterpillar traction device, which enables the fortress to move. A Frenchman took the idea from this and evolved the tank as an engine of war. The British first used the terrifying monster in actual fighting.

There is a common impression throughout America that the British Army invented the tank. The impression is wrong in two ways. The French government has recently awarded the ribbon of the Legion of Honor to the French ordnance officer who is officially hailed as the tank's inventor. His right to the honor, however, is disputed by a French civilian who possesses an impressive exhibition of drawings to prove that he and not the officer is the inventor. As this is written a lively controversy over the point is in progress in France. Wherever the credit for the invention belongs, the French were first to build tanks, building them only experimentally, however, and not using them until after the British had demonstrated their effectiveness.

In the second place, it was not the British Army which adopted them first in England, but the British Navy. The tank as an idea shared the experience of many another war invention in being skeptically received by the conservative experts. The British Navy, indeed, produced the first ones in England; but to the British Army goes the glory of having first used them in actual fighting and of establishing them in the forefront of modern offensive weapons.

Brought forth as a surprise, the tanks made an effective début in the great British drive for Cambrai. Later the enemy affected to scoff at their usefulness. The closing months of the tanks' brief history, however, found them in greater favor than ever, and they were used by both sides in increasing numbers.

Up to the beginning of the summer of 1917 there was little accurate information in this country regarding the tanks. Somewhat hazy specifications then began to come from Europe about the designs of the different tanks at that particular time in use on the battle front, but these specifications were exceedingly rough and sketchy, consisting in the main of merely the fact that the machines should be able to

THREE-TON TANK.

Weight, 5,800 pounds; crew, two men (one gunner, one driver); power plant, two Ford motors, geared together, each motor driving one track; speed, nine miles per hour; climbing ability, 45°

SIX-TON TANK.

This machine is practically a copy of the French Renault tank and carries two men (one driver, one gunner). About half of these tanks were equipped with 37-millimeter cannon and about half with machine guns. Certain of these tanks also made with wireless apparatus substituted for the turret of the fighting tanks. Power plant, one Buda 4-cylinder motor; speed, five to six miles per hour; grade capacity, 45°; weight, 15,000 pounds.

cross trenches about 6 feet wide, that each should carry one heavy gun and two or three machine guns, and that their protection should consist of armor plate about five-eighths of an inch thick.

With these facts as a guide, two experimental machines were decided upon, and work on them was begun immediately. With these machines it was determined to test the relative advantages of a specially articulated form of caterpillar tractor with wheeled traction, making use of very large wheels, and to develop the possibilities between the gas-electric and steam systems of propulsion.

In September, 1917, decision had been made to supply the American Army with two types of tanks—one the large size, typical of that used by the British and capable of containing a dozen men, and the other a smaller one patterned after the French two-man model and known as the Renault. In September one of our officers charged with tank production was dispatched to Europe for a more intimate study of the machines used abroad and for the purpose of getting more detailed information respecting the merits of the various types of tanks, as well as to make arrangements for sending specimens here.

The decision to equip the American forces in Europe with tanks of two sizes was made only after thorough and somewhat protracted conferences with British, French, and American officers in Europe. Complete drawings and samples of the small tank were obtained from the French and shipped to this country. As all of the drawings were made in accordance with the metric system of measurements, it was necessary before anything could be done toward actual production to remake the drawings, as the machine shops here were not equipped to use the metric system.

The large British tank had been successful in its operations on the battle front, but its very decided limitations, recognized by British authorities, caused our officers to think it best to redesign the large tank in preference to copying the existing big British tank with its limitations.

General "fighting" specifications for the big tank were laid down by the British general staff at the conference at British headquarters at which American officers were present. It was agreed that this big tank, known as the Mark VIII, should be of Anglo-American design and construction. Arrangements were made for producing 1,500 of this type. To do this, Great Britain and the United States entered into a working agreement that provided for England to furnish the hulls, guns, and ammunition, while the United States was to furnish the power plant and driving details of the monster. Roughly speaking, each tank would cost about $35,000, of which $15,000 represented the American part of the job, on which some 72 contractors were at once engaged. About 50 per cent of the work on these tanks had been completed when the armistice was signed, and the first units were undergoing trials.

It was confidently expected that all of the 1,500 contracted for would have been completed by March, 1919. While these Anglo-American tanks were in the process of construction there were also being built here 1,450 all-American tanks of the large English type, and for this all-American tank 50 per cent of the work had also been done at the signing of the armistice.

In December, 1917, a sample French tank of the Renault type reached this country along with detailed drawings and a French engineer. Much difficulty then ensued in getting American concerns to take on production of this machine, because of the difficult nature of its manufacture. Considerable time, too, was taken up in changing the drawings from the French metric dimensions to the American dimensions, and this involved redesigning many parts.

In the manufacture of the armor built for the Renault type of tank the French made no attempt to adhere to simple shapes, and for this reason practically a new source of supply for this kind of armor had to be developed. Contracts for 4,440 of the Renault type of tanks were finally made. The approximate cost of each one of these machines was $11,500. Manufacturing activities for the various parts had to be divided up among more than a score of plants, so that many plants were turning out parts for these machines, while the assembling was done at only three plants, which also made a portion of the parts.

The three assembly plants were the Van Dorn Iron Works, of Cleveland, Ohio; the Maxwell Motors Co., of Dayton, Ohio; and the C. L. Best Co., also of Dayton.

Finished machines of this type started to come through in October. When the armistice was signed 64 of these 6-ton Renault tanks, each designed to carry two men and a machine gun, were completed, while up to the end of December the number of those finished amounted to 209, with 289 in the process of assembly. There is every reason to believe that had the armistice not been signed, the entire original program would have been completed by April.

During the summer and fall of 1918 our tank program had been augmented by the development of two entirely new types of tanks. One was a two-man tank weighing 3 tons, built by the Ford Motor Co. and costing in the neighborhood of $4,000. This tank, mounting one machine gun, has a speed of about 8 miles an hour. Of this type 15 had been built; and, up to the 1st of January, 1919, 500 were to have been finished, after which they were to have been turned out by the Ford Co. at the rate of 100 a day.

The other new tank developed was a successor to the French Renault, designed for production in great volume. This tank was

SIX-TON TANK BEING MOUNTED ON A 10-TON TRAILER.

CHAPTER IX.

MACHINE GUNS.

The machine gun is typically and historically an American device. An American invented the first real machine gun ever produced. Another American, who had taken British citizenship, produced the first weapon of this type that could be called a success in war. Still a third American gave to the allies at the beginning of the great war a machine gun which revolutionized the world's conception of what that weapon might be; while a fourth American inventor, backed by our Ordnance Department, enabled the American forces to take into the field in France what is probably the most efficient machine gun ever put into action.

The machine gun as an idea is not modern at all. The thought has been engaging the attention of inventors for several centuries. The idea was inherent in guns which existed in the seventeenth and eighteenth centuries, but they should be called rapid-fire guns rather than machine guns, since no machine principle entered into their construction. They usually consisted of several gun barrels bound together and fired simultaneously.

The first true machine gun was the invention of Richard Jordon Gatling, an American, who in 1861 brought out what might be termed a revolving rifle. The barrels, from 4 to 10 in number, were placed parallel to each other and arranged on a common axis about which they revolved in such a manner that each barrel was brought in succession into the firing position. This gun was used to some extent in our Civil War and later in the Franco-Prussian War.

In 1866 Reffye, a French inventor, brought out the first mitrailleuse—a mounted machine gun of the Gatling type towing a limber and drawn by four horses. It had 25 rifled barrels and could fire 125 shots per minute. The weapon, however, during the Franco-Prussian War, turned out to be a failure for the reason that it proved an excellent target for the enemy's artillery and was not sufficiently mobile. Accordingly the French government abandoned it.

Sir Hiram S. Maxim, who was American born, in 1884 developed a machine gun which operated automatically by utilizing the force of the recoil. This gun was perfected and became a serviceable weapon for the British army in the Boer War. The Maxim gun barrel was cooled by the water-jacket system. When the water became hot it exhausted a jet of steam which could be seen for long distances across the South African veldt, making it a mark for the Boer sharpshooters. This defect was remedied in homemade fashion by carrying the exhaust steam through a hose into a bucket of water where it was condensed. This Maxim gun fired 500 shots a minute.

Meanwhile in this country the Gatling gun had been so improved that it became one of our standard weapons in the Spanish-American War. Later on it was used in the Russo-Japanese War.

The Colt machine gun also existed in 1898. This was the invention of John M. Browning, whose name has been prominently associated with the development of automatic firearms for the last quarter of a century.

In England the Maxim gun was taken up by the Vickers Co., eventually becoming what is known to-day as the Vickers gun. In 1903 or 1904 the American Government bought some Maxim machine guns which were then being manufactured by the Colt Co. at Hartford, Conn.

In no war previous to the one concluded in 1918 did the machine gun take a prominent place in the armaments of contending forces. The popularity of the earlier machine guns was retarded by their great weight. Some of them were so heavy that it took several men to lift them. All through the history of the development of machine guns the tendency has been toward lighter weapons, but it was not until the great war that serviceable machine guns were made light enough to give them great effectiveness and popularity. Such intense heat is developed by the rapid fire of a machine gun that unless the barrel can be kept cool the gun will soon refuse to function. The water jacket which keeps the gun cool proved to be the principal handicap to the inventors who were trying to remove weight from the device. The earliest air-cooled guns were generally unsuccessful, since the firing of a few rounds would make the barrel so hot that the cartridges would explode voluntarily in the chamber, thus rendering the weapon unsafe. The Benét-Mercié partly overcame this difficulty by having interchangeable barrels. As soon as one barrel became hot it could be quickly removed and its cool alternate inserted in its place.

These conditions led to the development of machine guns along two separate lines—the heavy type machine gun, which must be capable of long sustained fire, and the automatic rifle, whose primary requisite is extreme lightness. These requirements brought the ultimate elimination from ground use in France and in the United States of guns of the so-called intermediate weight as being incapable of fulfilling either of the above requirements to the fullest degree.

The machine gun produced by the American inventor, Col. I. N. Lewis, was a revelation when it came to the aid of the allies early in the great war. This was an air-cooled gun which could be fired for a considerable time without excessive heating, and it weighed only 25 pounds, no great burden for a soldier. The Lewis machine gun was hailed by many as the greatest invention brought into prominence by the war, although its weight put it in the intermediate class, with limitations as noted above.

Along in the first decade of the present century the Benét-Mercié automatic machine rifle was developed. This was an air-cooled gun of the automatic rifle type and weighed 30 pounds. Light as this gun was, it was still too heavy to be of great service as an automatic rifle, since a strong man would soon tire of holding 30 pounds up to his shoulder, and it was therefore in the intermediate class.

The Germans had apparently realized better than anyone else the value of machine guns in the kind of fighting which they expected to be engaged in, and therefore supplied them to their troops in greater numbers than did the other powers, having, an early report stated, 50,000 Maxim machine guns at the commencement of hostilities. The Austrian Army had adopted an excellent heavy type machine gun known as the Schwarzlose whose chief feature lay in the fact that it operated with only one major spring.

Such was the machine-gun situation, although incompletely set forth here, at the beginning of the great war. The nations, with the exception of Germany, had been slow to promote machine gunnery as a conspicuous phase of their military preparedness. In our Army we had a provisional machine-gun organization, but no special officers and few enthusiasts for machine guns. We were content with a theoretical equipment of four machine guns per regiment. The fact was that in no previous war had the machine gun demonstrated its tactical value. The chief utility of the weapon was supposed to lie in its police effectiveness in putting down mobs and civil disorders and in its value in other special situations, particularly defensive ones.

The three years of fighting in Europe before the United States was drawn in had demonstrated the highly important place which the machine gun held in modern tactics. Because of the danger of our position we had investigated many phases of armed preparedness, and in this investigation numerous questions had arisen regarding machine guns. The Secretary of War had appointed a board of five Army officers and two civilians to study the machine-gun subject, to recommend the types of guns to be adopted, the number of guns we should have per unit of troops, how these guns should be transported, and other matters pertaining to the subject. Six months before we declared war this board submitted a report strongly recommending the previously adopted Vickers machine gun and the immediate procurement of 4,600 of them. In December, 1916, the War

Department acted on this report by contracting for 4,000 Vickers machine guns from the Colt Co. in addition to 125 previously ordered.

The Vickers gun belongs to what is known as the heavy type of machine gun. The board found that the tests it had witnessed did not then warrant the adoption of a light-type machine gun, although the Lewis gun of the intermediate type was then being manufactured in this country. The board, however, recommended that we conduct further competitive tests of machine guns at the Springfield Armory, in Massachusetts, these tests to begin May 1, 1917, the interval being given to permit inventors and manufacturers to prepare equipment for the competition.

The war came to us before these tests were made. On the 6th day of April, 1917, our equipment included 670 Benét-Mercié machine rifles, 282 Maxim machine guns of the 1904 model, 353 Lewis machine guns, and 148 Colt machine guns. The Lewis guns, however, were chambered for the .303 British ammunition and would not take our service cartridges.

Moreover, the manufacturing facilities for machine guns in this country were much more limited in extent than the public had any notion of then or to-day. Both England and France had depended mainly upon their own manufacturing facilities for their machine guns, the weapons which they secured on order from the United States being supplementary and subsidiary to their own supplies. We had at the outbreak of the war only two factories in the United States which were actually producing machine guns in any quantity at all. These were the Savage Arms Corporation, which in its factory at Utica, N. Y., was nearing the completion of an order for about 12,500 Lewis guns for the British and Canadian Governments, and the Marlin-Rockwell Corporation, which had manufactured a large number of Colt machine guns of the old lever type for the Russian Government. The Colt factory in the spring of 1917 was equipping itself with machinery to produce the 4,125 Vickers guns, the order for 4,000 of which had been placed the previous December by the War Department on recommendation of the Machine Gun Board. None of these guns, however, had been completed when the United States entered the war. The Colt Co. also held a contract for Vickers guns to be produced for the Russian Government.

It was therefore evident that we should have to build up in the United States almost a completely new capacity for the production of machine guns. Nevertheless, we took advantage of what facilities were at hand; and at once, in fact within a week after the declaration of war, began placing orders for machine guns. The first of these orders came on April 12, when we placed a contract with the Savage Arms Corporation for 1,300 Lewis guns, which, as manufactured by that corporation, had by this time been overhauled in design and much improved. This order was subsequently heavily increased. On June 2 we placed an order with the Marlin-Rockwell Corporation for 2,500 Colt guns, these weapons to be used in the training of our machine-gun units.

In this connection the reader should bear continually in mind that throughout the development of machine-gun manufacture we utilized all existing facilities to the limit in addition to building up new sources of supply. In other words, whenever concerns were engaged in the manufacture of machine guns, whatever their make or type, we did not stop the production of these types in these plants and convert the establishments into factories for making other weapons; but we had them continue in the manufacture in which they were engaged, giving them orders which would enable them to expand their facilities in their particular lines of production. Then when it became necessary for us to find factories to build Browning guns and some of the other weapons on which we specialized, we found new capacity entirely for this additional production.

Since we sent to France the first American division of troops less than three months after the declaration of war, they were necessarily armed with the machine guns at hand, which in this case proved to be the Benét-Mercié machine rifles.

Meanwhile the development of machine guns in Europe had been going on at a rapid rate. The standard guns in use by the French Army were now the Hotchkiss heavy machine gun and the Chauchat light automatic rifle, both effective weapons. Upon the arrival of our first American division in France the French Government expressed its willingness to arm this division with Hotchkiss and Chauchat guns; and thereafter the French facilities proved to be sufficient to equip our troops with these weapons until our own manufacture came up to requirements.

The 1st of May, 1917, brought the tests recommended by the investigation board, these tests continuing throughout the month. To this competition were brought two newly developed weapons produced by the inventive genius of that veteran of small-arms manufacture, John M. Browning. Mr. Browning had been associated with the Army's development of automatic weapons for so many years that he was peculiarly fitted to produce a mechanism that could adapt itself to the quantity production which our forthcoming effort demanded. Both the Browning heavy machine gun and the Browning light automatic rifle which were put through these tests in May had been designed with the view of enormous production quickly attained, so that their simplicity of design was one of their chief merits. After the tests the board pronounced these weapons the most effective guns of their type known to the members. The

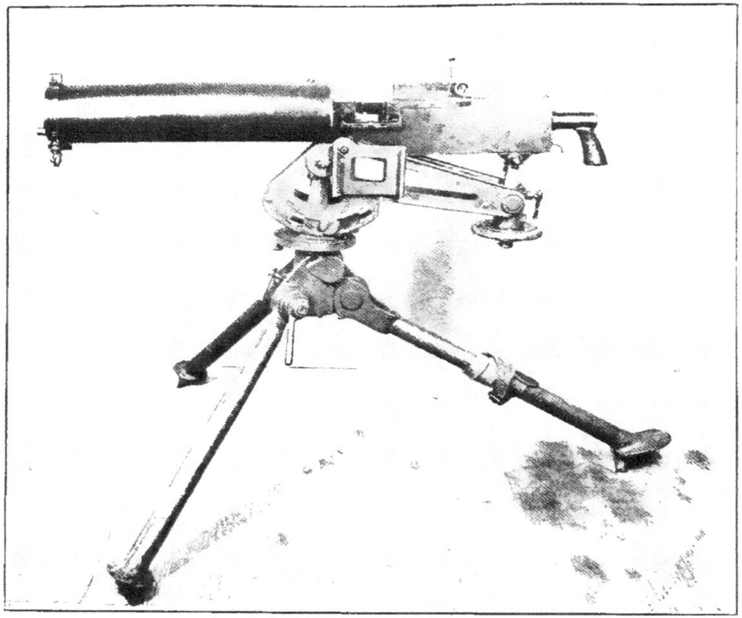

BROWNING MACHINE GUN, MODEL 1917.

MARLIN TANK MACHINE GUN.

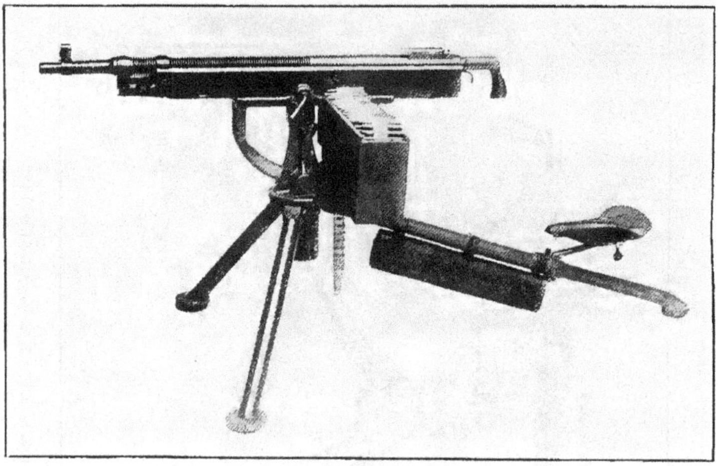

COLT MACHINE GUN, MODEL 1917, CALIBER .30.

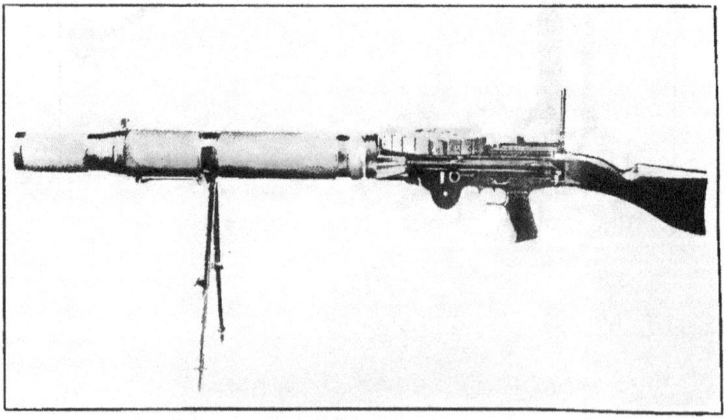

LEWIS MACHINE GUN, MODEL 1917, CALIBER .30.

Browning heavy gun with its water jacket filled weighs 36.75 pounds, whereas the Browning automatic rifle weighs only 15.5 pounds. These May tests also proved the Lewis machine gun to be highly efficient. The board recommended the production of large numbers of all three weapons; the two Brownings and the Lewis. The board also approved the Vickers gun, which weighs 37.50 pounds, and we accordingly continued it in manufacture.

The first act of the Ordnance Department after this report had been received was to increase greatly the orders for Lewis machine guns with the Savage Arms Corporation, and the second to make preparation for an enormous manufacture of Browning machine guns and Browning automatic rifles. Mr. Browning had developed these weapons at the plant of the Colt's Patent Firearms Manufacturing Co., of Hartford, Conn., which concern owned the exclusive rights to both these weapons under the Browning patents. This company at once began the development of manufacturing facilities for the production of Browning guns. In July, 1917, orders for 10,000 Browning machine guns and 12,000 Browning automatic rifles were placed with the Colt Co. It should be remembered that the Colt Co. was in the midst of preparations for the production of large numbers of Vickers machine guns; and the Government required that the Browning manufacture should be carried on without interference with the existing contracts for Vickers guns. This requirement necessitated an enormous expansion of the Colt plant to take care of its growing contracts for Browning guns. The concern prepared to make the Browning automatic rifle, the lighter gun, at a new factory at Meriden, Conn.

In its arrangements with the Colt Co. the Government recognized that its future demands for Browning guns would be far beyond the capacity of this one concern to supply. Consequently, for a royalty consideration, the Colt Co. surrendered for the duration of the war, its exclusive rights to manufacture these weapons, this arrangement being approved by the Council of National Defense. Mr. Browning, the inventor of the guns, was also compensated by the Government for weapons of his invention manufactured during the war. In the arrangement the Government acquired the right to manufacture during the period of the emergency all other inventions that might be developed by Mr. Browning—an important consideration, since at any time the inventor might add improvements to the original designs or bring out accessories that would add to the efficiency or effectiveness of the weapons.

It may also be added that throughout this period Mr. Browning's efforts were constantly directed toward the perfection of these guns and the development of new types of guns and accessories. His services along these lines were of great value to the War Department.

When these necessary preliminary matters had been settled the Ordnance Department made a survey of the manufacturing facilities of the United States to determine what factories could best be set to work to produce Browning guns and rifles, always with special care that no existing war contracts, either for the allies or for the United States, be disturbed.

By September this survey was complete, and also by this time we had definite knowledge of the rate of enlargement of our military forces and their requirements for machine guns. We were ready to adopt the program of machine-gun construction that would keep pace with our needs, no matter what numbers of troops we might equip for battle. As a foundation for the machine-gun program, in September, 1917, we placed the following orders: 15,000 water-cooled Browning machine guns with the Remington Arms-Union Metallic Cartridge Co., of Bridgeport, Conn.; 5,000 Browning aircraft machine guns with the Marlin-Rockwell Corporation, of New Haven, Conn.; and 20,000 Browning automatic rifles with the Marlin-Rockwell Corporation. In this connection it should be explained that the Browning aircraft gun is essentially the heavy Browning with the water-jacket removed. It was practicable to use it thus stripped, because in aircraft fighting a machine gun is not fired continuously, but only at intervals, and then in short bursts of fire too brief to heat a gun beyond the functioning point.

At the same time these orders were placed the Winchester Repeating Arms Co., of New Haven, Conn., was instructed to begin its preliminary work looking to the manufacture of Browning automatic rifles; and less than a month later, in October, an order for 25,000 of these weapons was placed with this concern. Then followed in December an additional order for 10,000 Browning aircraft guns to be manufactured by the Marlin-Rockwell Corporation. A contract for Browning aircraft guns was also given to the Remington Arms-Union Metallic Cartridge Co.

Before the year ended the enormous task of providing the special machinery for this practically new industry was well under way. The Hopkins & Allen factory, at Norwich, Conn., had been engaged upon a contract for military rifles for the Belgian government. Before this order was completed the Marlin-Rockwell Corporation took over the Hopkins & Allen plant and set it to producing parts for the light Browning automatic rifles. Even this concern, however, could not produce the parts in sufficient quantities for the Marlin-Rockwell order, and the latter concern accordingly acquired the Mayo Radiator factory, at New Haven, and equipped it with machine tools for the production of Browning automatic-rifle parts. Such expansion was merely typical of what went on in the other concerns engaged in our machine-gun production. Immense quantities of new machinery had to

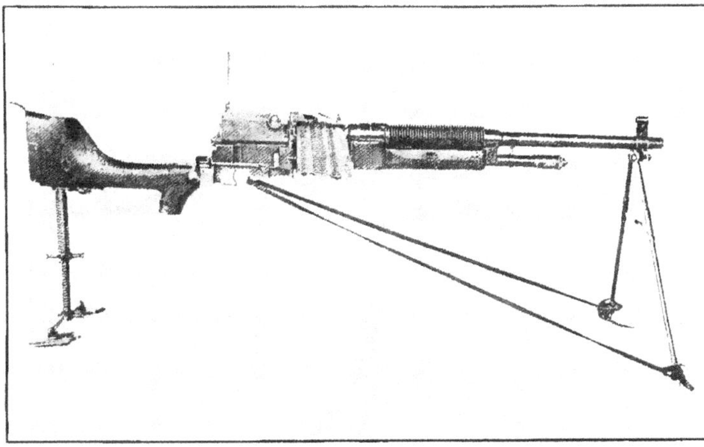

BENET-MERCIE MACHINE GUN.

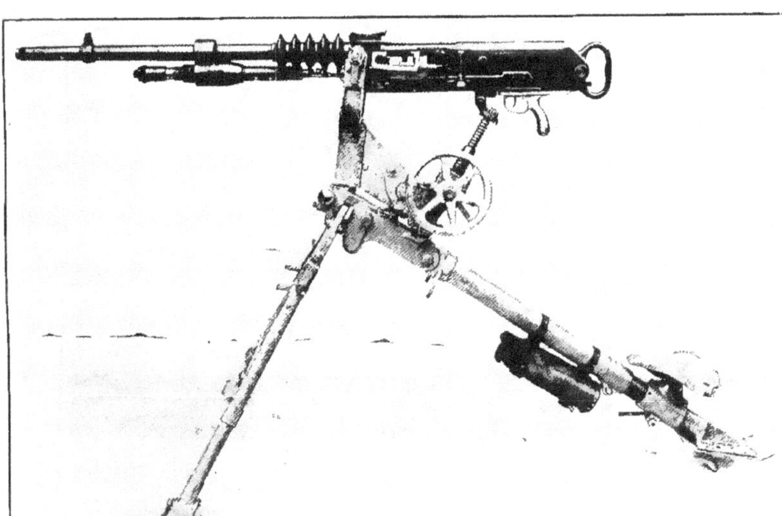

HOTCHKISS MACHINE GUN, MODEL 1914, 8-MILLIMETER.

This is the machine gun adopted by the French Army. This gun is of a heavy type, air-cooled, gas-operated, and fed from either a strip holding 30 cartridges or a metallic link belt. Its rate of fire is about 500 rounds per minute.

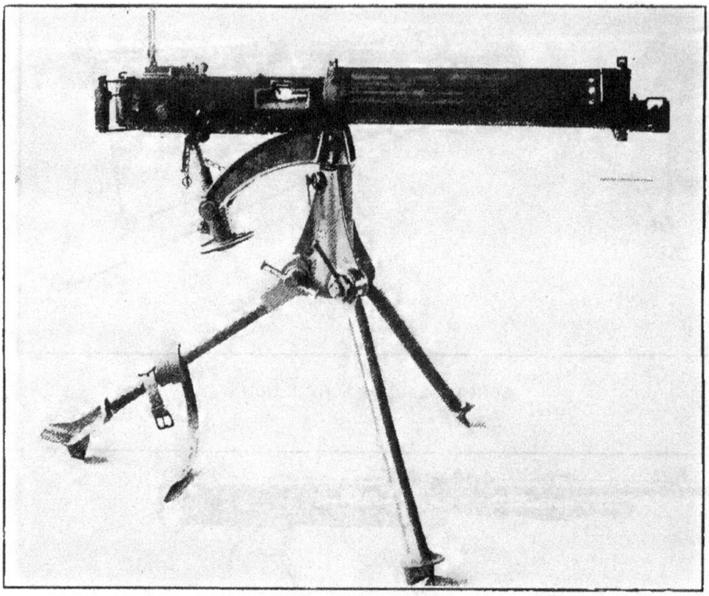

VICKERS MACHINE GUN, MODEL 1915, CALIBER .30.

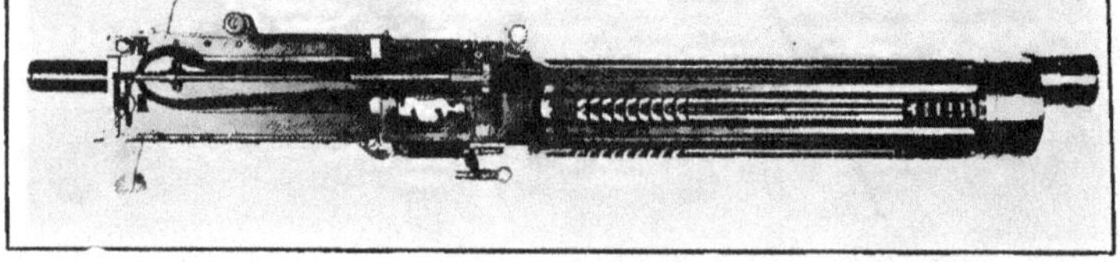

VICKERS AIRCRAFT MACHINE GUN, MODEL 1918, CALIBER .30.

be built and set up in all these factories. But still the Ordnance Department kept on expanding the machine-gun capacity. The New England Westinghouse Co., of Springfield, Mass., in January, 1918, completed a contract for rifles for the Russian government and was at once given an order for Browning water-cooled guns. For reasons which will be explained later, the original order for Browning aircraft guns, which had been placed with the Remington Arms Co., was later transferred to the New England Westinghouse Co. at their Springfield plant.

As soon as our officers in France could make an adequate study of our aircraft needs in machine guns, they discovered that in the three years of war only one weapon had met the requirements of the allies for a fixed machine gun that could be synchronized to fire through the whirling blades of an airplane propeller. This was the Vickers gun, which was already being manufactured in some quantity in our country, and for which three months before we entered the war we had given an order amounting to 4,000 weapons. On the other hand, the fighting aircraft of Europe were also finding an increased need for machine guns of the flexible type—that is, guns mounted on universal pivots, and which could be aimed and fired in any direction by the second man, or observer, in an airplane. The best gun we had for this purpose was the Lewis machine gun.

For technical reasons that need not be explained here, the Vickers gun was a difficult one to manufacture. The Colt Co., which was producing these weapons, in spite of their long experience in the manufacture of such arms and in spite of their utmost efforts, had been unable to deliver the finished Vickers guns on time, either to the Russian government or to this country. However, by expanding the facilities of this factory to the utmost, by the month of May, 1918, the concern achieved a production of over 50 Vickers guns per day. Doubtless, because of these same difficulties, neither the British nor the French governments had been able to procure Vickers guns as rapidly as they expanded the number of their fighting aircraft, and consequently when we entered the war we received at once a Macedonian cry from the allies to aid in equipping the allied aircraft with weapons of the Vickers type. An arrangement was readily reached in this matter. Our first troops in France needed machine guns for use on the lines. Our own factories had not yet begun the production of these weapons. Accordingly, in the fall of 1917, we arranged with the French high commissioner in this country to transfer 1,000 of our Vickers guns to the French air service, receiving in exchange French Hotchkiss machine guns for Gen. Pershing's troops.

Now while the demands of the allied service had brought forth only the Vickers machine gun as a satisfactorily synchronized weapon, we, shortly after our entry into the war, had succeeded in developing two additional types of machine guns which gave every promise of being satisfactory for use as fixed synchronized guns on airplanes. One of these, of course, was the heavy Browning gun, stripped of its water jacket; but because this was a new weapon, requiring an entirely new factory equipment for its production, the day when Brownings would begin firing at the German battle planes was remote, indeed, as time is reckoned in war.

On the other hand, our inventors had been improving a machine gun known as the Marlin, which was, in fact, the old Colt machine gun, Mr. Browning's original invention, but now of lighter construction and with a piston firing action instead of a lever control. In the face of considerable criticism at the time, we proposed to adapt this weapon to our aircraft needs as a stop-gap until Brownings were coming from the factory in satisfactory quantities. We took this course because we were prepared to turn out quantities of the Marlin guns in relatively quick time. As has been said, the Marlin resembled the Colt. The Marlin-Rockwell Corporation was already tooled up for a large production of Colt guns, and this machinery with slight modifications could be used to produce the Marlin.

We decided upon this course shortly after the declaration of war, and there followed a severe engineering and inventive task to develop a high-speed hammer mechanism and a trigger motor which would adapt the gun for use with the synchronizing mechanism. But then occurred one of those surprising successes that sometimes bless the efforts of harassed and hurried executives at their wits' end to meet the demand of some great emergency. The improvements added to the Marlin gun eventually transformed it in unforeseen fashion into an aircraft weapon of such efficiency that not only our own pilots but those of the French air forces as well were delighted with the result.

When it was proposed to adapt the Marlin gun for synchronized use on airplanes, the Ordnance Department detailed officers to cooperate with the Marlin company in its efforts. For technical reasons of design the original gun apparently had little or no adaptability to such use. Many new models were built only to be knocked to pieces after the failure of some feature to perform properly the work for which it was designed. Nevertheless the enthusiasm of the company for its project could not be chilled, and it continued the development until the gun finally became a triumph in gas-operated aircraft ordnance.

In the latter part of August we were using the Marlin gun at the front, and cablegram after cablegram told us of the surprisingly excellent performances of this weapon in actual service. It is

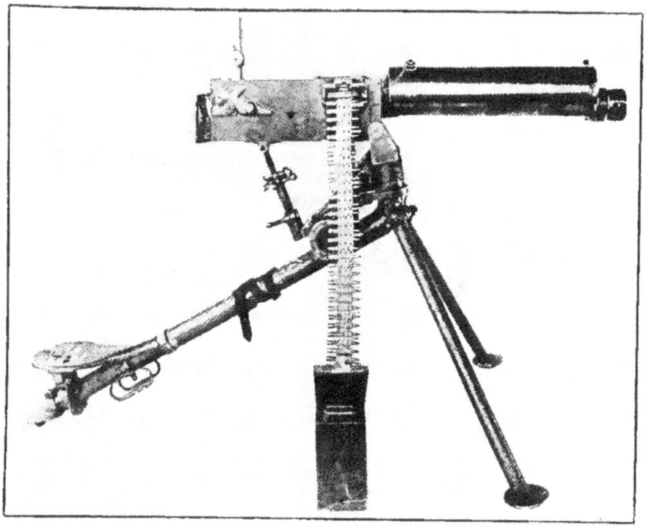

MAXIM MACHINE GUN AND TRIPOD (AMERICAN), MODEL 1904, CALIBER .30.

This was the first automatic machine gun to be developed. It is of the heavy type. recoil operated, water cooled, and belt fed. The gun is capable of sustained fire for long periods of time provided its water supply is properly maintained, and is adaptable to indirect barrage fire. It is used by the British and U. S. forces and in modified form by the Germans.

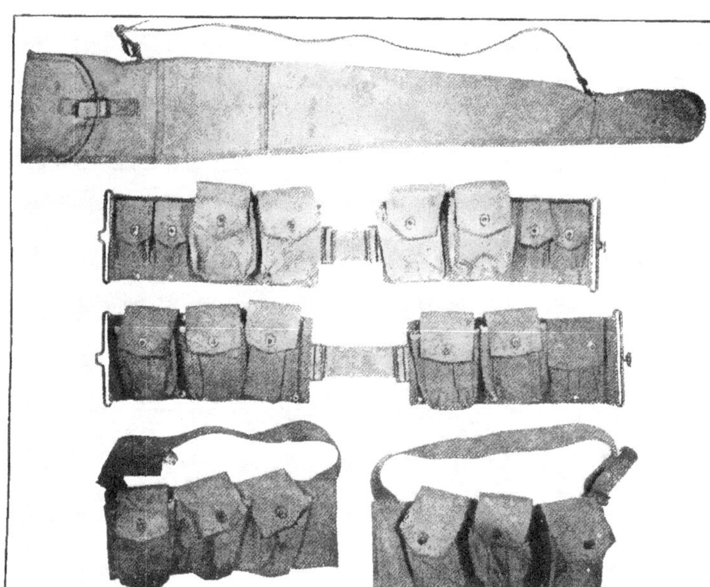

BROWNING AUTOMATIC RIFLE EQUIPMENT.

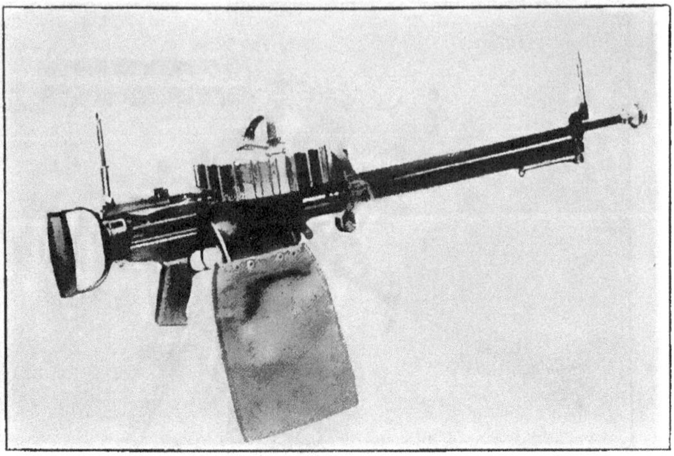

LEWIS AIRCRAFT MACHINE GUN, MODEL 1917, CALIBER .30.

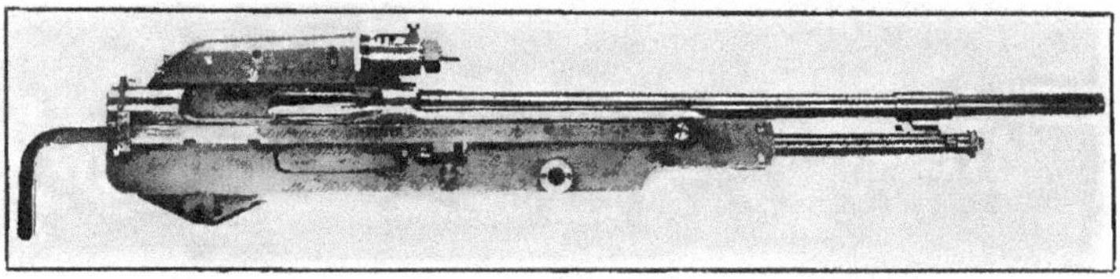

MARLIN AIRCRAFT MACHINE GUN, TYPE 8 M. G.

A fixed synchronized gun developed from the Colt gun solely for aircraft use. It is of the heavy type, gas operated, air cooled, and belt fed. It is the only gas-operated gun which has been successfully synchronized and has been found to give the closest grouping of shot in synchronized fire which has ever been obtained with any gun.

GERMAN MAXIM MACHINE GUN ON MOUNT.

sufficient here to quote one of these messages from Gen. Pershing, dated February 23, 1918:

> Marlin aircraft guns have been fired successfully on four trips 13,000, 15,000 feet altitude, and at temperature of minus 20° F. On one trip guns were completely covered ice. Both metallic links and fabric belts proved satisfactory.

(Cartridges are fed into the fixed aircraft guns inserted in belts made of metallic links which disintegrate as the guns are fired.)

On November 2, 1918, just before the armistice was signed, Gen. Pershing cabled as follows, in part:

> Marlin guns now rank as high as any with pilots, and are entirely satisfactory.

The French government tested the Marlin guns and declared them to be the equal of the Vickers. In order to meet the ever-increasing demands of the Air Service for machine guns capable of synchronization, the original order for 23,000 Marlin guns, placed in September, 1917, with the Marlin-Rockwell Corporation, was afterwards increased to 38,000. Along in 1918 the French tried to procure Marlins from this country, but by that time the Browning production was reaching great proportions, and the equipment at the Marlin plant was being altered to make Brownings.

The original order for Lewis guns, placed with the Savage Arms Corporation, had contemplated their use by our troops in the line; but when it became evident that the available manufacturing capacity of the United States would be strained to the utmost to provide enough guns for our airplanes, we diverted the large orders for Lewis guns entirely to the Air Service. This action was confirmed by cabled instructions from Gen. Pershing. For this flexible aircraft work the weapon was admirably adapted.

To the machine-gun tests, May, 1917, the producers of the Lewis gun brought an improved model, chambered for our own standard .30-caliber cartridges, instead of for the British .303 ammunition, with some 15 modifications in design in addition to those which had been presented to us before, and some added improvements in construction and in the metallurgical composition of its materials. From our point of view, this new model Lewis was a greatly improved weapon. The fact should be stated here that the Lewis gun, as so successfully made for the British service by the Birmingham Small Arms Co., had never been procurable by the United States, even in a single sample for test.

The Lewis accordingly became the standard flexible gun for our airplanes. The Savage Arms Corporation was able to expand its facilities to fulfill every need of our Air Service for this type of weapon, and therefore we made no effort to carry the manufacture of Lewis guns into other plants. Before 1917 came to an end the Savage company was delivering the first guns of its orders.

During the difficulties on the Mexican border the United States secured from the Savage Arms Co. several hundred Lewis guns made to use British ammunition. In order to be sure that the guns would be properly used, experts from the factory were sent out to instruct the troops who were to receive the guns. Ordnance officers also went out on this instruction work and established machine-gun schools along the border. The troops did not find the guns entirely satisfactory, in spite of expert instruction that they received from men from the factory. The trouble with the guns at this time was due to the fact that the company making them in the United States had been engaged in the manufacture of machine guns for a short time only and had run into several minor difficulties in the design and manufacture, difficulties which caused considerable trouble in operating the guns in the field, and which were subsequently corrected in the 15 changes mentioned above. The machine-gun schools which were established on the border taught not only the mechanism of the Lewis gun, but also those of the other types of guns with which the various troops were armed. The first thing that these schools developed was the fact that much of the trouble which had been encountered in machine guns was undoubtedly due to the fact that our soldiers were unfamiliar with the operation of the weapons. In fact, at that time we had few experts in the operation of any make of machine guns.

Soon after the establishment of machine-gun schools on the border it became apparent that the system of instruction devised by our ordnance officers had gone a long way toward overcoming the difficulties which the Army had encountered in the use of machine guns. The advantage of these schools was so marked that on the outbreak of the war with Germany the Ordnance Department established a machine-gun school at Springfield Armory. The first class of this school consisted of a large number of technical graduates from the Massachusetts Institute of Technology and other such schools. These men were employed as civilians, and were taught the mechanism of machine guns in a theoretical way in as thorough a manner as could possibly be done, and were given an opportunity to fire the guns and find out for themselves just what troubles were likely to occur. Many of these men were afterwards commissioned as officers in the Ordnance Department and were sent to the various cantonments throughout the United States to establish schools of instruction in the mechanism of the various machine guns.

After this class of civilians had been graduated from the Springfield school, a number of training-camp candidates were instructed and were afterwards commissioned. When the full success of this school was realized, it was enlarged and expanded, and it instructed not only civilians and training-camp candidates, but also officers of the Ordnance Department, who were trained as armament officers, instructors, etc. Later the school was still further expanded to include a large class of enlisted men for duty as armorers. In all, over 500 officers were instructed at the Springfield school.

When the war with Germany ceased, the graduates of the Springfield Armory machine gun school were found in almost every line of endeavor connected with arms, ammunition, and kindred subjects.

Now, let us look at the first results of the early effort in machine-gun production. Within a month after the first drafted troops reached their cantonments we were able to ship 50 Colt guns from the Marlin-Rockwell Corporation to each National Army camp, these guns to be used exclusively for training our machine-gun units. Before another 30 days passed we had added to the machine-gun equipment of each camp 20 Lewis guns of the ground type, and 30 Chauchat automatic rifles which we bought from the French. (The Lewis ground gun was almost identical with the aircraft type, except that its barrel was surrounded by an aluminum heat radiator for cooling, a device not needed on the guns of airplanes because of the latter's shorter periods of fire.) Also, in the autumn of 1917 we were able to issue to each National Guard camp a training equipment consisting of 30 Colt machine guns, 30 Chauchat automatic rifles, and some 50 to 70 Lewis ground guns.

At the beginning of 1918 our machine-gun manufacture was well under way. Such was the industrial situation at this time: the Savage Arms Corporation was producing Lewis aircraft machine guns of the flexible type; the Marlin-Rockwell Corporation was manufacturing large quantities of Marlin aircraft machine guns of the synchronizing type; the Colt's Patent Fire Arms Manufacturing Co. was building Vickers machine guns of the heavy, mobile type; and a number of great factories were tooling up at top speed for the immense production of Browning guns of all types soon to begin. Meanwhile we kept increasing our orders as rapidly as conditions warranted.

By May, 1918, the first 12 divisions of American troops had reached France. They were all equipped with Hotchkiss heavy machine guns and Chauchat automatic rifles—both kinds supplied by the French government. During May and June, 11 American divisions sailed, and the heavy machine-gun equipment of these troops was American built, consisting of Vickers guns. For their light machine guns these 11 divisions received the French Chauchat rifles in France. After June, 1918, all American troops to sail were supplied with a full equipment of Browning guns, both of the light and heavy types. Part of these Brownings were issued to the troops before they sailed, and the rest upon their arrival in France.

The Savage Arms Corporation built nearly 6,000 Lewis guns of the ground type before diverting their manufacture to the aircraft type exclusively. On May 11, 1918, this concern had built 16,000 Lewis guns for the American Government, of which more than 10,000 were for use on airplanes. By the end of July the company had turned out 16,000 aircraft Lewis guns, not to mention 6,000 of the same sort which it had built and supplied to the American Navy. By the end of September we had accepted over 25,000 Lewis aircraft guns. On the date of the signing of the armistice approximately 32,000 of these guns had been completed.

By the first of May, 1918, the Marlin-Rockwell Corporation had turned out nearly 17,000 Marlin aircraft guns with the synchronizing appliances. Thirty days later its total had reached 23,000. On October 1 the entire order of 38,000 Marlin guns had been completed, and the company began the work of converting its plant into a Browning factory.

On May 1, 1918, the Colt Co. had delivered more than 2,000 Vickers guns of the ground type. Before the end of July this output totaled 8,000, besides 3,000 Vickers guns which were later converted to aircraft use. In addition the Colt Co. had undertaken another machine-gun project of which nothing has been said before. This concern had completed manufacture of about 1,000 Vickers guns for the Russian government. At this time the aviators at the front began using machine guns of large caliber, principally against observation balloons and dirigible aircraft. The allies had developed an 11-millimeter Vickers machine gun for this purpose, which means a gun with a bore diameter of nearly one-half inch. The Ordnance Department undertook to change these Russian Vickers guns into 11-millimeter aircraft machine guns. This undertaking was successfully carried through by the Colt Co., which delivered the first modified weapon in July and had increased its deliveries to a total of 800 guns by November 11, 1918.

When the fighting ceased the Colt Co. had delivered 12,000 heavy Vickers guns and nearly 1,000 of the aircraft type. As was mentioned before, a considerable quantity of Vickers ground guns had been subsequently converted to aircraft use. The production of ground-type Vickers ceased on September 12, 1918, by which date the manufacture of Browning guns had developed sufficiently to meet all of our future needs. Thereafter the Colt plant produced the aircraft types of Vickers guns only. We shipped 6,309 Vickers ground guns overseas before the armistice was signed, besides equipping six France-bound divisions of troops with these weapons in this country, making a total of 7,653 American-built Vickers in the hands of the American Expeditionary Forces. Later, we planned to replace these weapons with Brownings, turning over the Vickers guns to the Air Service.

But America's greatest feat in machine-gun production was the development of the Browning weapons. These guns, as has been noted, were of three types: the heavy Browning water-cooled gun, weighing 37 pounds, for the use of our troops in the field; the light Browning automatic rifle, weighing 15.5 pounds, and in appearance similar to the ordinary service rifle, also for the use of our soldiers fighting on the ground; and, finally, the Browning synchronized aircraft gun of the rigid type, which was the Browning heavy machine gun made lighter by the elimination of its water-jacket, speeded up to double the rate of fire, and provided with the additional attachment of the synchronized firing mechanism. Let us take up separately the expansion of the facilities for manufacturing these types.

In the first place, the Colt Co., which owned the Browning rights, in September, 1917, turned over to the Winchester Repeating Arms Co. the task of developing the drawings and gauges for the manufacture of Browning automatic rifles on a large scale. The latter concern did a splendid job in this work. Early in March, 1918, the Winchester Co. had tooled up its plant and turned out the first Browning rifles. These were shipped to Washington and demonstrated in the hands of gunners before a distinguished audience of officers and other Government officials, and their great success assured the country that America had an automatic rifle worthy of her inventive and manufacturing prestige. By the first of May the Winchester Co. had turned out 1,200 Browning rifles.

The Marlin-Rockwell Corporation attained its first production of Browning rifles in June, 1918, by which time the Winchester Co. had built about 4,000 of them. Before the end of June the Colt Co. added its first few hundreds of Browning rifles to the expanding output. By the end of July the total production of Browning rifles had reached 17,000, produced as follows: 9,700 by Winchester; 5,650 by Marlin-Rockwell; and 1,650 by Colt's. Two months later this total had been doubled—the exact figure being 34,500 Browning rifles—and on November 11, 1918, when the flag fell on this industrial race, the Government had accepted 52,238 light Browning rifles. Of these in round numbers the Winchester Co. had built 27,000; Marlin-Rockwell, 16,000; and Colt's, 9,000.

But these figures give only an indication of the Browning rifle program as it had expanded up to the time hostilities ceased. When the armistice was signed our orders for these guns called for a production of 288,174, and still further large orders were about to be placed. As an illustration of the size which this manufacture would have attained, we had completed negotiations with one concern whereby its factory capacity was to be increased to produce 800 Browning rifles every 24 hours by June of 1919. After the armistice was signed we canceled orders calling for the manufacture of 186,000 Browning automatic rifles.

Of the 48,082 of these weapons sent overseas, 38,860 went in bulk on supply transports, while the rest constituted the equipment of 12 Yankee divisions which carried their automatic rifles with them.

The Colt Co. itself developed the drawings and gauges for the quantity manufacture of the Browning gun of the ground type. It will be remembered that the New England Westinghouse Co. was the first outside concern to begin the manufacture of these weapons. The New England Westinghouse Co. received its orders in January, 1918, and within four months had turned out its first completed guns, being the first company to deliver these weapons to the Government. By the first of May it had delivered 85 heavy Brownings.

By the middle of May the Remington Co. came into production of the heavy Brownings. The Colt Co., which was required to continue its production of Vickers guns, was also retarded by the duty of preparing the drawings for the other concerns who had contracted to make heavy Brownings; and this factory, the birthplace of the Browning gun, was not able to produce any until the end of June. By this time the Westinghouse Co. had turned out more than 2,500 heavy Brownings, and Remington over 1,600.

By the end of July the production of Browning machine guns at all plants had reached the total of 10,000; and two months later 26,000 heavy Brownings were in the hands of the Government. In the following six weeks this production was enormously increased, the total receipts by the Government up to November 11 amounting to about 42,000 heavy Browning guns. In round numbers Westinghouse produced 30,000 of these, Remington 11,000, and Colt about 1,000.

We shipped in all 30,582 heavy Brownings to the American Expeditionary Forces, 27,894 going on supply ships and the rest in the hands of 12 divisions of troops.

These shipments actually put in France before the armistice was signed enough heavy Brownings to equip completely all the American troops on French soil. However, at the time these supplies were arriving the fighting against the retreating German Army was at its height, and there was no time for the troops on the line to exchange their British-built and French-built machine guns for Brownings, nor to replace their Chauchat automatic rifles with light Brownings, of which there was also an ample supply in France.

A report of the Chief Ordnance Officer, American Expeditionary Forces, as of February 15, 1919, shows that, except for antiaircraft use, Vickers and Hotchkiss machine guns with troops had been almost entirely replaced by heavy Brownings on that date, and that Chauchat automatic rifles had been replaced by light Brownings.

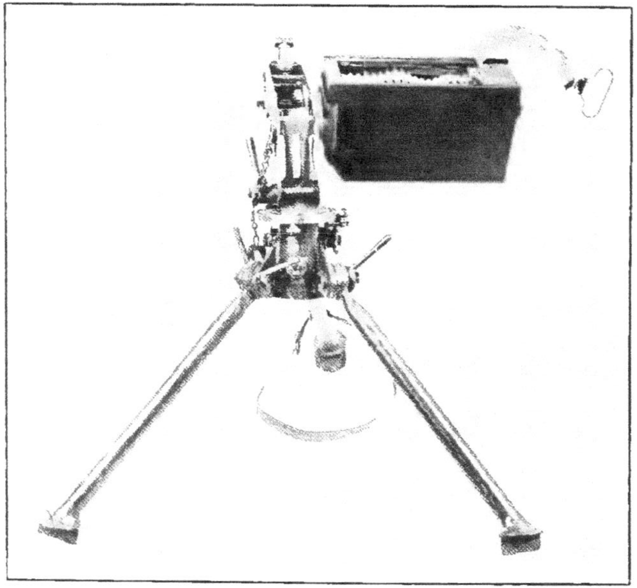

BROWNING EXPENDABLE CARTRIDGE BELT BOX, MARK I, MOUNTED ON A BROWNING MACHINE GUN TRIPOD, MODEL 1917.

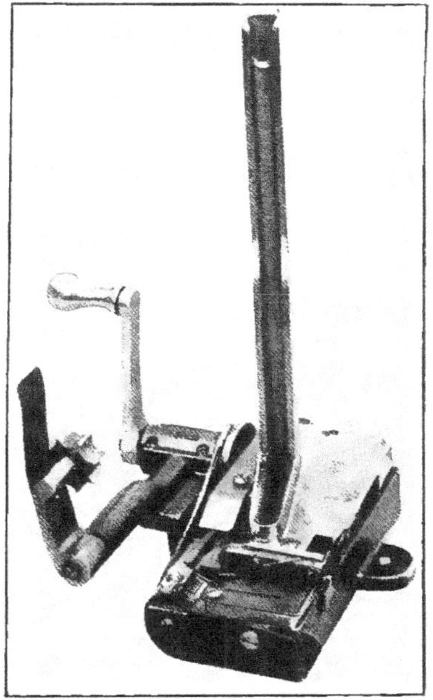

BELT-FILLING MACHINE FOR BROWNING MACHINE GUN, MODEL 1917.

BROWNING TANK MACHINE GUN, MODEL 1919, MOUNTED ON TANK.

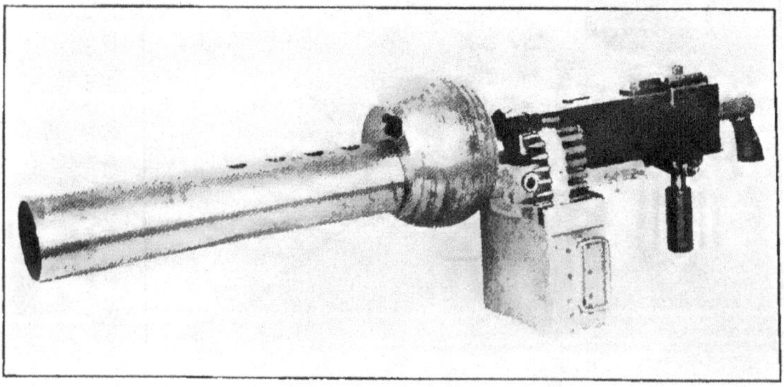

BROWNING TANK MACHINE GUN, MODEL 1919, ON BALL MOUNT, SHOWING CASING.

When the armistice was signed we had placed orders for 110,000 heavy Brownings and were contemplating still further orders. We later reduced these orders by 37,500 guns.

Because the Marlin aircraft gun had performed so satisfactorily, and because our facilities for the manufacture of this weapon were large, the production of the Browning aircraft guns had not been pushed to the limit, which latter action would have interfered with the production of the Marlin gun at a time when it was most essential to obtain an immediate supply of fixed synchronized aircraft guns. Only a few hundred Browning aircraft guns had been completed before the close of the fighting. In its tests and performances this weapon had been speeded up to a rate of fire of from 1,000 to 1,300 shots per minute, which far surpassed the performances of any synchronized gun then in use on the western front.

By the spring of 1918 it became evident that we would require a special machine gun for use in our tanks. Several makes of guns were considered for this purpose and finally discarded for one reason or another. The ultimate decision was to take 7,250 Marlin aircraft guns which were available and adapt them to tank service by the addition of sights, aluminum heat radiators, and handle grips and triggers. The rebuilding of these guns at the Marlin-Rockwell plant when the armistice was signed was progressing at a rate that insured the adequate equipment of the first American-built tanks.

Meanwhile the Ordnance Department undertook the production of a Browning tank machine gun. This gun was developed by taking a heavy Browning water-cooled gun, eliminating the water jacket and substituting an air-cooled barrel of heavy construction, and adding hand grips and sights. The work was begun in September, 1918, and the completed model was delivered by the end of October. Before the armistice was signed five sample guns had been built, demonstrated at the Tank Corps training camps, and unanimously approved by the officers of the Tank Corps designated to test it. After a test in France, the report stated: "The gun is by far the best weapon for tank use that is now known, and the Department is to be congratulated upon its development." An order for 40,000 Browning tank guns was given to the Westinghouse Co. This concern, already equipped for the manufacture of heavy Browning guns, was scheduled to start its deliveries in December, 1918, and to turn out 7,000 tank guns per month after January 1, 1919. After the signing of the armistice, however, the order was cut down to approximately 1,800 guns. By March 27, 1919, the company had delivered 500 Browning tank guns, and the order for the remaining 1,300 was thereafter canceled.

After the entrance of the United States in the war the armies on both sides developed a new type of machine-gun fighting, which consisted in indirect firing, or laying down barrages of machine-gun bullets. This required the development of special tripods, clinometers for laying angles of elevation, and other special equipment; and speedy progress was being made in the quantity production of this matériel when the war came to an end.

In a complete machine-gun program not only must the guns themselves be built, but they must be fully equipped with accessories, such as tripods, extra magazines, carts for carrying both guns and ammunition, feed belts of various types, belt-loading machines, observation and fire control instruments, and numerous other accessories the manufacture of which is absolutely essential but usually unseen by the public. The extent of our work in accessories is indicated by a few approximate figures of deliveries up to the signing of the armistice: nonexpendable ammunition boxes, 1,000,000; expendable ammunition boxes, 7,000; expendable belts, 5,000; nonexpendable belts, 1,000,000; belt-loading machines, 25,000; water boxes, 110,000; machine-gun carts, 17,000; ammunition carts, 15,000; tripods, 25,000.

The aircraft machine guns also required numerous accessories, some of them highly complicated in their manufacture. This special equipment consisted in part of special mounts for the guns, synchronizing attachments, metallic disintegrating link belts, electric heaters to keep the guns warm at the low temperatures at the high altitudes of the aviator's battle field, and many other smaller items.

Not only our own forces but the allied armies as well were enthusiastic about the Browing guns of both types, as soon as they had seen them in action. The best proof of this is that in the summer of 1918 the British, Belgian, and French Governments all made advances to us as to the possibility of the United States producing Browning automatic rifles for their respective forces. On November 6, a few days before the end of hostilities, the French high commissioner requested that we supply 15,000 light Browning rifles to the French Army. We would not make this arrangement at the time because we thought it inadvisable to divert any of our supplies of these guns from our own troops until the spring of 1919, when we expected that our capacity for making light Brownings would exceed the demands of our own troops. Our demand for the lighter guns, incidentally, was far greater than we had originally expected it to be. As soon as the Browning rifle was seen in action the General Staff of our Expeditionary Forces at once increased by 50 per cent the number of automatic rifles assigned to each company of troops, and we were manufacturing to meet this augmented demand when the war ended. By spring of 1919 we expected to be furnishing light Brownings to the British and French Armies as well as to our own.

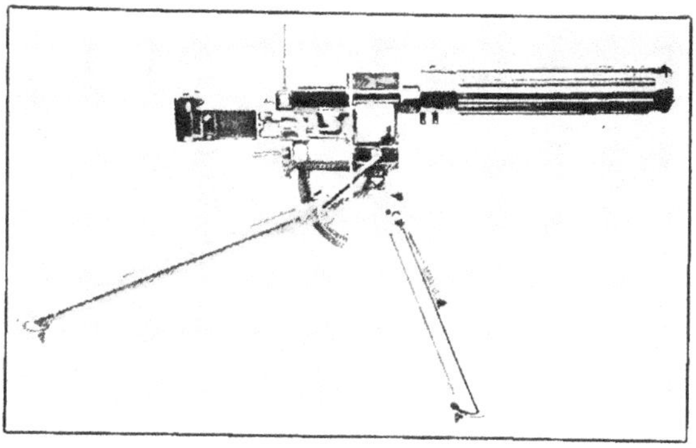

FIAT (ITALIAN) MACHINE GUN AND TRIPOD.

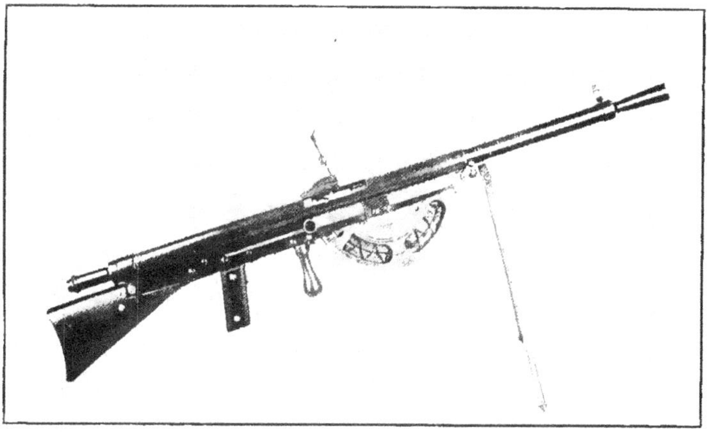

CHAUCHAT MACHINE RIFLE, MODEL 1915, CALIBER, 8 MILLIMETERS.

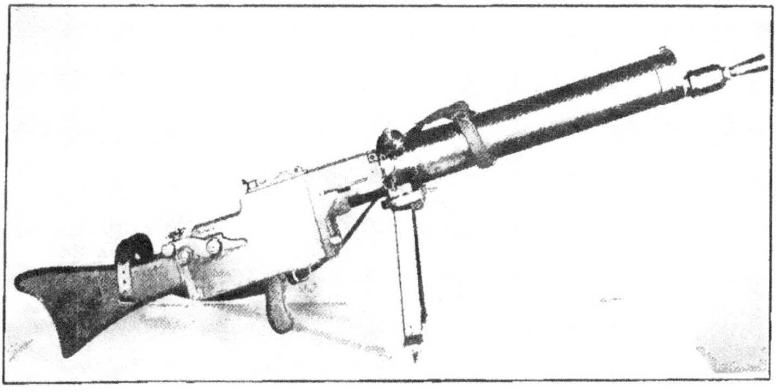

GERMAN 08/15 (SPANDAU) MACHINE GUN.

Both types of Browning guns proved to be unqualified successes in actual battle, as numerous reports of our Ordnance officers overseas indicated. The following report from an officer, in addition to carrying historical information of interest to those following our machine-gun development, is typical of numerous other official descriptions of these weapons in battle use:

The guns [heavy Brownings] went into the front line for the first time in the night of September 13. The sector was quiet and the guns were practically not used at all until the advance, starting September 26. In the action which followed, the guns were used on several occasions for overhead fire, one company firing 10,000 rounds per gun into a wood in which there were enemy machine-gun nests, at a range of 2,000 meters. Although the conditions were extremely unfavorable for machine guns on account of rain and mud, the guns performed well. Machine-gun officers reported that during the engagement the guns came up to the fullest expectations and, even though covered with rust and using muddy ammunition, they functioned whenever called upon to do so.

After the division had been relieved, 17 guns from one company were sent in for my inspection. One of these had been struck by shrapnel, which punctured the water jacket. All of the guns were completely coated with mud and rust on the outside, but the mechanism was fairly clean. Without touching them or cleaning them in any way, except to run a rod through the bore, a belt of 250 rounds was fired from each without a single stoppage of any kind.

It can be concluded from the try-out in this division that the gun in its operation and functioning when handled by men in the field is a success.

The Browning automatic rifles were also highly praised by our officers who had to use them. Although these guns received hard usage, being on the front for days at a time in the rain and when the gunners had little opportunity to clean them, they invariably functioned well.

On November 11 we had built 52,238 Browning automatic rifles in this country. We had bought 29,000 Chauchats from the French. Without providing replacement guns or reserves, this was a sufficient number to equip over 100 divisions with 768 guns to the division. This meant light machine guns enough for a field army of 3,500,000 men. In heavy machine guns at the signing of the armistice we had 3,340 of the Hotchkiss make, 9,237 Vickers, and 41,804 Brownings, or a total of 54,627 heavy machine guns—enough to equip the 200 divisions of an army of 7,000,000 men, not figuring in reserve weapons.

The daily maximum production of Browning rifles reached 706 before our manufacturing efforts were suddenly stopped, and that of Browning heavy machine guns 575. At the peak of our production a total of 1,794 machine guns and automatic rifles of all types was produced within a period of twenty-four hours.

Based upon our output in July, August, and September, 1918, we were producing monthly 27,270 machine guns and machine rifles of all types, while the average monthly production of France was at this time 12,126 and that of Great Britain 10,947.

In total production between April 6, 1917, and November 11, 1918, we had turned out 181,662 machine guns and machine rifles, as against 229,238 by France and 181,404 by England in that same period.

One of the important features which contributed to the success of the machine-gun program was the cordial spirit of cooperation which the War Department met from the machine-gun manufacturers. Competitive commercial advantages weighed not at all against the national need, and the Department found itself possessed of a group of enthusiastic and loyal partners with whom it could attack the vast problem of machine-gun supply. Without these partners and this spirit, the problem could not have been solved. The United States, starting almost from the zero point, developed in little more than a year a machine-gun production greater than that of any other country in the world, although some of those countries had been fighting a desperate war for three years and building machine guns to the limit of their capacity.

Acceptances of automatic arms, by months, in United States and Canada on United States Army orders only.

	To Jan. 1.	1918												Total.
		Jan.	Feb.	Mar.	Apr.	May.	June.	July.	Aug.	Sept.	Oct.	Nov.	Dec.	
Ground machine guns.														
Browning heavy					12	922	2,620	4,225	9,182	8,838	14,639	6,654	9,516	56,608
Vickers field	2,031	1,021	951	1,386	1,341	1,208	1,349	1,565	789	381	103			12,125
Colt	2,500						305	11						2,816
Lewis field	2,209	291												2,500
Lewis caliber .303	750				300									1,050
Aircraft machine guns.														
Browning											211	363	6	580
Marlin	12	3,134	3,850	3,419	5,750	6,250	219	6,356	7,269	1,691	50			38,000
Lewis flexible	6	540	1,085	1,568	1,333	2,629	4,342	4,338	5,595	3,973	5,857	3,792	4,142	39,200
Vickers caliber .30										307	575	373	1,221	2,476
Vickers 11-mm							72	263	95	254	117	161	276	1,238
Tank machine guns.														
Browning												3	1	4
Marlin									[1] 103	[1] 9	[1] 316	[1] 460	[1] 582	[1] 1,470
Automatic rifles.														
Browning light			15	548	363	1,822	3,876	8,196	12,517	6,896	13,687	11,368	10,672	69,960
Total	7,508	4,986	5,901	6,921	9,099	12,831	12,783	24,954	35,447	22,340	35,239	22,714	25,834	226,557

[1] Modified from aircraft, not included in total.

CHAPTER X.

SERVICE RIFLES.

Although in the 19 months of American belligerency in the great war we had sent to France upward of two million soldiers, each rifleman among them as he stepped aboard his transport carried his own gun. This weapon, which was to be his comrade and best friend in the perilous months to come, was an American rifle, a rifle at least the equal of any in use by soldiers of other nations, a rifle manufactured in an American plant. It may have been one of the dependable Springfield rifles. More likely, it was a modified 1917 Enfield, built from a design British in fundamental character, but modified for greater efficiency by American ordnance officers after the actual entry of the United States in the great struggle. When it is considered that even a nation of such military genius as France, especially skilled as she was in the construction of military weapons, was three years developing her full ordnance program, even though working at top speed, the rifle production of the United States stands out as one of the feats of the war.

The story of the modified 1917 Enfield, which was the rifle on which the American Expeditionary Forces based their chief dependence, is an inspiring chapter in our munitions history. To get this weapon we temporarily forsook the most accurate Army rifle the world had ever seen and straightway produced in great quantities another one, a new model, that proved itself to be almost, if not quite, as servicable for the kind of warfare in which we were to engage. It is the story of triumph over difficulties, of American productive genius at its best.

America, since the days of Daniel Boone a nation of crack shots, was naturally the home of good rifles. Hence, perhaps, it is not surprising that the United States should be the nation to produce the closest shooting military rifle known in its day. This was the United States rifle, model of 1903, popularly called the "Springfield."

The Springfield rifle had superseded in our Army the Krag, which we had used in the Spanish-American War. In that conflict the Spanish Army used a rifle of German design, the Mauser. Our ordnance officers at that time considered the Krag to be a more accurate weapon than the Mauser. Still we were not satisfied with the Krag, and, after several years of development, in 1903 we brought out the Springfield, the most accurate and quickest firing rifle that had ever come from an arsenal.

There was no questioning the superiority of the Springfield in point of accuracy. Time after time we pitted our Army shooting teams against those of the other nations of the earth and won the international competitions with the Springfield. We won the Olympic shoot of 1908 over England, Canada, France, Sweden, Norway, Greece, and Denmark. Again, in 1912, we won the Olympic shoot against England, Sweden, South Africa, France, Norway, Greece, Denmark, Russia, and Austria-Hungary. In 1912 the Springfield rifle, in the hands of Yankee marksmen, won the Pan American match at Buenos Aires, and in 1913 it defeated Argentina, Canada, Sweden, and Peru. In all of these matches the Mauser rifle was fired by various teams; but the Springfield never failed to defeat this German weapon, which it was to meet later in the fighting of the great war.

Altogether the Springfield rifle defeated the military rifles of 15 nations in shooting competitions prior to the war, and in 1912, at Ottawa, an American team firing Springfields set markmanship records for 800 yards, 900 yards, and 1,000 yards that have never been broken. Much is to be said for the men behind these guns, but due credit must be given to the rifles that put the bullets where the marksmen aimed.

Such was the history of this splendid arm when the United States neared the brink of the great conflict. But as war became inevitable for us and we began to have a realization of the scale on which we must prosecute it, our ordnance officers studying the rifle problem became persuaded that our Army could not hope to carry this magnificent weapon to Europe as its chief small-arms reliance. A brief examination of the industrial problem presented by the rifle situation in 1917 should make it clear even to a man unacquainted with machinery and manufacturing why it would be humanly impossible to equip our troops with the rifle in developing which our ordnance experts had spent so many years.

The Model 1903 rifle had been built in two factories and only two—the Springfield Armory, Springfield, Mass., and the Rock Island Arsenal at Rock Island, Ill. Our Government for several years prior to 1917 had cut down its expenditures for the manufacture of small arms and ammunition. The result was that the Rock Island Arsenal had ceased its production of Springfields altogether, while the output of rifles from the Springfield Armory had been greatly reduced.

This meant that the skilled artisans once employed in the manufacture of Springfield rifles had been scattered to the four winds. When in early 1917 it became necessary to speed up the production of rifles to the limit in these two establishments those in charge of the undertaking found that they could recover only a few of the old, trained employees. Yet even when we had restaffed these two factories with skilled men their combined production at top speed could not begin to supply the quantity of rifles which our impending Army would need. Therefore, it was obviously necessary that we procure rifles from private factories.

Why, then, was not the manufacture of Springfields extended to the private plants? Some ante bellum effort, indeed, had been made looking to the production of Springfields in commercial plants, but lack of funds had prevented more than the outlining of the scheme.

Any high-powered rifle is an intricate production. The 1917 Enfield is relatively simple in construction, yet the soldier can dismount his Enfield into 86 parts, and some of these parts are made up of several component pieces. Many of these parts must be made with great precision, gauged with microscopic nicety, and finished with unusual accuracy. To produce Springfields on a grand scale in private plants would imply the use of thousands of gauges, jigs, dies, and other small tools necessary for such a manufacture, as well as that of great quantities of special machines. None of this equipment for Springfield rifle manufacture had been provided, yet all of it must be supplied to the commercial plants before they could turn out rifles.

We should have had to spend preliminary months or even years in building up an adequate manufacturing equipment for Springfields, the while our boys in France were using what odds and ends of rifle equipment the Government might be able to purchase for them, except for a condition in our small-arms industry in early 1917 that now seems to have been well-nigh providential.

Among others, both the British and the Russian Governments in the emergency of 1914 and 1915 had turned to the United States to supplement their sources of rifle supply while they, particularly the British, were building up their home manufacturing capacity. There were five American concerns engaged in the production of rifles on these large foreign orders when we entered the war. Three of them were the Winchester Repeating Arms Co., of New Haven, Conn.; the Remington Arms-Union Metallic Cartridge Co., of Ilion, N. Y.; and the Remington Arms Co. of Delaware at its enormous war-contract factory at Eddystone, Pa., later a part of the Midvale Steel & Ordnance Co. These concerns had developed their manufacturing facilities on a huge scale to turn out rifles for the British Government. By the spring of 1917 England had built up her own manufacturing facilities at home, and the last of her American contracts were nearing completion.

Here, then, was at hand a huge capacity which, added to our Government arsenals, could turn out every rifle the American Army would require, regardless of how many troops we were to put in the field.

But what of the gun that these plants were making—the British Enfield rifle? As soon as war became a certainty for us the Ordnance Department sent its best rifle experts to these private plants to study the British Enfield in detail. They returned to headquarters without enthusiasm for it; in fact, regarding it as a weapon not good enough for an American soldier.

A glance at the history of the British Enfield will make clear some of our objections to it. Until the advent of the 1903 Springfield, the German Mauser had occupied the summit of military-rifle supremacy. From 1903 until the advent of the great war these two rifles, the Mauser and the Springfield, were easily the two leaders. The British Army had been equipped with the Lee-Enfield for some years prior to the outbreak of the great war, but the British ordnance authorities had been making vigorous efforts to improve this weapon. The Enfield was at a disadvantage principally in its ammunition. It fired a .303-caliber cartridge with a rimmed head. From a ballistic standpoint this cartridge was virtually obsolete.

In 1914 a new, improved Enfield, known as the Pattern '14, was brought out in England, and the British Government was on the point of adopting it when the great war broke out. This was to be a gun of .276 caliber and was to shoot rimless, or cannelured, cartridges similar to the standard United States ammunition. The war threw the whole British improved Enfield project on the scrap heap. England was no more equipped to build the improved Enfields than we were to produce Springfields in our private plants. The British arsenals and industrial plants and her ammunition factories were equipped to turn out in the quantities demanded by the war only the old "short Enfield" and its antiquated .303 rimmed cartridges.

Now England was obliged to turn to outside sources for an additional rifle supply, and in the United States she found the three firms named above willing to undertake large rifle contracts. Having to build up factory equipment anew in the United States for this work, England found that she might as well have the American plants manufacture the improved Enfield as the older type. To produce the 1914 Enfield without change in America and the older-type Enfield in England would complicate the British rifle-ammunition manufacture, since these rifles used cartridges of different sizes and types. Accordingly, the British selected the improved Enfield for the American manufacture, but modified it to receive the .303 rimmed cartridges.

This was the gun, then, that we found being produced at New Haven, Ilion, and Eddystone in the spring of 1917. The rifle had many of the characteristics of the 1903 Springfield, but it was not so good as the Springfield in its proportions, and its sights lacked some of the refinements to which Americans were accustomed. Yet even so it was a weapon obviously superior to either the French or Russian rifle. The ammunition which it fired was out of the question for us. Not only was it inferior, but, since we expected to continue to build the Springfields at the Government arsenals, we should, if we adopted the Enfield as it was, be forced to produce two sizes of rifle ammunition, a condition leading to delay and unsatisfactory output. The rifle had been designed originally for rimless ammunition and later modified; so it could be modified readily back again to shoot our standard .30-caliber Springfield cartridges.

It may be seen that the Ordnance Department had before it three courses open, any one of which it might take. It could spend the time to equip private plants to manufacture Springfields, in which case the American rifle program would be hopelessly delayed. It could get guns immediately by contracting for the production of British .303 Enfields, in which case the American troops would carry inferior rifles with them to France. Or, it could take a relatively brief time, accept the criticism bound to come from any delay, however brief such delay might be and however justified by the practical conditions, and modify the Enfield to take our ammunition, in which case the American troops would be adequately equipped with a good weapon.

The decision to modify the Enfield was one of the great decisions of the executive prosecution of the war—all honor to the men who made it.

The three concerns which had been manufacturing the British weapons conceded that it should be changed to take the American ammunition. Each company sent to the Springfield Armory on May 10, 1917, a model modified rifle to be tested. The test showed the weapons still to be unsatisfactory, principally because they had not been standardized. Standardization was regarded as an essential for two reasons, one of them a matter of practical tactics in the field and the other relating to production speed.

To begin with, the soldier on the battle field is his own rifle repairman. His unit usually has on hand a supply of weapons damaged or out of commission for one reason or another. If, therefore, any part of the soldier's rifle is broken or damaged, he can go to the stock of unused guns on hand and take from another rifle the part which he requires; and it will fit his gun, provided there has been standardization in the rifle manufacture at home. But if the guns have not been standardized and each weapon is a filing and tinkering job in the assembly room of the factory, then the soldier in the field is not likely to be able to find a part that will fit on his gun; and his rifle, if damaged, goes out of commission. Or, if he finds a part which fits but does not fit perfectly, his gun may break as he fires it, and he himself may suffer serious injury.

In the second place, standardization is essential to great speed in production. If one plant producing rifles encounters a shortage in any of the parts of the gun, it can send to another plant and secure a supply of these parts, a favorable condition in manufacture that is impossible if the weapon has not been standardized. The value of standardization in speeding up manufacture, however, is best shown in the actual records of rifle production during the war. The fastest mechanic in any of the three Enfield factories before 1917 had set an assembly record of 50 rifles in one working day for the British gun. After we had standardized the Enfield the high assembly record was 280 rifles a day, while the assemblers in the plants averaged 250 rifles a day per man when the work was well started.

The Enfields sent to the Springfield Armory test were not standardized at all, but were largely hand fitted. Little or no attempt had been made to obtain interchangeability of parts among the rifles turned out by the three plants. Even the bolt taken from one company's rifle would not enter the receiver of another company's.

The Ordnance Department was confronted with the dilemma of approving and issuing a weapon pronounced unsuitable by its own experts and thus obtaining speedy production, or of delaying until interchangeability was established. It chose the latter course.

On July 12 a second set of rifles had been tested. These came more nearly up to our ideas of standardization, but were still not entirely satisfactory. Nevertheless we decided to go ahead with production and improve the standardization as we went along. The Winchester and Ilion plants elected to start work on that understanding, but Eddystone preferred to wait for the final requirements. Ilion afterwards decided to postpone production until the final specifications were adopted. It would have been well if the same course had been followed at the Winchester plant, for word came later from Europe not to send over rifles of Winchester manufacture of that period. The final drawings of the standardized and modified Enfield did not come from the plants until August 18. Six days later the thousands of dimensions had been carefully checked and finally approved by the ordnance officers, and after that production started off in earnest.

The wisdom of adopting the Enfield rifle and modifying it to meet our requirements instead of extending the manufacture of Springfields was almost immediately apparent, for in August, almost as soon as the final drawings were approved, the first rifles were delivered to the Government. This was possible because the modifications which we adopted did not require any fundamental changing of machinery.

The principal equipment of these plants was in place and ready to begin manufacturing Enfields at once; and while the changes in the rifle were under discussion, the manufacturers were producing their gauges and small tools as each modification was decided upon.

While we did not succeed in attaining, nor did we attempt to attain, in fact, complete standardization and interchangeability of the parts of the Enfields, we did all that was practicable in this direction, several tests showing that the average of interchangeability was about 95 per cent of the total parts.

Meanwhile we were building up the working staffs of the Rock Island Arsenal and Springfield Armory and speeding the production of Springfields. Before the war ended the Rock Island Arsenal, which was making spare parts for Springfields, reached an output equaling 1,000 completed rifles a day, while the Springfield Armory attained a high average of 1,500 assembled rifles a day in addition to spare parts equaling 100 completed rifles daily.

The Eddystone plant finished its British contracts on June 1, Winchester produced its last British rifle on June 28, and Ilion on July 21, 1917. Winchester delivered the first modified Enfields to us on August 18, Eddystone on September 10, and Ilion about October 28.

The progress in the manufacture was thereafter steadily upward. During the week ending February 2, 1918, the daily production of military rifles in the United States was 9,247, of which 7,805 were modified Enfields produced in the three private plants, and 1,442 were Springfields built in the two arsenals. The total production for that week was 50,873 guns of both types, or nearly enough for three Army divisions. In spite of the time that went into the standardization of the Enfield rifle, all troops leaving the United States were armed with American weapons at the ports of embarkation.

Ten months after we declared war against Germany we were producing in a week four times as many rifles as Great Britain had turned out in a similar period after 10 months of war, and our production was then twice as large in volume as Great Britain had attained in the war up to that time. By the middle of June, 1918, we had passed the million and one-half mark in the production of rifles of all sorts, this figure including over 250,000 rifles which had been built upon original contracts placed by the former Russian government.

The production of Enfields and Springfields during the war up to November 9, 1918, amounted to a total of 2,506,307 guns. Of these 312,878 were Springfield rifles produced by the two Government arsenals. We had started the war with a reserve of 600,000 Springfield rifles on hand, and in addition stored in our armories and arsenals were 160,000 Krags. These latter had to be cleaned and repaired in large part before they could be used. From the

Canadian Government we purchased 20,000 Ross rifles. The deliveries of Russian rifles totaled 280,049. This gave us a total equipment of 3,575,356 rifles. Since approximately one-half of the soldiers of an army as actually organized carry rifles, the number of rifles procured in all by the Ordnance Department was sufficient to arm both for fighting and for training an army of 7,000,000 men, disregarding reserve and maintenance rifles.

The Enfield thus became the dominant rifle of our military effort. With its modified firing mechanism it could use the superior Springfield cartridges with their great accuracy. The Enfield sights, by having the peep sight close to the eye of the firer, gave even greater quickness of aim than the Springfield sights afforded. In this respect the weapon was far superior to the Mauser, which was the main dependence of the German Army. All in all to a weapon that made scant appeal to our ordnance officers in a few weeks we added improvements and modifications that made the 1917 Enfield a gun that for the short-range fighting in Europe compared favorably with the Springfield and was to the allied cause a distinct contribution which America substantially could claim to be her own.

Standardization not only made possible the ultimate speed with which our rifles were produced but, together with the care of the Government in purchasing raw materials and in drawing contracts, it saved a great deal of money in the cost of these weapons. The British had been paying approximately $42 apiece for Enfields produced in the United States. The modified Enfields cost the Government approximately $26 each. Thus in the total production of 2,202,429 modified Enfields, we saved $37,441,293 compared with what this weapon had cost in the past.

Both the Springfield and the 1917 Enfield rifle possessed advantages of accuracy and speed of fire over the German Mauser. It is true that the Mauser fired a heavier bullet than that of our standard ammunition and sent it with somewhat greater velocity; but at the longer fighting ranges the Mauser bullet is not so accurate as the United States bullet. Due to its peculiar shape, the Mauser bullet is apt to tumble end over end at long ranges—"key-holing," the marksmen call it—particularly when the wind blows across the range. Such tumbling causes a bullet to curve as a baseball thrown by a good pitcher, destroying its accuracy.

Early in our fighting with Germany we captured Mauser rifles and hastened to compare them with the Springfields and modified Enfields. We found in the American rifles a marked superiority in the rapidity of fire, the quickness and ease of sighting, and in the accuracy of shots fired. The accuracy was due not only to our standard Springfield ammunition, but also to the greater mechanical accuracy in the finish of the chamber and bore of the American rifles. The rapidity of fire of the American guns was due to the position and shape of the bolt handle, which is the movable mechanism on the rifle with which the soldier ejects a spent shell and throws in a fresh one.

How we developed this bolt handle is an interesting story in itself. In 1903, when we brought out the first Springfield rifle, we decided to abandon the old carbine which had been carried by our Cavalry regiments and, by making a rifle with a comparatively short barrel, furnish a gun which could be used by both Infantry and Cavalry. The original bolt handle of the Springfield, like the one on the present Mauser, had projected horizontally from the side of the chamber. It was found that this protuberance did not fit well in the saddle holster of the cavalryman, but jammed the side of the rifle against the leather of the holster, with frequent injury to the rifle sight. For this primary reason the rifle designers bent the bolt handle down and back. This modification incidentally brought the bolt handle much nearer to the soldier's hand as he fingered the trigger than it had been before. The Enfield design had carried this development even farther, so that the bolt handle was practically right at the trigger, and the rifleman's hand was ready to pull the trigger the instant after it had thrown in a new cartridge.

Let us see what effect this design of the bolt handle had in the recent war. The Mauser still clung to the old horizontal bolt handle well away from the trigger grip. Some of our best riflemen practiced with the captured Mausers and, firing at top speed with them, could not bring the rate of shooting anywhere near up to the marks set by the Enfields and Springfields. One enthusiast has even maintained that the speed of the Mauser is not over 50 per cent of that of the 1917 American rifle, but this may be an underestimate. On such a basis the result was that under battle conditions with equal numbers of men on a side the Americans had in effect two rifles to the Germans one.

To put it another way, by bending back the bolt handle we had placed two men on the firing line where there was only one before; but the added man required no shelter, nor any clothing, nor rations, nor water, nor pay. Although he sometimes needed repairing, he did not get sick, nor did he ever become an economic burden nor draw a pension. His only added cost to the Government was an increased consumption of cartridges.

When American troops were in the heat of the fighting in the summer of 1918, the German government sent a protest through a neutral agency to our Government asserting that our men were using shotguns against German troops in the trenches. The allegation was true; but our State Department replied that the use of such weapons was not forbidden by the Geneva Convention as the Germans had asserted. Manufactured primarily for the purpose of arming guards placed over German prisoners, these shotguns were undoubtedly in some instances carried into the actual fighting. The Ordnance Department procured some 30,000 to 40,000 shotguns of the short-barrel or sawed-off type, ordering these from the regular commercial manufacturers. The shell provided for these guns each contained a charge of nine heavy buckshot, a combination likely to have murderous effect in close fighting.

Such was the rifle record of this Government in the war. The Americans carried into battle the best rifles used in the war, and America's industry produced these weapons in the emergency at a rate which armed our soldiers as rapidly as they could be trained for fighting. Success in such a task looked almost impossible at the start; but that it was attained should forever be a source of gratification to the American people.

Rifle production to Nov. 9, 1918.

Months.	Eddystone.	Winchester.	Ilion.	Springfield Armory.	Rock Island Arsenal.	Total.
Before August, 1917	....	....	....	14,986	1,680	16,666
Aug. 1, 1917 to Dec. 31, 1917	174,160	102,363	26,364	89,479	22,330	414,696
1918.						
January	81,846	39,200	32,453	23,890	7,680	185,069
February	98,345	32,660	39,852	6,910	2,460	180,227
March	68,404	42,200	49,538	120	420	160,682
April	87,508	43,600	36,377	2,631	....	170,116
May	84,929	41,628	54,477	3,420	550	185,004
June	104,110	34,249	52,995	6,140	619	198,113
July	135,080	35,700	60,413	14,841	2,038	248,072
August	106,595	20,030	65,144	27,020	1,597	220,386
September	110,058	31,550	58,027	29,770	3,813	233,218
October	100,214	33,700	53,563	35,920	3,256	226,653
Nov. 1-9, 1918	30,659	9,100	16,338	10,500	808	67,405
Total	1,181,908	465,980	545,541	265,627	47,251	2,506,307

NOTE.—Eddystone, Winchester, and Ilion plants turned out the United States rifle, caliber .30, model 1917, popularly known as the Enfield, while the Springfield Armory and the Rock Island Arsenal produced the United States rifle, caliber .30, model 1903, popularly known as the Springfield rifle. The months marked by a drop in the production at Springfield and at Rock Island were months in which the components manufactured were not assembled but were used for spare parts.

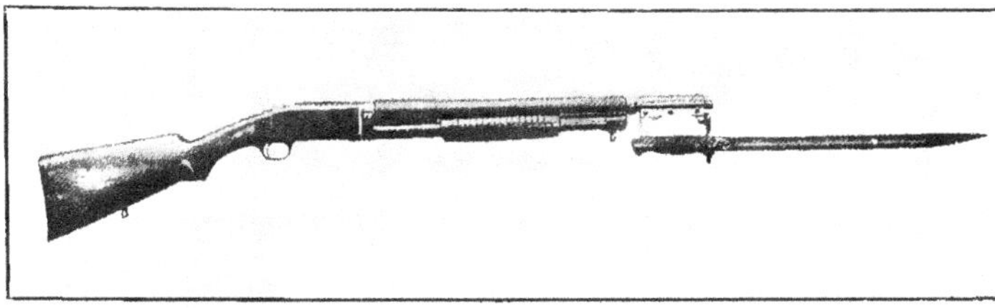

REMINGTON REPEATING SHOTGUN, 12-GAUGE, WITH BAYONET, MODEL 1917.

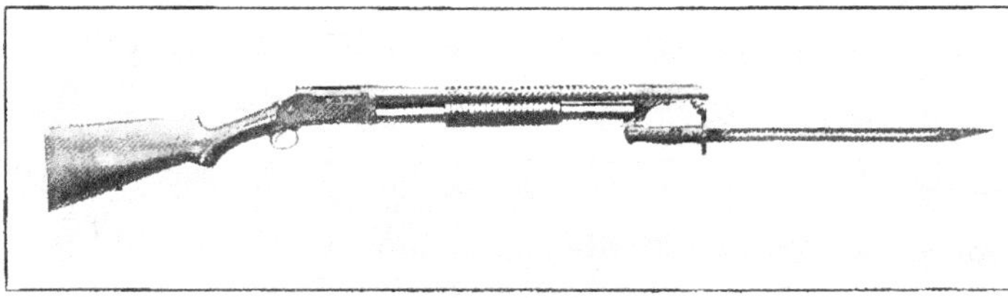

WINCHESTER REPEATING SHOTGUN, 12-GAUGE, WITH BAYONET, MODEL 1917.

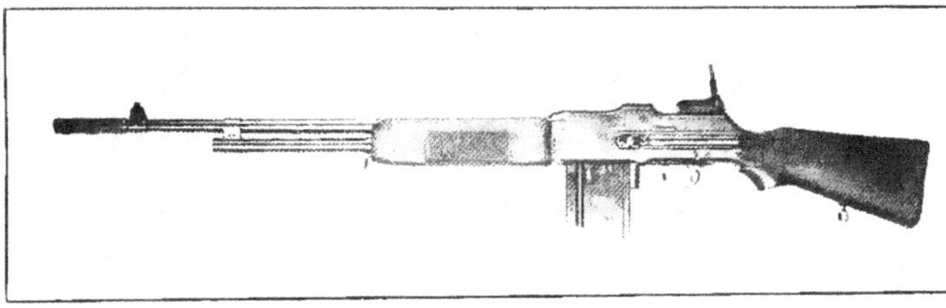

BROWNING AUTOMATIC RIFLE, MODEL 1918, CALIBER .30.

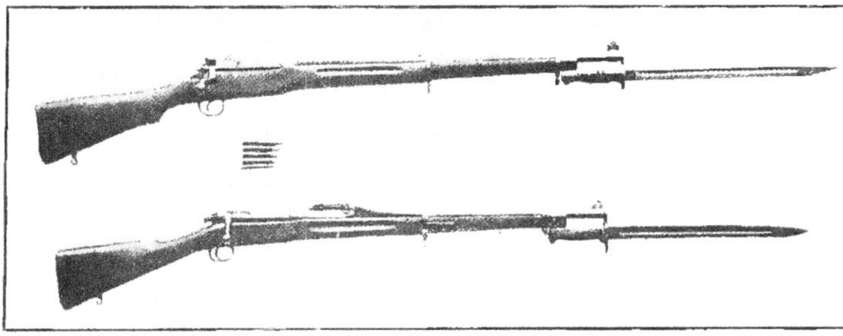

U. S. MODIFIED ENFIELD RIFLE, CALIBER .30, MODEL 1917, BAYONET, MODEL 1917.
U. S. MAGAZINE (SPRINGFIELD) RIFLE, CALIBER .30, MODEL 1903. BAYONET, MODEL 1905.

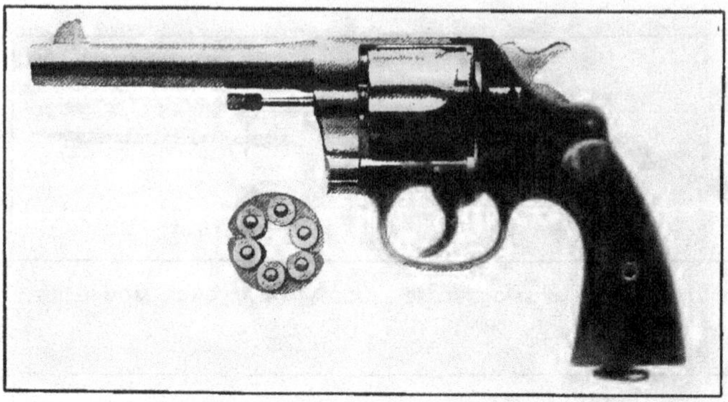

COLT DOUBLE ACTION REVOLVER, CALIBER .45, MODEL 1917.

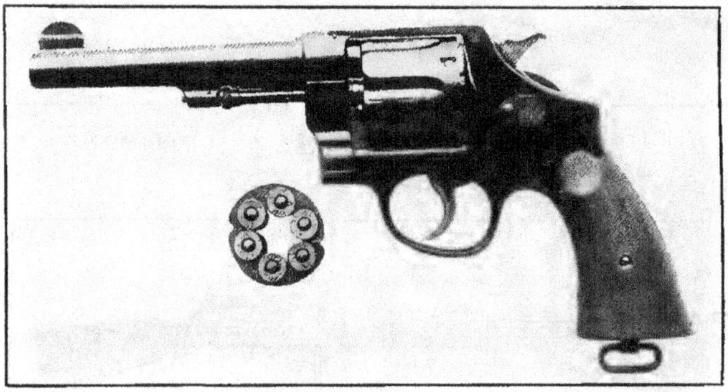

SMITH & WESSON DOUBLE ACTION REVOLVER, CALIBER .45, MODEL 1917.

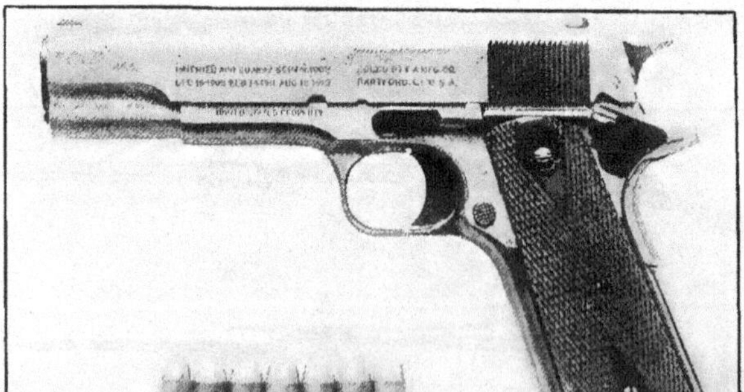

AUTOMATIC PISTOL, CALIBER .45, MODEL 1911.

CHAPTER XI.

PISTOLS AND REVOLVERS.

The American pistol was one of the great successes of the war. For several years before the war came the Ordnance Department had been collaborating with private manufacturers to develop the automatic pistol; but none of our officers realized until the supreme test came what an effective weapon the Colt .45 would be in the hand-to-hand fighting of the trenches. In our isolation we had suspected, perhaps, that the bayonet and such new weapons as the modern hand grenade had encroached upon the field of the pistol and revolver. We were soon to discover our mistake. In the hands of a determined American soldier the pistol proved to be a weapon of great execution, and it was properly feared by the German troops.

We had long been a nation of pistol shooters, we Americans, but not until the year 1911 did we develop a pistol of the accuracy and rapidity of fire demanded by our ordnance experts. The nations of Europe had neglected this valuable arm almost altogether, regarding it principally as a military ornament which only officers should carry. The result of Europe's neglect was that the small-caliber revolvers of the Germans and even of the French and English were toys in comparison with the big Colts that armed the American soldiers.

America owed the Colt .45 to the experiences of our fighters in the Philippines, and to the inventive genius of John Browning of machine-gun fame. In the earlier Philippine campaigns our troops used a .38-caliber pistol. Our soldiers observed that when the tough tribesmen were hit with these bullets and even seriously wounded they frequently kept on fighting for some time. What was needed was a hand weapon that would put the adversary out of fighting the instant he was hit, whether fatally or not. We therefore increased the caliber of the automatic pistol to .45 and slowed down the bullet so that it tore flesh instead of making a clean perforation. These improvements gave the missile the impact of a sledge hammer, and a man hit went down every time.

Moreover, in this development great improvement had been made in the accuracy of the weapon, the 1911 Colt being the straightest-shooting pistol ever produced in this country. Even the best of the older automatics and revolvers were accurate only in the hands of expert marksmen. But any average soldier with average training can hit what he shoots at with a Colt. The improvements in the automatic features brought it to the stage where it could be fired by a practiced man 21 times in 12 seconds. In this operation the recoil of each discharge ejects the empty shell and loads in a fresh one.

Only a few men of each infantry regiment carried pistols when our troops first went into the trenches. But in almost the first skirmish this weapon proved its superior usefulness in trench fighting. Such incidents as that of the single American soldier who dispersed or killed a whole squad of German bayoneteers which had surrounded him struck the enemy with fear of Yankee prowess with the pistol. The "tenderfoot's gun," as the westerners loved to call it, had come to its own.

By midsummer of 1917 the decision had been made to supply to the infantry a much more extensive equipment of automatic pistols than had previously been prescribed by regulations—to build them by hundreds of thousands where we had been turning them out by thousands. In February, with war in sight, realizing the limitations of our capacity then for producing pistols, the Colt automatic being manufactured exclusively by the Colt's Patent Firearms Manufacturing Co. at Hartford, Conn., and for a limited period by the Springfield Armory, we took up with the Colt Co. the proposition of securing drawings and other engineering data which would enable us to extend the production of this weapon to other plants. This work was in progress when in April, 1917, it was interrupted by the military necessity for calling upon every energy we had in the production of rifles.

In order to supplement the pistol supply, although the Colt automatic was the only weapon of this sort approved for the Army, the Secretary of War authorized the Chief of Ordnance to secure other small arms, particularly the double-action .45-caliber revolver as manufactured by both the Colt Co. and the Smith & Wesson Co. These revolvers had been designed to use the standard Army caliber-.45 pistol cartridges. The revolver was not so effective a weapon as the automatic pistol, but it was adopted in the emergency only to make it possible to provide sufficient of these arms for the troops at the outset.

At the start of hostilities the Colt Co. indicated that it could tool up to produce pistols at the rate of 6,000 per month by December, 1917, and could also furnish 600 revolvers a week beginning in April. As soon as funds were available we let a contract to the Colt Co. for 500,000 pistols and 100,000 revolvers, and to the Smith & Wesson Co. one for 100,000 revolvers. Although these contracts were not placed until June 15, in the certainty that funds would eventually be available both concerns had been working on the production of weapons on these expected contracts for many weeks.

When the order came from France to increase the pistol equipment, in addition to efforts to increase production at the plants of the two existing contractors we made studies of numerous other concerns which might undertake this class of manufacture. The proposal to purchase .38-caliber revolvers as a supplementary supply was abandoned for the reason that any expansion of this manufacture and of that for the necessary ammunition would be at the expense of the ultimate output of .45s and ammunition therefor.

In December, 1917, the Remington Arms-Union Metallic Cartridge Co. was instructed to prepare for the manufacture of 150,000 automatics, Colt model 1911, at a rate to reach a maximum production of 3,000 per day. Considerable difficulty was experienced in obtaining the necessary drawings and designs, because the manufacture of these pistols at the Colt Co. plant had been largely in the hands of expert veteran mechanics, who knew tricks of fitting and assembling not apparent in the drawings. The result was that the drawings in existence were not completely representative of the pistols. Finally complete plans were drawn up that covered all details and gave interchangeability between the parts of pistols produced by the Remington Co. and those by the Colt Co., which was the goal sought.

During the summer of 1918 in order to fill the enormously increased pistol requirements of the American Expeditionary Forces, contracts for the Colt automatic were given to the National Cash Register Co., at Dayton, Ohio; the North American Arms Co., Quebec; the Savage Arms Co., Utica, N. Y.; Caron Bros., Montreal; the Burroughs Adding Machine Co., Detroit, Mich.; the Winchester Repeating Arms Co., New Haven, Conn.; the Lanston Monotype Co., Philadelphia, Pa.; and the Savage Munitions Co., San Diego, Calif.

All of these concerns, none of which had ever before produced the .45-caliber pistol, were proceeding energetically with their preparations for manufacture when the armistice came to cancel their contracts. No pistols were ever obtained from any except the Colt's Patent Fire Arms Manufacturing Co. and the Remington Arms-Union Metallic Cartridge Co.

Difficulty was experienced in securing machinery to check the walnut grip for the pistols, and to avoid delay in production the Ordnance Department authorized the use of bakelite for pistol grips in all the new plants which were to manufacture the gun. Bakelite is a substitute for hard rubber or amber, invented by the eminent chemist Dr. Baekeland.

At the outbreak of the war the Army owned approximately 75,000 .45-caliber automatic pistols. At the signing of the armistice there had been produced and accepted since April 6, 1917, a total of 643,755 pistols and revolvers. The production of pistols was 375,404 and that of revolvers 268,351. In the four months prior to November 11, 1918, the average daily production of automatic pistols was 1,993 and of revolvers 1,233. This was at the yearly production rate of approximately 600,000 pistols and 370,000 revolvers. These pistols were produced at an approximate cost of $15 each.

Production of pistols and revolvers to Dec. 31, 1918.

	Pistols.			Revolvers.			Total pistols and revolvers.
	Colt.	Remington U. M. C.	Total pistols.	Colt.	Smith & Wesson.	Total revolvers.	
Apr. 6 to Dec. 29, 1917	58,500	....	58,500	20,900	9,513	30,413	88,913
January, 1918	11,000	....	11,000	8,700	7,500	16,200	27,200
February, 1918	14,500	....	14,500	8,800	8,550	17,350	31,850
March, 1918	21,300	....	21,300	11,800	12,400	24,200	45,500
April, 1918	22,400	....	22,400	10,400	10,650	21,050	43,450
May, 1918	35,000	....	35,000	11,100	12,150	23,250	58,250
June, 1918	37,800	....	37,800	11,100	14,250	25,350	63,150
July, 1918	39,800	....	39,800	11,600	11,555	23,155	62,955
August, 1918	40,400	....	40,400	11,300	13,358	24,658	65,058
September, 1918	32,100	640	32,740	11,100	12,650	23,750	56,490
October, 1918	42,300	3,881	46,181	13,500	16,675	30,175	76,356
November, 1918	45,800	4,102	49,902	11,900	12,660	24,560	74,462
December, 1918	24,600	4,529	29,129	9,500	11,400	20,900	50,029
Total	425,500	13,152	438,652	151,700	153,311	305,011	743,663

CHAPTER XII.

SMALL-ARMS AMMUNITION.

Prior to the war with Germany the Ordnance Department, in providing .30-caliber ammunition for our Army rifles and machine guns, had thought in terms of millions and had placed its ammunition orders on that scale. But when hostilities were at hand and steel and walnut were being assembled into rifles to arm the indefinitely increasing millions of Yankee soldiers that we would send and keep on sending to Europe until victory was ours, small-arms ammunition stepped out of the million class and became an industry whose units of production were reckoned by the billion.

The war increased the human strength of the American Army approximately thirty times. That ratio of increase was carried over into a production of ammunition for rifles and machine guns. The story of ammunition in the war is the story of a three-billion output forced from a hundred-million capacity. In this effort we find another of those frequent industrial romances which the war produced in America; for, when called upon to do more than an industrial possibility, as we regarded such things in 1917, the contriving executive and organizing ability and the skillful hands of the ammunition industry made good.

Our .30-caliber ammunition capacity in the United States prior to the war was about 100,000,000 cartridges per year. We actually produced in the war period the huge total of 3,507,023,300 small-arms cartridges. Pushed at feverish haste, such expansion naturally recorded its mistakes and its failures; but none of these was fatal or irremediable. The fact will always remain that a difficult art was enlarged in time to take care of every demand of the American Army for small-arms ammunition, and that no military operation on our part was held up by lack of this ammunition. Hence it is submitted that the production of small-arms cartridges was one of the genuine achievements of our Ordnance Department.

Let us consider first the production of the .30-caliber service ammunition, which may be regarded as the standard product of the ammunition industry. This was the ammunition used in our two service rifles, the Springfield or United States model of 1903 and the United States model of 1917, which is a modification of the British rifle, pattern 1914, and in most of the machine guns which we fired in France, although we used the 8-millimeter cartridge with the Chauchat machine rifle. When the war broke out we had on hand approximately 200,000,000 rounds of .30-caliber cartridges. Most of these had been manufactured by the Government at the Frankford Arsenal, which was, in fact, practically the only plant in the United States equipped to produce this ammunition in any appreciable quantities.

For some years prior to the war, however, the Government had adopted the policy of encouraging the manufacture of Army ammunition in private plants. This was done by placing with various concerns small annual orders for this type of ammunition. These orders were usually in the neighborhood of 1,000,000 rounds each. The purpose of such orders, insignificant as they were, was to scatter throughout the principal private ammunition factories the necessary jigs, fixtures, gauges, and other tooling required in the production of cartridges for Army rifles and machine guns. These small orders might also be expected to educate the operating forces of the private plants in this manufacture. By this means the Government hoped to have in an emergency a nucleus of skill and equipment which could be quickly expanded to meet war requirements.

As a further means of stimulating interest in this peace-time undertaking the Ordnance Department conducted each year a sort of competition among the private manufacturers of small-arms ammunition. The output of each factory accepting the Government orders was tested for proper functioning and accuracy; and those cartridges which won in this competition were used as the ammunition shot in the national rifle matches. Thus the winning concern could use its achievement in its advertising.

But these educational efforts on the part of the Government failed to create a capacity that was anywhere near to being adequate to meet the demands of such a war as that into which we were plunged in the year 1917. We had built up no large reserves of ammunition, and the orders placed with private manufacturers had been so small that they had resulted in virtually no factory preparation at all for great quantity production. To all practical purposes the entire ammunition manufacturing capacity of .30-caliber cartridges in 1917 was encompassed within the walls of the Frankford Arsenal.

There was, however, in the ammunition industry a fortunate condition existing when we entered the war. For some time numerous American concerns had been working on the manufacture of cartridges for both the British and the French Governments. The cartridges being turned out under these contracts were not suitable for our use, being of different caliber than those taken by American weapons, and this meant that the machinery in existence could not be converted to the production of American ammunition without radical and time-consuming alteration of tools, etc. However, cartridges are cartridges, regardless of their size; and the manufacture which was supplying France and England had resulted in educating thousands of mechanics and shop executives in the production of ammunition. Consequently, when we went into the war, we had the men and the skill ready at hand; we needed only to produce the tools and the machinery in addition to the raw materials.

Yet this in itself was a problem. How should we meet it? Three courses seemed to be possible for the Government. In the first place, we could build from the ground up an immense Government arsenal having an annual capacity of 1,000,000,000 rounds, or ten times that of the great Frankford Arsenal. Or we could interest manufacturers in a project of building a private cartridge factory capable of producing 1,000,000,000 rounds per year. Both of these methods were predicated on the assumption that the existing cartridge factories had their hands full with orders. The third plan was to place our cartridge demands with the existing ammunition plants and let them increase their facilities to take care of our orders.

As soon as the early orders had been given and all available capacity had been set going, this problem engaged the study and attention of the Ordnance Department. In the early fall of 1917 a meeting of the manufacturers of small-arms ammunition was held in Washington to discuss the matter. Principally on account of the difficulties in providing a trained working force for a new Government arsenal or private plant, the opinion was unanimous that the existing concerns should expand in facilities and trained personnel to handle the cartridge project. Out of this meeting grew the American Society of Manufacturers of Small Arms and Ammunition. Thereafter until the close of the war this society or its committees met about once every two weeks to discuss problems arising in the work. The officers of the Ordnance Department in charge of the ammunition project attended all of these meetings. The result of such cooperation was gratifyingly shown not only in the standardization of manufacturing processes in the various plants but also in the output of cartridges.

The success of this effort is best shown in the production figures in the period from April, 1917, to November 30, 1918. In that time the United States Cartridge Co. turned out 684,334,300 rounds of our caliber-.30 service ammunition; the Winchester Repeating Arms Co., 468,967,500 rounds; the Remington Arms-Union Metallic Cartridge Co., 1,218,979,300; the Peters Cartridge Co., 84,169,800; the Western Cartridge Co., 48,018,800; the Dominion Arsenal, 502,000; the Frankford Arsenal, 76,739,300; and the National Brass & Copper Tube Co., 22,700,400.

This production record to some extent was made possible by a leniency on the part of the Ordnance Department which we had not displayed before the war. When we could take plenty of time in ammunition manufacture our specifications for cartridges were extremely rigid. It soon became apparent that if we adhered to our earlier specifications we would limit the output of cartridges. It was found in a joint meeting of ordnance officers and ammunition manufacturers that certain increased tolerances could be permitted in our specifications without affecting the serviceability of the ammunition. Consequently new specifications for our war ammunition were drawn, enabling the plants to get into quantity production much more quickly than would have been possible if we had not relaxed our prewar attitude.

The ordinary service cartridge consists of a brass cartridge case, a primer, a propelling charge of smokeless powder, and a bullet made with a jacket or envelope of cupronickel inclosing a lead slug or core. Cupronickel is a hard alloy of copper and nickel. Steel would be the ideal covering for a bullet because of its cheapness and availability, but steel has not been used because it is liable to rust and to destroy the delicate rifling of the gun barrel. Cupronickel is a compromise, being strong enough to hold the interior lead from deforming, but not so hard as to wear down excessively the rifling in the gun barrel.

Even as we entered the war the long continued fighting in Europe had created a shortage in cupronickel, and by the time the armistice came it was apparent that this shortage would soon become so acute that we would have to find a substitute for cupronickel. This shortage had already occurred in Germany, where the enemy ordnance engineers had produced a bullet incased in steel which in turn was clothed with a slight covering of copper. The soft copper coating kept the steel from injuring the gun barrel. We ourselves were experimenting with copper-coated steel bullets when peace came, and would have been prepared to furnish a substitute had cupronickel failed us.

Some of the earliest ammunition sent to our forces in France developed a tendency to hang fire and to misfire; and a liberal quantity of it, amounting to six months' production of the Frankford Arsenal, was condemned and withdrawn from use. This matter was aired fully in the newspapers at the time it occurred. It developed that the faulty ammunition had been produced entirely in the Frankford Arsenal and that the cause of the trouble was the primer in the cartridge.

The primer in a cartridge performs the same function that the flint did on the old-fashioned squirrel guns—it touches off the explosive propellant charge. But whereas the flint sent only a spark into the powder, the modern primer produces a long, hot flame.

The primers in the ammunition manufactured at the Frankford Arsenal had given ordinarily satisfactory results in 12 years of peacetime use. The flame charge in this primer contained sulphur, potassium chlorate, and antimony sulphide. Produced under normal conditions, with plenty of time for drying, this primer was satisfactory. But sulphur when oxidized changes to an acid extremely corrosive to metal parts, and oxidized primers are liable not to function perfectly. Heat and moisture accelerate the change of sulphur to acid; and if there happens to be bromate in the potassium chlorate of the priming charge, the change is even more rapid.

An investigation of the Frankford Arsenal showed that these very elements were present. Because of the haste of production of cartridges, too much moisture had been allowed to get into the arsenal dry houses. The potassium chlorate was also found to contain appreciable quantities of bromate.

The condition was remedied by adopting another primer composition. And then, to play doubly safe, the Government specifications were amended to prevent the use of potassium chlorate containing more than 0.01 per cent of bromate.

However, this condemned ammunition was but a trifling fraction of the total output or even of the production then going on. The primers used by the various private manufacturers of ammunition functioned satisfactorily.

While we were not rigid in our specifications for the bulk of the service ammunition, in one respect we were most meticulous, and this was in respect to the ammunition used by the machine guns mounted on our airplanes. For these weapons we created an A-1 class of service .30-caliber cartridges, since it was highly important that there be no malfunctioning of ammunition in the air. Every cartridge of this class had to be specially gauged throughout its manufacture. This care resulted in a slower production of airplane cartridges than that of those for use on the ground, but we always had enough for our needs.

Until we went to war with Germany our Army had known only the cartridge firing the hard-jacketed lead bullet. But we entered a conflict in which several novel sorts of small-arms projectiles were in familiar use; and it became necessary for us to take up the manufacture of these strange missiles at once. These included such special types as tracer bullets to indicate the path of fire in the air, incendiary bullets for setting on fire observation balloons, hostile planes, and dirigible airships, and, finally, armor-piercing bullets for use against armor plate with which airplanes and tanks are equipped. We had developed none of these in this country before the war, except that in the Frankford Arsenal our designers had done some little experimental work with armor-piercing ammunition, in fact carrying it to the point of an efficient design.

One of the first acts of the Ordnance Department was to send an officer to visit the ammunition factories of France and England to study the methods of manufacturing these special types of bullets.

These friendly nations willingly gave us full information at first hand with respect to this complicated manufacture, which we were thus enabled to begin in September, 1917. Special machinery was required for loading the tracer bullet and also for producing the incendiary projectile. We adopted British practice for both of these. We ourselves were well equipped to begin the production of armor-piercing bullets, for which we had previously solved the problems of design; yet the production of metals to be used in this missile required some further experimental work. By February, 1918, however, our production of armor-piercing bullets was well under way and by the time the war came to an end we had produced nearly 5,000,000 of them.

The tracer bullet which we manufactured contained a mixture of barium peroxide and magnesium and in flight burned with the intensity of a calcium light. These bullets were principally used by machine gunners of aircraft, since in the air it is impossible to tell where machine-gun projectiles are going unless there is some device enabling the gunner to see the trajectory of the bullets. This is done by inserting tracer bullets at intervals in the belts of cartridges fed into the machine gun. The common conception of a tracer bullet is one that leaves a trace of smoke in its flight; whereas the truth is that our tracer and the British tracer were practically smokeless, the gunner observing the direction of aim by following the bright lights of the tracer bullets with his eye. These lights were plainly visible in the brightest sunlight. Although the slight quantity of the flaming mixture burned but a few seconds, it was sufficient to trace the flight for 500 yards or more from the muzzle of the machine gun.

The tracer bullet consisted of a cupronickel shell, the nose of which contained a leaden core to balance the bullet properly. The rear chamber of the bullet held a cup containing the mixture of barium peroxide and magnesium. The rear end of the bullet was left slightly open, and through this opening the mixture was ignited by the hot flame of the propelling powder discharge.

An entirely different principle was used in the construction of the incendiary bullet. This bullet was also incased in cupronickel; but the incendiary chemical, which was phosphorus, was contained in a chamber in the nose of the bullet. A serrated plug held the phosphorus in its chamber, and behind this plug was a solid plug of lead coming flush with the base of the bullet and soldered thereto. On one side of the missile was a hole drilled through the cupronickel into one of the grooves of the serrated plug. This hole was stopped by a special kind of solder. The heat of friction developed in the infinitesimal space of time while the projectile was passing through the gun barrel served the double purpose of melting out the solder

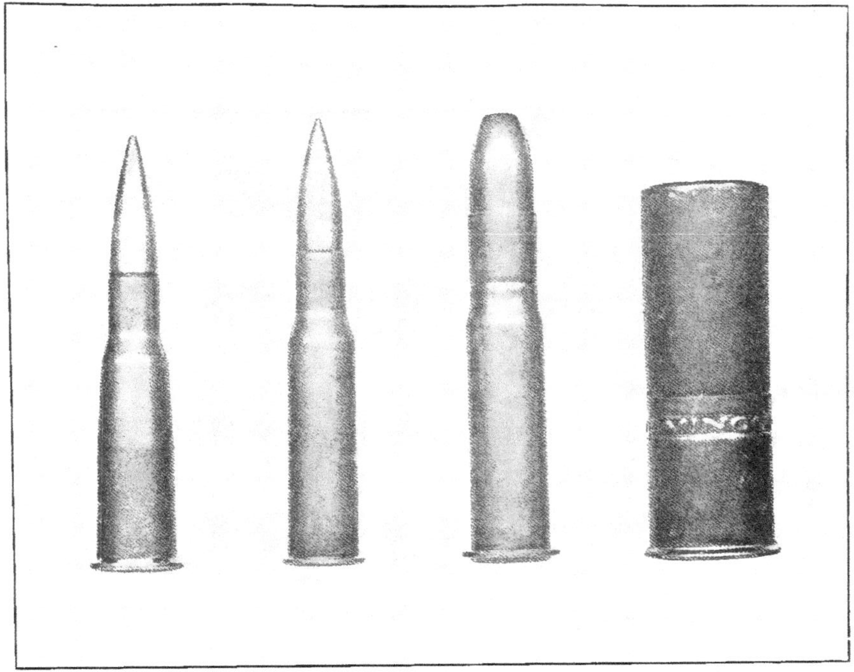

8-MILLIMETER FRENCH CARTRIDGE, .303 BRITISH CARTRIDGE—(MACHINE GUN), 11-MILLIMETER FRENCH INCENDIARY BULLET, AND A SHOT-GUN SHELL.

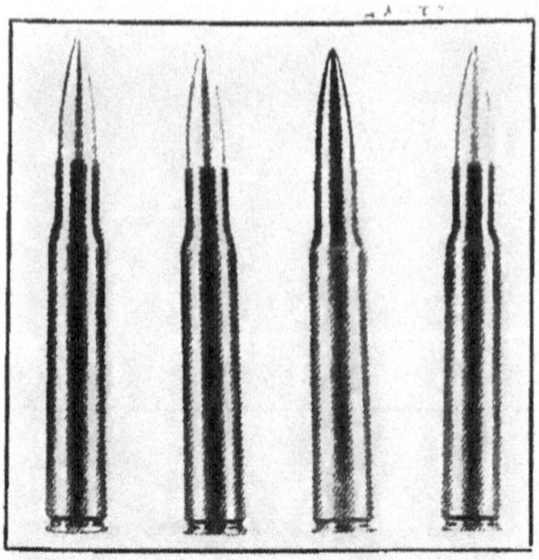

LEFT TO RIGHT—ARMOR PIERCING, TRACER, INCENDIARY, ORDINARY.

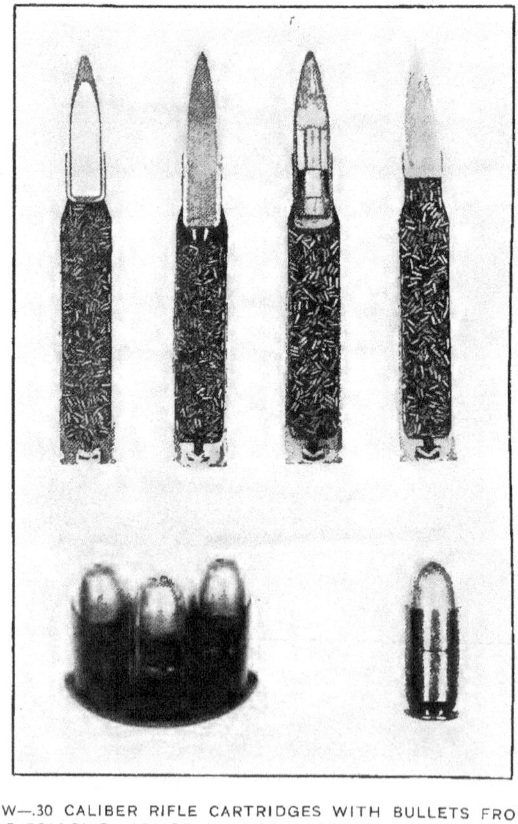

UPPER ROW—.30 CALIBER RIFLE CARTRIDGES WITH BULLETS FROM LEFT TO RIGHT, AS FOLLOWS: ARMOR PIERCING, TRACER, INCENDIARY, ORDINARY. LOWER ROW—.45 CALIBER AMMUNITION IN CLIP FOR REVOLVER; .45 CALIBER AMMUNITION FOR PISTOL.

from the hole and igniting the phosphorus within the chamber. Thereafter the centrifugal force of the revolving bullet whirled the burning phosphorus out through the unplugged hole. Seen in the air the fire of the phosphorus could not be discerned, but the burning chemical threw off considerable smoke, so that the eye of the gunner could follow the blue spiral to its mark. Our incendiary bullet had an effective range of 350 yards, after which distance the phosphorus was burned out.

Equally interesting was the construction of the armor-piercing bullet. Heavy and solid as the jacketed lead bullet used in our service guns seems to be, when fired against even light armor plate it leaves only a small mark upon its objective. As soon as the cupronickel jacket strikes the armor plate it splits and the lead core flattens out and flies into fragments. The armor plate may not even be dented by this impact. Yet change the core of this missile from lead to hardened steel and an entirely different result is produced. Our armor-piercing bullet was made with a cupronickel jacket for the sake of the gun barrel. The inner side of this jacket was lined with a thin coat of lead which was made thicker in the nose of the bullet. Finally a core of specially heat-treated steel completed the construction of the projectile. When this missile is fired against armor plate the jacket splits and the lead lining virtually disappears from the impact, but the pointed steel core keeps on and bores a hole through the plate as it might through soft wood.

The production figures show the degree of success which we attained in the manufacture of this special ammunition. Up to November 30, 1918, the E. I. du Pont de Nemours Co. had produced 6,057,000 tracer cartridges of .30 caliber and 1,560,000 incendiary cartridges of the same size. The Frankford Arsenal turned out 22,245,000 tracer cartridges of this size, 14,148,000 incendiary cartridges, and 4,746,900 armor-piercing cartridges. We placed an additional order for armor-piercing projectiles with the Dominion arsenals, which delivered to us 1,980,000 of such cartridges.

We also set out to develop new manufacturing facilities for the production of this special aircraft ammunition. Excellent tracer bullets were produced by the National Fireworks Co., of West Hanover, Mass., and that company was getting into a satisfactory production stride when the armistice was signed. The Hero Manufacturing Co., of Philadelphia, Pa., also was turning out an approved incendiary bullet when peace came. These various special bullets were loaded in cartridges at the Frankford Arsenal.

When the fighting ceased we were working on the development of armor-piercing bullets that would also be incendiary; and armor-piercing bullets that would also contain a tracing mixture. It was thought that bullets of these types would be particularly valuable for aircraft use. While we had done considerable experimenting along both lines, no satisfactory types had yet been developed.

There was another class of small arms for which we also had to produce ammunition on a war scale. Our automatic pistols and revolvers demanded .45-caliber ball cartridges. In normal times the Frankford Arsenal had been almost our sole producer of these cartridges, and it had attained an annual output of approximately 10,000,000 rounds of them. This quantity was nowhere nearly adequate for our war needs, especially after the decision to equip our troops much more numerously with pistols and revolvers than had formerly been the case.

Consequently it was necessary for us to develop additional manufacturing facilities for .45-caliber ammunition. We did this by placing orders with some of the same manufacturers who were developing the .30-caliber production. Because it was necessary for us to give preference always to the rifle and machine-gun ammunition, the manufacture of pistol cartridges was not carried through as rapidly as some other phases of the ammunition program. However, a satisfactory output was reached in time to meet the immediate demands of our forces in the field, and this production was expanding and keeping ahead of the increased needs for this sort of cartridges. The total war production of .45-caliber ammunition by the various factories was as follows:

United States Cartridge Co	75,500,000
Winchester Repeating Arms Co	46,446,800
Remington Arms-Union Metallic Cartridge Co	144,825,700
Peters Cartridge Co	55,521,000
Frankford Arsenal	12,349 200

Early in 1918 our Air Service field forces saw the need of a machine gun of larger caliber than the quick-firing weapons in general use. The flying service of the principal allies had developed an 11-millimeter machine gun for use in attacking the captive balloons of the enemy. This gun fired a projectile only slightly less than one-half inch in diameter. To meet this new demand our Ordnance Department found at the Colt factory about 1,000 Vickers machine guns which were being built on order for the former Russian Government. The department took over these guns and modified them to take 11-millimeter ammunition, and that step made it necessary for us to produce machine-gun cartridges for these new weapons.

We at once developed a modified French 11-millimeter tracer incendiary cartridge, which in later use proved to be highly satisfactory. In an experimental order the Frankford Arsenal turned out about 100,000 of these cartridges, while at the time the armistice was signed the Western Cartridge Co. was prepared to produce this class of ammunition on a large scale.

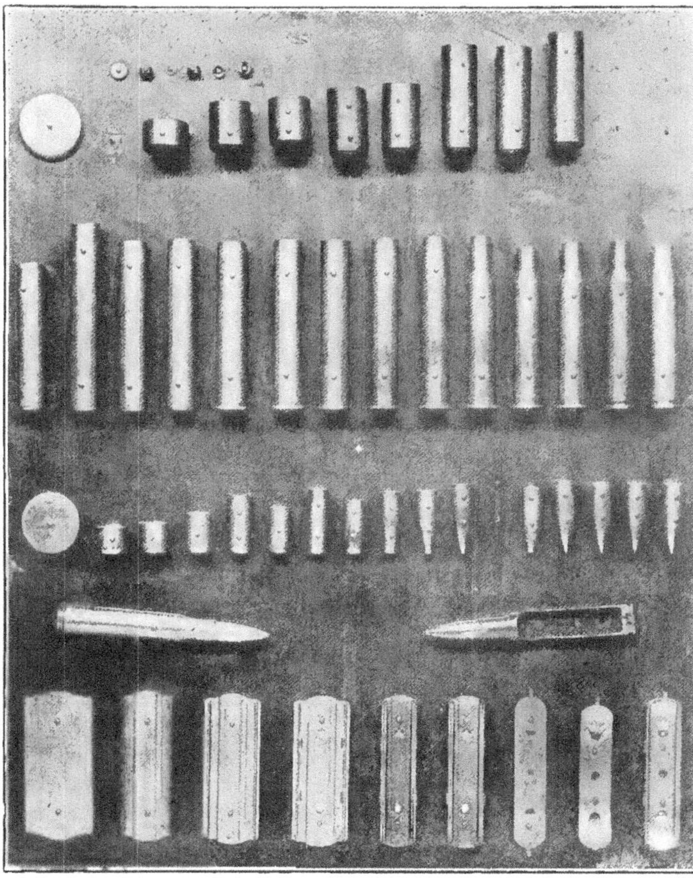

VIEW SHOWING SUCCESSIVE STAGES IN THE MANUFACTURE OF THE PRIMER BULLET AND CLIP FOR .30 CALIBER, MODEL OF 1906, AMMUNITION.

The top row shows the development of the primer cup and anvil. The second and third rows show the development in the manufacture of the cartridge case. The fourth and fifth rows show the development in the manufacture of the bullet jacket and the lead slug that fits into the jacket and finally the finished cartridge. The bottom row shows the development in the manufacture of the cartridge clip.

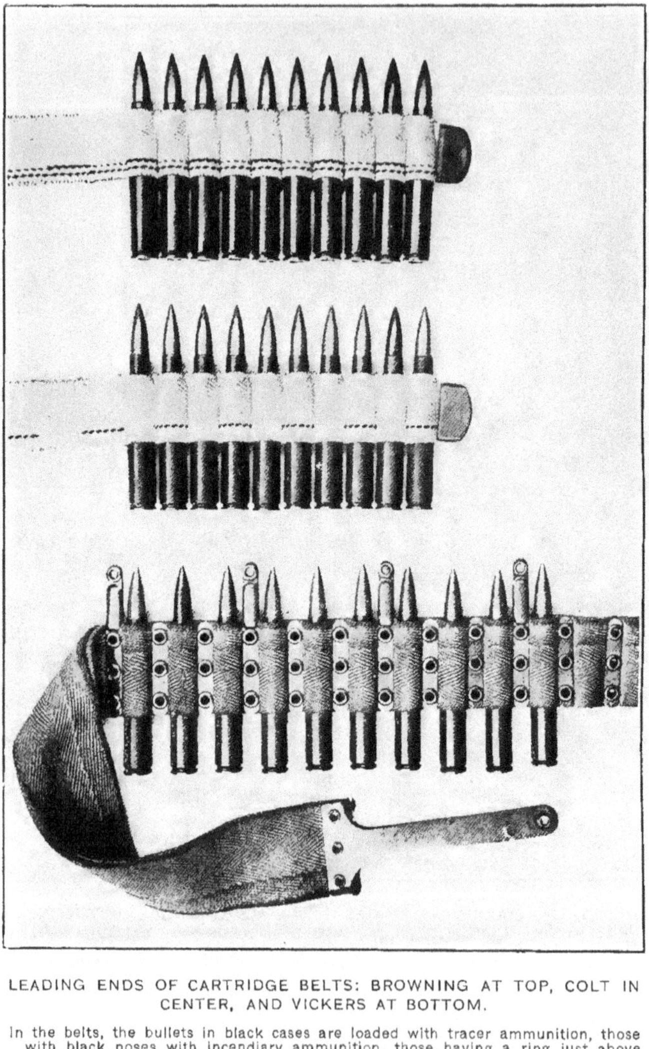

LEADING ENDS OF CARTRIDGE BELTS: BROWNING AT TOP, COLT IN CENTER, AND VICKERS AT BOTTOM.

In the belts, the bullets in black cases are loaded with tracer ammunition, those with black noses with incendiary ammunition, those having a ring just above the bullet casing with armor-piercing ammunition, while the rest are ordinary service cartridges.

Certain American concerns before April, 1917, had been producing 8-millimeter ammunition for the French government for use in its machine guns. When we entered the war our Ordnance Department found it necessary to continue the manufacture of these cartridges for the machine guns obtained from the French. Up to November 30, 1918, a total of 269,631,800 rounds had been produced under our supervision. These cartridges were manufactured by the Western Cartridge Co. and by the Remington Arms Co. at its Swanton plant.

How well and amply we were producing ammunition for our machine guns and rifles is indicated by the fact that our average monthly production, based upon our showing in July, August, and September, 1918, was 277,894,000 rounds as against a monthly average for Great Britain of 259,769,000 rounds and for France of 139,845,000.

Our total production of machine-gun and rifle ammunition during the 19 months of warfare was 2,879,148,000 rounds, while in that period England produced 3,486,127,000 rounds and France 2,983,675,000, but it must be remembered that they had been keyed up to that voluminous production by three years of fighting and that our monthly production rate indicated we would soon far surpass them in quantities.

The following table shows how our total production of ammunition for all small arms, including machine guns, rifles, pistols, and revolvers, grew month by month during the war:

	Rounds.		Rounds.
Nov. 30, 1917	156, 102, 792	July 31, 1918	2, 306, 999, 284
Dec. 31, 1917	351, 117, 928	Aug. 31, 1918	2, 623, 847, 546
Jan. 31, 1918	573, 981, 712	Sept. 30, 1918	2, 942, 875, 786
Feb. 28, 1918	760, 485, 688	Oct. 31, 1918	3, 236, 396, 100
Mar. 31, 1918	1, 021, 610, 956	Nov. 30, 1918	3, 507, 023, 300
Apr. 30, 1918	1, 318, 298, 492	Dec. 31, 1918	3, 741, 652, 200
May 31, 1918	1, 616, 142, 052	Jan. 31, 1919	3, 940, 682, 744
June 30, 1918	1, 958, 686, 784		

CHAPTER XIII.

TRENCH-WARFARE MATERIAL.

Like many of the other war implements produced by the Ordnance Department for use in France, the weapons employed in fighting from the trenches were entirely novel to American industry; and in the production of them we find the same story of the difficulties in the adoption of foreign designs, of the development of our own designs, of the delays encountered and mistakes made in equipping a new industry from the ground up, but, finally, of the triumphant arrival at quantity production in a marvelously brief time, considering the obstacles which had to be overcome.

When the movements of armies in the great war ceased and they were held in deadlock in the trenches, the fighters at once began devising weapons with which they could kill each other from below ground. For this purpose they borrowed from the experience of man running back to time immemorial. They took a leaf from the book of the Roman fire-ball throwers and developed the hand grenade beyond the point to which it had been brought in the European warfare of the last century. They called upon an industry which had once existed solely for the amusement of the people, the fireworks industry, for its golden rain and rainbow-hued stars for signals with which to talk to each other by night. Other geniuses of the trenches took empty cannon cartridges and, setting them up as ground mortars, succeeded in throwing bombs from them across No Man's Land into the enemy ranks. They even for a time resurrected the catapult of Trojan days, although this device attained no great success. But from all such activities new weapons of warfare sprang, crude at first, but later refined as only modern science and manufacture could perfect them.

America entered the war when this development of ordnance novelties had reached an advanced state. It became necessary for us, then, to make a rapid study of what had been done and then go ahead with our own production either from foreign designs or with inventions of our own.

To this end in April, 1917, a few days after we declared war with Germany, the Trench Warfare Section was organized within the Ordnance Department and given charge of the production of these novelties. The section did not entirely confine itself to trench-warfare materials, since one of its chief production activities was concerned with the manufacture of the various sorts of bombs to be dropped from airplanes. Also, at the start of its existence it had charge of the production of implements for fighting with poison gas and flame. Although in large part this phase of its work was taken away from it in the summer of 1917 and was later placed under the jurisdiction of the newly organized Chemical Warfare Service, the Trench Warfare Section continued to conduct certain branches of gas-warfare manufacture, in particular the production of the famous Livens projectors of gas and also the manufacture of the portable toxic-gas sets for producing gas clouds from cylinders.

All in all, the Trench Warfare Section was charged with the responsibility of producing some 47 devices, every one of them new to American manufacture and some extremely difficult to make. The backbone of the program consisted of the production of grenades, both of the hand-thrown and the rifle-fired variety, trench mortars, trench-mortar ammunition, pyrotechnics of various sorts, and bombs for the airplanes, with their sighting and release mechanisms.

In the production of these new devices there arose a new form of cooperation between Government and private manufacturers under the tutelage of the Trench Warfare Section. The manufacturers engaged in the production of various classes of these munition novelties joined in formal associations. There was a Hand Grenade Manufacturers' Association, under the capable leadership of William Sparks, president of the Sparks-Withington Co., of Jackson, Mich.; the Drop Bomb Manufacturers' Association, headed by J. L. Sinyard, president of A. O. Smith Corporation, Milwaukee; the Six-inch Trench-mortar Shell Manufacturers' Association, R. W. Millard, president of Foster-Merriam Co., Meriden, Conn.; the Rifle Grenade Manufacturers' Association, under the leadership of F. S. Briggs, president of the Briggs & Stratton Co., Milwaukee, Wis.; and the Livens Projector Manufacturers' Association. A similar association of manufacturers engaged in army contracts existed in the production of small-arms ammunition; but in no other branch of the Ordnance Department was the development of such cooperation carried on to the extent of that fathered by the Trench Warfare Section.

The existence of these associations was of inestimable benefit in securing the rapid development, standardization for quantity manufacture, and production of these strange devices. Each association had its president, its other officers, and its regular meetings. These meetings were attended by the interested officers of the Trench Warfare Section. In the meetings the experiments of the manufacturers and the short-cut methods developed in their shops were freely discussed; and, if modifications of design were suggested, such questions were thrashed out in these meetings of practical technicians, and all of the contractors simultaneously received the benefits.

The Trench Warfare Section produced its results under the handicap of being low in the priority ratings, many other items of ordnance being considered in Washington of more importance than the trench-fighting materials and therefore entitled to first call upon raw materials and transportation. In the priority lists the leader of 47 trench-warfare articles, the 240-millimeter mortars, stood twenty-second, and the others trailed after.

GRENADES.

The first of the trench-warfare weapons with which the rookie soldier became acquainted was the hand grenade, since this, at least in its practice or dummy form, was supplied to the training camps in this country. To all intents and purposes the hand grenade was a product of the war against Germany, although grenades had been more or less used since explosives existed. All earlier grenades had been crude devices with only limited employment in warfare, but in the three years preceding America's participation in the war the grenade had become a carefully built weapon.

The extent of our production of hand grenades may be seen in the fact that when the effort was at its height 10,000 workers were engaged exclusively in its manufacture. The firing mechanism of the explosive grenades which we built was known as the Bouchon assembly. In the production of this item 19 of every 20 workers were women. In fact no other item in the entire ordnance field was produced so exclusively by women. Incidentally, at no time during the war was there a strike in any grenade factory.

For a long time in the trenches of France only one type of hand grenade was used. This was the so-called defensive grenade, built of stout metal which would fly into fragments when the interior charge exploded. As might be expected, such a weapon was used only by men actually within the trenches, the walls of which protected the throwers from the flying fragments. But, as the war continued, six other distinct kinds of grenades were developed, America herself contributing one of the most important of them; and during our war activities we were engaged in manufacturing all seven.

The defensive, or fragmentation, type grenade was the commonest, most numerous, and perhaps, the most useful of all of them. Another important one, however, was that known as the offensive grenade, and it was America's own contribution to trench warfare. The body of the offensive grenade was made of paper, so that the deadly effect of it was produced by the flame and concussion of the explosion itself. It was quite sure to kill any man within 3 yards of it when it went off, but it was safe to use in the open offensive movements, since there were no pieces of metal to fly back and hit the thrower.

A third development was known as the gas grenade. It was built of sheet metal, and its toxic contents were effective in making enemy trenches and dugouts uninhabitable. A fourth, a grenade of similar construction, was filled with phosphorus, instead of gas, and was known as the phosphorus grenade. This grenade scattered burning phosphorus over an area 3 to 5 yards in diameter and released a dense cloud of white smoke. In open attacks upon machine-gun nests phosphorus grenades were thrown in barrages to build smoke screens for the attacking forces.

As a fifth class there was a combination hand and rifle grenade, a British device adopted in our program. The sixth class of grenades was known as the incendiary type. These were paper bombs filled with burning material and designed for use against structures intended to be destroyed by fire. Finally, in the seventh class were the thermit grenades, built of terneplate and filled with a compound containing thermit, which develops an intense heat while melting. Thermit grenades were used principally to destroy captured guns. One of them touched off in the breech of a cannon would fuse the breech-block mechanism and destroy the usefulness of the weapon.

All of these grenades except the incendiary grenades used the same firing mechanism, and the incendiary grenade firing mechanism was the standard one modified in a single particular.

The earliest American requirement in this production was for defensive grenades, of the fragmentation type. Our first estimate was that we would need 21,000,000 of these for actual warfare and 2,000,000 of the unloaded type for practice and training work. But, as the war continued and the American plans developed in scale, we saw we would require a much greater quantity than this; and orders were finally placed for a total of 68,000,000 live grenades and over 3,000,000 of the practice variety.

By August 20, 1917, the Trench Warfare Section had developed the design and the drawings for the defensive grenade. The first contract—for 5,000 grenades—was let to the Caskey-Dupree Co. of Marietta, Ohio. This concern was fairly entitled to such preference, because the experimentation leading up to the design for this bomb was conducted almost entirely at its plant in Marietta.

Next came an interesting industrial development by a well-known American concern which had previously devoted its exclusive energy to the production of high-grade silverware, but which now, as a patriotic duty, undertook to build the deadly defensive grenades. This was the Gorham Manufacturing Co. of Providence, R. I. This firm contracted to furnish complete, loaded grenades, ready for shipment overseas, and was the only one to build and operate a manufacturing and loading plant. Elsewhere contracts were let for parts only, these parts to converge at the assembling plants later; and such orders were rapidly placed until by the middle of December, 1917, various industrial concerns were tooling up for a total production of 21,000,000 of these missiles.

The grenade which these contractors undertook to produce was an American product in its design, although modeled after grenades already in use at the front. Its chief difference was in the firing mechanism, where improvements, or what were then thought to be improvements, had been installed to make it safer in the hands of the soldier than the grenades then in use at the front. This firing mechanism with its pivoted lever was, in fact, a radical departure from European practice. The body of this grenade was of malleable iron, and the grenade exploded with a force greater than that of any in use in France.

The remodeling of factories, the building of machines, and the manufacture of tools for this undertaking, pushed forward with determined speed, was completed in from 90 to 120 days, and by April almost all of the companies had reached the stage of quantity production.

And then, on May 9, 1918, came a cablegram from the American Expeditionary Forces that brought the entire effort to an abrupt halt. The officers of the American Expeditionary Forces in no uncertain terms condemned the American defensive grenade. The trouble was that in our anxiety to protect the American soldier we had designed a grenade that was too safe. The firing mechanism was too complicated. In the operation required to touch off the fuse five movements were necessary on the part of the soldier, and in this the psychology of a man in battle had not been taken sufficiently into consideration. The well-known story of the negro soldier who, in practice, threw his grenade too soon because he could feel it "swelling" in his hand, applies to most soldiers in battle. In using the new grenade the American soldier would not go through the operations required to fire its fuse. Cases came to light, too, showing that in the excitement of battle the American soldier forgot to release the safety device, thus giving the German an opportunity to hurl back the unexploded grenade.

As the result of this discovery all production was stopped in the United States and the ordnance engineers began redesigning the weapon. The incident meant that 15,000,000 rough castings of grenade bodies, 3,500,000 assembled but empty grenades and 1,000,000 loaded grenades had to be salvaged, and that on July 1, 1918, the production of live fragmentation grenades in this country was represented by the figure zero. Some of the machinery used in the production of the faulty grenades was useless and had to be replaced by new, while the trained forces which had reached quantity production in April had to be disbanded or transferred to other work while the design was being changed.

By August 1 the new design had been developed on paper and much of the new machinery required had been produced and installed in the plants, which were ready to go ahead immediately with the production. It is a tribute to the patriotism of the manufacturers who lost time and money by this change that little complaint was heard from them by the Government.

In the production of hand grenades the most difficult element of manufacture and the item that might have held up the delivery of completed mechanisms was the Bouchon assembly. There was an abundant foundry capacity in the United States for the production of gray-iron castings for grenade bodies, and so this part of the program gave no anxiety. The Bouchon assembly threatened to be the choke point. In order to assure the success of defensive-grenade production, the Precision Castings Co. of Syracuse, N. Y., and the Doehler Die Castings Co. of Toledo, Ohio, and Brooklyn, N. Y., worked their plants 24 hours a day until they had built up a reserve of Bouchons and screw plugs and removed all anxiety from that source. The total production of Bouchons eventually reached the figure 64,600,000.

The first thought of the Ordnance Department was to produce grenades by the assembling and quantitative method; that is, by the production of parts in various plants and the assembling of those parts in other plants. But, due to delay in railway shipments and difficulties due to priorities, it was discovered that this method of manufacture, however adaptable it might be to other items in the ordnance program, was not a good thing in grenade production; and when the war ended the tendency was all in the direction of having the assembly contractors produce their own parts either by purchase from subcontractors or by manufacture in their own plants.

The orders for the redesigned grenades called for the construction of 44,000,000 of them. So rapidly had the manufacturers been able to reach quantity production this time that a daily rate of 250,000 to 300,000 was attained by November 11, 1918, and by December 6, less than a month after the fighting stopped, the factories had turned out 21,054,339 defensive grenades.

It should be remembered that the great effort in ordnance production in this country was directed toward the American offensive expected on a tremendous scale in the spring of 1919. Had the war continued the fragmentation grenade program, in spite of the delays encountered in its development, would have produced a sufficient quantity of these weapons.

Special consideration is due the following-named firms for their efforts in developing the production of defensive grenades:

Caskey-Dupree Co., Marietta, Ohio.
Spacke Machine & Tool Co., Indianapolis, Ind.
Stewart-Warner Speedometer Co., Chicago, Ill.
Miami Cycle & Manufacturing Co., Middletown, Ohio.
American Radiator Co., Buffalo, N. Y.
International Harvester Co., Chicago, Ill.
Doehler Die Castings Co., Brooklyn, N. Y.
Precision Castings Co., Syracuse, N. Y.

The American offensive grenade was largely the production of the Single Service Package Corporation of New York, both in the development of its design and in its manufacture. The body of this grenade was built of laminated paper spirally wound and waterproofed by being dipped in paraffine. The top of this body was a die casting, into which the firing mechanism was screwed. Practically no changes were made in the design of this weapon from the time it was first produced, and the production record is an excellent one.

Our earliest thought was that we would need some 7,000,000 of these grenades and orders for that quantity of bodies were placed in January and March, 1918, with the Single Service Package Corporation. Then it became necessary to discover factories which could produce the metal caps. The orders for these were first placed with the Acme Die Castings Co. and the National Lead Casting Co. for 3,375,000 castings from each concern. But these companies failed to make satisfactory deliveries, and in May, 1918, a contract for 5,000,000 caps was let to the Doehler Die Castings Co. which reached quantity production in August. After that the Single Service Package Corporation, the chief contractor, forged ahead in its work and on November 11 was producing the bodies for offensive grenades at the rate of 55,000 to 60,000 daily. By December 6, 1918, the Government had accepted 6,179,321 completed bodies. The signing of the armistice brought to an end a project to build 17,599,000 additional grenades of this type.

The production of gas grenades offered some peculiar difficulties. We set out at first to produce 3,684,530 of them. By January, 1918, the engineers of the Ordnance Department had completed the plans and specifications for the American gas grenade, and on February 12, an order for 1,000,000 of them was placed with the Maxim Silencer Co., of Hartford, Conn.

The gas grenades were to be delivered at the filling plants complete except for the detonator thimbles, which seal both gas and phosphorus grenades and act as sockets for the firing mechanism. It was seen that the construction of these thimbles might be a choke point in the construction of grenades of both types, and orders were early placed for them—1,500,000 to be delivered by the Maxim Silencer Co. and an equal quantity by the Bassic Co., of Bridgeport, Conn. On December 6, 1918, these concerns had produced 1,982,731 detonator thimbles.

The body of the gas grenade is built of two sheet-metal cups welded together to be gas-tight. Since, when we started out on this production, we did not know what kind of gas would be used or at what pressure it would be held within the grenade, we set the specifications to make grenade bodies to hold an air pressure of 200 pounds. The welding of the cups frequently failed to hold such pressure, so that the rejections of gas-grenade bodies under this test ran as high as 50 per cent. But in June, 1918, the gas for the grenades had been developed, and we were thereupon able to reduce the pressure of the standard test to 50 pounds. Under such a test the bodies readily passed inspection.

In September, 1918, we let additional contracts for gas grenades—500,000 to the Evinrude Motor Co., of Milwaukee; 500,000 to the John W. Brown Manufacturing Co., of Columbus, Ohio; and 400,000 to the Zenite Metal Co., of Indianapolis.

On November 11 gas grenade bodies were being produced at the rate of 22,000 per day, and the total production up to December 6 was 936,394.

The phosphorus grenade was similar to the gas grenade in construction. The plans and specifications for this weapon were ready in January, 1918. In February the following contracts were let: Metropolitan Engineering Co., Brooklyn, N. Y., 750,000; Evinrude Motor Co., Milwaukee, 750,000; Zenite Metal Co., Indianapolis, 500,000. On December 6, 1918, these concerns had delivered a total of 521,948 phosphorus grenade bodies.

The difficulties which had been experienced in the production of gas grenades were repeated in this project. The Evinrude Co. was especially quick in getting over the obstacles to quantity production. The Metropolitan Engineering Co. was already engaged with large orders for adapters and boosters in the heavy-gun ammunition manufacture for the Ordnance Department and found that the order for phosphorus grenades conflicted to a considerable extent with its previous war work. The matter was thrashed out in the Ordnance Department, which gave the priority in this plant to the adapters and boosters, with the result that this firm was able to make only a small contribution to the total production of phosphorus grenade bodies.

The development of thermit grenades was still in the experimental stage when the armistice was signed. There was no actual production in this country of grenades of this character. In October, however, the development of the grenade in design had reached a stage where we felt justified in letting a contract for 655,450 die-casting parts to the Doehler Die Castings Co., at its Toledo plant, and for an equal number of bodies with firing-mechanism assemblies to the Stewart-Warner Speedometer Corporation at Chicago.

The incendiary grenade not only did not get out of the development stage, but even a perfected model was regarded as of doubtful value by the officers of the American Expeditionary Forces. Nevertheless, the Chemical Warfare Service was of the opinion that such a grenade should be worked out, and an order for 81,000 had been given to the Celluloid Co., of Newark, N. J. Experimental work was progressing satisfactorily when the armistice was signed.

When the war ended, we were adapting to American manufacture a combination hand and rifle phosphorus grenade, borrowed from the English. The body of this grenade was built of terneplate and it had a removable stem, so that it could be thrown by hand or fired from the end of a service rifle. The American Can Co. built 1,000 of these to try out the design and strengthen the weak features.

Article.	Completed to Nov. 8, 1918.	Completed to Feb. 1, 1919.	Sent overseas.
Dummy hand grenade	415,870	415,870	....
Practice hand grenade	3,605,864	3,605,864	....
Defensive hand grenade	17,477,245	25,312,794	516,533
Offensive hand grenade	5,359,321	7,000,000	173,136
Gas hand grenade	635,551	1,501,176	249,239
Phosphorus hand grenade	505,192	521,948	150,600
Thermit hand grenade	....	....	....

NOTE.—In above figures all grenades are unloaded with the exception of those sent overseas, which were loaded.

RIFLE GRENADES.

In the construction of our rifle grenades there was another unfortunate experience due to a faulty design. The rifle grenade fits in a holder at the muzzle of an ordinary service rifle. When the rifle is fired the bullet passes through a hole in the middle of the grenade, and the gases of the discharge following the bullet throw the grenade approximately 200 yards. Any man within 75 yards of an exploding rifle grenade is likely to be wounded or killed. The rifle grenade is used both as a defensive and offensive weapon, since the firer is well out of range of the exploding missile.

In developing a rifle grenade for American manufacture our engineers adopted the French Viven-Bessiere type. The French service ammunition is larger than ours, and it was therefore necessary to design our grenade with a smaller hole. But in the anxiety to produce this weapon in the shortest time possible the models were not sufficiently tested, and no consideration was taken of the difference in design between a French bullet and an American bullet. The result was that the French grenade did not function well with our

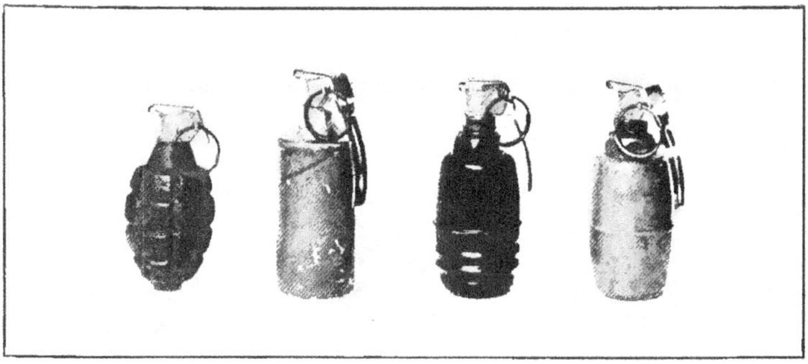

LEFT TO RIGHT—DEFENSIVE HAND GRENADE, OFFENSIVE HAND GRENADE, GAS HAND GRENADE, PHOSPHORUS HAND GRENADE.

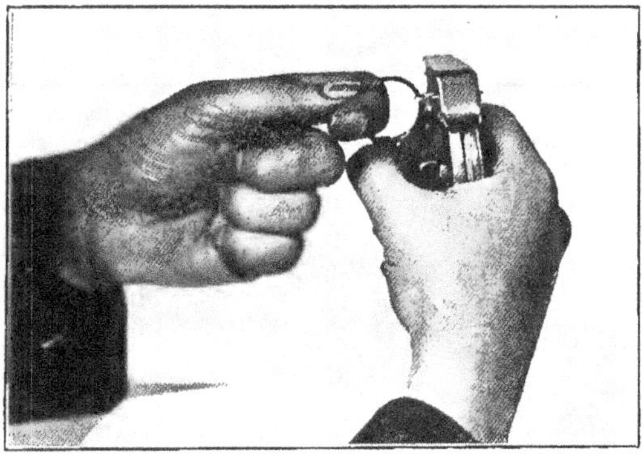

THROWING A HAND GRENADE. FIRST OPERATION: WITHDRAWING THE COTTER PIN.

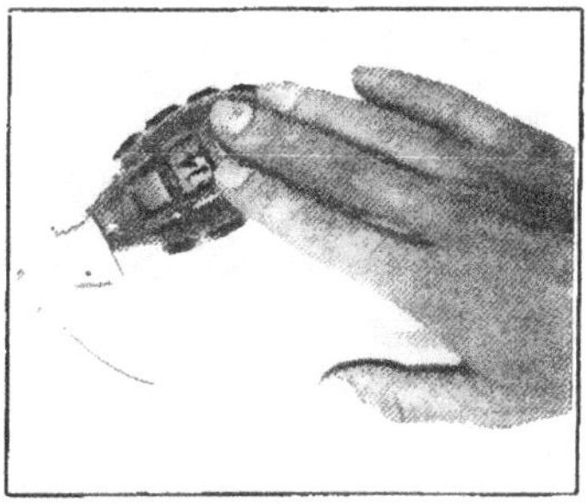

THROWING A HAND GRENADE. SECOND OPERATION: RELEASING THE GRENADE AND FIRING LEVER.

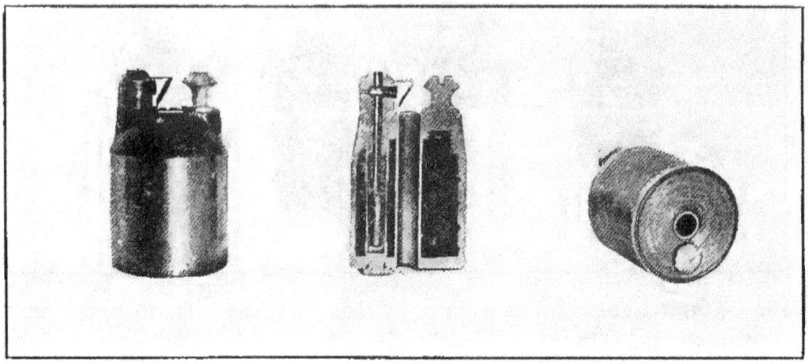

V. B. RIFLE GRENADE, LIVE, MARK I.

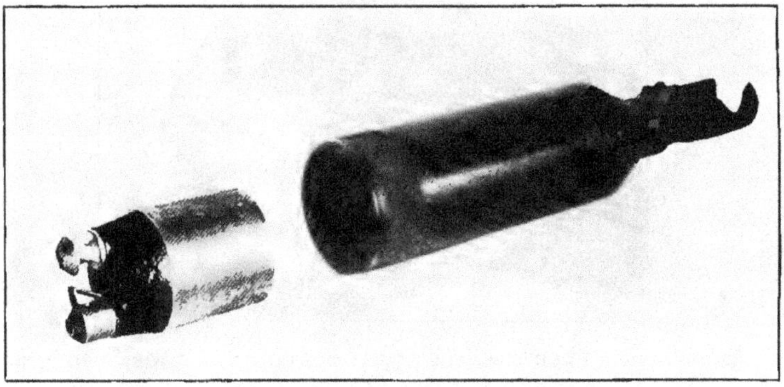

V. B. RIFLE GRENADE, MARK I, AND DISCHARGER.

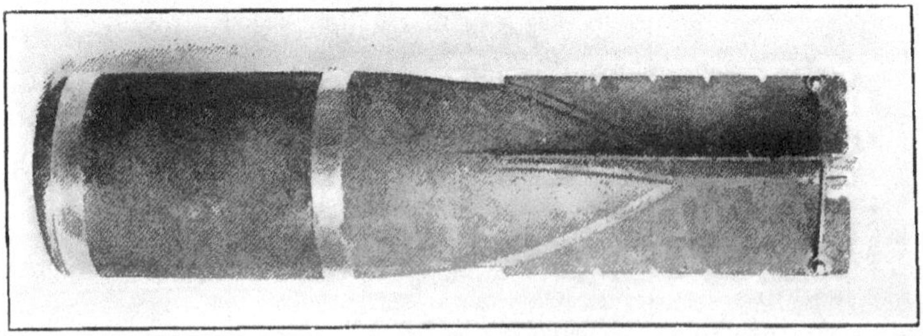

6-INCH TRENCH MORTAR SHELL.

ammunition, due to the splitting of the Springfield bullet as it passed through the grenade. The result was that in May, 1918, several months after the manufacture of this grenade had been in progress, the entire undertaking was canceled pending the development of new designs; and 3,500,000 completed grenades had to be salvaged.

The original contract for rifle grenades had been let to the Westinghouse Electric & Manufacturing Co. of Pittsburgh. This called for the production of all parts by the Westinghouse Co. and the assembling of them in the Westinghouse plant to the number of 5,000,000 grenades. But there was such a diversity of material employed in the manufacture of rifle grenades that succeeding contracts were let for parts and for assembling separately.

After the rifle grenade had been redesigned new contracts were let for a total of 30,115,409 of them. In August, a few weeks later, the daily production of these grenades in the various plants had reached a total of 130,000 and by the end of October the daily production was 250,000. The goal toward which this production was aiming was the expected spring offensive of the American Expeditionary Forces in 1919. We should have met this event adequately because, while only 685,200 American rifle grenades had actually been shipped overseas when the fighting ceased, we had 20,000,000 of them ready for loading at that time and the production was already heavy and constantly increasing.

Special consideration is due the following-named firms for their efforts in developing the production of rifle grenades:

Westinghouse Electric & Manufacturing Co., Pittsburgh, Pa.
Briggs & Stratton Co., Milwaukee, Wis.
Holcomb & Hoke, Indianapolis, Ind.
Stewart-Warner Speedometer Corporation, Chicago, Ill.
Cutler-Hammer Manufacturing Co., Milwaukee, Wis.
American Radiator Co., Buffalo, N. Y.
Link-Belt Co., Indianapolis, Ind.
Doehler Die Castings Co., Brooklyn, N. Y.

TOXIC GAS EQUIPMENT.

America entered the war nearly two years after the Germans had made their first gas attack. In those intervening months gas warfare had grown to be a science in itself, requiring special organizations with each army to handle it.

The employment of toxic gas had developed along several lines. The original attack by the Germans upon the maskless Canadians at Ypres had been in the form of a gas cloud from projectors, these latter being pressure tanks with nozzle outlets. For some time the Germans continued the use of gas solely by this method. Retaliation on the part of the allies quickly followed. However, the employment of gas cloud attacks involved great labor of preparation and was absolutely dependent upon certain combinations of weather conditions. In consequence, the launching of a gas attack in this form could not be timed with regard to other tactical operations. Therefore the allies were put to the necessity of developing other means of throwing toxic gases, and this they did by inclosing the gas in shell shot from the big guns of the artillery, in grenades thrown by hand from the trenches, and—most effectively of all—by the agency of an ingenious invention of the British known as the Livens projector.

The Livens projector was deadly in its effect, since it could throw suddenly and in great quantity gas bombs, or drums, into the enemy's ranks. It is notable that although the British used this device with great success throughout much of the latter period of the war, and though the French and Americans also adopted it and used it freely, the Germans were never able to discover what the device was that threw such havoc into their ranks, nor were they ever able to produce anything that was similar to it. The Livens projector remained a deep secret until the close of hostilities, and the Government offices in Washington, where the design was adapted to American manufacture, and the American plants producing the parts, were always closely guarded against enemy espionage.

Without going into details of the construction of the Livens projector it may be said that it was usually fired by electricity in sets of 25 or multiples thereof. The drums, which were cylindrical shell about 24 inches long and 8 inches in diameter, were ejected from long steel tubes, or barrels, buried in the ground resting against pressed-steel base plates. At the throwing of an electric switch a veritable rain of these big shell, as many as 2,500 of them sometimes, with their lethal contents, would come hurtling down upon the enemy. The Livens projectors could throw their gas drums nearly a mile.

The projector was entirely a new type of munition for our manufacturers to handle. The Trench Warfare Section of the Ordnance Department took up the matter late in 1917 and by May, 1918, had designed the weapon for home manufacture. Early in June the contracts were allotted for barrels and gas drums, or shell. The production of barrels was exclusively in the hands of the National Tube Co., of Pittsburgh, Pa., and the Harrisburg Pipe & Pipe-Bending Co., of Harrisburg, Pa. These companies reached the production stage in August, 1918, and completed about 63,000 barrels before the armistice was signed. Their respective plants reached a daily production rate of approximately 600 barrels per day.

Somewhat later in the spring of 1918 the contracts for base plates, on which the barrels rest when ready for firing, muzzle covers, and for various other accessories were closed. Over 100,000 base plates were produced by the Gier Pressed Steel Co., of Lansing, Mich., and

EIGHT-INCH LIVENS PROJECTOR, MARK II, WITH POWDER CHARGE AND SHELL.

Vertical cross section as laid in the ground ready for firing at 45° elevation.

EFFECTS OF LIVENS PROJECTOR SHELL.

the American Pulley Co., of Philadelphia, Pa. The Perkins-Campbell Co., of Philadelphia, built the muzzle covers, 66,180 of them. Cartridge cases were manufactured by Art Metal (Inc.), of Newark, N. J., and the Russakov Can Co., of Chicago, the former producing 288,838 and the latter 47,511.

The Ensign-Bickford Co., of Simsbury, Conn., produced 334,300 fuses for Livens shell; the Artillery Fuse Co., of Wilmington, Del., assembled 26,000 firing mechanisms; the E. I. du Pont Co., at its Pompton Lakes (N. J.) plant, manufactured 20,000 detonators, and 487,350 detonators were produced by the Aetna Explosives Co., at Port Ewan, N. Y.; while the American Can Co., at Lowell, Mass., assembled 256,231 firing mechanisms.

Shear wire pistols were used in the operation of the Livens projector. The Edison Phonograph Co., of Orange, N. J., produced 181,900 of these, and the Artillery Fuse Co., of Wilmington, Del., 11,747. The adapters and boosters of the shell were all built by the John Thompson Press, of New York. The Waterbury Brass Goods Co., of Waterbury, Conn., made the fuse casing. Adapters and boosters to the number of 334,500 were turned out by the former, and 299,900 fuse casings by the latter.

The manufacture of gas drums for the projectors was delayed for some time because of difficulties in welding certain parts of the drums. Acetylene and arc welding processes were tried out, and a good many shell were made by such welding; but the lack of expert welders for these processes, and the rejections of shell due to leakage in the welded joints, caused the manufacturer to turn to fire welding, the process for which had been developed by the Air-tight Steel Tank Co., of Pittsburgh, Pa. At the time the armistice was signed the welding problem had been overcome and the production was going forward at a rate to meet the requirements of the expected fighting in the spring of 1919. The shell delivered were produced as follows:

By the Federal Pressed Steel Co., of Milwaukee, Wis., 5,609; by the Pressed Steel Tank Co., also of Milwaukee, 20,536; by the Air-tight Steel Tank Co., of Pittsburgh, Pa., 600; by the National Tube Co., of Pittsburgh, 27,098; by the Truscon Steel Co., of Youngstown, Ohio, 19,880. The entire Livens shell program, as it existed in November, 1918, called for the production of 334,000 shell.

TRENCH MORTARS.

The production of trench mortars was not only an important part of our ordnance program but it was an undertaking absolutely new to American experience. Not only did we have to produce mortars, but we had to supply them with shell in great quantities, this latter in itself an enterprise of no mean proportions.

Some seven different types of mortars were in use when we came into the war. Our ordnance program contemplated the manufacture of all seven of them, but we actually succeeded in bringing only four types into production. These four were the British Newton-Stokes mortars of the 3-inch, 4-inch, and 6-inch calibers, and the French 240-millimeter mortar, which had also been adopted by the British. As usual in the adoption of foreign devices, we had to redesign these weapons to make them adaptable to American shop methods. We encountered much difficulty throughout the whole job, largely because of insufficient information furnished from abroad, and because in spite of this handicap we had to produce mortars and ammunition that would be interchangeable with French and British munitions stocks.

The first one of these weapons which we took up for production here was the 3-inch Newton-Stokes. The first contract for the manufacture of mortars of this size was placed with the Crane Co., of Chicago, on November 8, 1917, for 1,830 mortars. This concern at once arranged with the Ohio Seamless Tube Co., of Shelby, Ohio, for the drawing of steel tubes for the mortar barrels. This latter concern, however, was already handling large contracts for the Navy and for the aircraft program, and these operations took priority over the mortar contracts. But the Crane Co. took advantage of the interim to build the accessories for the weapons—the tripods, clinometers, base plates, and tool boxes. In the spring of 1918 the company received the first barrel tubes and began producing completed weapons. But when these mortars were sent to the proving ground the test-firing deformed the barrels and broke the metal bases. Finally it was decided that the propelling explosive used was not a suitable one for the purpose. Another was substituted. The new propellant permitted as great a range of fire without damage to the mortar in firing.

The Crane Co. was eventually able to reach a production of 33 of the 3-inch mortars a day, and up to December 5, 1918, it had built 1,803 completed weapons, together with the necessary tools and spare parts. In the early fall of 1918 an additional contract for 677 of these mortars was placed with the Crane Co. and another for 2,000 mortars of this size with the International Harvester Co., of Chicago. Neither of these two latter contracts ever came to the production stage.

A few days after the original contract for 3-inch mortars was let the Trench Warfare Section took up the matter of producing ammunition for these weapons. Two sorts of shell were to be required—live shell filled with high explosive and practice shell made of malleable iron. The original program adopted in November, 1917, called for the production of 5,342,000 live shell for the 3-inch mortars and 1,500,000 practice shell.

3-INCH STOKES TRENCH MORTAR.

4-INCH STOKES TRENCH MORTAR AND AMMUNITION.

6-INCH SUTTON TRENCH MORTAR.

11-INCH SUTTON TRENCH MORTAR.

The plan was adopted of building these shell of lap-welded, 3-inch steel tubing, cut into proper lengths. The contracts for the finished machined and assembled shell were placed with the General Motors Corporation at its Saginaw (Mich.) plant, with H. C. Dodge (Inc.), at South Boston, Mass., and with the Metropolitan Engineering Co., of Brooklyn, N. Y. In order to facilitate production, the Government agreed to furnish the steel tubing. For this purpose it ordered from the National Tube Co., of Pittsburgh, Pa., 1,618,929 pieces of steel tubing, each 11 inches in length, and from the Allegheny Steel Co., at Brakenridge, Pa., 2,332,319 running feet of tubing. These tube contracts were filled by the early spring of 1918.

The railroad congestion of February and March, 1918, held up the delivery of tubing, but the assembly plants utilized the time in tooling up for the future production. All the plants thereafter soon reached a quantity production, the General Motors Corporation in particular tuning up its shop system until it was able to reach a maximum daily production in a 10-hour shift of 35,618 completed shell.

The casting of malleable iron bodies for the practice shell of this caliber was turned over to the Erie Malleable Iron Co., of Erie, Pa., and to the National Malleable Castings Co., with plants at Cleveland, Chicago, Indianapolis, and Toledo. The former concern cast 196,673 bodies and the latter 1,015,005. The Gorham Manufacturing Co., of Providence, R. I.; the Standard Parts Co., of Cleveland, Ohio; and the New Process Gear Corporation, of Syracuse, N. Y., machined and assembled the practice shell. When the armistice was declared, these three contracts were approximately seven-tenths complete.

We were dissatisfied with our 3-inch shell, for the reason that they tumbled in air and were visible to the eye. The French had developed a mortar shell on the stream-line principle which was invisible in flight and had twice the range of ours. Had the war continued the Trench Warfare Section would have produced stream-line shell for mortars.

The second mortar project undertaken was the manufacture of the 240-millimeter weapon. This was the largest mortar which we produced, its barrel having a diameter of approximately 10 inches. It proved to be one of the toughest nuts to crack in the whole mortar undertaking. The British designs of this French weapon we found to be quite unsuited to our factory methods, and for the sake of expediency we frequently modified them in the course of the development. The total contracts called for the production of 938 mortars.

It was obvious that the manufacture of this and of other larger mortars would fall into three phases. The forging of barrels, breechblocks, and breech slides was a separate type of work, and we allotted the contracts for this work to the Standard Forging Co., of Indiana Harbor, Ind. The machining of these parts to the fine dimensions required by the design was an entirely separate phase of manufacturing, and we placed this work with the American Laundry Machine Co., of Cincinnati. Still a third class of work was that of assembling the completed mortars, and this contract went to the David Lupton Sons Co., of Philadelphia, who also engaged to manufacture the metal and timber bases and firing mechanisms. These big mortars had to have mobile mountings, and the contract for the mortar carts we placed with the International Harvester Co., of Chicago. These contracts were signed in December, 1917.

The Lupton plant had difficulty in securing the heavy machinery it needed for this and for other mortar contracts, its machinery being held up by the freight congestion. Early in 1918 the American Expeditionary Forces advised us to redesign the 240-millimeter mortar to give it a stronger barrel. Consequently all work was stopped until this could be done. The first mortars of the new design to be tested were still unsatisfactory with respect to the strength of the barrels; and as a consequence the Standard Forging Co. urged that nickel steel be substituted for basic open-hearth steel as the material for the barrels. This change proved to be justified.

There was also trouble at the shops of the American Laundry Machine Co., its equipment not having the precision to do machining of the type required in these weapons. Accordingly a new machining contract was made with the Symington-Anderson Co., of Rochester, N. Y., which concern was eventually able to reach a production of 20 machined barrels per week.

In all we produced 24 of the 240-millimeter mortars in this country. Certain of the parts were manufactured up to the total requirements of the contracts, but others were not built in such numbers. The International Harvester Co. built all 999 carts ordered.

The production of shell for these big mortars was another difficult undertaking. After consultation with manufacturers we designed shell of two different types. One of these was a shell of pressed plates welded together longitudinally; and a contract for the production of 283,096 of these was placed with the Metropolitan Engineering Co. The other form was that of two steel hemispheres welded together. The Michigan Stamping Co., of Detroit, undertook to build 50,000 of these.

These shell contracts were placed in December, 1917. The Michigan Stamping Co. had to wait five months before it could secure and install its complete equipment of machinery. It was September before all of the difficulties in the Detroit plant's project could be overcome and quantity production could be started. The concern eventually, before and after the signing of the armistice, built 9,185 shell of this type at a maximum rate of 56 per day.

Greater promise seemed to be held forth by the Metropolitan Engineering Co.'s project to build shell of pressed-out plates, electrically welded. The Government undertook to furnish the steel plates for this work and secured from the American Rolling Mills Co., of Middletown, Ohio, a total production of 6,757 tons of them. The Metropolitan Engineering Co. had great difficulty in perfecting a proper welding process; and the concern lost a great deal of money on the contract, yet cheerfully continued its development without prospect of recompense in order that we might have in this country the knowledge of how to build such shell. In all, including production after the armistice was signed, the Metropolitan Engineering Co. built 136,189 shell bodies of this size at a maximum rate of 987 per day.

During the summer of 1918 a single-piece shell body of the 240-millimeter size, produced by a deep-drawing process, was worked out. A contract for 125,000 of them was given to the Ireland & Matthews Manufacturing Co., of Detroit, Mich. The armistice brought this contract to an end before it had produced any shell of this new and most promising type.

Early in 1918 we received the first samples of the 6-inch trench mortar. By April all the plans were ready for American production. Again this work was divided by types. The National Tube Co., of Pittsburgh, contracted to build 510 rough forgings of mortar barrels at its Christy Park plant. The Symington-Anderson Co. undertook to machine these barrels. The David Lupton Sons Co. agreed to assemble the mortars, as well as to produce the metal and timber bases for them.

The first machined barrels reached the Lupton plant in June and found bases ready for them. But, as the assembling was in progress, the American Expeditionary Forces cabled that the British producers of mortars had changed their designs, and that we must suspend our manufacture until we also could adopt the changes. The altered plans reached us some weeks later; yet, nevertheless, we were able to make good our original promise to deliver 48 of the 6-inch Newton-Stokes mortars at the port of embarkation in October, 1918.

Meanwhile we had increased the contracts by an additional requirement of 1,577 mortars of this size. The National Tube Co. eventually reached a maximum daily production of 60 barrel forgings. The Symington-Anderson Co. machined the barrels finally at a 33-per-day clip. As many as 11 proof-fired guns per day came from the David Lupton Sons Co.

An interesting fact in connection with the production of shell for the 6-inch mortars is that they were built principally by American makers of stoves. The 6-inch mortar-shell bodies were of cast iron instead of steel, and thus were adaptable to manufacture in stove works. Each shell weighed 40 pounds without its explosive charge. Such shell were used at the front for heavy demolition purposes.

The contracts for these shell were placed in March, 1918. The Trench Warfare Section was immediately called upon to secure favorable priority for the pig iron required for this purpose. The various stove works did not have the necessary machinery for building these shell, and so a special equipment in each case had to be built. At the tests the first castings which came through the foundry were found to leak, and this required further experiments in the design, holding up production until July, 1918.

Because of the many troubles encountered in this work the various stove makers in the summer of 1918 formed an association which they called the Six-inch Trench-mortar Shell Manufacturers' Association. This association held monthly meetings and its members visited the various plants where shell castings were being made. The United States Radiator Corporation, the Foster-Merriam Co., and the Michigan Stove Co., were especially active in improving methods for making these shell.

The various concerns producing 6-inch mortar shell and the amounts turned out were as follows:

Foster Merriam Co., Meriden, Conn	33,959
U. S. Radiator Corporation, Detroit, Mich	240,700
Globe Stove & Range Co., Kokomo, Ind	17,460
Rathbone, Sard & Co., Albany, N. Y	97,114
Michigan Stove Co., Detroit, Mich	100,000

The following concerns shortly before the armistice was signed received contracts for the production of 6-inch mortar shell, orders ranging in quantity from 50,000 to 150,000, but none of these concerns started production:

Wm. Crane Co., Jersey City, N. J.
Frontier Iron Works, Buffalo, N. Y.
Henry E. Pridmore, (Inc.), Chicago, Ill.
Best Foundry Co., Bedford Ohio.
McCord & Co., Chicago, Ill.

It was not until July, 1918, that the plans were ready for the 4-inch Newton-Stokes trench mortars. The American Expeditionary Forces estimated that they would require 480 of these weapons. A total of 500 drawn barrel tubes was ordered from the Ohio Seamless Tube Co., of Shelby, Ohio. This concern was able to ship one-fifth of its order within 10 days after receiving it. The barrels were sent to the Rock Island Arsenal for machining. The Crane Co., of Chicago, held the contract for building the bases, tripods, spare parts, and tools, and also for the assembling of the completed mortars. This factory was already equipped with tools for this work, since it had been building similar parts for 3-inch mortars. Consequently, the Crane Co., in August, almost within a month of receiving its contract, was producing completed 4-inch mortars and sending them to the Rock Island Arsenal for proof firing. The Ohio Seamless Tube Co. reached

240-MILLIMETER TRENCH MORTAR.

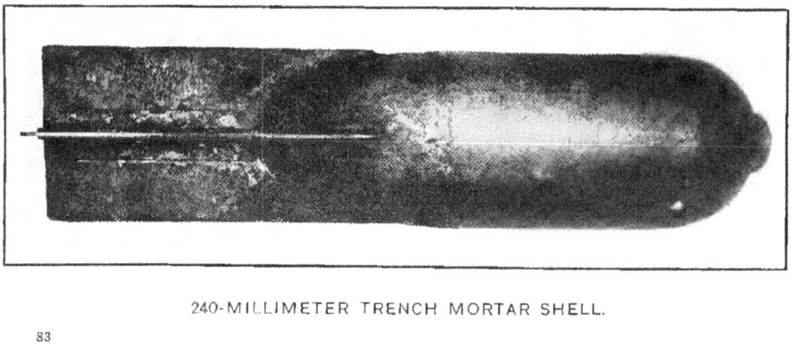

240-MILLIMETER TRENCH MORTAR SHELL.

a high daily production of 83 barrel forgings per day; the Rock Island Arsenal, 10 machined barrels per day; and the Crane Co., 19 assembled mortars per day.

We planned to build only smoke shell and gas shell for the 4-inch mortars. Large contracts for various parts of these shell were placed and the enterprise was gaining great size when the armistice was declared, but no finished smoke shell and only a few gas shell for 4-inch trench mortars had been produced. The contracts for the smoke shell were let in October, 1918, and work had not started further than the procurement of raw material before the armistice came. A large number of contractors expected to produce the parts for the 4-inch gas shell, and considerable of the raw materials were actually produced; but only one of the machining and assembling contractors, the Paige-Detroit Motor Car Co., actually completed any of these shell, and production at this plant did not start until December 5, 1918.

Production of trench mortars and trench-mortar ammunition.

TRENCH MORTARS.

Character.	Completions to Nov. 11, 1918.	Completions to Feb. 1, 1919.	Shipped overseas.
3-inch	1,609	1,830	843
4-inch	444	778	....
6-inch	368	500	48
240-millimeter (9.45 inches)	29	30	....

TRENCH-MORTAR AMMUNITION.

Character.	Completions to Nov. 11, 1918 (unloaded).	Completions to Feb. 1, 1919 (unloaded).	Shipped overseas (loaded).
	Rounds.	*Rounds.*	*Rounds.*
3-inch live	3,136,275	3,741,237	157,785
3-inch practice	607,178	782,340	....
4-inch gas	....	212	....
4-inch smoke	....	....	....
6-inch live	292,882	492,404	....
240-millimeter (9.45 inches)	67,829	131,124	....

TOXIC GAS SETS.

Another extensive project in the trench-warfare program was the manufacture of the so-called toxic gas sets. Each set consisted of a one-man portable cylinder equipped with a nozzle and a firing mechanism. Each set was ready for firing as soon as it was placed in position.

In August, 1918, the toxic-gas-set project was taken up by the Trench Warfare Section. Contracts for cylinders were awarded to the Ireland-Matthews Manufacturing Co., of Detroit, Mich., who produced 13,642 cylinders, and to the American Car & Foundry Co. at its Milton, Pa., plant, which concern turned out 11,046 cylinders.

The Pittsburgh Reinforcing, Brazing & Machine Co. produced 9,765 valves for the cylinders in two months after receiving the contract. The Yale & Towne Manufacturing Co., of Stamford, Conn., which received the contract for nozzles on September 5, 1918, manufactured 20,501 of them before the armistice was signed; and J. N. Smith & Co., of Detroit, Mich., who did not receive their contract until September 26, built 3,252 nozzles before the fighting stopped. The Liquid Carbolic Co., of Chicago, and the Ruud Manufacturing Co., of Pittsburgh, had the contracts for the firing mechanism; but none of these was produced because at the time the armistice was signed the firing mixture to be used with the cylinders had not been developed.

In connection with the production of materials for gas warfare the Ordnance Department also designed several types of containers for the shipment of poison gas, these including not only the portable cylinders but larger tanks and even tank cars.

PYROTECHNICS.

A few years ago, when we allowed the adventurous American boy to blow off his fingers and hands by the indiscriminate use of explosives in celebrating the Nation's birthday, we had an extensive fireworks industry in this country. But the spread of the sane Fourth reform had virtually killed this manufacture, so that when we entered the war there were only three or four plants in the United States making fireworks. These concerns kept the trade secrets closely guarded. However, as we approached the brink of hostilities it was evident we would have to build up a large production capacity for the pyrotechnics demanded by the various new types of fighting which had sprung into existence since 1914. Fireworks were extensively used principally for signaling at night and as an aid to aviators in the dark.

One of the men to foresee this need was Lewis Nixon, who had long been in the public eye and was known especially for his advocacy of an American merchant marine. He organized a pyrotechnics concern known as the Nixon Fulgent Products Co., built a plant at Brunswick, N. J., and was ready to talk business with the Government when the war began.

Also there had long been in existence that perennial delight of children and adults alike known as Paine's Fireworks, whose spectacular exhibitions are familiar to most city dwellers in the United States. This concern had its own manufacturing plant, which was ready to expand to meet Government war requirements.

In addition, two other concerns of the formerly declining industry were ready to increase their facilities and produce pyrotechnics for war purposes. These were the Unexcelled Manufacturing Co., of New York, and the National Fireworks Co., of West Hanover, Mass.

The four concerns proved to be able to meet every war requirement we had.

Prior to the war some few military pyrotechnics had been procured by the Signal Corps, the Coast Artillery, the Engineer Corps, and also by the Navy; but on September 27, 1917, the design of all Army pyrotechnics was centralized in the Trench Warfare Section.

Much experimentation was necessary before specifications could be prepared, since the entire fire-signaling field had long been in confusion. We had made our own designs and were proceeding with production in the spring of 1918, when the American Expeditionary Forces made the positive recommendation that the entire French program of pyrotechnics be adopted by the United States. This meant a fresh start in the business, but nevertheless pyrotechnic devices were developed to meet all of our needs. These devices included signal rockets, parachute rockets, signal pistols and their ammunition, position and signal lights, flares, smoke torches, and lights to be thrown by the V. B. discharger, the French device attached to the end of the rifle in which a rifle grenade fits.

At the outset of our efforts we started to build signal rockets, position lights, rifle lights, signal lights, and lights for use with the Very signal pistol. The Very signal pistol, which we adopted first, had the caliber of a 10-gauge shotgun, and its cartridges resembled shotgun shells in appearance, although containing Roman candle balls of various colors instead of leaden shot. The orders from abroad in the spring of 1918 changed the caliber of the Very pistol to 25 millimeters and brought into our requirements some 16 different styles of star and parachute cartridges. In addition to these, there were required about 20 styles of star and parachute cartridges for the French V. B. discharger. The recommendations from France brought in 13 new styles of signal rockets, as well as smoke torches, wing-tip flares for airplanes, parachute flares for lighting the ground under bombing airplanes, and also 12 styles of cartridges for a new 35-millimeter Very pistol for the use of aviators.

After we received these instructions there was great uncertainty here as to the quantity of each item that should be produced; and this matter was not settled until August 5, 1918, when an enormous program of requirements was issued. At first it seemed that the Government itself must build new factories to take care of these needs, but a careful examination showed that the existing facilities could be expanded to take care of the production. The placing of contracts in this undertaking was under way when the armistice stopped the work.

The following table indicates the size of the pyrotechnic undertaking and also what was accomplished. All of this production came from the plants of the four companies which have been named. In addition to the fireworks themselves, accessories were produced by a number of other concerns. The Japan Paper Co., New York City, manufactured and imported from Japan approximately 3,000,000 paper parachutes. The Remington Arms Co., New Haven, Conn., built about 2,500,000 10-gauge signal-pistol cartridges, except for the stars they contained. The Empire Art Metal Co., College Point, N. Y., produced nearly 2,000,000 Very pistol cartridge cases. The Winchester Repeating Arms Co., Bridgeport, Conn., supplied nearly 5,000,000 primers for these cartridges. Rose Bros. & Co., Lancaster, Pa., produced 65,600 silk parachutes for Very cartridges. Cheney Bros., South Manchester, Conn.; D. G. Dery (Inc.), Allentown, Pa.; Stehli Silk Corporation, New York City; Sauquoit Silk Co., Philadelphia; Lewis Roessel & Co., Hazleton, Pa.; Schwarzenback-Huber Co., New York City; and the Duplane Silk Corporation, Hazleton, Pa., produced a total of 1,231,728 yards of silk for parachutes to float airplane flares. The parachutes themselves for the airplane flares, a total of 28,570 of them, were manufactured by the Duplane Silk Corporation; Folmer-Clogg Co., Lancaster, Pa.; and Jacob Gerhardt Co., Hazleton, Pa. The Edw. G. Budd Manufacturing Co., Philadelphia, built 41,020 metal cases for the airplane flares.

Articles.	Ordered.	Completed to Nov. 8, 1918.	Completed to Feb. 1, 1919.
Signal rockets	615,000	437,101	544,355
Position lights	2,072,000	1,187,532	1,670,070
Rifle lights	55,000	55,000	55,000
Signal lights	3,110,000	2,661,008	2,710,268
V. B. cartridges	1,215,000	110,000	673,200
Very cartridges, 25-millimeter	300,000	....	....
Smoke torches	500,000	31,000	188,102
Wing-tip flares	112,000	70,000	100,865
Airplane flares	50,083	2,100	8,000

We also contracted for the production of many thousands of Very signal pistols. Before the original program was canceled the Remington Arms Co. had produced 24,460 of the 10-gauge pistols in contracts calling for a total output of 35,000.

In August, 1918, we let contracts for 135,000 of the 25-millimeter pistols and for approximately 30,000 of the 35-millimeter pistols. The A. H. Fox Gun Co. completed 4,193 of the smaller pistols and the Scott & Fetzger Machine Co. turned out 7,750 of them. Other concerns which had taken contracts but had not come into production when the armistice was signed were the National Tool & Manufacturing Co., the Doehler Die Castings Co., the Hammond Typewriter Co., and Parker Bros.

Considerable experimental work of an interesting nature was carried out looking toward the development of incendiary devices. Three types of flame projectors, flaming bayonets, an airplane destroyer, incendiary darts, and the smoke knapsack were among the projects undertaken. Owing in large measure to changes in requirements by the American Expeditionary Forces none of these devices was actually turned out on any considerable scale.

CHAPTER XIV.

MISCELLANEOUS ORDNANCE EQUIPMENT.

The miscellaneous ordnance equipment of the American soldier in the recent war—that is, articles which he carried with him and which added to his comfort, his safety, or his efficiency as a fighter—while in many respects identical with the equipment used by our troops for many years, at the same time contained several novelties.

In the novelty class were helmets and armor. There is a widespread impression that helmets and body armor passed away with the invention of gunpowder and because of that invention. This impression is not at all true. Body armor came to its highest development long after gunpowder was in common use in war. The sixteenth century witnessed the most extensive use of armor; yet at that time guns and pistols formed an important part of the equipment of every army, and even a weapon which is generally fancied to be ultramodern, the revolver, had been invented.

The fact is that not gunpowder but tactics caused the decline of armor. Not that armor was unable to stop many types of projectiles shot from guns, but that its weight hampered swift maneuvering, caused it to be laid aside by the soldier. The decline of armor may be said to date from the Thirty Years' War. The armies in that period, and particularly that of the Swedes, began making long marches for surprise attacks, and the body armor of the troops was found to be a hindrance in such tactics. Thereafter armor went out of fashion.

Yet it never completely disappeared in warfare. Gen. Rochambeau is said to have worn body armor at the siege of Yorktown. Great numbers of corselets and headpieces were worn in the Napoleonic wars. The corselet which John Paul Jones wore in his fight with the *Serapis* is preserved at the Metropolitan Museum of Art in New York. The Japanese army was mailed with good armor as late as 1870. Breastplates were worn to some extent in the Civil War in the United States, and an armor factory was actually established at New Haven, Conn., about 1862. In the museum at Richmond, Va., is an equipment of armor taken from a dead soldier in one of the trenches at the siege of that city. There was a limited use of armor in the Franco-Prussian War. Some of the Japanese troops carried shields at Port Arthur. Helmets were worn in the Boer War. A notorious Australian bandit in the eighties for a long time defied armed posses to capture him because he wore armor and could stand off entire squads of policemen firing at him with Martini rifles at close range.

Thus it can not be said that armor, in coming into use again in the great war, was resurrected; it was merely revived. In its static condition during most of the four-year period, the war against Germany was one in which armor might profitably be used. This opportunity could scarcely be overlooked, and indeed it was not. Everybody knows of the helmets that were in general use; yet body armor itself was coming into favor again, and only the welcome but unexpected end of hostilities prevented it, in all probability, from becoming again an important part of the equipment of a soldier.

As a consequence of the attenuated but persistent use of armor by soldiers during the past two centuries and of the demand of the aristocratic for helmets and armor as ornaments, the armorer's trade had been kept alive from the days of Gustavus Adolphus to the present. The war efforts of the United States in 1917 and 1918 demanded a wide range of human talents and special callings; but surely the strange and unusual seemed to be reached when in the early days of our undertaking the Engineering Division of the Ordnance Department sought the services of expert armorers.

Through the advice of the National Research Council, which had established a committee of armor experts, the Ordnance Department commissioned in its service Maj. Bashford Dean, a life-long specialist in armor, curator in the Metropolitan Museum of Art, an institution which, learning of the Government's need, at once placed at its disposal its wonderful specimens of authentic armor, its armor repair shop where models could at once be made, and the services of Maj. Dean's assistant there whom he had brought from France, Daniel Tachaux, one of the few surviving armorers, who had inherited lineally the technical side of the ancient craft.

It may be said that there were but two nations in the great war which went to the Middle Ages for ideas as to protective armor—ourselves and Germany. The Germans, who applied science to almost every phase of warfare, did not neglect it here. Germany at the start consulted her experts on ancient armor and worked along lines which they suggested. The German helmet used in the trenches was undoubtedly superior to any other helmet given a practical use.

The first helmets to be used in the great war were of French manufacture. They were designed by Gen. Adrien, and 2,000,000 of them were manufactured and issued to the French Army. These helmets were the product of hasty pioneer work, but the fact that they saved from 2 to 5 per cent of the normal casualties of such a war as was being fought at once impelled the other belligerents to adopt

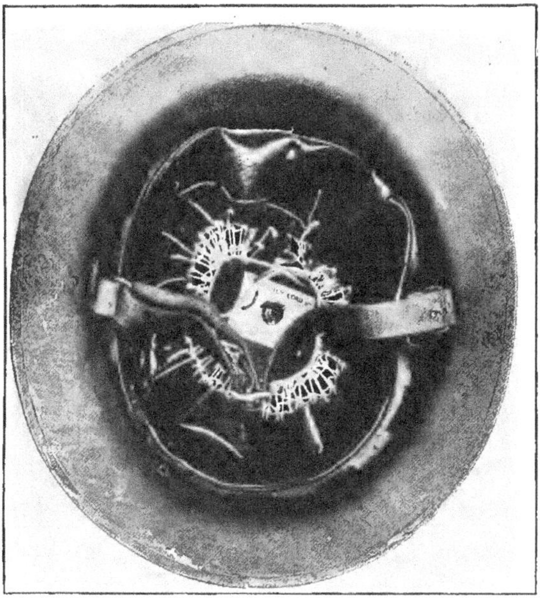

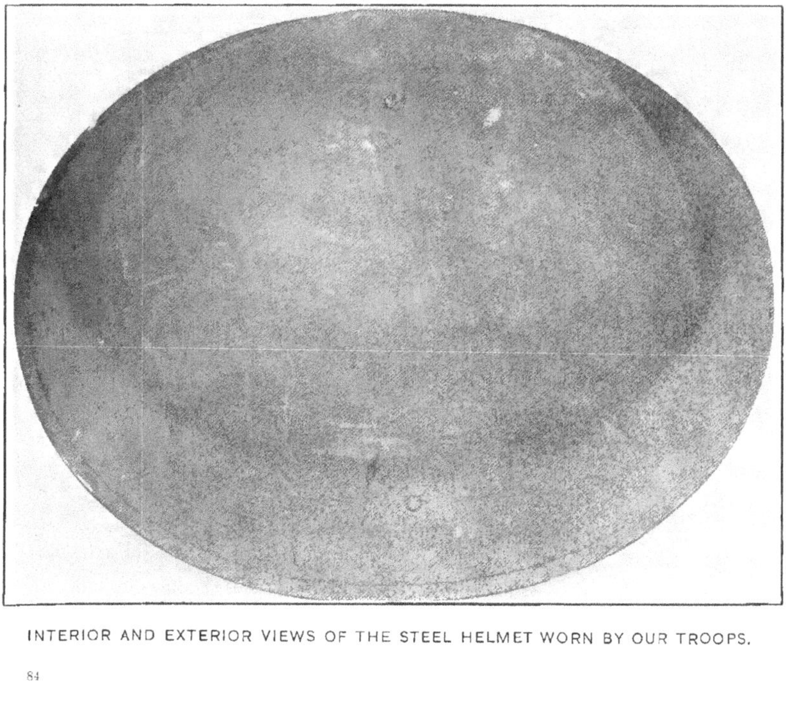

INTERIOR AND EXTERIOR VIEWS OF THE STEEL HELMET WORN BY OUR TROOPS.

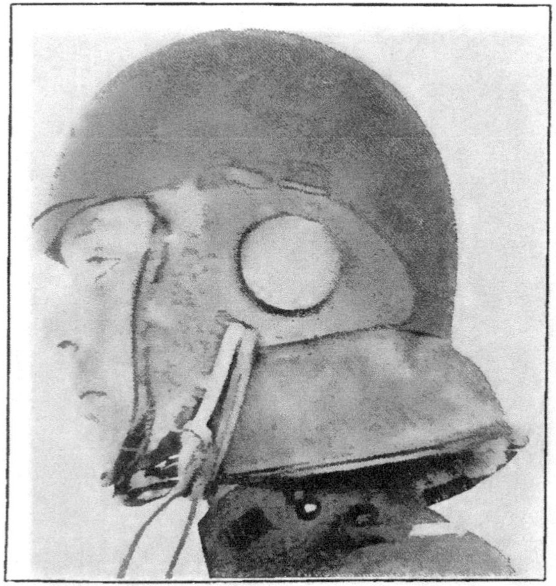

AVIATOR'S HELMET

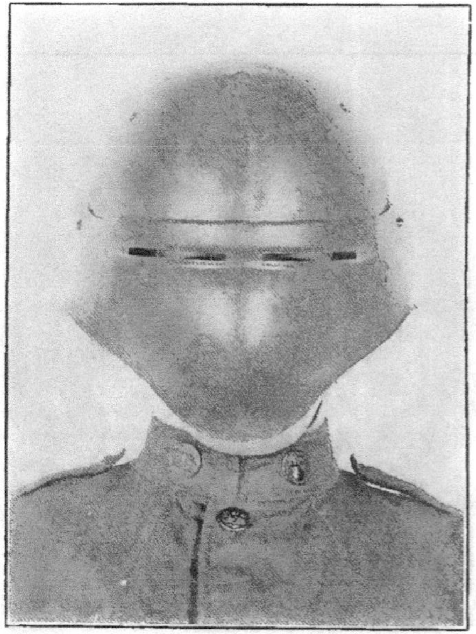

VISOR HELMET FOR SNIPERS AND MACHINE GUNNERS.

the idea. Great Britain, spurred by the necessity of producing quickly a helmet in quantity, designed the most simple helmet to manufacture, which could be pressed out of cold metal.

When America entered the war she had, naturally, no distinctive helmet; and the English type, being easiest to make, was adopted to fill the gap until we could design a more efficient one ourselves. Consequently 400,000 British helmets were bought in England and issued to the vanguard of the American Expeditionary Forces. Our men wore them, became accustomed to them, and came to feel that they were the badge of English-speaking troops. The British helmet thus became a habit with our men, one difficult to change, a fact which mitigated against the popularity of the more advanced and scientific models which we were to bring out.

Now, the British helmet possessed some notable defects. It did not afford a maximum of protective area. The center of gravity was not so placed as to keep the helmet from wobbling. The lining was uncomfortable and disregarded the anatomy of the head. It was vulnerable at the concave surface where bowl and brim joined.

It is not an astonishing circumstance that some of the earlier helmets worn by the men-at-arms of the days of knighthood possessed certain of these same defects, notably, that they were apt to be top-heavy and uncomfortable. Only by centuries of constant application and improvement were the armorers of the Middle Ages able to produce helmets which overcame these defects and which embodied all of the principles of defense and strength which science could put into them. The best medieval helmets stand at the summit of the art. It was the constant aim of the modern specialist, aided by the facilities of the twentieth century industries, to produce helmets as perfect technically as those rare models which are the pride of museums and collectors.

Certainly in one respect we had the advantage of the ancients in that we have nowadays at our disposal the modern alloy-steels of great resistance. An alloy of this kind having a thickness of 0.036 of an inch is able to stop at a distance of 10 feet a jacketed, automatic pistol ball, .45 caliber, traveling at the rate of 600 feet a second. This was important not only from the standpoint of helmet production, but from the further inference that body armor of such steel might still be profitably used. The records of the hospitals in France show that 7 or 8 of every 10 wounded soldiers were wounded by fragments of shell and other missiles which even thin armor plate would have kept out. The German troops used body armor in large numbers, each set weighing from 19 to 24 pounds. In this country we believed it possible to produce body armor which would not be difficult to carry and which would resist the impact of a machine-gun bullet at fairly close range.

The production of helmets, however, was our first concern; and in order to be sure of a sufficient quantity of these protective headpieces, we adopted the British model for production in the United States and went ahead with it on a large scale. For the metal we adopted after much experimentation a steel alloy with a high percentage of manganese. This was practically the same as the steel of the British helmet. Its chief advantage was that it was easy to work in the metal presses in existence and it required no further tempering after leaving the stamping presses. Its hardness, however, wore away the stamping tools much more quickly than ordinary steel sheets would do.

While we adopted the British helmet in design and substantially in metal used, we originated our own helmet lining. The lining was woven of cotton twine in meshes three-eighths of an inch square. This web, fitting tightly upon the wearer's head, evenly distributed the weight of the two-pound helmet, and in the same way distributed the force of any blow upon the helmet. The netting, together with small pieces of rubber around the edge of the lining, kept the helmet away from the head, so that even a relatively large dent could not reach the wearer's skull.

It is an interesting fact that the linings for the American helmets were produced by concerns whose ordinary business was the manufacture of shoes. There were 10 of these companies taking such contracts. Steel for the helmet was rolled by the American Sheet & Tin Plate Co. The helmets were pressed and stamped into shape by seven companies which had done similar work before the war. These concerns were:

Contractor.	Delivered.
Edward G. Budd Manufacturing Co., Philadelphia	1,150,775
Sparks, Withington Co., Jackson, Mich	473,469
Crosby Co., Buffalo, N. Y	469,968
Bossett Corporation, Utica, N. Y	116,735
Columbian Enameling & Stamping Co., Terre Haute, Ind	268,850
Worcester Pressed Steel Co., Worcester, Mass	193,840
Benjamin Electric Co., Des Plaines, Ill	33,600
Total	2,707,237

The metal helmets and the woven linings were delivered to the plant of the Ford Motor Co. at Philadelphia, where they were painted and assembled. The helmets were painted in the olive-drab shade for protective coloring. While on dull days such objects could not be discerned at a great distance, in bright weather their rounded surfaces might catch and reflect sunbeams, thus betraying the positions of their wearers. To guard against this, as soon as the helmets were treated to a first coat of paint fine sawdust was blown upon the wet surface. When this had dried, another coat of paint was applied, and a nonreflective, gritty surface was thus produced.

AMERICAN EXPERIMENTAL MODELS OF HELMETS, LIGHT BREASTPLATES, AND ARM GUARDS.

American Helmet. Experimental Model No. 2.

American Helmet No. 8. (Visor up.)

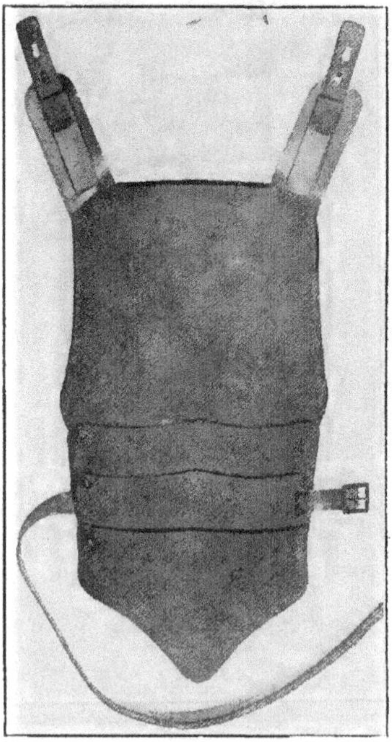

AMERICAN LIGHT BACK PLATE. EXPERIMENTAL MODEL.

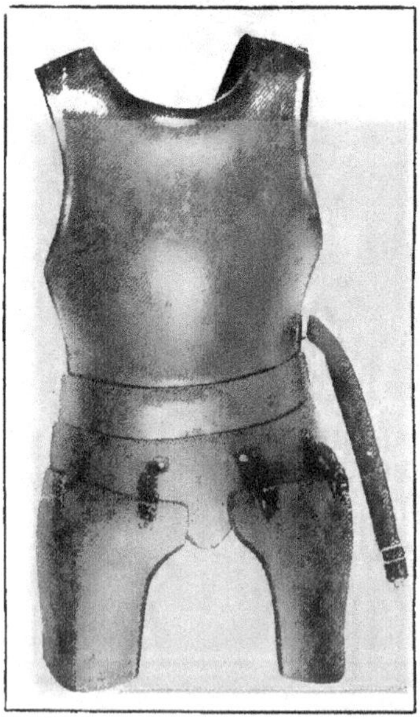

AMERICAN HEAVY ARMOR FOR MACHINE GUNNERS. EXPERIMENTAL MODEL 1917.

We began receiving substantial quantities of finished helmets by the end of November of the first year of the war. On February 17, 1918, practically 700,000 had been shipped abroad or were ready for shipment at the ports of embarkation. Later in the spring of 1918, when we began sending men to France much beyond our earlier expectations, the orders for helmets were greatly expanded. In July the total orders reached 3,000,000, in August 6,000,000, and in September 7,000,000. This would give us enough to meet all requirements until June, 1919.

When the armistice was signed the factories were producing more than 100,000 helmets every four days, and were rapidly approaching the time when their daily output would be 60,000. The Government canceled all helmet contracts as soon as the fighting ceased, having received up to that time a total of 2,700,000 of them.

While this manufacture was going on we were developing helmets of our own. Major Dean went to France to collect information dealing with the actual needs of the service and to present numerous experimental models of helmets for the comment and criticism of the General Staff. In numerous cases these models were accepted for manufacture here in experimental lots.

In all we developed four models which seemed to have merits recommending their adoption. The first distinctive American helmet was known as model No. 2. The Ford Co. at Detroit pressed about 1,200 of these helmets. The helmet, however, was similar in appearance to the German helmet, and for that reason was disapproved by the American Expeditionary Forces.

Helmet model No. 3 was of a deep-bowl type, but it was rejected when the Hale & Kilburn Co., of Philadelphia, after a great deal of experimentation, found that the helmet was too deep for successful manufacture by pressing.

Model No. 4 was designed by the master armorer of the Metropolitan Museum of Art. It was also found too difficult to manufacture.

Helmet No. 5 was strongly recommended by American experts, but was not accepted by the General Staff. It was designed by the armor committee at the Metropolitan Museum of Art in conjunction with the Engineering Division of the Ordnance Department. Hale & Kilburn undertook to manufacture these helmets, which were to be painted, assembled, and packed by the Ford Motor Co. at its Philadelphia plant. Various component parts of the helmet were sublet in experimental quantities to numerous manufacturers.

The No. 5 helmet, complete, weighed 2 pounds, 6½ ounces. It combined the virtues of several types of helmets. It gave a maximum of protection for its weight. It was comparatively easy to produce. This helmet, with slight variations, was later adopted as the standard helmet of the Swiss Army. The latest German helmet, it is interesting to note, was approaching similar lines.

We also produced helmets for special services—one with a visor to protect machine gunners and snipers, and another, known as model 14, for aviators, it being little heavier than the leather helmet which airmen wore in the war and twenty times as strong a defense for the head. A third special helmet, known as model 15, was for operators of tanks. It was provided with a neck guard of padded silk to stop the lead splash which penetrated the turret of the tank. The Ordnance Department turned out 25 of these in 10 days and sent them by courier to France for a test.

The Germans issued body armor only to troops holding exposed positions under heavy machine-gun and rifle fire; but such use was distinctly valuable, as was shown by captured German reports.

The Engineering Division of the Ordnance Department developed a body defense including a light front and body plate, these together weighing 9½ pounds. One lot of 5,000 sets was manufactured by the Hale & Kilburn Corporation. The linings of these plates were of sponge rubber, and they were made by the Miller Rubber Co., of Akron, Ohio. All of these sets were shipped abroad for testing; but the report was not favorable, as the American soldier did not wish to be hampered with armor. He had learned to wear his helmet, but he had yet to be convinced of the practical value of body armor.

We developed a heavy breast plate with thigh guards, weighing 27 pounds, which stopped machine gun bullets at 150 yards. An experimental lot of these were completed in 26 days by the Mullins Manufacturing Co., of Salem, Ohio. These were also shipped abroad for test.

A few defenses for arms and legs were prepared which, although light in weight, would protect the wearer from an automatic-pistol ball at 10 feet. About 70 per cent of the hospital cases in France were casualties caused by wounds in the arms and legs. These defenses, however, were rejected on account of their impeding to a certain degree the movements of the wearer.

Our development in armor also produced an aviator's chair weighing 60 pounds. It would protect the pilot from injury from below and from the back, withstanding armor-piercing bullets fired at a distance of 50 yards. Since the piercing of the gas mask canister by a bullet might result in the death of the soldier by admitting gas directly into the breathing system of his mask, the Ordnance Department designed an armored haversack for the gas mask and its canister, this haversack incidentally serving as a breast defense.

BAYONETS AND TRENCH KNIVES.

Another large ordnance operation was the production of bayonets for the service rifles. The British bayonet had proved to be highly satisfactory in the war; and, since it was already designed to fit the Enfield rifle, which we had adopted for our own, we took the British bayonet as it was and, with only one slight alteration, set out to produce it in this country.

The Government found both the Remington Arms-Union Metallic Cartridge Co. at its Bridgeport, Conn., works, and the Winchester Repeating Arms Co. building these bayonets for the English Government. The latter's bayonet needs by 1917 were being well supplied by home manufacture, and this permitted us to buy approximately 545,500 bayonets which had already been manufactured for the British.

The Ordnance Department at once started out these two concerns on contracts for bayonets for the American Government, Remington with total orders for 2,820,803 bayonets and Winchester with orders for 672,500. Remington delivered in all 1,565,644 bayonets and Winchester 395,894. This was a total of 1,961,500 bayonets.

The total production of 1917 rifles was about 2,520,000. These figures indicate that we were short over 500,000 bayonets at the time hostilities ceased; and as a matter of fact this shortage had already become acute, especially in the training camps.

The bayonets had not come as rapidly as we had expected, because to produce them at the rate originally planned would have interfered with the more essential production of rifles by these same companies. Accordingly in 1918 additional contracts for bayonets were made. Landers, Frary & Clark, of New Britain, Conn., engaged to manufacture 500,000 bayonets, and the National Motor Vehicle Co., 255,000. These latter contracts, however, were suspended after the armistice was signed. The additional orders had made it certain that there would be no bayonet shortage by the spring of 1919.

While this production was under way we were also manufacturing bayonets for the model 1903 Springfield rifle. The Springfield Armory produced 347,533 of these and the Rock Island Arsenal 36,800. In addition the Springfield Armory delivered 50,000 bayonet blades as spare parts.

We not only had to provide bayonets but also the scabbards to hold them in. The scabbard of the 1917 bayonet was of simple manufacture and there were no difficulties in securing sufficient quantities. The Jewell Belt Co. delivered 1,810,675 of them; Graton & Knight delivered 1,669,581; while the Rock Island Arsenal produced 3,000. This gave us a total of 3,480,000 scabbards, a quantity greatly in excess of the production of either bayonets or rifles.

A new weapon which had come into use during the great war as part of the soldier's individual equipment was the trench knife. The question of making such knives was taken up by the Government with various manufacturers throughout the country and they were given a general idea of what was required and, in conjunction with the Ordnance Department, were requested to develop details. The design submitted by Henry Disston & Sons, of Philadelphia, received the most favorable consideration. This knife was manufactured and known as model 1917. It was a triangular blade 9 inches long. The triangular blade was deemed the most efficient because of the ease with which it would pierce clothing and even leather. This knife was slightly changed as regards handle and given a different guard to protect the man's knuckles, and was known as model 1918. These knives were sent abroad in large quantities to be used by the American Expeditionary Forces. Landers, Frary & Clark produced 113,000 of these knives and the Oneida Community (Ltd.), Oneida, N. Y., 10,000.

On June 1, 1918, the American Expeditionary Forces made an exhaustive test, comparing the various trench knives used abroad. The four knives tested were as follows; United States, model 1917; Hughes; French; and British knuckle knife. These tests were made to determine the merits of the different knives as to the following points:

(*a*) Serviceability—ability to carry in hand and function other arms.

(*b*) Quickness in action.

(*c*) If the soldier were knocked unconscious, would knife drop from hand?

(*d*) Suitability to carry in hand when crawling.

(*e*) Probability of being knocked out of hand.

(*f*) Weight, length, shape of blade.

(*g*) Shape of handle.

It was found that the model 1917, although a satisfactory knife, could be improved. Therefore the trench knife known as Mark I was developed partially by the American Expeditionary Forces and partially by the Engineering Division of Ordnance. This knife was entirely different from the model 1917, having a flat blade, metal scabbard, and a cast-bronze handle. It was a combination of all the good points of all the knives used by the foreign armies.

The Government placed orders for 1,232,780 of the new knives. Deliveries were to have begun in December, but before that time peace had come and the orders had been reduced to 119,424. The new model knives were to have been manufactured by A. A. Simons & Son, Dayton, Ohio; Henry Disston & Son, Philadelphia; Landers, Frary & Clark, and the Oneida Community (Ltd.). All contracts were canceled except the one with Landers, Frary & Clark.

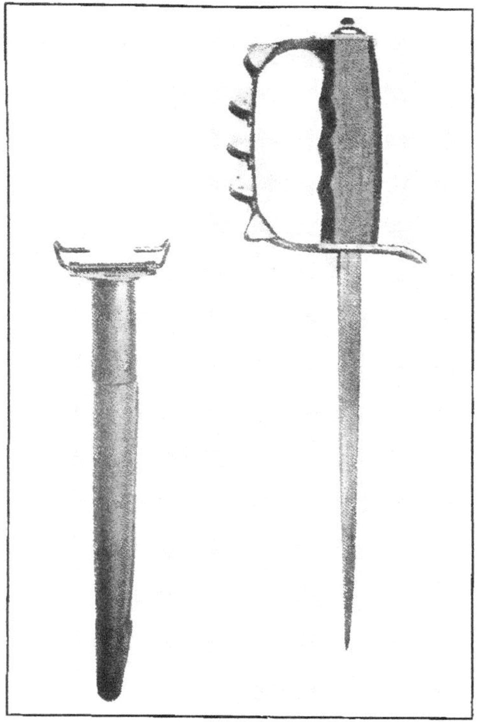

1917 MODEL OF TRENCH KNIFE AND SCABBARD.

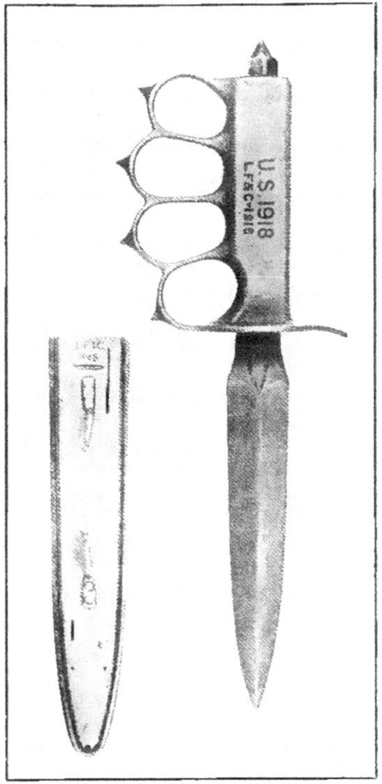

MARK I TRENCH KNIFE WITH FLAT BLADE, DESIGNED BY A. E. F.

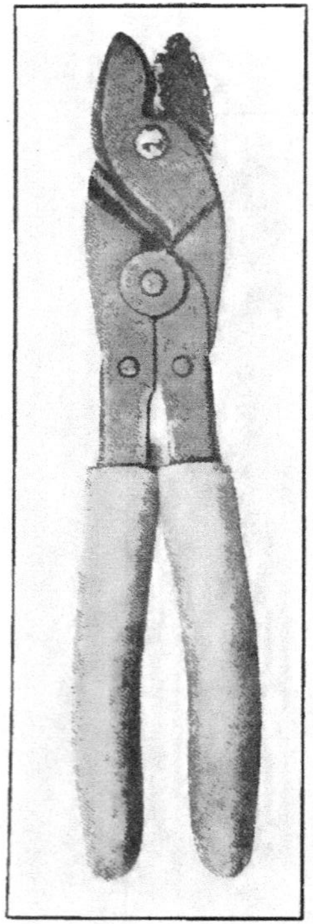

WIRE CUTTER (ONE HAND).

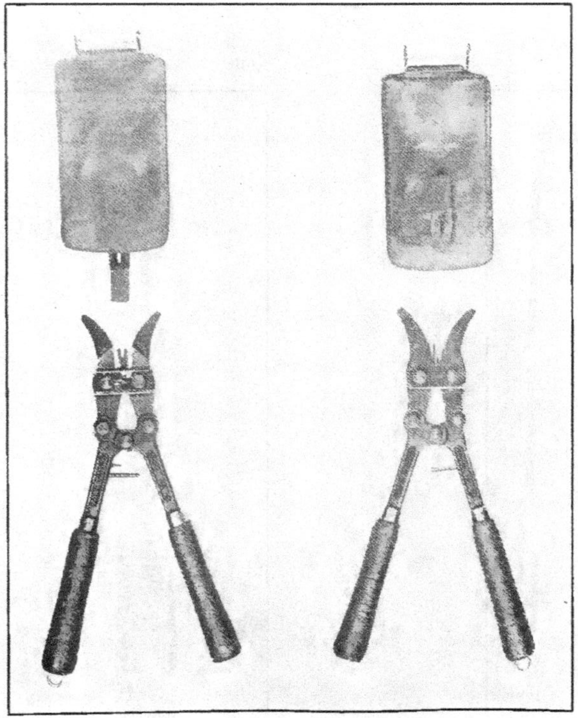

FRENCH WIRE CUTTERS.

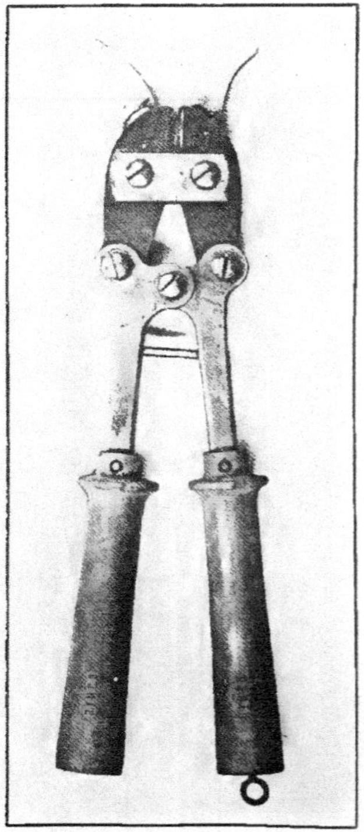

WIRE CUTTER, MODEL 1918, SHOWING SPECIAL RUBBER HANDLES. TESTED AT 10,000 VOLTS.

PERISCOPES, BELTS, ETC.

Another new article in the equipment of our soldiers was the trench periscope, a device enabling a man to look over the edge of the trench without exposing himself to fire. The ordinary periscope was merely a wooden box 2 inches square and 15 inches long, with an inclined mirror set at each end. Production was commenced in October, 1917, by two companies, and 81,000 were delivered by the middle of January. In August, 1918, an additional lot of 60,000 was ordered, but the deliveries of these were slow.

An even simpler periscope was merely a mirror about three inches long and an inch and a half wide which could be placed on a bayonet or a stick and set up over the trench so that it gave a view of the ground in front. A total of 100,000 of these was delivered before the end of July, 1918, and 50,000 additional ones before November. Further facts about periscopes are set down in the chapter in this report relating to sights and fire-control apparatus.

At the beginning of the war all textile equipment, such as cartridge belts, bandoleers to carry ammunition, haversacks, pack carriers, pistol holsters, canteen covers and similar material were supplied in woven material. Only two concerns in this country could manufacture articles of this quality. They were the Mills Woven Cartridge Belt Co., Worcester, Mass., and the Russell Manufacturing Co., Middletown, Conn. Although these two concerns practically doubled their output and worked day and night to supply the material, the demand was too great, and belts and carriers were designed to be stitched and sewn and not woven. Equipment made in this manner is inferior to the woven article. However, the Mills Woven Cartridge Belt Co. produced approximately 3,200,000 of these articles and the Russell Manufacturing Co., 1,500,000. Large producers of the stitched and sewn material were the Plant Brothers Co., Boston, Mass.; R. H. Long Co., Framingham, Mass.; L. C. Chase Co., Watertown, Mass.

For the Browning automatic rifle and the Browning machine gun there were specially designed belts and bandoleers. The rifleman had his own special belt, and his first and second assistants had their own individual belts, and the assistants also had two bandoleers each, one right and one left, which were carried across their shoulders. These were manufactured in quantities by the following manufacturers:

R. H. Long Co., Framingham, Mass.	175,000
Plant Bros., Boston, Mass.	75,000
L. C. Chase Co., Watertown, Mass.	20,000

Many small articles of textile equipment were produced in immense quantities. There were approximately four and a half million canteen covers produced prior to November 1. Large contracts were placed with the following concerns: Perkins-Campbell Co., Cincinnati, Ohio; Brauer Bros., St. Louis, Mo.; L. C. Chase Co., Watertown, Mass.; Miller-Hexter Co., Cleveland, Ohio; Powers Manufacturing Co., Waterloo, Iowa; R. H. Long Co., Framingham, Mass.; Bradford Co., St. Joseph, Mich.; Galvin Bros., Cleveland, Ohio; Progressive Knitting Works, Brooklyn, N. Y.

Approximately four and a half million haversacks were produced and delivered prior to November 1, 1918. Large manufacturers producing these were as follows: Canvas Products Co., St. Louis, Mo.; Rock Island Arsenal, Rock Island, Ill.; Plant Bros., Boston, Mass.; Simmons Hardware Co., St. Louis, Mo.; R. H. Long Co., Framingham, Mass.; Liberty, Durgin (Inc.), Haverhill, Mass.; Wiley, Bickford & Sweet, Hartford, Conn.

It is impossible here to enumerate the entire range of ordnance munitions produced, outside of the development of guns and their ammunition; but their manufacture, in orders that ordinarily amounted to the millions of individual pieces, engaged the activities of a large number of manufacturers of the United States.

The Government ordered about 1,200,000 axes to be used in trench operations, of which 661,690 were delivered. Bags of all sorts for horse feed, grain, rations, and supplies totaled in their deliveries about 2,250,000. The Government received 809,541 saddle blankets; about 3,750,000 carriers for entrenching shovels, axes, and picks; nearly 4,450,000 covers for the breech locks of rifles; over 1,000,000 currycombs; 76,230 lariats; 727,000 entrenching picks; nearly 4,750,000 first-aid pouches, and over 2,000,000 pouches for small articles; 234,689 Cavalry saddles; 134,092 Field Artillery saddles; 15,287 mule saddles; 482,459 saddle bags; nearly 1,800,000 entrenching shovels; 2,843,092 spur straps; 70,556 steel measuring tapes each 5 feet long.

These figures selected at random from thousands of miscellaneous items indicate to some extent the scale on which America went into the war.

The old model 1910 American wire cutter, although efficient in times past, was not capable of cutting specially constructed manganese wire which the Germans used. Therefore it became necessary for this country to develop a better cutter. A meeting of the plier manufacturers of the country was called and the question was put before them. The spirit of cooperation of the American manufacturers was evident, inasmuch as over 90 per cent of the manufacturers attended the meeting.

The model submitted by Kraeuter & Co., Newark, N. J., was adopted and 5,000 were manufactured and sent to France. Although this was the best cutter developed in this short time, it was evident

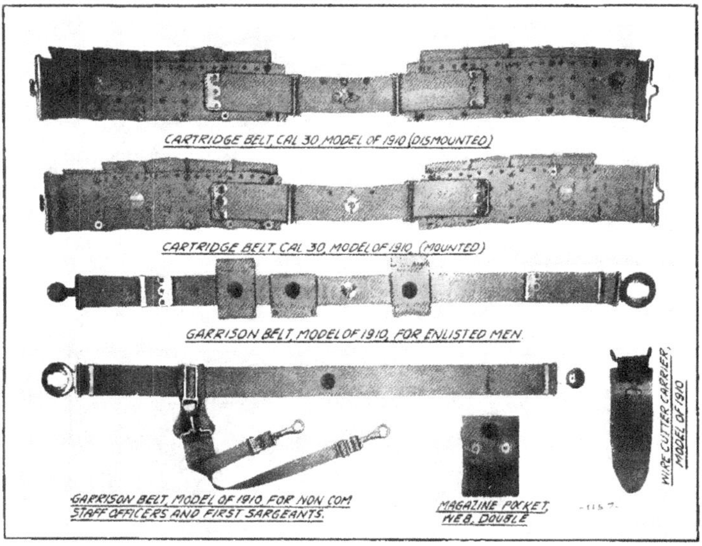
CARTRIDGE BELT, CAL 30, MODEL OF 1910 (DISMOUNTED)
CARTRIDGE BELT, CAL 30, MODEL OF 1910 (MOUNTED)
GARRISON BELT, MODEL OF 1910, FOR ENLISTED MEN.
WIRE CUTTER CARRIER, MODEL OF 1910
GARRISON BELT, MODEL OF 1910, FOR NON COM STAFF OFFICERS AND FIRST SARGEANTS.
MAGAZINE POCKET, WEB, DOUBLE

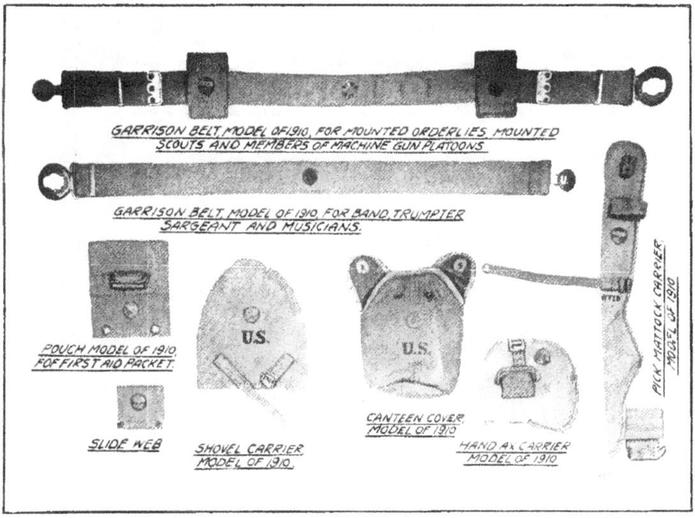
GARRISON BELT, MODEL OF 1910, FOR MOUNTED ORDERLIES, MOUNTED SCOUTS AND MEMBERS OF MACHINE GUN PLATOONS.
GARRISON BELT, MODEL OF 1910, FOR BAND, TRUMPTER SARGEANT AND MUSICIANS.
PICK MATTOCK CARRIER, MODEL OF 1910
POUCH MODEL OF 1910, FOR FIRST AID PACKET.
U.S.
U.S.
SLIDE WEB
SHOVEL CARRIER MODEL OF 1910.
CANTEEN COVER, MODEL OF 1910
HAND AX CARRIER MODEL OF 1910

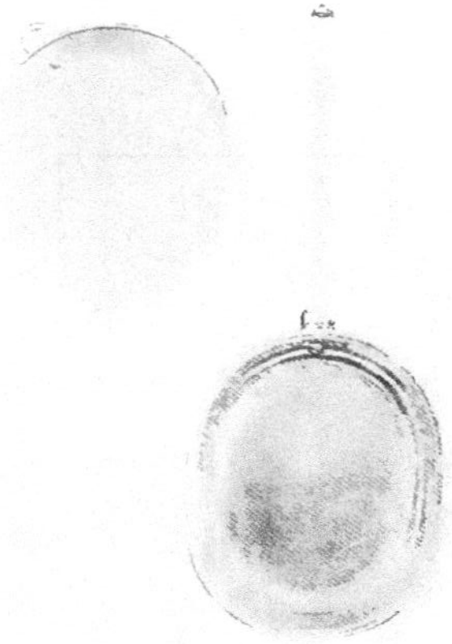

MEAT CAN, MODEL 1918.

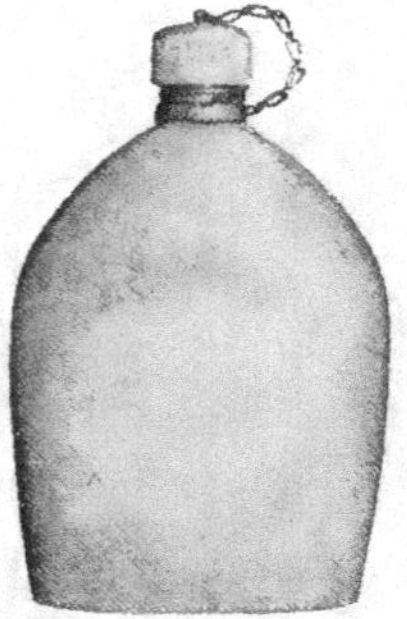

CANTEEN, MODEL 1910.

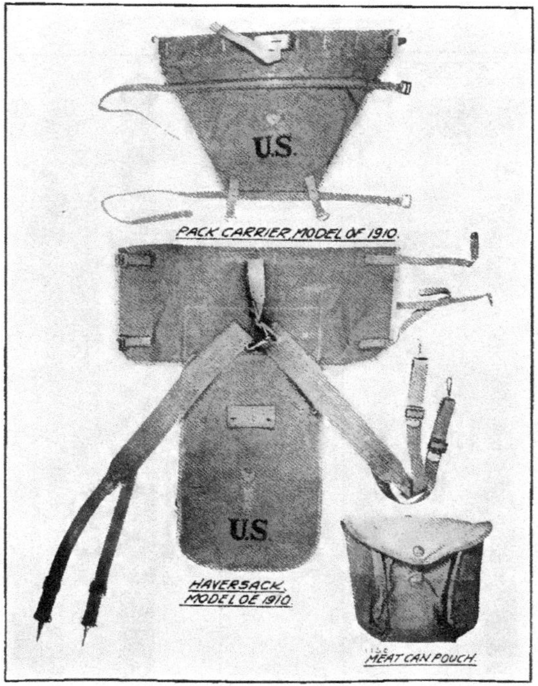

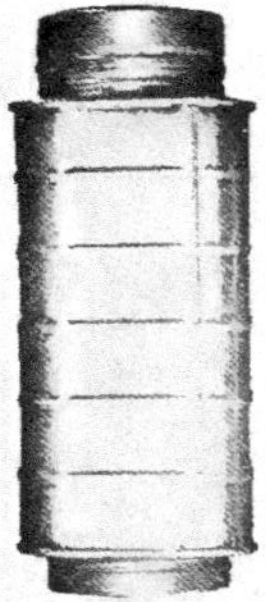

CONDIMENT CAN, MODEL 1910.

KNIFE, MODEL 1910.

FORK, MODEL 1910.

that it was not the right article, and the Engineering Division of Ordnance continued experimenting to make a more satisfactory one. In this connection a one-hand wire cutter was developed by the William Schollhorn Co., of New Haven, Conn. This cutter was a very efficient and satisfactory article, and, although it was never adopted by the American Army during the war, it is worthy of consideration. The American Expeditionary Forces eventually sent back drawings and sample of the French wire cutter, which was developed abroad and known as model 1918. This was a large, two-handed cutter. Production was started. The article was found difficult to manufacture, but the manufacturers undertook it with a will and production was well under way when the armistice was signed.

The mess equipment of the soldier included the following items: meat can, condiment can, canteen and cup, knife, fork, and spoon. These articles were practically the same as the Army had always used, with one exception—the meat can. Advice was received from the American Expeditionary Forces that the meat cans in which the soldiers' food was placed by the cooks of the various organizations were not large enough to hold the portions that the American doughboys needed when they were fighting at the front. Although production was well under way with various American manufacturers on the old model, a new model can was designed which was half an inch deeper. The American manufacturers immediately, with a great deal of trouble to themselves, changed their dies and tools and manufactured a new meat can which was larger than the old. Thousands of cans were turned out daily.

Production data.

CONDIMENT CANS.

Contractor.	Contract.	Completed and delivered.
American Can Co., New York City	3,553,940	3,553,940
Tin Decorating Co., Baltimore, Md	2,003,640	2,003,640
Gotham Can Co., Brooklyn, N. Y	500,000	500,000
Total	6,057,580	6,057,580

BACON CANS.

Sturgis & Burns, Chicago	2,303,800	1,731,000
Landers, Frary & Clark, New Britain, Conn	534,360	534,360
Rock Island Arsenal, Rock Island, Ill	1,658,000	1,358,570
Wisconsin Metal Products Co., Racine, Wis	50,000	50,000
Acklin Steel Co., Toledo, Ohio	250,000	250,000
Cleveland Metal Products Co., Cleveland, Ohio	300,000	21,750
Whittaker, Glessner Co., Wheeling, W. Va	500,000	131,880
Total	5,596,160	4,077,560

Production data—Continued.

MEAT CANS.

Contractor.	Contract.	Competed and delivered.
Aluminum Co. of America, Pittsburgh	3,385,955	3,385,955
Landers, Frary & Clark, New Britain, Conn	3,000,000	3,000,000
J. W. Brown & Co., Columbus, Ohio	641,945	641,945
Wheeling Stamping Co., Wheeling, W. Va	940,812	940,812
Edmunds & Jones Co., Detroit, Mich	138,360	138,360
Rock Island Arsenal	138,852	138,852
Total	8,245,924	8,245,924

CANTEENS.

Aluminum Co. of America, New York	3,470,000	3,470,000
Landers, Frary & Clark	2,862,150	2,862,150
Aluminum Goods Co., Manitowoc, Wis	2,370,000	2,370,000
J. W. Brown Co	861,471	861,471
Buckeye Aluminum Co., Wooster, Ohio	776,014	776,014
Rock Island Arsenal	361,000	361,000
Total	10,700,635	10,700,635

KNIVES.

American Cutlery Co., Chicago	2,865,910	2,865,910
Landers, Frary & Clark	7,286,550	7,286,550
Rock Island Arsenal	527,600	527,600
International Silverware Co	473,000	473,000
Hinckley Manufacturing Co	130,000	130,000
Total	11,283,060	11,283,060

FORKS.

R. Wallace & Co.,	8,585,000	8,585,000
Wallace Bros	367,810	367,810
Rock Island Arsenal	200,000	200,000
Charles Parker Co., Meriden, Conn	810,000	810,000
Wm. B. Durgin Co., Concord, N. H	500,000	500,000
Total	10,462,810	10,462,810

SPOONS.

R. Wallace & Co	8,037,600	8,037,600
National Enameling & Stamping Co	906,400	906,400
Wm. B. Durgin Co	500,000	500,000
Charles Parker Co	902,000	902,000
Total	10,346,000	10,346,000

BOOK II.

THE AIR SERVICE.

CHAPTER I.

THE AIRCRAFT PROBLEM.

When the United States entered the war against Germany in 1917 there was no phase of her forthcoming industrial effort from which so much was expected as from the building of airplanes and equipment for aerial warfare; yet there was no phase of the immense undertaking in which the United States was so utterly unprepared. In many other branches of the work of providing matériel for a modern army, however inadequately acquainted America might be with the developments which had gone on in Europe since 1914, yet she had splendid resources of skill and equipment which could quickly turn from the pursuits of peace to the arts attending warfare. But there was no large existing industry in the United States which could turn easily to the production of airplanes, since such airplanes as were known in Europe in 1917 had never been built in the United States.

It seems difficult now for us to realize how utterly unlearned we were, both in official and technical quarters, in the design, the production, or the use of aeronautical equipment in those early days of 1917. Here in America mechanical flight had been born; but we had lived to see other nations develop the invention into an industry and a science that was a closed book to our people. In the three years of warfare before American participation, the airplane had been forced through a whole generation of normal mechanical evolution. Of this progress we were aware only as nontechnical and distant observers. Such military study of the progress as we had conducted was casual. It had, in fact, brought to America scarcely a single basic fact on which we could build our contemplated industry.

When the United States became a belligerent no American-built airplane had ever mounted a machine gun or carried any other than the simplest of necessary instruments. Such things as oxygen apparatus, electrically heated clothing for aviators, radio-communication with airplanes, landing and bombing flares, electric lighting systems for planes, bomb-dropping devices, suitable compasses, instruments for measuring height and speed, and the like—in short, all the modern paraphernalia that completes the efficiency of combat airplanes—these were almost entirely unknown to us.

The best of the prewar activities of America in this line had produced some useful airplane engines and a few planes which the countries then at war were willing to use only in training of aviators.

Within the Army itself there was small nucleus of skill around which could be built an organization expert and sophisticated. We had in the official files no adequate information as to sizes, capacities, and types of planes or engines, or character of ordnance, armament, or aeronautical appliances demanded by the exacting service in which our young birdmen were soon to engage. Even the airplanes on order in April, 1917 (over 350 of them), proved to be of such antiquated design that the manufacturers of them, in the light of their increased knowledge of war requirements a few months later, asked to be released from their contracts.

Nor was there in the United States any industry so closely allied to airplane manufacture that its engineers and designers could turn from one to the other and take their places at once abreast of the progress in Europe. There was little or no engineering talent in the United States competent to design fully equipped military aircraft which could compete with Europe. Our aircraft producers must first go to France and England and Italy, and ground themselves in the principles of a new science before they could attempt to produce their own designs or even before they could be safe in selecting European designs for reproduction in this country.

The first consideration of the whole program by the Joint Army and Navy Technical Board indicated a figure of 22,000 as the number of airplanes, including both training and battle types, which should be furnished for the use of the Army during the 12 months following July 1, 1917. This figure represented the determination of America to play a major part in aerial warfare. It was not possible for the board to realize at that time all of the problems which would be encountered, and the figures indicated confidence in the ability of the industrial organizations of the United States to meet a difficult situation, rather than an exact plan under which such production might be developed.

It is probable that it was not fully realized that the production of this program, with the proper proportion of spare parts required for military operations, meant the manufacture of the equivalent of about 40,000 airplanes.

Without an industry then, and with little knowledge or understanding of the problems of military aerial equipment, we faced the task of securing the equivalent of 40,000 airplanes in 12 brief months beginning July, 1917.

In one respect we were in a degree prepared in professional skill and mechanical equipment to go ahead on broad lines. This was in the matter of producing engines. The production of aviation engines in America had, indeed, been comparatively slight, but in the automobile industry had been developed a vast engine-building capacity. The detail equipment of automobile shops was not entirely suited to

250 AIRPLANES IN THE AIR AT ONE TIME OVER SAN DIEGO, CALIF.

TWO ROWS OF AIRPLANES LINED UP AT A TEXAS FLYING FIELD.

aviation engines, but, nevertheless, it furnished the basis for the future successful production of the Liberty engine and the other engines called for by the air program.

America succeeded, once the requirements were known, in producing the various accessories of aerial warfare. It was necessary first to learn from foreign sources what these accessories were and how they should be built; but as a rule it was possible to adapt American production resources to the problem, and the difficulties experienced were rather those of determining requirements and the exact adaptation of the various articles to specific airplanes.

The achievements of America in aircraft production during the war period may be summarized as follows:

In our 19 months of warfare we outdid any one of the belligerent nations in Europe in the production of airplanes in its first 19 months of intensive production. In our second year of war we nearly equaled the record of England in her third.

At the end of the effort, after our designers had saturated themselves in the science and were abreast of the developments of Europe, they produced several typical American airplanes which gave promise of being superior to any that Europe was turning out.

We created one of the three or four best airplane engines, if not the best of all, that the world had seen, and produced it in great quantities. We took a standard but complicated aero-engine from Europe and not only duplicated it in quantity here but turned out a finer product than the original French makers had been able to obtain with their careful and more leisurely methods.

In the steel cylinders of all the aero-engines we built was a capacity for producing some seven or eight million horsepower, an energy equivalent to one-fifth of the commercially practicable water power of the United States. The Liberty engines built could alone do the work of the entire flood of Niagara and have a million horsepower to spare.

In three years of warfare the allies had been able to develop only a single machine gun that could be successfully synchronized to fire through a revolving airplane propeller. In 12 months of actual effort America produced two others as good, both susceptible of factory quantity production.

We developed new airplane cameras. We carried to new stages the science of clothing aviators. We developed in quantity the wireless airplane telephone that stilled in the ears of the pilot the bedlam of wind and machine guns and engine exhaust and placed him within easy speaking radius of his ground station and his commander in the air.

We built balloons at a rate to supply more than our own needs.

When the shortage of linen threatened the entire airplane output of the nations opposing Germany, we developed cotton wing fabric that not only substituted for linen, but proved to be better; and in producing a liquid filler to make this fabric wind-tight we established on a large scale an entirely new chemical industry in the United States.

Such were the high points in the history of America's aircraft production for war. The details of the developments which led to these results are set forth on the following pages.

VIEW OF AIRPLANE IN FLIGHT TAKEN FROM ANOTHER MACHINE.

PARACHUTE JUMP FROM AN ALTITUDE OF 7,900 FEET, NOVEMBER 12, 1918.

Picture taken at Love Field, Tex., from an airplane.

CHAPTER II.

AIRPLANE PRODUCTION.

Sketchy and incomplete as was our knowledge of airplane construction in the early days of 1917, it was no more hazy than our notion of how many planes to build. What would constitute overwhelming superiority in the air?

As an indication of the rapidity with which history has moved, it may be stated that in January and February of 1917 the Signal Corps discussed the feasibility of building 1,000 planes in a year of construction. This seems now to us a ridiculously low figure to propose as representative of American resources, but in the early weeks of 1917 the construction of a thousand airplanes appeared to be a formidable undertaking. In March, when war was inevitable, we raised this number to 2,500 planes within 12 months; in April, when war was declared, we raised it again to 3,700.

But once we were in the war, through the exchange of military missions our designers were taken into the confidence of the aviation branches of the French, British, and Italian Armies and shown then for the first time a comprehensive view of the development of the war plane, both what had been done in the past and what might be expected in the future. As a result our Joint Army and Navy Technical Board in the last week of May and the early part of June, 1917, recommended to the Secretaries of War and the Navy that a building program be started at once to produce the stupendous total of 19,775 planes for our own use and 3,000 additional ones, if we were to train foreign aviators, or approximately 22,000 in all. This was a program worthy of America's industrial greatness. Of these proposed planes, 7,050 were for training our flyers, 725 for the defense of the United States and insular possessions, and 12,000 for active service in France.

Such was the task assigned to an industry that in the previous 12 months had manufactured less than 800 airplanes, and those consisting principally of training planes for foreign governments.

The expanding national ambition for an aircraft industry was also shown by the mounting money grants. On May 12 Congress voted $10,800,000 for military aeronautics. On June 15 an appropriation of $43,450,000 was voted for the same purpose. Finally on July 24, 1917, the President signed the bill appropriating $640,000,000 for aircraft. This was the largest appropriation ever made by Congress for one specific purpose, and this bill was put through both Houses within the period of a little more than a week.

The figure 22,000, however, scarcely indicates the size of this undertaking, as we were to realize before long. We little understood the infinite complications of fully equipping battle planes. Lacking that invaluable experience which Europe had attained in three years of production, we had no practical realization of the fact that for each 100 airplanes an equivalent of 80 additional airplanes must be provided in spare parts. In other words, an effective fighting plane delivered in France is not one plane, but it is one plane and eight-tenths of another; which means that the program adopted in June, 1917, called for the production in 12 months of not 22,000 airplanes but rather the equivalent of 40,000 airplanes.

Let us set down the inventory of the Government's own resources for handling this project.

The American Air Service, which was then part of the Signal Corps, had had a struggling and meager existence, working with the old pusher type of planes until in 1914 an appropriation of $250,000 was made available for the purchase of new airplanes and equipment. Shortly after this appropriation was granted, five officers were sent to the Massachusetts Institute of Technology for a course in aeronautics. When the war broke out in Europe in August, 1914, these men constituted the entire technically trained personnel of the Air Service of the United States. By April 6, 1917, we had 65 officers in the Air Service, an enlisted and civilian personnel of 1,330, two flying fields, and a few serviceable planes of the training type.

This equipment may be compared with that of Germany, France, and England at the time they went to war. Germany is believed to have had nearly 1,000 airplanes in August, 1914; France had about 300; and England barely 250. America's 224, delivered up to April 6, 1917, were nearly all obsolete in type when compared with the machines then in effective service in France.

No sooner had the United States embarked upon the war than the agents of the European manufacturers of airplanes descended upon the Aircraft Board in swarms. France and Italy had both adopted the policy of depending upon the private development of designs for their supplies of airplanes, with the result that the builders of each country had produced a number of successful types of flying machines and an even greater number of types of engines. On the assumption that the United States would adopt certain of these types and build them here, the agents for the Sopwiths, the Capronis, the Handley-Pages, and many others proceeded to demonstrate the particular excellences of their various articles. Out of this confusion of counsel stood one pertinent fact in relief—the United States would have to pay considerable royalties for the use of any of these European devices.

As to the relative merits of types and designs, it was soon apparent that no intelligent decision could be reached in Washington or anywhere but in Europe. Because of our distance from the front and the length of time required to put the American industrial machine into operation on a large scale, it was necessary that in advance we understand types and tendencies in aircraft construction, so that we might anticipate aircraft development in such special designs as we might adopt. Otherwise, if we accepted the types of equipment then in use in Europe, by the time we had begun producing on a large scale a year or so later we would find our output obsolete and out of date, so rapidly was the aircraft art moving.

Consequently, in June the United States sent to Europe a commission of six civilian and military experts, headed by Maj. R. C. Bolling, part of whose duties was to advise the American War Department as to what types of planes and engines and other air equipment we should prepare to manufacture. Also, in April the Chief of the Signal Corps had cables sent to England, France, and Italy, requesting that aviation experts be sent at once to this country; and shortly after this we dispatched to Europe more than 100 skilled mechanics to work in the foreign engine and airplane plants and acquire the training that would make them the nucleus of a large mechanical force for aircraft production in this country.

But while these early educational activities were in progress, much could be done at home that need not await the forthcoming reports from the Bolling mission. We had, for instance, in this country several types of planes and engines that would be suitable for the training fields which were even then being established. The Signal Corps, therefore, bent its energies upon the manufacture of training equipment, leaving the development of battle aircraft to come after we should know more about that subject.

It was evident that we could not equip an airplane industry and furnish machines to our fliers abroad before the summer of 1918; and so we arranged with France for this equipment by placing orders with French factories for 5,875 planes of regular French design. These were all to be delivered by July 1, 1918.

In the arrangement with the French factories we agreed to supply from the United States a great deal of the raw materials for these machines, and the contract for furnishing these supplies was given to J. G. White & Co. of New York City. This concern did a creditable job, shipping about 5,000,000 feet of lumber, much necessary machinery, and a multitude of items required in the fabrication of airplanes, all to the value of $10,000,000.

The total weight of the shipments on this contract was something like 23,000 tons, this figure including 7,500 tons of lumber. The other tonnage consisted of tubing of steel, brass, copper and aluminum; sheets of steel, copper, lead, and aluminum; as well as bar steel, tool steel, structural steel, ball bearings, crank shafts, turnbuckles, radiator tubes, wire, cable, bolts, nuts, screws, nails, fiber cloth, felt, and rubber. All of this was in addition to approximately 1,000 machine tools, such as motors, lathes, and grinders.

The orders for French planes were divided as follows: 725 Nieuport training planes, 150 Spad training planes, 1,500 Breguet service planes; 2,000 Spad service planes; and 1,500 New Spad or Nieuport service planes. The decision between the New Spad or Nieuport service planes was to be made as soon as the New Spad could be tested. These planes were to be delivered in specified monthly quantities increasing in number until the total of 1,360 planes should be placed in our hands during the month of March, 1918, alone. The contracts were to be concluded in June with the delivery of the final 1,115 planes. We also contracted for the manufacture of 8,500 service engines of the Renault, Hispano and Gnome makes, all of these to be delivered by the end of June.

When the armistice ended the fighting, we had produced a total of 11,754 airplanes in America, together with most of the necessary spare parts for about one-third of them.

While a large part of the American airplanes built in the war period were of the training type rather than the service, or battle, type, nevertheless it was necessary that we have a large equipment of training planes in order to prepare the swiftly expanding personnel of the Air Service for its future activity at the front. The nations associated with us in the war, however, had produced their training equipment prior to our participation as a belligerent, and at the time we entered the war the French, British, and Italians were producing only enough training planes to maintain their training equipment and were going in heavily with the rest of their airplane industries for the production of service planes.

With these considerations in mind, the reader may make an interesting comparison of British and American plane production, the British figures being for both the British Army and the British Navy, whereas the American figures are for the American Army alone. In the following table of comparison the British figures are based on the Lockhart Report of November 1, 1918:

Comparative rate of airplane production—British and United States Army.

Calendar year.	British Army and Navy.	United States Army.
1915, Jan. 1 to Dec. 31	2,040	20
1916, Jan. 1 to Dec. 31	6,000	[1] 83
1917, Jan. 1 to Dec. 31	14,400	[2] 1,807
1918, Jan. 1 to Dec. 31	30,000	[3] 11,950

[1] Experimental.
[2] 1,476 built in last seven months only.
[3] Inclusive of 135 secured by Engineering Department. American total 12,837 if October production had continued through November and December.

Broadly stated, and without reference to types of planes produced, these figures mean that the United States in her second year of the war produced for the American Army alone almost as many airplanes as Great Britain in her third year of the war built for both her army and navy. In October, 1918, factories in this country turned out 1,651 planes, which, without allowing for the monthly expansion in the production, was at the rate of 20,000 planes per year. Assuming no increase in the October rate of production, we would have attained the 22,000 airplanes in 23 months after July 1, 1917, the date on which the production effort may be said to have started. Our production of fighting planes in the war period was 3,328.

On the day the armistice was signed we had received from all sources 16,952 planes. Of these 5,198 had been produced for us by the allies. We had 48 flying fields, 20,568 Air Service officers, and 174,456 enlisted men and civilian personnel. These figures do not mean that we had more than 17,000 planes on hand at that time, because the mortality in airplanes is high from accidents and ordinary wear and tear.

THE PROBLEM OF MATERIAL.

Once we had started out on this enterprise we soon discovered that the production of airplanes was something more than a mere manufacturing job. With almost any other article we might have made our designs, given orders to the factories, and rested in the security that in due time the articles would be forthcoming. But with airplanes we had to create the industry; and this meant not only the equipping of factories, but the procurement and sometimes the actual production of the raw materials.

For instance, the ideal lubricant for the airplane motor is castor oil. When we discovered that the supply of castor oil was not nearly sufficient for our future needs, the Government itself secured from Asia a large quantity of castor beans, enough to seed more than 100,000 acres in this country and thus to provide for the future lubrication for our motors. This actual creation of raw materials was conducted on a much larger scale in the cases of certain other commodities used in airplane construction, particularly in the production of lumber and cotton and in the manufacture of the chemicals for the "dope" with which the airplane wings are covered and made air-tight.

An airplane must have wings and an engine with a propeller to make it go; and, like a bird, it must have a tail to make it fly straight and a body (fuselage) to hold all together. Part of the tail (the rudder) moves sideways and steers the airplane from left to right; part moves up and down (the elevators) and makes the airplane go up or down, and parts of the wings (the ailerons) move up and down and make the airplane tip from side to side. All of these things must be connected to the controls in the hands of the pilot. The front edges of the wings are raised above the line of flight; and when the propeller driven by the engine forces the wings through the air, the airplane is lifted and flies.

All of the airplanes built for the United States during the war were tractor biplanes. In a plane of the tractor type the propeller is in front and pulls the machine. The biplane is so called because it has two planes or wings, one above the other. The biplane has been the most largely used of all types in war for two reasons: first, the struts and wires between the planes form a truss structure, and this gives the needed strength; and second, there is less danger of enemy bullets wrecking a biplane in the air because its wing support is greater than that of the monoplane or single-winged machine.

Since the airplane can lift only a limited weight, every part of the mechanism must be as light as possible. An airplane engine weighs from 2 to 3 pounds per horsepower, whereas an automobile motor weighs from 8 to 10 pounds per horsepower. The skeleton of the airplane is made of wood, mostly spruce, with sheet-steel fittings to join the wood parts together, and steel wires and rods to make every part a truss. This skeleton is covered with cloth, and the cloth is stretched and made smooth by dope.

Wood, sheet steel, wire, cloth, varnish—these are the principal components of an airplane. As raw materials they all seem easy to obtain in America. And so they are in peace times and for ordinary purposes. But never before had quality been so essential in an American industry, from the raw material up to the finished product—quality in the materials used, and quality in the workmanship which fashions the parts. But combined with this quality we were forced to produce in quantities, bounded only by our own physical limitations, and these quantities must include not only the materials for our own air program but also some of the principal raw materials used by the airplane builders in France and England, specifically, all of the spruce which the allies would require and, later, much of the wing fabric and dope for their machines.

Quite early it was apparent to us that we had on our hands a problem in spruce production which the Government itself must solve, if the airplane undertaking were not to fail at the outset. When we entered the war linen was exclusively used for the covering of wings; and it developed almost immediately that the United Kingdom was practically the sole source of linen. But the Irish looms could not begin to furnish us with our needs for this commodity. Later on came up the question of supplying dope and castor oil. Finally, during the last months of the war, it became necessary for us to follow up the production of all classes of our raw material, particularly in working out a suitable supply of steel tubing. But our great creative efforts in raw materials were confined to spruce, fabric, and dope.

The lumber problem involved vast questions of an industrial and technical character. We had to conduct a campaign of education in the knowledge of aircraft requirements that reached from the loggers themselves in the woods to the saw-mill men, to the cut-up plants, and then followed through the processes of drying and sawing to the proper utilization of the lumber in the aircraft factories.

In working out these problems, while we drew heavily upon the experience in Europe, yet we ourselves added our own technical skill to the solution. The Signal Corps was assisted by the forest products laboratory at Madison, Wis., and by the wood section of the inspection department of the Bureau of Aircraft Production. The United States Forest Service contributed its share of technical knowledge. At the end of the war we considered that our practice in the handling of aircraft lumber was superior to that of either France or England.

THE SPRUCE PROBLEM.

Each airplane uses two distinct sorts of wood—first, the spruce or similar lumber for the wing beams or other plane parts; and second, mahogany, walnut, or other hardwoods for propellers. In each case the Army production authorities were involved both in securing the lumber and in educating manufacturers to handle it properly.

In an ordinary biplane there are two beams for each lateral wing, eight beams to the plane. These form the basis of strength for the wings. Because of the heavy stresses put upon the airplanes by battle conditions, only the most perfect and straight-grained wood is suitable for these beams. All cross-grained or spiral-grained material, or material too coarse in its structure, is useless.

Spruce is the best of all woods for wing beams. Our problem was to supply lumber enough for the wing beams, disregarding the other parts, as all other wood used in the manufacture of planes could be secured from cuttings from the wing-beam stock. At the beginning we built each beam out of one piece of wood; and this meant that the lumber must be extra long, thick, and perfect. Until we learned how to cut the spruce economically we found that only a small portion of the lumber actually logged was satisfactory for airplanes. An average sized biplane uses less than 500 feet of lumber. In the hands of skilled cutters this quantity can be worked out of 1,000 feet of rough lumber. But in the earlier days of the undertaking as high as 5,000 feet of spruce per plane were actually used because of imperfections in the lumber, lack of proper inspection at the mills, and faulty handling in transit and in the factories.

We also used certain species of fir in building training planes. This wood is, like spruce, light, tough, and strong. The only great source of supply of these woods was in the Pacific Northwest, although there was a modest quantity of suitable timber in West Virginia, North Carolina, and New England.

While at first we expected to rely upon the unaided efforts of the lumber producers, labor difficulties almost immediately arose in the Northwest to hinder the production of lumber. The effort, too, was beset with difficulties of a physical nature, since the large virgin stands of spruce occurred only at intervals and often at long distances from the existing railroads. By the middle of October, 1917, it became evident that the northwestern lumber industry unaided could not deliver the spruce and fir; and the Chief of Staff of the Army formed a military organization to handle the situation. On November 6, 1917, Col. Brice P. Disque took command of the Spruce Production Division of the Signal Corps, this organization later being transferred to the Bureau of Aircraft Production.

When Col. Disque went into the Northwest he found the industry in chaotic condition. The I. W. W. was demoralizing the labor forces. The mills did not have the machinery to cut the straight-grained lumber needed and their timber experts were not sufficiently skilled in the selection and judging of logs to secure the maximum footage. The whole industry was organized along lines of quantity production and desired to avoid all high quality requirements insisted upon by the Government.

One of the first acts of the military organization was to organize a society called the Loyal Legion of Loggers and Lumbermen, the "L. L. L. L.," to offset the I. W. W. propaganda, its platform being, no strikes, fair wages, and the conscientious production of the Government's requirements. On March 1, 1918, 75,000 lumbermen and operators agreed without reservation to give Col. Disque power to decide all labor disputes. The specifications for logs were then standardized and modified as far as practicable to meet the manufacturers' needs. We arranged financial assistance that they might equip their mills with the proper machinery. We instituted a system of instruction for the personnel. Finally, the Government fixed a price for aircraft spruce that stabilized the industry and provided against delays from labor disputes.

While these basic reforms were being instituted our organization had energetically taken up the physical problems relating to the work. We surveyed the existing stands of spruce timber, built railroads connecting them with the mills, and projected other railroads far into the future. We began and encouraged logging by farmers in small operations. By these and other methods employed, the efficiency of this production effort gradually increased.

In all, we took 180,000,000 feet of aircraft lumber out of the northwestern forests. To the allies went 120,000,000 feet; to the United States Army and Navy, 60,000,000 feet.

Yet when we had resolved the difficulties in the forests only part of the problem had been met. Next came the intricate indus-

ASSEMBLING DE HAVILAND-4 WINGS AT THE DAYTON-WRIGHT PLANT.

SEWING FABRIC ON AIRPLANE WINGS.

APPLYING THE DOPE TO AIRPLANE WINGS.

trial question of how to prepare this lumber for aircraft use. We possessed little knowledge as to the proper methods of seasoning this timber. The vast majority of woodworking plants in this country, such as furniture and piano makers, had always dried lumber to the end that it might keep its shape. We now were faced with the technical question of drying lumber so as to preserve its strength. The forest products laboratory worked out a scientific method for this sort of seasoning. Incidentally they discovered that ordinary commercial drying had seldom been carried on scientifically. The country will receive a lasting benefit from this instruction carried broadcast over the industry.

In the progress of our wood studies we discovered a method of splicing short lengths of spruce to make wing beams and in the later months of the production used these spliced beams exclusively at a great saving in raw materials. The use of laminated beams would probably have become universal in another year of warfare.

COTTON FABRIC AND DOPE.

The flying surfaces of an airplane are made by stretching cloth over the frames. When we came into war it was supposed that linen was the only common fabric with sufficient strength for this use, and linen was almost exclusively used by the airplane builders, although Italian manufacturers were trying to develop a cotton fabric. Of the three principal sources of flax, Belgium had been cut off from the allies, Russia was isolated entirely after the revolution there, and Ireland was left as the sole available land from which flax for airplane linen could be obtained.

As late as August, 1917, England assured us that she could supply all of the linen that would be needed. It rapidly became evident that England had underestimated our requirements. An average airplane requires 250 yards of fabric, while some of the large machines need more than 500 yards. And these requirements do not take into consideration the spare wings which must be supplied with each airplane. This meant a demand for millions of yards put upon the Irish supply, which had no such surplus above allied needs.

For some time prior to April 6, 1917, the Bureau of Standards at Washington had been experimenting with cotton airplane cloths. Out of the large variety of fabrics tested several promising experimental cloths were produced. The chief objection to cotton was that the dope which gave satisfactory results on linen failed to work with uniformity on cotton. Therefore, it became apparent that if we were to use cotton fabric, we should have to invent a new dope.

Two grades of cotton airplane cloth were finally evolved—A, which had a minimum strength of 80 pounds per inch, and B, with a minimum strength of 75 pounds per inch. Grade A was later universally adopted. This cloth weighed four and one-half ounces per square yard.

We placed our first orders for cotton airplane fabric in September, 1917—orders for 20,000 yards—and from that time on the use of linen decreased. By March of 1918 the production of cotton airplane cloth had reached 400,000 yards per month. In May the production was about 900,000 yards; and when the war ended this material was being turned out at the rate of 1,200,000 yards per month. Starting with a few machines, our cotton mills had gradually brought 2,600 looms into the enterprise, each loom turning out about 120 yards of cloth in a week. A total of 10,248,355 yards of cotton fabric was woven and delivered to the Government—over 5,800 miles of it, nearly enough to reach from California to France. The use of cotton fabric so expanded that in August, 1918, we discontinued the importations of linen altogether.

There was, however, danger that we would be limited in our output of cotton fabric if there were any curtailment in the supply of the long-staple sea-island and Egyptian cotton of which this cloth is made. To make sure that there would be no shortage of this material the Signal Corps in November, 1917, went into the market and purchased 15,000 bales of sea-island cotton. This at all times gave us an adequate reserve of raw material for the new fabric.

Cotton proved to be not only an admirable substitute for linen but even a better fabric than the original cloth which had been used. No matter how abundant the supply of flax may be, it is unlikely that linen will ever again be used in large quantities for the manufacture of airplane wings.

Thus, as the airplane situation was saved by the prompt action of the Signal Corps in organizing and training the spruce industry, so again its decision to produce cotton fabric and its prompt action in cornering the necessary cotton supply made possible the uninterrupted expansion of the allied aviation program.

The wings of an airplane must not only be covered with fabric, but the fabric must be filled with dope, which is a sort of varnish. The function of the dope is to stretch the cloth tight and to create on it a smooth surface. After the dope is on the fabric the surface is protected further by a coat of ordinary spar varnish.

We found in the market two sorts of dope which were being furnished to airplane builders of all countries by various chemical and varnish manufacturers. One of these dopes was nitrate in character and was made from nitrocellulose and certain wood-chemical solvents including alcohol. This produced a surface similar to that of a photographic film. The other kind of dope had an acetate base and was made from cellulose-acetate and such wood chemical solvents as acetone.

The nitrate dope burned rapidly when ignited but the acetate type was slow burning. Thus in training planes not subject to attack by enemy incendiary bullets the nitrate dope would be fairly satisfactory, but in the fighting planes the slow-burning acetate dope was a vital necessity. Up to our participation in the war the dopes produced in the United States were principally nitrate in character.

It was evident that we should make our new dope acetate in character to avoid the danger of fire. But for this we would require great quantities of acetone and acetate chemicals, and a careful canvass of the supply of such ingredients showed that it would be impossible for us to obtain these in anything like the quantities we should require without developing absolutely new sources of production.

Already acetone and its kindred products were being absorbed in large quantities by the war production of the allies. The British Army was absolutely dependent upon cordite as a high explosive. Acetone is the chemical basis of cordite; and therefore the British Army looked with great concern upon the added demand which the American aviation program proposed to put upon the acetone supply.

We estimated that in 1918 we would require 25,000 tons of acetone in our dope production. The British war mission in this country submitted figures showing that the war demands of the allies, together with their necessary domestic requirements, would in themselves be greater than the total world production of acetone.

There was nothing then for us to do but to increase the source of supply of these necessary acetate compounds; and this was done by encouraging, financially and otherwise, the establishment of 10 large chemical plants. These were located in as many towns and cities, as follows: Collinwood, Tenn.; Tyrone, Pa.; Mechanicsville, N. Y.; Shawinigan Falls, Canada; Kingsport, Tenn.; Lyle, Tenn.; Freemont, Mo.; Sutton, W. Va.; Shelby, Ala.; and Terre Haute, Ind.

But it was evident that before these plants could be completed the airplane builders would be needing dope; and therefore steps were taken to keep things going in all the principal countries fighting Germany until the acetate shortage could be relieved. In December, 1917, we commandeered all the existing American supply of acetate of lime, the base from which acetone and kindred products are made. Then we entered into a pool with the allied governments to ration these supplies of chemicals, pending the era of plenty. Our agency in this pool was the wood-chemical section of the War Industries Board, whereas the allies placed their demands in the hands of the British war mission. These two boards allocated the acetate chemicals among the different countries according to the urgency of their demands. Since it was evident there might be financial losses incurred as the result of the commandeering order or in the project of the new Government chemical plants, the British war mission agreed that any deficit should be shared equally by the American and British Governments. It was also agreed we should not have any advantage in prices paid for acetates of American origin. Under this arrangement we were able to produce 1,324,356 gallons of fabric dope during the period of hostilities, without upsetting any of the European war-production projects. Had the war continued, the output from the 10 chemical plants in which the Government was a partner would have cared for all American and allied requirements, allowing the production of private plants to go exclusively for the ordinary commercial purposes.

THE TRAINING PLANES.

The actual building of the airplanes gave a striking example of the value of previous experience, either direct or of a kindred nature, in the quantity production of an article. What airplanes we had built in the United States—and the number was small, being less than 800 in the 12 months prior to April, 1917—had been entirely of the training type. These had been produced principally for foreign governments. But this slight manufacture gave us a nucleus of skill and equipment that we were able to expand to meet our own training needs almost as rapidly as fields could be equipped and student aviators enlisted. The training-plane program can be called a success, as the final production figures show. Of the 11,754 airplanes actually turned out by American factories, 8,567 were training machines. This was close to the 10,000 mark set as our ambition in June, 1917.

There are two types of training planes—those used in the primary instruction of students and those in the advanced teaching, the latter approaching the service planes in type. The primary plane carries the student and the instructor. Each occupant of the fuselage has before him a full set of controls which are interconnected so that the instructor at will can do the flying himself, or correct the student's false moves, or allow the student to take complete charge of the machine. These primary planes fly at the relatively slow speed of 75 miles per hour on the average and require engines so reliable that they need little attention.

For our training planes we adopted the Curtiss JN-4, with the Curtiss OX-5 engine, and as a supplementary equipment the Standard Aero Corporation's J-1 plane, with the Hall-Scott A-7-A engine. Both of these planes and both engines had been previously manufactured here. The Curtiss equipment, which was the standard at our training camps, gave complete satisfaction. The J-1 plane was later withdrawn from use, partly because the plane itself was not liked, partly because of the vibration resulting from this Hall-Scott engine, it having only four cylinders, and partly because of the uncertainty of the engine in cold weather.

CURTISS JN4-D, USED AS A PRIMARY TRAINING MACHINE. ENGINE, CURTISS OX-5.

This machine has a dual control and is used solely for training purposes.

V. E. 7. EQUIPPED WITH 180-HORSEPOWER HISPANO-SUIZA ENGINE.

An American designed training plane.

It was evident that at the start we must turn our entire manufacturing capacity to the production of training planes, since we would need these first in any event, and we were not yet equipped with the knowledge to enable us to make intelligent selections of service types.

In taking up the manufacturing problem the first step was to divide the existing responsible airplane plants between the Army and Navy, following the general rule that a single plant should confine its work to the needs of one Government department only. There were, of course, exceptions to this rule. This division gave to the Army the plants of the—

Curtiss Aeroplane & Motor Corporation, Buffalo, N. Y.
Standard Aircraft Corporation, Elizabeth, N. J.
Thomas-Morse Aircraft Corporation, Ithaca, N. Y.
Wright-Martin Aircraft Corporation, Los Angeles, Calif.
Sturtevant Aeroplane Co., Boston, Mass.

The factories which fell to the Navy were those of the—

Curtiss Aeroplane & Motor Corporation, Buffalo, N. Y.
The Burgess Co., Marblehead, Mass.
L. W. F. (Lowe, Willard & Fowler) Engineering Co., College Point, Long Island.
Aeromarine Engineering & Sales Co., New York.
Gallaudet Aircraft Corporation, New York.
Boeing Airplane Co., Seattle, Wash.

Of these concerns, Curtiss, Standard, Burgess, L. W. F., Thomas-Morse, and Wright-Martin were the only ones which had ever built more than 10 machines.

These factories were quite insufficient in themselves to carry out the enterprise. Therefore it became necessary to create other airplane plants. Two new factories thereupon sprang into existence under Government encouragement. The largest producer of automobile bodies was the Fisher Body Co., at Detroit, Mich. The manufacture of automobile bodies is akin to the manufacture of airplanes to the extent that each is a fabrication of accurate, interchangeable wood and sheet-steel parts. The Fisher organization brought into the enterprise not only machinery and buildings but a skilled organization trained in such production on a large scale.

At Dayton, Ohio, the Dayton-Wright Airplane Corporation was created. With this company was associated Orville Wright, and its engineering force was built up around the old Wright organization. A number of immense buildings which had been recently erected for other purposes were at once utilized in this new undertaking.

As an addition to these two large sources of supply, J. G. White & Co. and J. G. Brill & Co., the well-known builders of street cars, formed the Springfield Aircraft Corporation at Springfield, Mass.

Also certain forward-looking men on the Pacific coast created in California several airplane plants, some of which ultimately became satisfactory producers of training planes.

At this point in the development we were not aware of the great production of spare parts that would be necessary. Yet we did understand that there must be a considerable production of spares; and in order to take the burden of this manufacture from the regular airplane plants, and also to educate other factories up to the point where they might undertake the construction of complete airplanes, we placed many contracts for spare parts with widely scattered concerns. Among the principal producers of spares were the following:

The Metz Co., Waltham, Mass.
Sturtevant Aeroplane Co., Jamaica Plains, Mass.
Wilson Body Co., Bay City, Mich.
West Virginia Aircraft Corporation, Wheeling, W. Va.
The Rubay Co., Cleveland, Ohio.
Engel Aircraft Co., Niles, Ohio.
Hayes-Ionia Co., Grand Rapids, Mich.

For a long time the supply of spare parts was insufficient for the needs of the training fields. This was only partly due to the early lack of a proper realization of the quantity of spares that would be required. The production of spares on an adequate scale was hampered by numerous manufacturing difficulties incident to new industry of any sort in shops unacquainted with the work, and by a lack of proper drawings for the parts.

As to the training planes themselves, with all factories in the country devoting themselves to this type exclusively at the start, the production soon attained great momentum. The Curtiss Co. particularly produced training planes at a pace far beyond anything previously obtained. The maximum production of JN-4 machines was reached in March, 1918, when 756 were turned out.

Advanced training machines are faster, traveling at about 105 miles per hour; and they carry various types of equipment to train observers, gunners, photographers, and radio men. For this machine we adopted the Curtiss JN-4H, which was substantially the same as the primary training plane, except that it carried a 150-horsepower Hispano-Suiza engine. We also built a few "penguins," a kind of half airplane that never leaves the ground; but this French method of training with penguins we never really adopted.

The finishing school for our aviators was in France, where the training was conducted in Nieuports and other fighting machines.

In July, 1918, we reached the maximum production of the advanced training machines, the output being 427. As the supply of primary training planes met the demands of the fields, the production was reduced, since the original equipment, kept up by only enough manufacture to produce spares and replacement machines, would suffice to train all of the aviators that we would need.

The actual production of training planes by months was as follows:

	Primary training planes, SJ-1, JN-4D, Penguin.	Advanced training planes, JN-4 and 6H, S-4B and C, E-1, SE-5.		Primary training planes, SJ-1, JN-4D, Penguin.	Advanced training planes, JN-4 and 6H, S-4B and C, E-1, SE-5.
1917.			1918—Continued.		
April	....	....	March	756	178
May	....	....	April	645	81
June	9	....	May	419	166
July	56	....	June	126	313
August	103	....	July	236	427
September	193	....	August	296	193
October	340	....	September	233	132
November	331	1	October	212	320
December	423	20	November	186	297
			December	162	259
1918.					
January	700	29	Total	5,952	2,615
February	526	199			

THE SERVICE PLANES.

It was not until we took up the production of fighting, or service, airplanes that we came into a full realization of the magnitude of the engineering and manufacturing problems involved.

We had perhaps a dozen men in the United States who knew something about the designing of flying machines, but not one in touch with the development of the art in Europe or who was competent to design a complete fighting airplane. We had the necessary talent to produce designs and conduct the manufacture of training planes; but at the outset, at least, we were unwilling to attempt designs for service planes on our own initiative. At the beginning we were entirely guided by the Bolling mission in France as to types of fighting machines.

In approaching this, the more difficult phase of the airplane problem, our first act was to take an inventory of the engineering plants in the United States available for our purposes. With the Curtiss Co. was Glenn Curtiss, a leader of airplane design, and several competent engineers. The Curtiss Co. had been the largest producers in the United States of training machines for the British and had had the benefit of assistance from British engineers, and so possessed more knowledge and experience to apply to the service-plane problem than any other company. For this reason we selected this plant to duplicate the French Spad plane, the story of which undertaking will be told further on.

Orville Wright, the pioneer of flying, was not in the best of health, but was devoting his entire time to experimental work in Dayton.

Willard, who had designed the L. W. F. airplane and was then with the Aeromarine Co.; Chas. Day, formerly with the Sloane Manufacturing Co., and then with the Standard Aero Corporation; Starling Burgess, with the Burgess Co., of Marblehead, Mass.; Grover C. Loening, of the Sturtevant Co.; and D. D. Thomas, with the Thomas-Morse Co., were all aviation engineers on whom we could call. One of the best experts of this sort in the country was Lieut. Commander Hunsaker, of the Navy. In the Signal Corps we had Capt. V. E. Clark, who was also an expert in aviation construction, and he had several able assistants under him.

The Burgess factory at Marblehead, the Aeromarine plants at Nutley and Keyport, N. J., and the Boeing Airplane Co. at Seattle were to work exclusively for the Navy, according to the mutual agreement, taking their aeronautical engineers with them. This gave the Army the engineering resources of the Curtiss, Dayton-Wright, and Thomas-Morse companies.

Quite early we decided to give precedence in this country to the observation type of service plane, eliminating the single-place fighter altogether and following the observation planes as soon as possible with production of two-place fighting machines. This decision was based on the fact, not always generally remembered, that the primary purpose of war flying is observation. The duels in the air that occurred in large numbers, especially during the earlier stages of the war, were primarily to protect the observation machines or to prevent observation by enemy machines.

The first service plane which we put into production and which proved to be the main reliance of our service-plane program was the De Haviland-4, which is an observation two-place airplane propelled by a Liberty 12-cylinder engine. As soon as the Bolling mission began to recommend types of service machines, it sent samples of the planes thus recommended. The sample De Haviland was received in New York on July 18, 1917. After it had been studied by various officers it was sent to Dayton. It had reached us without engine, guns, armament, or many other accessories later recommended as essential to a fighting machine. Before we could begin any duplication the plane had to be redesigned to take our machine guns, our instruments, and our other accessories, as well as our Liberty engine.

The preliminary designing was complete, and the first American-built De Haviland model was ready to fly on October 29, 1917.

Figure 11 does not tell quite the complete story of De Haviland production, since in August and September, 204 De Haviland planes which had been built were shipped to France without engines and were there knocked down to provide spare parts for other De Havilands in service. These 204 machines, therefore, do not appear in the production total. Adding them to the figures above, we find

DE HAVILAND-4. USED FOR OBSERVATION, RECONNAISSANCE, COMBAT, DAY BOMBING, AND DEFENSIVE FIGHTING.

Engine, Liberty 12-cylinder, 400-horsepower. Weight, empty, 2,391 pounds. Weight, full load, 3,582 pounds. Ground speed is 124.7 miles per hour. Speed at 10,000 feet, 117 miles per hour. Speed at 15,000 feet, 113 miles per hour. 10,000 feet is reached in 14 minutes with full load. Ceiling, 19,500 feet.

UNITED STATES DE HAVILAND 9–A.

This is the American development of the British DH–4

that the total output of De Haviland airplanes up to the end of December, 1918, was in number 4,587.

The production of the model machine only served to show us some of the problems which must be overcome before we could secure a standard design that could go into quantity production. Experimental work on the De Haviland continued during December, 1917, and January and February, 1918. The struggle, for it was a

De Haviland-4 Airplanes Produced Each Month During 1918.

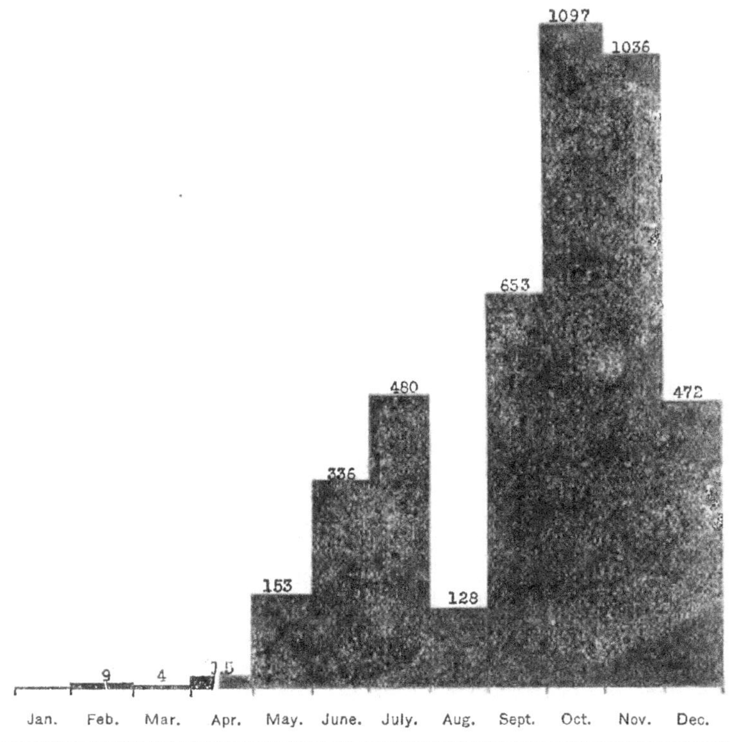

Figure 11.

struggle, to secure harmony between this English design and the American equipment which it must contain ended triumphantly on the 8th day of April, 1918, when the machine known as No. 31 was completely finished and established as the model for the future De Havilands. The characteristics of the standard American De Haviland-4 were as follows:

Endurance at 6,500 feet full throttle, 2 hours 13 minutes.
Endurance at 6,500 feet half throttle, 3 hours 3 minutes.
Ceiling, 19,500 feet.

Climb to 10,000 feet (loaded) 14 minutes.
Speed at ground level, 124.7 miles per hour.
Speed at 6,500 feet, 120 miles per hour.
Speed at 10,000 feet, 117 miles per hour.
Speed at 15,000 feet, 113 miles per hour.
Weight, bare plane, 2,391 pounds.
Weight, loaded, 3,582 pounds.

Endurance here means the length of time the fuel supply will last. The ceiling is the maximum altitude at which the plane can be maneuvered in actual service. Ground level means only far enough above the ground to be clear of obstructions.

The first De Havilands arriving in France were immediately put together, such remediable imperfections as existed were corrected then and there, and the machines were flown to the training fields. The changing and increasing demands of the service indicated the advisability of certain changes of design. The foreign manufacturers had brought out a covering for the gasoline tanks, making them nearly leak-proof, even when perforated by a bullet. In the first De Havilands the location of the principal gas tanks between the pilot and the observer was not the best arrangement in that the men were too far apart from each other so that, if the machine went down, the pilot would be crushed by the gas tank. Also the radius of action was not considered to be great enough, even though the later machines of this type carried 88 gallons of gasoline.

As a result the American aircraft designers brought out an improved De Haviland known as the 9–A. This carried a Liberty–12 engine; and the main differences between it and the De Haviland–4 were new locations for pilot and tanks, their positions being changed about, increased gasoline capacity, and increased wing surface. The machine was a cleaner, more finished design, showed slightly more speed, and had a greater radius of action than the De Haviland–4 which it was planned to succeed. We ordered 4,000 of these new machines from the Curtiss Co., but the armistice cut short this production.

The difficulties in the way of producing new service planes on a great scale without previous experience in such construction is clearly shown in the attempts we made to duplicate other successful foreign planes. On September 12, 1917, we received from the aviation experts abroad a sample of the French Spad. We had previously been advised to go into a heavy production of this model and had made arrangements for the Curtiss Co. at Buffalo to undertake the work. This development was well under way when in December a cablegram was received from Gen. Pershing advising us to leave the production of all single-place fighters to Europe. As a result we canceled the Spad order, and after that we attempted to build no single-place pursuit planes.

THE LEPERE CORPS OBSERVATION PLANE.

THE LEPERE (CAMOUFLAGED). THE ENGINE IS A LIBERTY 12-CYLINDER, 400-HORSEPOWER.

This plane was developed in the United States.

At the time this course seemed to be justified. The day of the single seater seemed to be over. The lone occupant of the single seater can not keep his attention on all directions at once; and as the planes grew thicker in the air, the casualties among flyers increased.

But the development of formation flying restored the single-place machine to favor. The formation had no blind spot, thus removing the principal objection to the single seater. The end of the war found the one-man airplane more useful than ever.

Our concentration here, however, was upon two-place fighters. On August 25, 1917, we received from abroad a sample of the Bristol fighting plane, a two-seat machine. The Government engineers at once began redesigning this machine to take the Liberty-12 engine and the American ordnance and accessories. The engine which had been used in the Bristol plane developed 275 horsepower. We proposed to equip it with an engine developing 400 horsepower.

The Bristol undertaking was not successful. The fact that later in the airplane program American designers successfully developed two-seater pursuit planes around the Liberty–12 engine shows that the engine decision was not the fault in the Bristol failure. There were repeated changes in the engineering management of the Bristol job. First the Government engineers alone undertook it; then the Government engineers combined with the drafting force of the airplane factory; finally the Government placed on the factory the entire responsibility for the job, without, however, permitting the manufacturer to correct any of the basic principles involved. All in all, the development of an American Bristol was most unsatisfactory, and the whole project was definitely abandoned in June, 1918.

The fundamental difficulty in all of these attempts was that we were trying to fit an American engine to a foreign airplane instead of building an American airplane around an American engine. It was inevitable that this difficulty should arise. We had skill to produce a great engine and did so, but for our earliest models of planes for this engine we relied upon the foreign models until we were sufficiently advanced in the art to design for ourselves. We were successful in making the adaptation only in the case of the De Haviland and then only after great delay.

But eventually we were to see some brilliantly successful efforts to design a two-place fighter around the Liberty–12. We had need of such a mechanism to supplement the De Haviland observation-plane production and make a complete service-plane program.

On January 4, 1918, Capt. Lepere, a French aeronautical engineer, who had formerly been with the French Government at St. Cyr, began experimental work on a new plane at the factory of the Packard Motor Car Co. By May 18 his work had advanced to a stage where

109287°—19——17

the Government felt justified in entering into a contract with the Packard Co. to provide shop facilities for the production of 25 experimental planes under Capt. Lepere's direction. The result of these efforts was a two-place fighting machine built around a Liberty engine. From the start this design met with the approval of the manufacturer and engineers because of its clean-cut perfection.

The performance of the Lepere plane in the air is indicated by the following figures:

R. P. M.=revolutions made by propellers in a minute.	Climb.		Speed.	
Altitude.	Time.	R. P. M.	Miles an hour.	R. P. M.
	min. sec.			
Ground	0 0	1,500	136	1,800
10,000 feet	10 35	1,520	132	1,740
15,000 feet	19 15	1,500	118	1,620
20,000 feet	41	1,480	102	1,550

Here at last was a machine that performed brilliantly in the air and contained great possibilities for quantity production, because it was designed from the start to fit American manufacturing methods. We placed orders for 3,525 Lepere machines. None of the factories, however, had come into production with the Lepere on November 11, 1918. Seven sample machines had been turned out and put through every test. It was the belief of those in authority that at last the training and technique of the best aeronautical engineers of France had been combined with the Liberty, probably the best of all aerial engines; and it was believed that the spring of 1919 would see the Yankee fliers equipped with American fighting machines that would be superior to anything they would be required to meet.

Nor were these expectations without justification. The weeks and months following the declaration of the armistice and extending through to the spring of 1919 were to witness the birth of a whole brood of new typically American designs of airplanes of which the Lepere was the forerunner. In short, when the armistice brought the great aviation enterprise to an abrupt end, the American industry had fairly caught that of Europe, and America designers were ready to match their skill against that of the master builders of France, Great Britain, Italy, and the central powers.

The Lepere 2-seated fighter was quickly followed by two other Lepere models—one of them, known as the Lepere C-21, being armored, and driven by a Bugatti engine, and the other a triplane driven by two Liberty engines and designed to be a day bomber. Then the first American designed single-seat pursuit planes began making their appearance—the Thomas-Morse pursuit plane, its 164 miles an hour at ground level, making it the fastest airplane ever tested by our Gov-

THE LOENING MONOPLANE.

This is one of the new distinctively American planes

THE LOENING TWO-PLACE PURSUIT PLANE.

ernment, if it were not the speediest plane ever built; the Ordnance Engineering Corporation's Scout, an advanced training plane; and several others. In two-seater fighting planes there was the Loening monoplane, an extremely swift and advanced type. There were several other new two-seaters designed experimentally in some instances and some of them giving brilliant promise.

Perhaps the severest and most exacting critic of aviation material is the aviator who has to fly the plane and fight with the equipment at the front. Brig. Gen. William Mitchell, then a colonel, was sent to France in 1917. He became in succession chief of the air service of the First Army Corps, chief of the air service of the First Army, and finally chief of the air service of the American group of armies in France. He commanded the aerial operations at the reduction of the St. Mihiel salient, where he gained the distinction of having commanded more airplanes in action than were ever assembled before under a single command. At St. Mihiel there were 1,200 allied planes in action, including, with our own, French, English, and Italian planes.

Gen. Mitchell, therefore, is a high authority as to the relative merits of air equipment from the airman's standpoint. In the spring of 1919, after a thorough investigation of the latest types of American planes and aerial equipment at the Wilbur Wright Field at Dayton, he sent to the Director of Air Service, Washington, D. C., the following telegram under date of April 20, 1919:

> I recommend the following airplanes in the numbers given be purchased at once: 100 Lepere 2-place corps observation, 50 Loening 2-place pursuit, 100 Ordnance Engineering Corporation 1-place pursuit, 100 Thomas-Morse 1-place pursuit, 50 USD9-A day bombardment, 700 additional Hispano-Suiza 300-horsepower engines, 2,000 parachutes. All of the above types are the equal of or better than anything in Europe.
>
> MITCHELL.

Now, let us see some of the specifications and performances of these new models. The USD9-A, being the redesigned and improved De Haviland 4, may be given a place as a latest model. It is a two-place bombing plane of the tractor biplane type, equipped with a Liberty 12 engine and weighing 4,872 pounds, loaded with fuel, oil, guns, and bombs, and with its crew aboard. With this weight its performance record in the official tests at Wilbur Wright Field in Dayton was as follows:

Speed (miles per hour):	
At ground	121.5.
At 6,500 feet	118.5.
At 10,000 feet	115.5.
At 15,000 feet	95.5.
Climb:	
To 6,500 feet, time	11 minutes 40 seconds.
To 10,000 feet, time	19 minutes 30 seconds.
To 15,000 feet, time	49 minutes.
Service ceilings (feet)	14,400.

The Lepere C–11, a tractor biplane equipped with a Liberty 12 engine, Packard make, weighing with its load aboard 3,655 pounds, performed as follows in the tests at the Wilbur Wright Field:

Speed (miles per hour):	
At ground	136.
At 6,500 feet	130.
At 10,000 feet	127.
At 15,000 feet	118.
Climb:	
To 6,500 feet, time	6 minutes.
To 10,000 feet, time	10 minutes 35 seconds.
To 15,000 feet, time	19 minutes 15 seconds.
Service ceiling (feet)	21,000.
Endurance at full speed at ground (hours)	2.5.

The Lepere carries two Marlin guns synchronized with the propeller and operated by the pilot and two Lewis guns operated by the observer. A total of 1,720 rounds of ammunition is carried.

The Loening monoplane, a tractor airplane equipped with an Hispano-Suiza 300-horsepower engine and representing, loaded, a gross weight of 2,680 pounds, its military load including two Marlin and two Lewis machine guns, performed as follows at the Wilbur Wright Field:

Speed (miles per hour):	
At ground	143.5.
At 6,500 feet	138.2.
At 10,000 feet	135.
At 15,000 feet	127.6.
Climb:	
To 6,500 feet, time	5 minutes 12 seconds.
To 10,000 feet, time	9 minutes 12 seconds.
To 15,000 feet, time	18 minutes 24 seconds.
Service ceiling (feet)	18,500.

The Ordnance Scout with a Le Rhone 80-horsepower engine, weighing, loaded, 1,117 pounds, is an advanced training plane. In its official test at Wilbur Wright Field it performed as follows:

Speed (miles per hour):	
At 6,500 feet	90.
At 10,000 feet	83.7.
At 15,000 feet	69.8.
Climb:	
To 6,000 feet, time	8 minutes 30 seconds.
To 10,000 feet, time	17 minutes 40 seconds.
To 14,000 feet, time	43 minutes 20 seconds.

The Thomas-Morse MB–3 pursuit plane, a tractor biplane equipped with an Hispano-Suiza 300-horsepower engine, weighing, including its crew but without military load, 1,880 pounds, in unofficial tests at Wilbur Wright Field, performed as follows:

Speed, at ground level (miles per hour)	163.68.
Climb, to 10,000 feet	4 minutes 52 seconds.

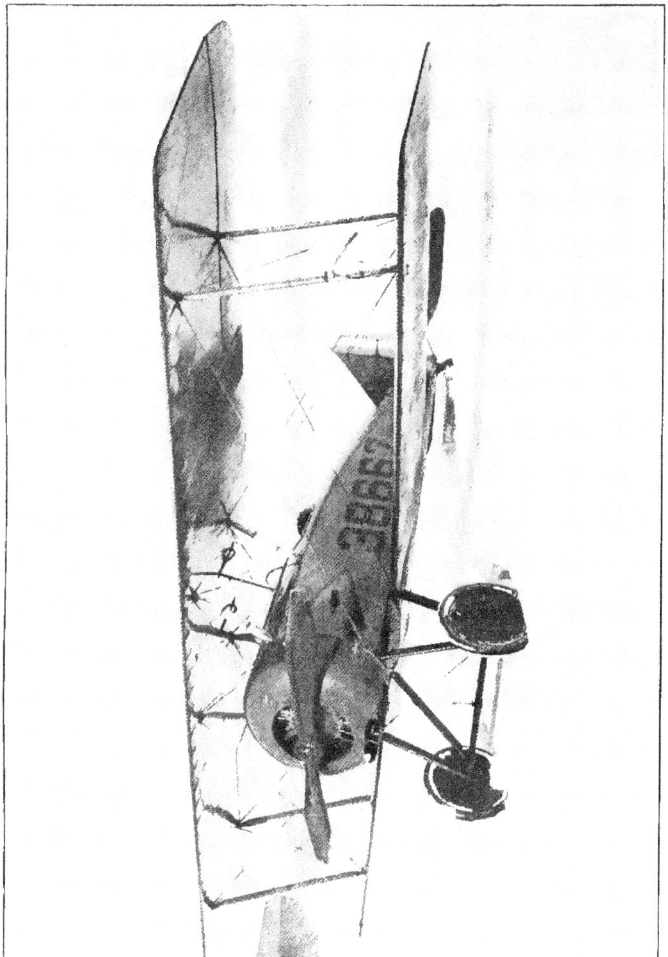

THE THOMAS-MORSE PURSUIT PLANE.

S. E. 5. EQUIPPED WITH 180-HORSEPOWER HISPANO-SUIZA ENGINE.

The Thomas-Morse pursuit plane is armed with two Browning machine guns synchronized with the propeller and carries 1,500 rounds of ammunition.

Uncertain as we were originally as to types of pursuit and observation planes to produce in this country, we were still more uncertain as to designs of night-bombing machines. These relatively slow weight-carrying planes were big and required the motive power of two or three engines, with the complications attendant upon double or triple power plants. They really presented the most difficult manufacturing problem which we encountered. Until the summer of 1918 there were only two machines of this type which we could adopt, the Handley-Page and the Caproni. We put the Handley-Page into production, not because it was necessarily as perfect as the Caproni, but because we could get the drawings for this machine and could not get the drawings for the Caproni, owing to complications in the negotiations for the right to construct the Italian airplane.

We were not entirely satisfied with the decision to build Handley-Pages, because the ceiling, or maximum working altitude which could be attained by this machine, was low; and, 12 months later, when we were in production, we might find the Handley-Pages of doubtful value because of the ever-increasing ranges of antiaircraft guns.

We secured a set of drawings, supposed to be complete, for the Handley-Page in August, 1917; but twice during the following winter new sets of drawings were sent from England, and few, if any, of the parts as designed in the original drawings escaped alteration. The Handley-Page has a wing spread of over 100 feet. Therefore, it was evident from the start that such machines could not have the fuselage, wings, and other large parts assembled in this country for shipment complete to Europe. We decided to manufacture the parts in this country and assemble the machines in England, the British air ministry in London having entered into a contract for the creation of an assembling factory at Oldham, England, in the Lancashire district. When it is realized that each Handley-Page involves 100,000 separate parts, the magnitude of the manufacturing job alone may be somewhat understood. But after they were manufactured, these parts, particularly the delicate members made of wood, had to be carefully packed so as to reach England in good condition. The packing of the parts was in itself a problem.

We proposed to drive the American Handley-Pages with two Liberty 12 engines in each machine. The fittings, which were extremely intricate pieces of pressed steel work, were practically all to be produced by the Mullins Steel Boat Co. at Salem, Ohio. Contracts for the other parts were placed with the Grand Rapids Airplane Co., a concern which had been organized by a group of furniture makers at Grand Rapids, Mich.

All of the parts were to be brought together previously to ocean shipment in a warehouse built for the purpose at the plant of the Standard Aero Corporation at Elizabeth, N. J. The Standard Aero Corporation was engaged under contract to set up 10 per cent of the Handley-Page machines complete in this country. These were to be used at our training fields.

Again, in the case of the Handley-Page, the engineering details proved to be a serious cause of delay. We found it difficult to install the Liberty engines in this foreign plane. When the armistice cut short operations, 100 complete sets of parts had been shipped to England, and seven complete machines had been assembled in this country.

None of the American-built Handley-Page machines saw service in France. There had been great delay in the construction of the assembling plant in England, and the work of setting up the machines had only started when the armistice was signed. The performance table of the Handley-Page shows its characteristics as follows:

Speed at ground level, 97 miles per hour.
Climb to 7,000 feet, 18 minutes 10 seconds.
Climb to 10,000 feet, 29 minutes.
Ceiling 14,000 feet, 60 minutes.

On its tests 390 gallons of gasoline, 20 gallons of oil, and 7 men were carried, but no guns, ammunition, nor bombs.

After a long delay, about January 1, 1918, tentative arrangements had been made with the Caproni interests looking toward the production of Caproni biplanes in this country. These machines had a higher ceiling and a greater speed than the Handley-Page. Capt. d'Annunzio with 14 expert Italian workmen, bringing with him designs and samples, came to this country and initiated the redesigning of the Caproni machine to accommodate three Liberty engines. The actual production of Caproni planes in this country was limited to a few samples which were being tested when the armistice was signed. The factories had tooled up for the production, however, and in a few months Capronis would doubtless have been produced in liberal quantities.

The performance of the sample planes in two tests is shown by the following figures:

	Test 1.	Test 2.
Speed at ground level	100 miles per hour	103.2 miles per hour.
Climb to 6,500 feet	16 minutes 18 seconds	14 minutes 12 seconds.
Climb to 10,000 feet	33 minutes 18 seconds	28 minutes 42 seconds.
Climb to 11,200 feet	49 minutes	
Climb to 13,000 feet		46 minutes 30 seconds.

ONE OF THE SMALL THOMAS-MORSE SCOUTS BESIDE A GIANT HANDLEY-PAGE MACHINE.

ARMORED GERMAN AIRPLANE SHOT DOWN ON THE WESTERN FRONT.

NIEUPORT SCOUT BESIDE A LOENING MONOPLANE.

As we had produced fighting planes built around the Liberty motor, so, too, in the night-bombing class American invention, with the experience of several months of actual production behind it, was able to bring out an American bombing plane that promised to supersede all other types in existence. This machine was designed by Glen L. Martin in the fall of 1918. It was a night-bomber equipped with two Liberty 12-cylinder engines. The Martin spread of 75 feet gave it a carrying capacity comparable with that of the Handley-Page. Its speed of 118 miles an hour at ground level far exceeded that of either the Caproni or Handley-Page, and it was evident that its ceiling would be higher than that of the Caproni, the estimated ceiling of the Martin being 18,000 feet. The machine never reached the state of actual quantity production, but several experimental models were built and tested. Being built around its engine it reflected clean-cut principles of design, and its performances in the air were truly remarkable for a machine of its type. The following table shows the results of the preliminary tests of the Martin bomber:

	Test 1.	Test 2.
Speed at ground level	113.3 miles per hour	118.8 miles per hour.
Climb to 6,500 feet	10 minutes 45 seconds	7 minutes.
Climb to 10,000 feet	21 minutes 20 seconds	14 minutes.
Climb to 15,000 feet		30 minutes 30 seconds.
Total weight	9,663 pounds	8,137 pounds.

The total delivery of airplanes to the United States during the period of the war was 16,952. These came from the following sources: United States contractors, 11,754; France, 4,881; England, 258; Italy, 59.

FIGURE 12.

U. S. SQUADRONS AT THE FRONT.

A squadron is equipped with from 15 to 25 planes.

Estimates of aircraft strength on the front were always uncertain, due to variations in the estimates of the number of planes in a squadron. The standing of the United States in aeroplanes at the front is indicated in the estimate of the American Air Service as of November 11, 1918. The figures of this estimate are as follows:

France	3,000
Great Britain	2,100
United States	860
Italy	600
Total	6,560

These figures represent fighting planes equipped ready for service, but do not include replacement machines at the front or in depots or training machines in France.

FIGURE 13.

COMPARISON ENEMY PLANES BROUGHT DOWN BY U. S. FORCES AND U. S. PLANES BROUGHT DOWN BY THE ENEMY.

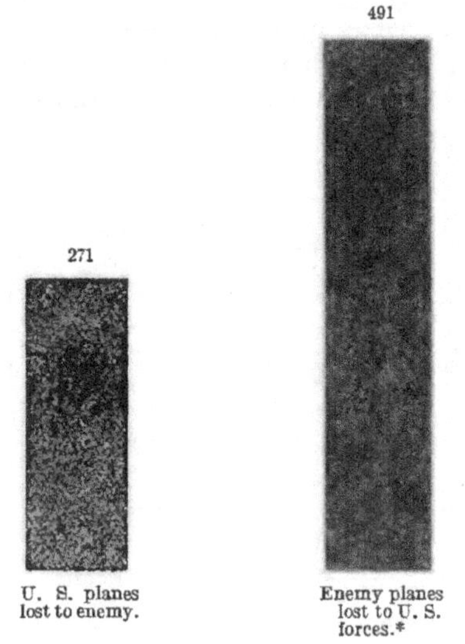

*Confirmed losses; in addition there were 354 unconfirmed.

The actual strength of the central powers in the air is at this time not definitely known to us. Such figures as we have are viewed with suspicion because of the two methods of observation in reporting an enemy squadron. This may be 24 planes to a squadron, that number representing the planes in active service in the air. But each squadron had a complement of replacement planes equalling the number of active planes, so that the squadron could be listed with 48 planes.

However, as some indication of the relative air strengths of the central powers we have a report from the chief of the Air Service of the American Expeditionary Forces showing that on July 30, 1918, Germany had 2,592 planes on the front and Austria 717.

THE GLENN MARTIN BOMBER.

The gross weight of this machine is 9,663 pounds. It can be equipped with five Lewis machine guns. Its ground speed is 113 miles an hour and its service ceiling is 12,800 feet. It climbs to a height of 6,500 feet in 10 minutes 45 seconds and to 10,000 feet in 21 minutes 20 seconds.

THE CAPRONI, EQUIPPED WITH THREE LIBERTY 12-CYLINDER ENGINES.

CHAPTER III.

THE LIBERTY ENGINE.

The Liberty engine was America's distinctive contribution to the war in the air, and her chief one. The engine was developed in those first chaotic weeks of preparation of 1917, when our knowledge of planes, instruments, and armament as then known in Europe was still a thing of the future. The manufacture of engines for any aeronautical purpose was one which we might approach with confidence. We possessed in the United States motor engineering talent at least as great as any in Europe, while in facilities for manufacture—in plants which had built our millions of automobile engines—no other part of the world could compare with us. Therefore, while waiting word from Europe as to the best type of wings, fuselages, instruments and the like, we went ahead to produce for ourselves a new, typically American engine which would uphold the prestige of America in actual battle.

Many Americans have doubtless wondered why we built our own engine instead of adopting one or more of the highly developed European engines then at hand; and no doubt our course in this vital matter has sometimes been set down to mere pride in our ability and to an unwillingness to follow the lead of other nations in a science in which we ourselves were preeminent—the science of building light internal-combustion engines. But national pride, aside from giving us confidence that our efforts in this direction would be successful, had little other weight in the decision. There were other reasons, and paramount ones, reasons leading directly from the necessity for the United States to arrive at her maximum aerial effort in a minimum of time, that irresistibly compelled the aircraft production organization to design a standard American engine. Let us examine some of these considerations.

If there was anything to be observed from this side of the Atlantic with respect to the tendencies of aircraft evolution in Europe it was that the horsepowers of the engines were continually increasing, these expansions coming almost from month to month as newer and newer types and sizes of engines were brought out by the European inventors. It was evident to us that there was not a single foreign engine then in use on the western front that was likely to survive the test of time. Each might be expected to have its brief day of supremacy, only to be superseded by something more modern and more powerful.

Yet time was an element to which in this country we must give grave consideration. To produce in quantities such as we were capable of producing would ordinarily require a year of maximum industrial effort to equip our manufacturing plants with the machines, tools, and skilled workmen necessary for the production of parts. The finished articles would under normal circumstances begin coming in quantity during the second year of our program. It would have been fatal to "tool up" our plants for the manufacture of equipment that would be out of date by the time we began producing it a year later.

The obvious course for the United States to adopt, not only with engines but with all sorts of aeronautical equipment, was to come into the manufacturing competition not abreast with European progress but several strides ahead of it, so that when we appeared on the field it would be with equipment a little in advance in type and efficiency of anything the rest of the world had to offer.

This factor of time was a strong element in the decision to produce a standard American engine, since with the possible exception of the Rolls-Royce there was no engine in Europe of sufficient horsepower and proved reliability to guarantee that it would retain its serviceability for the necessary two years upon which we must reckon. There was no other course that we could safely adopt.

But there were other conditions that influenced our conclusion. We believed that we could design and produce an engine much more quickly and with much better results than we could copy and produce any approved foreign model. This proved to be true in actual experience. Along with the production of Liberty engines we went into the quantity manufacture of a number of European engines in this country; and the experience of our engineers and factory executives in this work was anything but pleasant. Among others we produced in American factories the Gnome, the Hispano-Suiza, Le Rhone, and the Bugatti engines.

Now European manufacture of mechanical appliances differs from ours largely in the degree to which the human equation is allowed to enter the shop. In continental practice much of the metallurgical specifications and also of the details of mechanical measurements, limits of requisite accuracy, variations which can be allowed, etc., are not put on paper in detail for the guidance of operators, but are confided to the recollections of the individual workmen. A machine comes in its parts to the assembly room of a foreign factory, and after that it is subject to adjustments on the part of the skilled workmen before its operation is successful. It must be tinkered with before it will go, so to speak. Nothing of the sort is known in an American factory. When standard parts come together for assembly the calibrations must have been so exact that the machine will function perfectly when it is brought together; and assembling becomes mere routine. Thus when we came to adopt foreign plans and attempt to adapt them to our practices, we encountered trouble and delay.

Thirteen months were required to adapt the Hispano-Suiza 150-horsepower engine to our factory methods and to get the first engine from production tools, while eight months were similarly spent in producing the Le Rhone 80-horsepower engines. Both of these engines had been in production in European factories for a long time, and we had the advantage of all the assistance which the foreign manufacturers could give us.

These experiences merely confirmed the opinions of American manufacturers that the preparations for the production of any aviation engine of foreign design—if any such suitable and adequate engine could be found—would require at least as much time as to design and tool up for the production of an American engine. When to this was added the necessity of waiting for several weeks or months for a decision on the part of our aviation authorities, either in the United States or in Europe, as to which of the many types of engines then in use by the allies should be put into production here, procuring and shipping to this country suitable samples, drawings, and specifications, negotiating with foreign owners for rights to manufacture, etc., there was but one answer to be made on this score, and that was to design and build an all-American engine.

Another factor in the decision was that of our distance from France, a fact making it necessary for us to simplify as much as possible the problem of furnishing repair parts. At the time we entered the war the British air service was using or developing 37 different makes of engines, while France had 46. Should we be lured into any such situation it might have disastrous results, if only because of the difficulties of ocean transportation. Germany was practically concentrating upon not more than 8 engines. The obvious thing for us to do was to produce as few types of engines as possible, thus making simpler the problem of manufacturing repair parts and shipping them to the front.

With these considerations in mind, the Equipment Division of the Signal Corps in May of 1917 determined to go ahead with the design and production of a standard engine for the fighting forces of the aviation branch of the Army. In the engineering field two men stood out who combined in themselves experience in designing internal-combustion engines which approached nearest to combat engines, with experience also in large quantity production.

J. G. Vincent, with the engineering staff of the Packard Motor Car Co., had for approximately two years been engaged in research work, developing several types of 12-cylinder aviation engines of approximately 125 to 225 horsepower, which, however, were not suitable for military purposes because of their weight per horsepower. This work had resulted in the acquirement of a large amount of data and information which would be invaluable in the design of such an engine as the one proposed; and also had resulted in the upbuilding of an efficient experimental organization. He had also had wide experience in designing internal-combustion motors for quantity production.

E. J. Hall, of the Hall-Scott Motor Car Co., for eight years had been developing and latterly producing several types of aeronautical engines, which he had delivered into the service of several foreign governments, including Russia, Norway, China, Japan, Australia, Canada, and England. He had also completed and tested a 12-cylinder engine of 300 horsepower, which, however, was of too great weight per horsepower to be suitable in its form at that time for military purposes. He had thus acquired a large experience and fund of information covering the proper areas and materials for engine parts, and proper methods of tests to be applied to such engines, and in addition he had general experience in quantity production. All of this information and experience was of invaluable assistance not only in designing the new engine, but in determining its essential metallurgical and manufacturing specifications.

These two men were thus qualified in talent and in practice to lay down on paper the lines and dimensions of the proposed engine, an engine that would meet the Army's requirements and still be readily capable of prompt quantity production. They had in their hands the power to draw freely upon the past experience and achievement of practically the entire world for any features they might decide to install in the model power plant to be produced. And this applied not only to the patented features of American motors, but also of foreign engines; for each man had exhaustively studied the leading European engines, including the Mercedes upon which Germany largely pinned her faith up to the end of the war.

With respect to American motor patents, an interesting situation had arisen in the automobile industry. The leading producers of motor cars were in an association which had adopted an arrangement known as the cross-licensing agreement. Under this agreement all patents taken out by the various producers (with a few exceptions) were thrown into a pool upon which any producer at will was permitted to draw without payment of royalties.

A similar arrangement was adopted with respect to the Liberty engine, except that the Government pledged itself to pay an agreed royalty for the use of patents. Thus the engineers designing the engine might reach out and take what they pleased regardless of patent rights. The result was likely to be a composite type embracing the best features of the best engines ever built. Theoretically, at least, a superengine ought to result from such an effort.

The ideal aviation engine should produce a maximum of power with a minimum of weight; it must run at its maximum power during a large proportion of its operating time, a thing that an automobile motor seldom, if ever, does for more than a few minutes at a time; and it should consume oil and fuel economically to conserve space and weight on the airplane.

Such was the problem, the design of an engine to meet these requirements, that confronted these two engineers when they were called to Washington and asked to undertake the work.

There have been so many versions of the story of how the Liberty engine was designed and produced in its experimental models that it is fitting that the exact history of those memorable weeks should be set down here.

The engine was put on paper in the rooms occupied by Col. E. A. Deeds at the Willard Hotel in Washington. Col. Deeds had been the man of broad vision who, by taking into consideration the elements of the problems enumerated above, determined that America could best make her contribution to the aviation program by producing her typically own engine. He had proposed the plan to his associate, Col. S. D. Waldon, who had thereupon studied the matter and agreed entirely with the plan. The two officers persuaded Messrs. Hall and Vincent to forego further efforts on their individual developments and to devote their combined skill and experience to the creation of an all-American engine. The project was further taken up with the European authorities in Washington, and it was supported unanimously.

In these conferences it was decided to design two lines of combat engines. Each should have a cylinder diameter of 5 inches and a piston stroke 7 inches long; but one type should have 8 cylinders and the other 12. The 8-cylinder engine should develop 225 horsepower, as all the experts believed then, in May, 1917, that such a motor would anticipate the power requirements as of the spring of 1918, while the 12-cylinder engine should develop 330 horsepower, as it was believed that this would be the equal of any other engine developed through 1919 and 1920. Every foreign representative in Washington with aeronautical experience agreed that the 8-cylinder 225-horsepower engine would be the peer of anything in use in the spring of 1918; yet, so rapidly was aviation history moving that inside of 90 days it became equally clear that it was the 12-cylinder engine of 330 horsepower, and not the 8-cylinder engine, upon which we should concentrate for the spring of 1918.

With these considerations in mind Messrs. Hall and Vincent set to work to lay out the designs on paper. With them were Col. Deeds and Col. Waldon, the officers to insist that nothing untried or experimental be incorporated in the engines, the engineers to direct their technical knowledge by this sine qua non. The size of the cylinders, 5 by 7 inches, was adopted not only because the Curtiss and the Hall-Scott Companies, the largest producers of aviation engines in the United States, had had experience with engines of this size, but also because a new and promising French engine, the Lorraine-Dietrich, had just made its appearance in experimental form, and it was an engine approximately of that size.

On May 29, 1917, Messrs. Vincent and Hall set to work. Within two or three days they had outlined the important characteristics of the engine sufficiently to secure—on June 4—the approval of the Aircraft Production Board and of the Joint Army and Navy Technical Board to build five experimental models each of the 8 cylinder and the 12 cylinder sizes.

The detail and manufacturing drawings of the two engines were made partly by the staff of the Packard Motor Car Co., under Mr. O. E. Hunt, and partly by an organization recruited from various automobile factories and put to work under Mr. Vincent at the Bureau of Standards at Washington. Due credit must here be given to Dr. S. W. Stratton, the director of that important governmental scientific bureau. The Liberty engine pioneers woke him up at midnight and told him of their needs. He promptly tendered all the facilities of the Bureau of Standards, turning over to the work an entire building for use the following morning. Thereafter Dr. Stratton gave the closest cooperation of himself and his assistants to the work.

While the detail drawings were being made, the parts for the 10 engines were at once started through the tool rooms and experimental shops of various motor car companies. This work centered in the plant of the Packard Co., which gave to it its entire energy and wonderful facilities.

Every feature in the design of these engines was based on thoroughly proven practice of the past. That the engine was a composite is shown by the origin of its various parts:

Cylinders: The Liberty engine derived its type of cylinders from the German Mercedes, the English Rolls-Royce, the French Lorraine-Dietrich, and others produced both before and during the war. The cylinders were steel inner shells surrounded by pressed-steel water jackets. The Packard Co. had developed a practical production method of welding together the several parts of a steel cylinder.

Cam shafts and valve mechanism above cylinder heads: The design of these was based on the general arrangement of the Mercedes and Rolls-Royce, and improved by the Packard Motor Car Co. for automatic lubrication without wasting oil.

Cam-shaft drive: The general type as used on the Hall-Scott, Mercedes, Hispano-Suiza, Rolls-Royce, Renault, Fiat, and others.

WELDING JACKET ON CYLINDER FOR LIBERTY ENGINE. CADILLAC MOTOR CAR CO., DETROIT, MICH.

DRILLING CYLINDER FLANGES WITH MULTIPLE DRILL. PACKARD MOTOR CAR CO.

MACHINING THE CONNECTING RODS FOR THE LIBERTY ENGINE. CADILLAC MOTOR CO.

GAUGING VALVES AND PISTONS FOR THE LIBERTY ENGINE. LINCOLN MOTOR CO.

Angle between cylinders: In the Liberty the included angle between the cylinders is 45°. This angle was adopted to save head resistance, to give greater strength to the crank case, and to reduce periodic vibration. This decision was based on the experience of the Renault and Packard engines.

Electric generator and ignition: The Delco system was adopted, but specially designed for the Liberty to provide a reliable double ignition.

Pistons: The die-cast aluminum-alloy pistons of the Liberty were based on development work by the Hall-Scott Co. under service conditions.

Connecting rods: These were of the forked or straddle type as used on the DeDion and Cadillac automobile motors and also on the Hispano-Suiza and other aviation engines.

Crank shaft: A design of standard practice, every crank pin operating between two main bearings, as in the Mercedes, Rolls-Royce, Hall-Scott, Curtiss, and Renault.

Crank case: A box section carrying the shaft in bearings clamped between the top and bottom halves by means of long through bolts, as in the Mercedes and Hispano-Suiza.

Lubrication: The system of lubrication was changed, this being the only change of design made in the Liberty after it was first put down on paper. The original system combined the features of a dry crank case, such as in the Rolls-Royce, with pressure feed to the main crank-shaft bearings and scupper feed to the crank-pin bearings, as in the Hall-Scott and certain foreign engines. The system subsequently adopted added pressure-feed to the crank-pin bearings, as in the Rolls-Royce, Hispano-Suiza, and other engines.

Propeller hub: Designed after the practice followed by such well-known engines as the Hispano-Suiza and Mercedes.

Water pump: The conventional centrifugal type was adapted to the Liberty.

Carburetor: The Zenith type was adapted to the engine.

As the detailed and manufacturing drawings were completed in Washington and Detroit they were taken to various factories where the parts for the first engine were built.

The General Aluminum & Brass Manufacturing Co., of Detroit, made the bronze-back, babbitt-lined bearings.

The Cadillac Motor Car Co., of Detroit, made the connecting rods, the connecting-rod upper-end bushings, the connecting-rod bolts, and the rocker-arm assemblies.

The L. O. Gordon Manufacturing Co., of Muskegon, Mich., made the cam shafts.

The Park Drop Forge Co., of Cleveland, made the crank-shaft forgings. These forgings, completely heat treated, were turned out in three days, because Mr. Hall gave the Cleveland concern permission to use the Hall-Scott dies.

The Packard Motor Car Co. machined the crank shafts and all parts not furnished or finished elsewhere.

The Hall-Scott Motor Car Co., of Berkeley, Calif., made all the bevel gears.

The Hess-Bright Manufacturing Co., of Philadelphia, made the ball bearings.

The Burd High-Compression Ring Co., of Rockford, Ill., made the piston rings.

The Aluminum Castings Co., of Cleveland, made the die-cast alloy pistons and machined them up to grinding.

The Rich Tool Co., of Chicago, made the valves.

The Gibson Co., of Muskegon, Mich., made the springs.

The Packard Co. made all the patterns for the aluminum castings, which were produced by the General Aluminum & Brass Manufacturing Co., of Detroit.

The Packard Motor Car Co. used many of its own dies in order to obtain suitable drop forgings speedily, and also made all necessary new dies not made elsewhere.

As these various parts were turned out they were hurried to the tool room of the Packard Co., where the assembling of the model engines was in progress.

Before the models were built, however, extraordinary precautions had been taken to insure that the mechanism would be as perfect as American engineering skill could make it. The plans as developed were submitted to H. M. Crane, the engineer of the Simplex Motor Car Co. and of the Wright-Martin Aircraft Corporation, who had made a special study of aviation engines in Europe, and who for upward of a year had been working on the production of the Hispano-Suiza 150-horsepower engine in this country. He looked the plans over, and so did David Fergusson, chief engineer of the Pierce-Arrow Motor Car Co. Many other of the best experts in the country in the production of internal-combustion motors constructively criticized the plans, these including such men as Henry M. Leland and George H. Layng, of the Cadillac Motor Car Co., and F. F. Beall and Edward Roberts, of the Packard Car Co.

When the engineers were through, the practical production men were given their turn. The plane and engine builders examined the plans to make sure that each minute part was so designed as to make it most adaptable to quantity production. The scrutiny of the Liberty plans went back in the production scale even farther than this; for the actual builders of machine tools were called in to examine the specifications and to suggest modifications, if necessary, that would make the production of parts most feasible in machine tools either of existing types or of easiest manufacture.

Thus scrutinized and criticized, the plans of the engine were the best from every point of view which American industrial genius could produce in the time which was available. It was due to this exhaustive preliminary study that no radical changes were ever made in the original design. The Liberty engine was not the materialization of magic nor the product of any single individual or company, but it was a well-considered and carefully prepared design based on large practical aviation-engine experience.

On July 4, 1917, the first 8-cylinder Liberty engine was delivered in Washington. This was less than six weeks after Messrs. Hall and Vincent drew the first line of their plans. The same procedure was even then being repeated in the case of the 12-cylinder engine. By the 25th day of August the model 12-cylinder Liberty had successfully passed its 50-hour test. In this test its power ranged from 301 to 320 horsepower.

As an achievement in speed in the development of a successful new engine this performance has never been equaled in the motor history of any country. No successful American automobile motor was ever put in production in anything under a year of trial and experimentation. We may well believe that in the third year of war the European aviation designers were working at top speed to improve the motive power of airplanes; yet in 1917 the British war cabinet report contains the following language:

> Experience shows that as a rule, from the date of the conception and design of an aero engine, to the delivery of the first engine in series by the manufacturer, more than a year elapses.

But America designed and produced experimentally a good engine in six weeks and a great one in three months, and began delivering it in series in five months. This was due to the fact that we could employ our best engineering talent without stint, to the further fact that there were no restrictions upon our use of designs and patents proved successful by actual experience, and to the fact that the original engine design produced under such conditions stood every expert criticism and test that could be put upon it and emerged from the trial without substantial modification.

As soon as the first Liberty models had passed their official tests plans were at once made to put them in manufacture.

The members of the Aircraft Production Board chose for the chief of the engine production department Harold H. Emmons, an attorney and manufacturer of Detroit, Mich., who, as a lieutenant in the Naval Reserve Force, was just being called by the Navy Department into active service.

The production of all aviation engines, for both Army and Navy, was in his hands throughout the rest of the war. He placed orders for 100,993 aviation engines of all types, which involved the expenditure of $450,000,000 and more of Government funds. Of these 31,814 were delivered ready for service before the signing of the armistice. The United States reached a daily engine production greater than that of England and France combined.

In August, 1917, it was intended to manufacture both engines, the 8-cylinder and the 12-cylinder, and an agreement was reached with the Ford Motor Co. of Detroit to produce 8-cylinder Liberty engines to the number of 10,000. But before this contract could be signed the increasing powers of the newest European air engines indicated to our commission abroad that we should concentrate our manufacturing efforts upon the 12 alone, that being the engine of a power then distinctly in advance in the rapid evolution of aviation engines. The engine production department, therefore, entered into contracts for the construction of 22,500 of the 12-cylinder Liberties, and the first of these contracts was signed in August, a few days after the endurance tests had demonstrated that the 12-cylinder engine was a success.

Of this number of Liberty engines the Packard Motor Car Co. contracted to build 6,000; the Lincoln Motor Co., 6,000; the Ford Motor Co., 5,000; Nordyke & Marmon, 3,000; the General Motors Corporation (Buick and Cadillac plants), 2,000; while an additional contract of 500 engines was let to the Trego Motors Corporation.

Early in the Liberty engine project it became apparent that one of the great stumbling blocks to volume production would be the steel cylinder, if it were necessary to machine it out of a solid or partially pierced forging such as is used for shell making. This problem was laid before Henry Ford and the engineering organization of the Ford Motor Co., at Detroit, and they developed the unique method of making the cylinders out of steel tubing. One end of the tube was cut obliquely, heated, and in successive operations closed over and then expanded into the shape of the combustion chamber, with all bosses in place on the dome. The lower end was then heated and upset in a bulldozer until the holding-down flange had been extruded from the barrel at the right place. By this method a production of 2,000 rough cylinders a day was reached.

The final forging was so near to the shape desired that millions of pounds of scrap were saved over other methods, to say nothing of an enormous amount of labor thus done away with. The development of this cylinder-making method was one of the important contributions to the quantity production of Liberty engines.

It was evident that in the actual production of the Liberty engine there would continually arise practical questions of manufacturing policy that might entail modifications of the manufacturing methods, while our aviation authorities in Europe could be expected to advance

EXPERIMENTAL WORK ON NEW IDEAS FOR LIBERTY ENGINE.

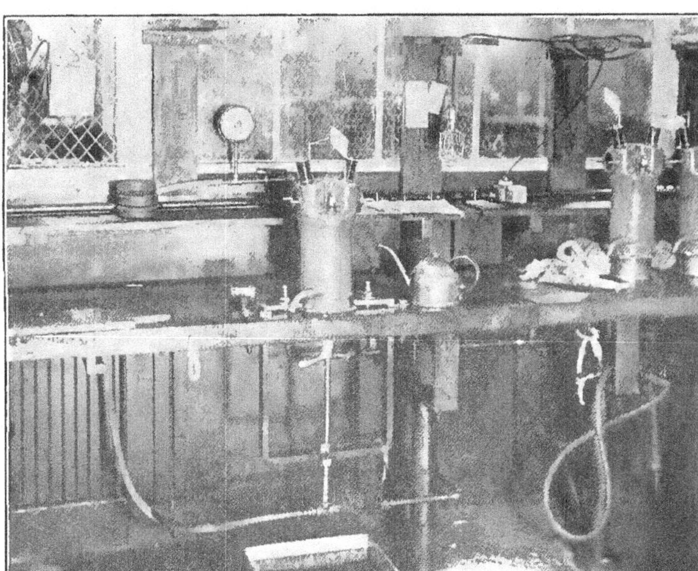

TEST CYLINDERS FOR THE LIBERTY ENGINE AT THE PLANT OF NORDYKE & MARMON, INDIANAPOLIS, IND.

CRANK-SHAFT DIE FOR THE LIBERTY ENGINE. BUICK ENGINE CO., DETROIT.

ACCEPTED LIBERTY ENGINES BEING BOXED FOR SHIPMENT.

suggestions from time to time that might need to be embodied in the mechanism. Consequently it was necessary to create a permanent development and standardization administration for the Liberty engine. Nor could this supervision be located in Washington, because of the extreme need for haste, but it must exist in the vicinity of the plants doing the manufacturing.

For this reason the production of the Liberty engine was centered in the Detroit manufacturing district, since in this district was located the principal motor manufacturing plant capacity of the United States. James G. Heaslet, formerly general manager of the Studebaker Corporation and an engineer and manufacturer of wide experience, was installed as district manager. The problems incident to the inspection and production of the Liberty engine were placed in charge of a committee consisting of Maj. Heaslet (chairman); Lieut. Col. Hall, one of the designers of the engine; Henry M. Leland; C. Harold Wills, of the Ford Motor Co.; and Messrs F. F. Beall and Edward Roberts, of the Packard Motor Car Co. With them were also associated D. McCall White, the engineer of the Cadillac Motor Co., and Walter Chrysler, of the Buick Co.

The creation of this committee virtually made a single manufacturing concern of the several, previously rival, motor companies engaged in producing the Liberty engine. To these meetings the experts without reservation brought the trade secrets and shop processes developed in their own establishments during the preceding years of competition. Such cooperation was without parallel in the history of American industry, and only a great emergency such as the war with Germany could have brought it about. But the circumstance aided wonderfully in the development and production of the Liberty engine.

Moreover, the Government drew heavily upon the talent of these great manufacturing organizations for meeting the special problems presented by the necessity of filling in the briefest possible time the largest aviation engine order ever known. Short-cuts that these firms might have applied effectively to their own private advantage were devised for the Liberty engine and freely turned over to the Government. The Packard Co. gave a great share of its equipment and personnel to the development. The most conspicuous success in the science of quantity production in the world was the Ford Motor Co., which devoted its organization to the task of speeding up the output of Liberty engines. In addition to the unique and wonderfully efficient method of making rough engine cylinders out of steel tubing, the Ford organization also perfected for the Liberty a new method of producing more durable and satisfactory bearings. Messrs. H. M. and W. C. Leland, whose names were indissolubly linked with the Cadillac automobile, organized and erected the enormous plant of the Lincoln Motor Co. and equipped it for the production of the Liberty, at a total expense of approximately $8,000,000.

Balanced against these advantages brought by highly trained technical skill and unselfish cooperation were handicaps such as perhaps no other great American industrial venture had ever known. In the first place, an internal-combustion engine with cylinders of a 5-inch bore and pistons of a 7-inch stroke—the Liberty measurements—was larger than the automobile engines then in use in this country. This meant that while we apparently had an enormous plant—the combined American automobile factories—ready for the production of Liberty engines, actually the machinery in these plants was not large enough for the new work, so that new machinery therefore must be built to handle this particular work. In some cases machinery had to be designed anew for the special purpose.

To produce every part of one Liberty engine something between 2,500 and 3,000 small jigs, tools, and fixtures are employed. For large outputs much of this equipment must be duplicated over and over again. To provide the whole joint workshop with this equipment was one of the unseen jobs incidental to the construction of Liberty engines—unseen by the general public, that is—yet it required the United States to commandeer the capacity of all available tool shops east of the Mississippi River and devote it to the production of jigs and tools for the Liberty engine factories.

Then there was the question of mechanical skill in the factories. It soon developed that an automobile motor is a simple mechanism compared with an intricate aviation engine. The machinists in ordinary automobile plants did not have the skill to produce the Liberty engine parts successfully. Consequently it became necessary to educate thousands of mechanics, men and women alike, to do this new work.

It was surprising to what extent unfriendly influence in the United States, much of it probably of a pro-German character, cut a figure in the situation. This was particularly true in the supply factories furnishing tools to the Liberty engine plants. Approximately 85 per cent of the tools first delivered for this work were found to be inaccurate and incorrect. These had to be remade before they could be used. Such tools as were delivered to the Liberty plants would mysteriously disappear, or vital equipment would be injured in unusual ways; in several instances cans of explosives were found in the coal at power plants; fire-extinguishing apparatus was discovered to be rendered useless by acts of depredation; and from numerous other evidences the builders of Liberty engines were aware that the enemy had his agents in their plants.

Difficulty was also experienced in the production of metals for the new engines. The materials demanded were frequently of a much higher grade than the corresponding materials used in ordinary auto-

LIBERTY ENGINE READY FOR TEST AT THE LINCOLN MOTOR CO., DETROIT, MICH.

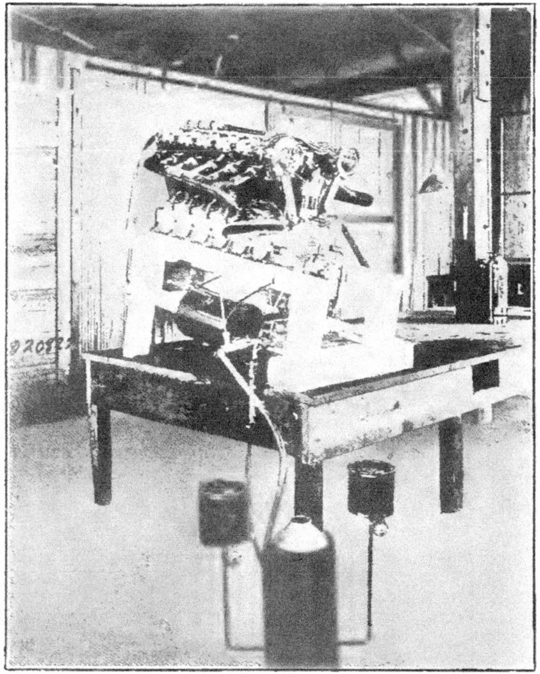

LIBERTY ENGINE ON INSTRUCTION STAND. WILBUR WRIGHT FIELD.

VIEW SHOWING LIBERTY ENGINE WITH PROPELLER HUB ATTACHED.

TEST SHED AND STAND WITH LIBERTY ENGINE MOUNTED WITH TEST PROPELLER. PACKARD MOTOR CAR CO.

mobile motors. Here was another unseen phase of development which had to be worked out patiently by the producers of raw materials.

Difficulties in transportation during the winter of 1917–18 added their share to the perplexing problems of the engine builders, while at times the scarcity of coal threatened the complete shutdown of some of the plants.

Under such obstacles the engine-production department forced the manufacture of the Liberty engine at a speed never before known in the automotive industry. In December, 1917, the Government received the first 22 Liberty engines of the 12-cylinder type, durable and dependable, a standardized, concrete product, only seven months after the Liberty engine existed merely as an idea in the brains of two engineers. These first engines developed a strength of approximately 330 horsepower, and this was true also of the first 300 Liberty engines delivered, these deliveries being completed in the early spring of 1918.

When the Liberty engine was designed our aviation experts believed that 330 horsepower was so far in advance of the development of aero engines in Europe that we could safely go ahead with the production of this type on a quantity basis. But again we reckoned without an accurate prophetic knowledge of the course of engine development abroad. We were building the first 300 Liberty engines at 330 horsepower when our aviation reports informed us from overseas that an even higher horsepower would be desirable. Therefore our engineers "stepped up" the power of the Liberty 12-cylinder engine to 375 horsepower. Several hundred motors of this power were in process of completion when again our observers in France advised us that we could add another 25 horsepower to the Liberty, making it 400 horsepower in strength, and be sure of leading all of the combatant nations in size and power of aviation engines during 1918 and 1919. This last step, we were assured, was the final, definite one. But to anticipate possible extraordinary development of engines by other nations, our engineers went even further than the mark advised by our overseas observers and raised the power of the Liberty engine to something in excess of 400 horsepower.

This enormous increase over the original power of the Liberty engine required changes in the construction, notably in increasing the strength of practically all of the working parts, including the crank shaft, the connecting rods, and the bearings. The change also resulted in making scrap iron of a large quantity of the jigs and special tools employed in making the lighter engines. A still further change had to come in the character of some of the steel used in some of the parts, and this went back to the smelting plants, where new and better methods of producing steel and aluminum for the Liberty engine had to be developed.

Thus while there were no fundamental changes in the design of the engine, the increase of its power required a considerable readjustment in the engine plants. Yet so rapidly were these changes made that on the first anniversary of the day when the design of the Liberty engine was begun—May 29, 1918—the Signal Corps had received 1,243 Liberty engines. In this achievement motor history was written in this country as it had never been written before.

From a popular standpoint it may seem that the Liberty engine was radically changed after its inception, but such an assertion is entirely unwarranted; for in the fundamental thing, the design, there was but one change made after the engine was laid down on paper in May, 1917, namely, in the oiling system. The original Liberty engine was partially fed with oil by the so-called scupper system, whereas this later was changed to a forced feed under compression. The scupper feed worked successfully, but the forced feed is foolproof and was therefore installed upon the advice of the preponderance of expert criticism.

It is also true that in working out certain practical manufacturing processes some of the original measurements were altered. But this is a common experience in the manufacture of any internal-combustion engine, and alterations made for factory expediency are not regarded as design changes, nor are they important.

The delivery of 22 motors in December of 1917 was followed by the completion of 40 in January, 1918. In February the delivery was 70. In March this jumped to 122; then a leap in April to 415; while in May deliveries amounted to 620.

The quantity production of Liberties may be said to have started in June, 1918, one year after the engine's conception in Washington. In that month 1,102 motors of the most powerful type were delivered to the service. In July the figure was 1,589; in August, 2,297; in September, 2,362. Then in October came an enormous increase to the total of 3,878 Liberty engines. During the month before the armistice was signed the engine factories were producing 150 engines a day.

In all, up to November 29, 1918, 15,572 Liberty engines were produced in the United States. In the disposal of them the American Navy received 3,742 for its seaplanes; the plants manufacturing airplanes in this country took 5,323 of them; 907 were sent to various aviation fields for training purposes; to the American Expeditionary Forces in France, in addition to the engines which went over installed in their planes, we sent 4,511 Liberty engines; while 1,089 went to the British, French, and Italian air services.

Some of the earliest Liberties were sent to Europe. In January, 1918, we shipped 3 to our own forces in France. In March we sent 10 to the British, 6 to the French, and 5 to the Italians. By June 7 the English tests had convinced the British air minister that the Liberty engine was in the first line of high powered aviation engines

INSTALLING LIBERTY ENGINE IN THE LEPERE FUSELAGE AT PACKARD PLANT, SHOWING PROGRESSIVE ASSEMBLY.

and a most valuable contribution to the allied aviation program. The British air minister so cabled to Lord Reading, the British ambassador in Washington. Again on September 26 the British air ministry reported that in identical airplanes the Liberty engine performed at least as well as the Rolls-Royce engine. Birkight, who

FIGURE 14.

LIBERTY ENGINES PRODUCED EACH MONTH DURING 1918.

designed the Hispano-Suiza engine in France, declared that the Liberty engine was superior to any high-powered aviation engine then developed on the Continent of Europe.

A more concrete evidence of the esteem in which this American creation was held by the European expert lies in the size of the orders which the various allied Governments placed with the United States for Liberty engines. The British took 1,000 of them immediately and declared that they wished to increase this order to 5,500 to be delivered by December 31, 1918. The French directed inquiries as to the possibility of taking one-fifth of our complete output of Liberty engines. The Italians also indicated their intention of purchasing heavily for immediate delivery.

This increased demand for the engine had not been anticipated in our original plans, as we had no idea that the allied Governments would turn from their own highly developed engines to ask for Liberty engines in such quantities. The original program of 22,500 engines was only sufficient for our own Army and Navy requirements. As soon as the foreign Governments, however, came in with their demands we immediately increased the orders placed with all the existing Liberty engine builders, and in addition contracted to take the entire manufacturing facilities of the Willys-Overland Co. at its plants in Toledo and Elyria, Ohio, and Elmira, N. Y. We also engaged the entire capacity of the Olds Motor plant at Lansing, Mich. In addition we had subsequently contracted for the production of 8,000 of the 8-cylinder engines. Thus the number of engines which would have been delivered under contract, if peace had not cut short the production, would have been 56,100 engines of the 12-cylinder type and 8,000 of the 8's.

The foreign Governments associated with us in the war against Germany showered their demands upon us for great numbers of the American engines, not only altogether because of the excellence of the Liberty, but because partially their plane production exceeded their output of engines. Mr. John D. Ryan, Director of Aircraft Production, verbally agreed to deliver to the French 1,500 Liberty engines by December 31, and further agreed to deliver motors to the French at the rate of 750 per month during the first six months of 1919. The British had already received 1,000 Liberty motors, and this order was increased with Mr. Ryan personally by several thousand additional engines to be delivered in the early part of 1919. When the armistice was signed the Liberty engine was being produced at a rate which promised to make it the dominant motive power of the war in the air before many months had passed.

The engine was originally named the "United States Standard 12-cylinder Aviation Engine." In view of the service which it promised to render to the cause of civilization, Admiral D. W. Taylor, the chief construction officer of the Navy, suggested during the early part of the period of production that the original prosaic name be discarded and that the engine be rechristened the "Liberty." Under this name the engine has taken its place in the history of the war as one of the most efficient agencies which was developed and employed by this country.

CHAPTER IV.

OTHER AIRPLANE ENGINES.

The production of the Liberty engine so captured popular attention that the public never fairly understood nor appreciated the extent of another production enterprise on the part of those providing motive power for our war airplanes. This was the supplementary manufacture of aero engines other than those which bore the proud appellation of "Liberty."

Let the production figures speak for themselves. In those 19 months, starting with nothing, we turned out complete and ready for service 32,420 aero engines. Of these thousands of engines less than one-half—the exact figure being 15,572—were Liberty engines. The rest were Hispano-Suizas, Le Rhones, Gnomes, Curtisses, Hall-Scotts, and some others, a total of 16,848 in all—built largely for the training of our army of the air.

This production would have been even more notable had the war continued, for at the date of the signing of the armistice the United States had contracted for the construction of 100,993 aircraft engines. Of these 64,100 were to be Liberty engines, so that the total plan of construction of engines other than the Liberty would have produced about 37,000 of them. The total cost of carrying through the combined engine project would have been in the neighborhood of $450,000,000.

While at the outbreak of the war American knowledge of military aviation may have been meager, still it was evident from the start that we would be able to go ahead with certain phases of production on a huge scale without waiting for the precise knowledge of requirements that would come only from an exhaustive study of the subject in Europe. In the first place we knew that we must train our aviators. For this purpose there was at the start no particular need of the highly-developed machinery then in use on the western front. The first aircraft requirement of the early training program was for safe planes, regardless of their type, and motive power to drive them. Later on, when we were better prepared, would come the training that would afford our aviators experience with the fighting equipment. So at the start there was no reason why we should not proceed at once with the construction of such training machines as we knew how to build.

An aviation program for war falls into these two divisions—the equipment required for training and that required for combat. While our organization, particularly through the Bolling commission which we had sent to Europe, was making a study of our combat requirements and while we were pushing forward the design and production of the Liberty engine, we forthwith developed on an ambitious scale the manufacture of training planes and engines in this country.

The training of battle aviators, on the other hand, also separates into two parts, the elementary training and the advanced training. The elementary training merely teaches the cadet the new art of maintaining himself in the air. Later, when he has mastered the rudiments of mechanical flight, he goes into the advanced training, the training in his fighting plane, where he requires equipment more nearly of the type used at the front.

For the elementary training we had some good native material to start with. The Curtiss Airplane Co. had been building training planes and engines both for the English and Canadian air authorities. This was evidently the most available American airplane for our first needs. The Curtiss plane was known as the "JN-4" and it was driven by a 90-horsepower engine called the Curtiss "OX." In the production of this equipment on the scale planned by the Signal Corps, the embarrassing feature, the choke point, was evidently to be the manufacture of the engine. The Curtiss plant at Buffalo for the manufacture of planes could be quickly expanded to meet the Government demands; but the Curtiss engine plant at Hammondsport, N. Y., could not develop the production of "OX" engines up to our needs and at the same time complete the orders which it was filling for the English and Canadian air services.

Consequently, contracts were awarded to the Curtiss Co. for its capacity in the production of "OX" engines, and then the American aviation authorities came to an agreement with the Willys-Morrow plant at Elmira, N. Y., for an additional 5,000 of these motors. Ordinarily it would require from five to six months to equip a plant with the large machine tools and the smaller mechanical appliances necessary for such a contract as this. But the Willys-Morrow plant tooled up in three months and was ready to start on the "OX" manufacturing job.

If speed in production was required at any point in the aviation development it was here in the manufacture of the elementary training planes and engines. Without training material, no matter how many aviation fields we set in order nor how many student aviators we enlisted, the movement of our flying forces toward the front could not even begin. And here entered an interesting engineering and executive problem that had to be worked out quickly by those in

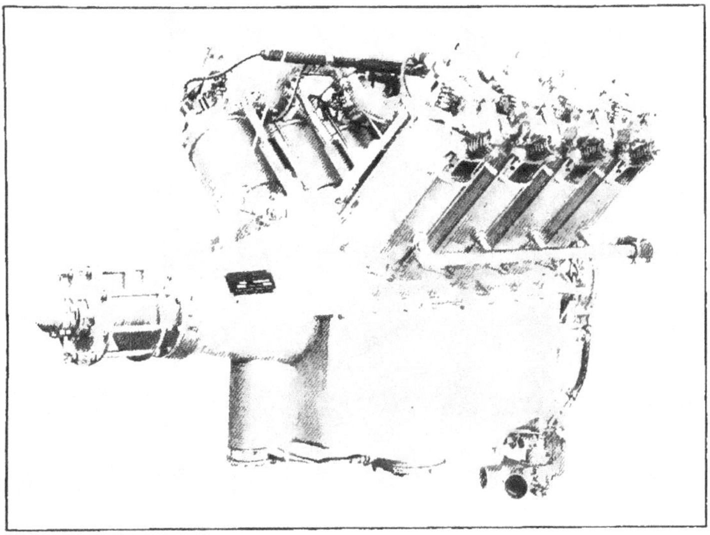

CURTISS ENGINE, MODEL OX-5.

HALL-SCOTT ENGINES BEING INSTALLED IN AIRPLANE FUSELAGES.

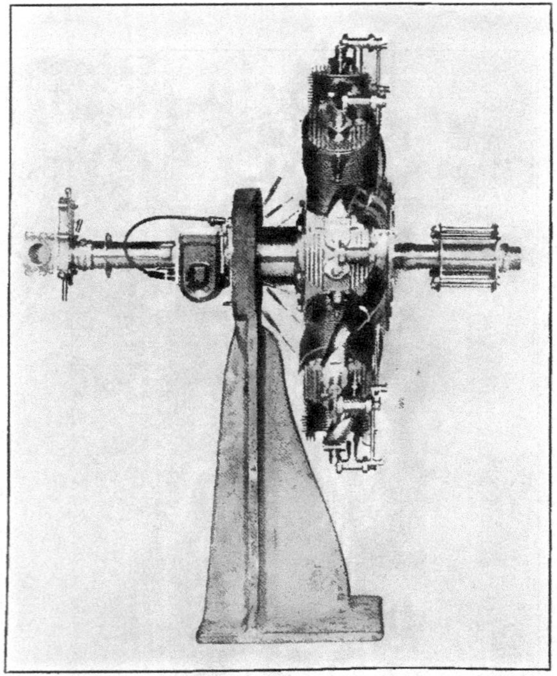

TWO VIEWS OF LE RHONE 80-HORSEPOWER ENGINE.

This is one of the successful rotary engines.

charge of our aircraft construction. If it were plotted on paper, the curve of requirements for aircraft training material would climb swiftly to its peak during the first six or eight months of the war and then decline with almost equal swiftness until it reached a low level. In other words, we must produce the great number of training machines in the shortest time possible in order to put our thousands of student aviators into the air at once over the training fields; but when this training equipment had been brought up to initial requirements, thereafter our needs in this direction could be met by only a small production, since the rate of wastage of such material is relatively low. Once our fields were fully equipped, the same apparatus could be used over and over again as the war went on, with little regard to the improvements of the type of battle planes, so that the ultimate manufacture need be large enough only to keep this equipment in condition.

It soon became evident that the production of Curtiss planes and engines, even under the heavy contracts immediately placed, would not be sufficient to take care of our elementary training needs; and the aviation administration began looking around for other types of aircraft that would fit into our plans. The experts in all branches of war flying which the principal allied nations had sent to the United States, warned us against the temptation to adopt many types of material in order to secure a quick early production. If the training equipment were not closely standardized in types, it would result in confusion and delay, both in training the aviator to fly and in preparing him for actual combat. Such had been the experience in Europe; and we were now given the benefit of this experience, so that we might avoid the mistakes which others had made. We were advised to adopt a single type of equipment for each class of training; but if that were not consistent with the demands for speed in getting our service in the air, then at the most we should not have more than two types either of planes or engines.

In the elementary training program it was evident that we could not equip ourselves with a single type of plane, except at considerable expense in time. Consequently we went ahead to develop another.

We found a training airplane being produced by the Standard Aero Corporation and known as the "Standard-J." The company had been developing this machine for approximately a year, and its plant could be expanded readily to meet a large contract. For the engine to drive this plane we adopted the Hall-Scott "A7A." This was a four-cylinder engine. It had the fault of vibration common to any four-cylinder engine, but it was regarded otherwise by experts as a rugged and dependable piece of machinery. The Hall-Scott Co. was equipped to produce this motor on an extensive scale, since at the time this concern was probably the largest manufacturer of aviation engines in the United States, with the possible exception of the Curtiss Co. The engine had been used in airplanes built by the Standard Aero Corporation, the Aero Marine Co., and the Dayton-Wright Co. Therefore the Joint Army and Navy Technical Board recommended the Standard-J plane and the Hall-Scott A7A engine as the elementary training equipment to alternate with the Curtiss plane and engine.

The Government placed contracts with the Hall-Scott Co. for 1,250 engines, its capacity. But, since a large additional number would be required, a supplementary contract for 1,000 A7A's was given to the Nordyke & Marmon Co. The Hall-Scott Co. cooperated with this latter concern by furnishing complete drawings, tools, and other production necessities.

When it came to the advanced training for our aviators, more highly developed mechanical equipment was required. There must be two sorts of this equipment. The advanced student must become acquainted with rotary engines such as were used by the French and others to drive the small, speedy chasse planes, while he must also come to be familiar with the operation of fixed cylinder engines, possessing upwards of 100 horsepower. These latter were the engines in commonest use on observation and bombing planes. For each type, the rotary and fixed, we were permitted by our policy to have two sorts of engines in order to get into production as quickly as possible, but not more than two.

Here again we had to survey the field of engine manufacture and select closely, at the same time making in point of speed approximately as good a showing as if we had adopted every engine with claims for our consideration and had told manufacturers of them to produce as many as they could.

In this case of rotary engines, our aviation representatives in Europe advised the production here of Gnome and Le Rhone motors. There were two models of the Gnome engine, one developing 110 horsepower and the other 150. The Le Rhone engine produced 80 horsepower. The Bolling commission had recommended that the Gnome 150 be used in some of our combat planes.

In the spring of 1917 we were producing a few Gnome 110 horsepower engines in this country. The General Vehicle Co. at some time previously had taken a foreign order for these engines. But neither the Gnome 150 nor the Le Rhone 80 had been built in the United States, both of these having been developed and used exclusively in France. The first recommendations from our observers in France advised us to produce 5,000 of the more powerful Gnome 150's and 2,500 Le Rhone 80's.

The production of Gnome engines in this country forms a good illustration of the manner in which aircraft requirements at the front were constantly shifting, due to the rapid evolution of the science of mechanical flight. Our officers did not hesitate to overrule their previous decisions, if such a course seemed to be justified, even at the cost of rendering useless great quantities of work already done and material already produced. This has been shown in the case of the Liberty engine. At the start we set out to build Liberty 8-cylinder engines on a large scale, only to discontinue this work before it was fairly started; but later on we again took up a Liberty 8-cylinder project on almost as great a scale as had been planned originally.

So with the production of the heavy 150-horsepower Gnome engine. Our European advisors were first of the opinion that we should go heavily into this production. Consequently the equipment end of the Signal Corps projected a program of 5,000 of the large Gnome engines. Such a contract was entirely beyond the capacity of the General Vehicle Co., which had been building the lighter Gnomes. So the Government entered into negotiations with the General Motors Co. to assume the greater burden of this undertaking. Under the pilotage of the aircraft authorities, an agreement was reached for the industrial combination of the General Motors Co. and the General Vehicle Co. The former concern brought its vast resources and numerous factories into the consolidation; while the latter furnished the only skilled knowledge and experience there was in the United States in the art of making rotary engines. This seemed to be a great step in our progress and an achievement in itself; but just as the undertaking of the construction of large Gnome engines was about to be started, events in Europe had caused our observers there to revise their first judgment, and we received cabled instructions recommending that we discontinue the development of the Gnome 150.

The entire program for Gnome 150's was canceled, and thereafter the General Vehicle Co., with its relatively small capacity, was called upon to produce as many of the small Gnome 110's as it could. As a matter of record the production of these engines amounted to 280 in number.

The Signal Corps found it difficult to induce manufacturers in this country to undertake the construction of foreign designed engines at all. The plans and specifications of mechanical appliances furnished by foreign engineers and manufacturers are so different from ours that trouble is invariably experienced in attempts to use them here. Successful concerns in this country naturally hesitated to pick up contracts on which they might fail and thus tarnish their reputations. Our advisors in Europe were insistent that we should produce Le Rhone engines in quantity in the United States, yet it was hard to find any manufacturing concern willing to undertake such a development. Nevertheless, the production of Le Rhone engines proved to be one of the most successful phases of the whole aircraft program. Its story illustrates the obstacles encountered in adapting a foreign device to American manufacture, and it also shows how American production genius can overcome these handicaps.

It was only after strenuous efforts on our part that the Union Switch & Signal Co., of Swissvale, Pa., a member of the Westinghouse chain of factories, was induced to take up the Le Rhone contract. This project called for the production of 2,500 rotary Le Rhones of 80 horsepower each. Let us see how the manufacturers took this totally unfamiliar machine and went about it to reproduce it in this country.

One might think that it would be necessary only to take the French drawings, change the metric system measurements to our own scale of feet and inches, and proceed to turn out the mechanism. But it was not so simple as that. We did receive the drawings, the specifications, the metallurgical instructions and the like, but these we found to be unreliable and unsatisfactory from our point of view. For instance, according to the French instructions the metallurgical requirements for the engine crank-shaft called for mild steel. This was obviously incorrect; and if an error had crept into this part of the plans there was no telling how faulty the rest of them might be. So from the metallurgical standpoint alone this became a laboratory job of analysis and investigation. A sample engine had been sent to us from France. Every piece of metal in this engine was examined by the chemists to determine its proper constituents, and from this original investigation new specifications were made for the steel producers.

The drawings of the engine were quite unsatisfactory from the point of view of American mechanics. They were found to be incorrect, and there were not enough of them. Consequently this required another study on the part of engineers and a new set of drawings to be made up. All of this fundamental work monopolized the time of a large force of draughtsmen and engineers for several months, working under the direction of E. J. Hall and Frank M. Hawley. The engine could not be successfully built without this preliminary study, yet this is a part of manufacture of which the unitiated have little knowledge.

The production of the Le Rhone engine might have been materially delayed by these difficulties, except for the organizing ability of the executives handling the contract. While the metallurgists were specifying the steel of the engine parts and the engineers were drafting correct plans, the factory officials, with the assistance of the engine production division of the Air Service, were procuring machinery and tooling up the plant for the forthcoming effort. By the

THREE VIEWS OF BUGATTI 410-HORSEPOWER ENGINE.

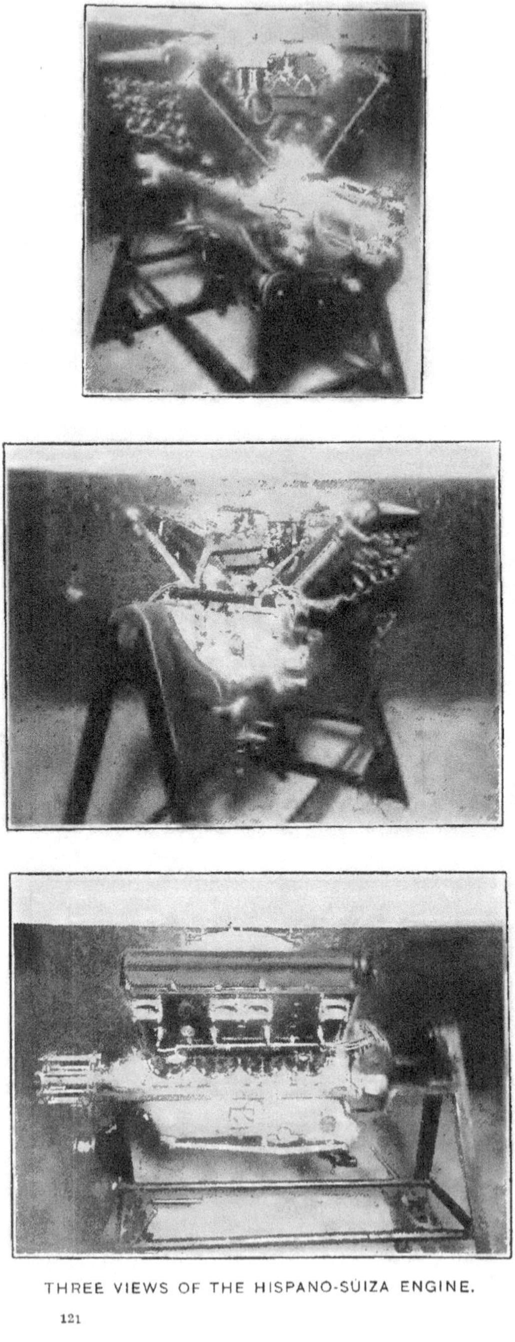

THREE VIEWS OF THE HISPANO-SUIZA ENGINE.

time this equipment was installed the plans were ready, the steel mills were producing the proper qualities of metal, and all was ready for the effort. The Gnome-Le Rhone factories in France sent one of their best engineers, M. Georges Guillot, and he assisted in the work at the Union Switch & Signal Co. So rapidly was the whole development carried out that the first American Le Rhones were delivered to the Government in May, 1918, considerably less than a year after the project was assumed by the Union Switch & Signal Co., which concern had not received the plans of the engine until September, 1917. By the time the armistice was signed the company had delivered 1,057 Le Rhone engines. Subsequent contracts had increased the original order to 3,900 Le Rhones, all of which would have been delivered before the summer of 1919, had the coming of peace not terminated the manufacture. Although France is the home of the rotary aviation engine, M. Guillot has certified to the Aircraft Board that these American Le Rhones were the best rotary engines ever built.

When it came to the selection of fixed cylinder engines for our advanced training program, all of the indications pointed to a single one, the Hispano-Suiza engine of 150 horsepower. This was a tried and true engine of the war, tested by a wealth of experience and found dependable. France had used the engine extensively in both its training and combat planes. In 1916 it had been brought to the United States for production for the allies, and when we entered the war the Wright-Martin Aircraft Corporation was producing Hispano-Suizas in small quantity. By the early summer of 1917, however, the motor had fallen behind in the development of combat engines because of the increasing horsepowers demanded by the fighting aces on the front, but it was still a desirable training engine and could, if necessary, be used to a limited extent in planes at the front.

The plane adopted by the American aircraft authorities for this type of advanced training was known as the Curtiss "JN 4H." It was readily adapted for the use of the Hispano-Suiza 150-horsepower engine. Contracts for several thousand of these engines were placed with the Wright-Martin Aircraft Corporation, and up to the signing of the armistice 3,435 engines were delivered. Before we could start the production of this engine it was necessary for the Government to arrange with the Hispano-Suiza Co. for the American rights to build it, this arrangement including the payment of royalties. Incidentally it is interesting to note that royalty was the chief beneficiary of the royalties paid by the American Government, King Alfonso of Spain being the heaviest stockholder of the Hispano-Suiza Co.

Although our policy permitted us to produce a second training engine of the fixed cylinder type, no engine other than the Hispano-Suiza was taken up by us. A number presented their claims for consideration, but they were one and all rejected. Among these were the Curtiss engines "OXX" and "V." A few of both of these had been used by the Navy, but neither one seemed to the Signal Corps to meet the requirements. The Sturtevant Co. had developed a 135-horsepower engine and built a few of them, while Thomas Bros., at Ithaca, N. Y., had taken the Sturtevant engine and modified it in a way that they claimed improved it, although the changes had not substantially increased the horsepower. This engine was rejected on the ground that it was too low in horsepower to endure as a useful machine through any considerable period of manufacture, and also because it was too heavy per horsepower to accomplish the best results.

To sum it up, our training program was built around the above named engines—the Curtiss "OX" and the Hall-Scott "A7A" for the elementary training machines; the Gnome and Le Rhone, for the rotary engine types of planes in the advanced training; and the Hispano-Suiza 150-horsepower, for the advanced training in fixed-cylinder-engine machines. Between the dates of September 1, 1917, and December 19, 1918, we sent to 27 fields 13,250 cadets and 9,075 students for advanced training. They flew a total of 888,405 hours and suffered 304 fatalities, or an average of 1 fatalty for every 2,922.38 flying hours. At one field the training fliers were in the air 19,484 hours before there was 1 fatality; another field increased this record to 20,269 hours; while a third made the extraordinary record of 1 casualty in 30,982 flying hours.

Although we do not possess the actual statistics, the best unofficial figures show that the British averaged 1 fatality for each 1,000 flying hours at their training camps, the French 1 for each 900 flying hours, while the Italian training killed 1 student for each 700 flying hours. These figures are significant, although varying conditions in the types of training programs may account to some extent for the wide differences in numbers of casualties at American as compared with allied training camps.

But while we were producing engines for the training airplanes, both elementary and advanced, we were not staking our whole combat program on the Liberty engine alone, although we expected that engine to be our main reliance in our battle machines. Our organization, both at home and abroad, was on the alert continually for other engines that might be produced in Europe or the United States and which would be so far in advance of anything in use by the air fighters in Europe in 1917 as to justify our production of them on a considerable scale. One of these motors which seemed to promise great results for the future was the Rolls-Royce, which had even then, in 1917, taken its place at the head of the British airplane engines.

Considerable difficulty was experienced in reaching a satisfactory arrangement with the Rolls-Royce Co. We expected to duplicate this engine at the plant of the Pierce-Arrow Motor Car Co., at Buffalo, N. Y., but the British concern objected to this arrangement on the ground that the Pierce-Arrow people were commercial competitors.

It was several months before we could agree on a factory and arrive at a contract satisfactory to both sides. Meanwhile the Liberty engine had scored its great success, and the expected enormous production of Liberties tended to cool the enthusiasm of our aircraft authorities for the Rolls-Royce, as it was evident that the Liberty itself would be as serviceable and as advanced in type as the British product.

The Rolls-Royce Co. wished to manufacture here its "190," an engine developing from 250 to 270 horsepower; and for this effort it was prepared to send to the United States at once a complete set of jigs, gauges, and all other necessary tooling of a Rolls-Royce plant. With this equipment ready at hand the company expected to produce about 500 American-built Rolls-Royce engines before the 1st of July, 1918.

But so rapidly was the evolution of aircraft engines going ahead that even during the time of these negotiations it became evident that something more than 250 horsepower would soon be needed in the fighting planes on the Western front. We therefore abandoned the Rolls-Royce model 190 and started negotiations for the 270-horsepower engine, the latest and most powerful one produced by the Rolls-Royce Co. But for this engine the British concern could not furnish the tooling, which would have to be made new in this country, and this would reduce the schedule of deliveries. As a result no American-built Rolls-Royce engine was ever made.

Another disappointing experience in attempting to produce a foreign designed motor in this country was the project to bring the manufacture of Bugatti engines to the United States. When our European aircraft commission arrived in France, the first experimental Bugatti engine had just made its appearance. It was apparently a long step in advance of any other motor that had been produced. This French mechanism was a geared 16-cylinder engine. It weighed approximately 1,100 pounds and was expected to develop 510 horsepower. It seemed to be the motor to supplement our own Liberty engine construction. Although heavier than a Liberty, it was much more powerful. The first Bugatti engine built in France was purchased by the Bolling commission and hurried to the United States with the urgent recommendation that we put it into production immediately and push its manufacture as energetically as we were pushing that of the Liberty engine.

The Signal Corps acted immediately upon this advice and prepared to proceed with the Bugatti on a scale that promised to make its development as spectacular as that of the Liberty. The Dusenberg Motor Corporation, of Elizabeth, N. J., was even then tooling up for the production of Liberty engines. We took this concern from its Liberty work and directed it to assume leadership in the production of Bugattis. The Liberty engine construction had been centered in the Detroit district. We now prepared to establish a new aviation engine district in the East, associating in it such concerns as the Fiat Plant at Schenectady, N. Y., the Herschell-Spillman Co., of North Tonawanda, N. Y., and several others. For a time the expectation for the Bugatti production ran almost as high as the enthusiasm for the Liberty engine, but the whole undertaking ended virtually in failure, a failure again due to the tremendous difficulty in adapting foreign engineering plans to American factory production.

This was the story of it. In due time the sample Bugatti engine arrived, and with it were several French engineers and expert mechanics. But, once set up, the Bugatti motor would not function, nor was it in condition to run; for, as we discovered, during its test in France a soldier had been struck by its flying propeller. His body had been thrown twice to the roof of the testing shed, and the shocks had bent the engine's crank shaft. Then, too, we learned for the first time that the design and development of this engine had not been carried through to completion and that a great deal of work would be required before the device could be put into manufacture. The tests in France had developed that such a fundamental feature as the oiling system needed complete readjustment, and this was only indicative of the amount of work yet to be done on the engineering side of the production. We did our best with this engine; but to redesign it and develop it so that it could pass the severe 50-hour test demanded by our Joint Army and Navy Technical Board was the work of months, and after that the tooling up of plants had to be accomplished. The American Bugatti was just getting into production when the armistice was signed, a total of only 11 having been delivered.

As we have seen, we were already building several hundred Hispano-Suiza 150-horsepower engines for our training planes. Soon after the arrival of our aircraft commission in France we were advised to go into the additional manufacture of the latest Hispano-Suiza geared engine of 220-horsepower. Consequently the Washington office at once arranged with the Wright-Martin Aircraft Corporation, which was building the smaller Hispano-Suizas, to undertake the production of this newer model also. The preparations for this manufacture had gone on in the Wright-Martin plant for a considerable period of time when further advice from Europe informed us that the Hispano-Suiza 220 was not performing successfully on account of trouble with the gearing. This fact, of course, canceled the new contract with the Wright-Martin Co., the incident being another of those ups and downs with which the undertaking was replete.

Along in the summer of 1918 the Hispano-Suiza designers in Europe brought out a 300-horsepower engine. By this date the development of military flying had made it apparent that engines of such great horsepower could be used advantageously on the smaller planes. However, the engine plants of the allied countries were already taxed to their capacities by their existing contracts, and the demands of these countries for high-powered engines could not be supplied unless we in America could increase our maufacturing facilities even further.

In following out this ambition, we placed contracts for the production of 10,000 Hispano-Suiza 300-horsepower engines. Of these, 5,000 were to be built by the Wright-Martin Aircraft Corporation. To enable this company to fulfill the new contract we leased to it the plant owned by the Government in Long Island City which had formerly been owned by the General Vehicle Co. The other 5,000 of these engines were to be built by the Pierce-Arrow Motor Car Co. at Buffalo. We also contracted for the entire manufacturing facilities of the H. H. Franklin Co., of Syracuse, N. Y., to aid both the Wright-Martin Corporation and the Pierce-Arrow Co., in this contract. The first of these high-powered Hispano-Suiza engines were expected to be delivered in January, 1919, but this project, of course, was interrupted by the armistice.

To summarize the complete engine program of the aviation development, the total contracts for engines provided for the delivery of 100,993 engines. These were divided as follows:

OX	9,450	Hispano-Suiza: 150-horsepower	4,000
A7A	2,250	Hispano-Suiza: 300-horsepower	10,000
Gnome	342	Bugatti	2,000
Le Rhone	3,900	Liberty-12	56,100
Lawrence	451	Liberty-8	8,000
Hispano-Suiza: 180-horsepower	4,500		

The delivery of aviation engines of all types to the United States Government, engines produced as part of our war program, were as follows, by months:

July, 1917	66	April, 1918	2,214
August, 1917	139	May, 1918	2,517
September, 1917	190	June, 1918	2,604
October, 1917	276	July, 1918	3,151
November, 1917	638	August, 1918	3,625
December, 1917	596	September, 1918	3,802
January, 1918	704	October, 1918	5,297
February, 1918	1,024		
March, 1918	1,666	Total	28,509

The production by types was as follows to November 29, 1918:

OX	8,458	Gnome	280
Hispano-Suiza	4,100	A7A	2,250
Le Rhone	1,298	Bugatti	11
Lawrence	451	Liberty	15,572

At the signing of the armistice the United States had produced about one-third of the engines projected in its complete aviation program.

Of the output of training engines to November 29, 1918, the various airplane plants took 9,069 for installation in planes, 325 (all of these being Le Rhone rotaries) went to the American Expeditionary Forces in France, 515 (all of which were Hispano-Suizas) were taken by the Navy, a single A7A model was sent to one of the allied countries, while 6,376 engines were sent directly to the training fields.

Of the combat engines produced to November 29, 1918 (which classification includes all of the Liberties, the two more powerful types of the Hispano-Suiza, and the Bugatti engine), 5,327 went to the various airplane plants for installation in planes, 5,030 of them were sent directly to the American Expeditionary Forces, 3,746 were turned over to the Navy, 1,090 went to the several allied nations, and 941 were taken by the training fields.

The shipment of aviation engines to Europe, however, does not imply the immediate use of them by our airplane squadrons at the front. In this report shipment to the American Expeditionary Forces means the shipment of engines from the American factories producing them. As a matter of fact several months usually elapsed from the dispatch of an engine from an American shop until it actually reached the Air Service in France, and even then another month might be required to put the engine into actual service. As a result, of the 5,000 and more aviation engines sent to France by the American engine producers, outside of those installed in their planes, less than 3,000 are recorded in the annals of the American Expeditionary Forces as having been received by them up to the end of December, 1918, the missing 2,000 being in that period either somewhere in transit or in warehouses on the route to their destination.

It is of interest to note what makes of foreign engines were used by our airmen in the war operations. An appended table shows the list of those received, their names, their rated powers, the numbers received month by month, and the totals. The records of the American Expeditionary Forces show that the squadrons in all received from all sources 4,715 aviation engines up to the end of the year 1918, but it should be borne in mind that this figure does not include more than 2,000 engines, principally Liberties, recorded on this side of the Atlantic as having been shipped to the Army abroad. Of the 4,715 engines noted as received, 2,710 were Liberties.

None of the foreign engines used by our pilots even approached the Liberty in power. The nearest in power were a Renault and an Hispano-Suiza, both rated at 300 horsepower.

Table of engines received from foreign sources in American Expeditionary Forces monthly.

Name and horsepower.	Jan.	Feb.	Mar.	Apr.	May.	June.	July.	Aug.	Sept.	Oct.	Nov.	Dec.	Total.
Hispano Suiza 180	......	......	......	......	......	......	......	......	8	......	11	......	19
Hispano Suiza 220	......	......	......	3	......	......	17	164	134	66	15	......	399
Hispano Suiza 300	......	......	......	......	......	......	......	......	1	......	......	......	1
Renault 190	......	......	4	......	4	......	18	......	......	......	......	......	26
Renault 300	......	......	......	......	......	4	10	14	3	32	20	......	83
Le Rhone 80	......	......	......	......	......	......	......	......	10	19	85	......	114
Le Rhone 120	......	......	......	6	......	8	14	43	......	43	......	......	114
Clerget 130	......	......	......	3	......	6	12	8	......	14	29	......	72
Clerget 140	......	......	......	......	......	......	......	......	......	......	10	......	10
Salmson 230	......	......	......	4	6	2	23	95	92	92	8	......	322
Fiat 300	......	......	......	......	......	......	......	......	23	10	150	......	183
Gnome 150	......	......	12	......	20	......	66	......	86	22	200	......	406
Peugeot 220	......	......	......	......	......	2	......	......	......	......	......	......	2
Beardmore 160	......	......	......	......	......	......	......	......	......	14	......	......	14
Total	......	......	16	16	30	22	160	324	357	312	528	......	1,765

CHAPTER V.

AVIATION EQUIPMENT AND ARMAMENT.

On one of the early days in the great war a Russian aviator, aloft in one of the primitive airplanes of that time, was engaged in locating the positions of the enemy when he chanced upon a German birdman engaged in a similar mission.

In those ancient times—for they seem ancient to us now, although less than five years have elapsed—actual fighting in the air was unknown. The aviators had no equipment for battle; indeed, it was doubtful if the thought had occurred to either side to keep down the enemy's aircraft by the use of armed force borne upon wings. In the first months of aviation in the great war the fliers of both sides recognized a sort of *noblesse oblige* of the air, which, if it did not make for actual friendship or fraternizing between the rival air services, at least amounted to a respect for each other often evidenced by an innocuous waving of hands as hostile flying machines passed each other.

But now the wounds of war had begun to smart; and when the Russian saw the German flier going unhindered upon a work that might bring death to thousands of soldiers in the Czar's army, a sudden rage filled his heart, and he determined to bring down his adversary, even at the cost of his own life. Maneuvering his craft, presently he was flying directly beneath the German and in the same direction and was but a short distance below his enemy's plane. Then, with a pull on his control lever, the Russian shot his machine sharply upward, hoping to upset the German and to escape himself. The result was that the machines collided, and both crashed to the ground. This was probably the first aerial combat of the war.

It seems strange to us to-day that the highly complicated and standardized art of fighting with airplanes was developed entirely during the great war and, indeed, was only started after the war had been in progress for several months. Yet such was the case. At the beginning of the war there was no such thing as armament in aircraft, either of the offensive or defensive sort. It is true that a small amount of experimentation in this direction had occurred prior to the war and also in the early months of fighting, but it was not until the summer of 1915 that air fighting, as it is so well known to the entire world to-day, was begun.

In this country we had successfully fired a machine gun from an airplane in 1912, while at the beginning of the war the French had a few heavy airplanes equipped to carry machine guns. Yet in August, 1915, Maj. Eric T. Bradley, of the United States Air Service, but then a flight sublieutenant in the Royal Flying Corps, frequently flew over the lines hunting for Germans; and his offensive armament consisted of a Lee-Enfield rifle or sometimes a 12-gauge double-barreled shotgun.

The aviators in those pioneer days usually carried automatic pistols, but the danger to one side or the other from such weapons was slight, owing to the great difficulty of hitting an object moving as swiftly as an airplane travels. The earlier planes also packed a supply of trench grenades for dropping upon bodies of troops. Another pioneer offensive weapon for the airplane was the steel dart, which was dropped in quantities upon the enemy's trenches. Great numbers of these darts were manufactured in the United States for the allies, but the weapon proved to be so ineffective that it had but a brief existence.

It is said that before the pilots carried any weapons at all the first war aviators used to shoot at each other with Very pistols, which projected Roman candle balls. The start of air fighting may be said to have come when the Lewis machine guns were brought out for use in the trenches. Presently these ground guns were taken into the planes and fired from the observers' shoulders. Then for the first time war flying began to be a hazardous occupation so far as the enemy's attentions were concerned.

It was soon discovered that the machine gun was the most effective weapon of all for use on an airplane, because only with rapid firers could one hope to hunt successfully such swiftly moving prey as airplanes. It had become patent to the strategists that it was of supreme importance to keep the enemy's aircraft on the ground. Hence invention began adapting the machine gun to airplane use.

The swiftest planes of all were those of the single-seater pursuit type. It was obviously impossible for the lone pilot of one of these to drop his controls and fire a machine gun from his shoulder. This necessitated a fixed gun that could be operated while the pilot maintained complete control of his machine, and such necessity was the mother of the invention known as the synchronizing gear.

This ingenious contrivance, however, did not come at once. Most of the war planes were of the tractor type; that is, that they had the engine and propeller in front, this arrangement giving them better maneuvering and defensive powers in the air than those possessed by planes with the rear, pushing propellers. The first fixed machine gun was carried on the upper plane of the biplane so as to shoot over the arc described by the propeller. With the gun thus attached parallel to the line of flight, the pilot needed only to point the airplane itself directly at the target to have the gun trained on its objective. But such an arrangement proved to be unsatisfactory. A single belt or magazine of cartridges could, indeed, be fired from the gun, but there was no more firing on that trip, because the pilot could not reach up to the upper plane to reload the weapon.

So the fixed gun was brought down into the fuselage and made to fire through the whirling propeller. At first the aviators took their chances of hitting the propeller blades, and sometimes the blades were armored at the point of fire, being sheathed in steel of a shape calculated to cause the bullets to glance off. This system was not satisfactory. Then, since a single bullet striking an unprotected propeller blade would often shatter it to fragments, attempts were made to wrap the butts of the blades in linen fabric to prevent this splintering, and this protection actually allowed several shots to pierce the propeller without breaking it.

This was the state of affairs on both sides early in 1915. The French Nieuports had their fixed guns literally shooting through the propellers, the bullets perforating the blades, if they did not wreck them. As late as February, 1917, Maj. Bradley, who was by that time a flight commander in the British service, worked a Lewis gun over the Bulgarian lines with the plane propellers protected only by cloth wrappings.

All of this makeshift operation of fixed machine guns was changed by the invention of the synchronizing device. This is an appliance for controlling the fire of the fixed gun so that the bullets miss the blades of the flying propeller and pass on in the infinitesimal spaces of time when the line of fire ahead of the gun is clear of obstruction. The term "synchronizing" is not accurate, since that word implies that the gun fires after each passage of a propeller blade across the trajectory. Such is not the truth. The propeller revolves much more rapidly than the gun fires. The device is also called an "interrupter," another inexact term, since the fire of the gun is not interrupted, but only caused at the proper moments. Technicians prefer the name "gun control" for this mechanism.

Who first invented the synchronizer is a matter of dispute, but all observers agree that the Germans in the Fokker monoplanes of 1915 were the first to use it extensively. Not until some time after this did the allies generally install similar devices. Some have attributed the original invention to the famous French flier, Roland Garros.

Two types of synchronizers were developed, one known as the hydraulic type and the other as the mechanical. In operation they are somewhat similar. In each case there is a cam mounted on the engine shaft so that each impulse of the piston actuates a plunger.

The plunger passes on the impulses to the rest of the mechanism. In the mechanical control the impulse is carried through a series of rods to the gun, causing the latter to fire at the proper moments. In the hydraulic control the impulse is transmitted through oil held at a pressure in a system of copper tubes. The hydraulic synchronizer is known as the Constantinisco control, commonly called the "C. C." after the military fashion of using initials. This was the device copied for American planes in the war.

In April, 1917, we knew practically nothing about the use or manufacture of aircraft guns. We had used airplanes at the Mexican border, but not one of them carried a machine gun. The Lewis gun, which is a flexible type of aircraft weapon pointed on a universal pivot by the observer in a two-place plane, was being manufactured by the Savage Arms Corporation for the British Government; but we had never made a gun of the fixed type in this country, nor did we know anything about the construction or manufacture of synchronizers.

One special requirement of the aircraft machine gun is that it must be reliable in the extreme. It is bad enough to have a gun jam on the ground, but in the air it may be fatal, for little can be done there to repair the weapon. A jam leaves the gunner to the mercy of his adversary, so in the production of aircraft armament there must be not only special care in the manufacture of the guns, but the ammunition, too, must be as perfect as human accuracy can make it. The cartridges must be either hand-picked and specially selected from the run of service ammunition, or else manufactured slowly and expressly for the purpose, with minute gauging from start to finish of the process.

Another requirement for the aircraft gun is that it must function perfectly in any position. On the ground a machine gun is fired essentially in a horizontal position, but the airman dives and leaps in his maneuvering and must be able to shoot at any instant.

Aircraft guns are subject to extreme variations of temperature, and so they must be certain to function perfectly in the zero cold of the high altitudes, regardless of the contraction of their metal parts.

Then, too, such guns must be able to fire at a much greater rate than those of the ground service. Five hundred shots per minute is regarded as sufficient for a ground gun, but aircraft guns have been brought up to a rate of fire as high as 950 to 1,000 shots per minute. The Browning aircraft gun, never used by us, but in process of development when the armistice was signed, had been speeded up to 1,300 shots per minute, with all shots synchronized to miss the blades of the propeller.

The rate of fire in the air can not be made too swift. Suppose an airplane were flying past a long, stationary target, such as a billboard, at the relatively slow speed of 100 miles an hour. Assume on this plane a flexible machine gun aimed at the billboard at right angles to the line of flight. If this is a fast machine gun, it may shoot 880 times a minute, at which rate the shots will come so fast that the explosions will merge into a continuous roar. Yet the bullets fired at such a rate from a machine moving at even such low speed will be spaced out along the billboard at intervals of 10 feet. But most of the fighting planes traveled much faster than 100 miles per hour. Thus it is entirely possible for two antagonists in the air to aim with complete accuracy at each other and both to pass unscathed through the lines of fire. The faster, therefore, the aircraft gun fired, the better the chances of bringing down the enemy plane.

The Lewis gun, invented by Col. Lewis, of the United States Army, was the weapon most generally used by the allies as the flexible gun for their airplanes, operated on a universal mount which permitted it to be pointed in any direction. The Lewis aircraft gun was the ground gun modified principally by stripping it of the cooling radiator and by the addition of a gas check to reduce the recoil. The Lewis was fed by a drum magazine, a more desirable feed for flexible guns than any belt system. The German flexible gun, the Parabellum, had the unsatisfactory belt feed.

The Vickers gun was the only successful weapon of the fixed type developed in the war before we became a belligerent. We were manufacturing Vickers guns in the United States prior to April, 1917; but when the Signal Corps faced the machine-gun problem, in September, 1917, it found that the Infantry branches of the Army had contracted for the entire Vickers production in this country.

Accordingly, the equipment division of the Signal Corps, in the face of marked opposition, took up the development of the Marlin gun as an aircraft gun of the fixed type. This gun, however, proved to be extraordinarily successful and was regarded by our Flying Service and by the aviators of the allies to be the equal of the Vickers in efficiency. Because of this development, when there came the need of tank guns, in June, 1918, the Aircraft Board, which had succeeded the Signal Corps as the director of aerial activities, was able to supply 7,220 Marlin machine guns within two weeks for this purpose.

The first order for Marlin guns was placed on September 25, 1917; and over 37,500 of them had been produced before December, 1918. The Marlin-Rockwell factory began producing 2,000 guns per month in January, 1918, and increased this rapidly until as many as 7,000 guns were built in one month. The Marlin gun shoots at the rate of 600 to 650 shots per minute and is fed by a belt of the disintegrating metal-link type.

As to Lewis guns, which we adopted as our flexible weapon, more than 35,000 of them were delivered to the Air Service up to December, 1918. In February, 1918, the Savage Arms Corporation built 1,500 of them, increasing their monthly deliveries until in the month of October, 1918, they turned out 5,448 of these weapons. The Lewis gun which the British had been using carried 47 cartridges in its magazine. A notable accomplishment of the manufacture of Lewis guns for our use was to increase the capacity of the magazine to hold 97 cartridges.

In our De Haviland-4 planes we installed two Marlin fixed guns, each firing at the rate of 650 shots per minute, equipping the weapons with Constantinisco controls to give the plane a maximum fire of 1,300 shots per minute through the blades of a propeller whirling at a rate as high as 1,600 revolutions per minute. Four fixed guns have also been successfully fitted to one plane and timed so that none of the bullets struck the propeller blades.

At the time the armistice was signed the rate of production of special aircraft ammunition, a classification including tracer bullets, incendiary bullets, and armor-piercing bullets, exceeded 10,000,000 rounds per month.

The original estimate for the quantity of ammunition our Flying Service should have was later greatly increased because the squadrons at the front began installing as many as four guns on a single observation plane.

Although different aviators had their own notions about the loading of ammunition belts, certain sequences in the use of the three types of special ammunition were usually observed. First usually came the tracer cartridge, which assists the gunner in directing his aim; then two or three armor-piercing cartridges, relied upon to injure the hostile engine or tap the gasoline tank; and finally one or two incendiary cartridges to ignite the enemy's gasoline as it escaped, sending him down in flames. Such a sequence would be repeated throughout the ammunition belt or magazine container.

The belts for the fixed guns carry a maximum of 500 rounds of cartridges. The belt which we furnished to our fliers at the front was made of small metallic links fastened together by the cartridges themselves. As the gun was fired and the cartridges ejected, the links fell apart and cleared the machine through special chutes. The total production of such belting in this country amounted to 59,044,755 links. Although the links are extremely simple in design, the great accuracy required in their finish made production of them a difficult manufacturing undertaking. The production and inspection of each link involved over 36 separate operations. It actually cost more to inspect belt links than to manufacture them.

We produced 12,621 British unit sights for airplane guns and sent 1,550 of them overseas. We also bought an adequate number of small electric heaters to keep the gun oil from congealing in the cold of high altitudes.

A novel undertaking for our photographic manufacturers was the production of the so-called gun cameras which are used to train airplane gunners in accuracy of fire. Target practice with a machine gun in an airplane is dangerous to the innocent bystander; and it was found to be impracticable, moreover, to tow suitable targets for actual machine-gun fire. Consequently, quite early in the war, the air services of the allies adopted the practice of substituting cameras for the machine guns on the practice planes.

One of these gun cameras, invented by Thornton Pickard, of Altringham, England, imitated in design a Marlin aircraft machine gun; and in order to make a picture with it, the gunner must go through the same movements that he would employ in firing a Marlin gun. Thus, if the gun were pointed directly on the target, the target would appear squarely in the center of the picture taken; and this showed the gunner's accuracy as well as if he had fired cartridges from the actual weapon.

These gun cameras were of two sorts. One type took a single picture each time the trigger was pulled. Those of the other sort took a number of pictures automatically at a speed approximately that of the firing of a machine gun. This latter type was much the same as a moving picture camera, the resulting film being a string of silhouettes of the target, each exposure showing whether the aim of the gunner was exact at the instant the picture was taken.

In September, 1917, the Eastman Kodak Co. began the development of a camera gun of the "burst" or automatic moving-picture type. After our authorities had seen the model, the Navy ordered a number of them, while the Air Service placed increasing orders for these instruments until 1,057 had been produced and delivered to the Government by November, 1918. This camera was not used in the fixed airplane guns, but was designed to train the operators of the flexible Lewis gun. The camera exactly replaced the ammunition magazine on a Lewis gun.

Of the single-shot gun cameras 150 were delivered during the hostilities. This design was obtained from Canada and duplicated here.

The use of the so-called Bromotype, paper in gun cameras was one of the interesting phases of this development. As everyone acquainted with photography knows, a picture is made ordinarily by exposing a sensitized plate or film, developing the latter to make a negative, then exposing sensitive print paper to the light that comes through the negative, thus reversing the lights and shadows and creating a positive in the exact semblance of the subject photographed. A concern in Cleveland, Ohio, the Posetype Co., produced Bromotype paper which could be exposed directly in the camera, coming out of the developing process as a positive without the intervention of a film or plate negative.

Bromotype paper is much more highly sensitized than ordinary print paper, so that it may be adequately exposed in an instantaneous, high-speed snapshot. The exposure is then developed in the ordinary way in the dark room, the familiar negative image appearing on the surface in the ruby light of the lantern. At this point the special developing process enters. The paper negative, without being fixed, is immersed in a bath of chemicals that dissolves away the sensitized surface that has been oxidized by the light from the camera lens—that is, the image—leaving on the paper only the unoxidized, or unexposed, parts of the sensitization. The paper now presents an unbroken white surface. It is then redeveloped by a special solution, and the picture in its true values of light and shade thus comes into existence. The entire development and finishing of thir paper requires only 2½ to 3 minutes.

Under this system, of course, only one finished print of each exposure can be made; but the airplane gunners needed only one print to show their aim. Positype paper was thus admirably adapted for use in the airplane gun cameras; and because of its cheapness and the simplicity and rapidity of its use, it was rapidly supplanting film at the training camps in this country when the armistice was signed.

AIRPLANE BOMBS.

The American production of bombs to be dropped from airplanes was not started so soon as production in some of the other branches of ordnance development, due to numerous difficulties encountered in working up the design of this new matériel. Although aerial bombing was steadily increasing in effectiveness and magnitude when hostilities ended, yet this kind of fighting was a development that came relatively late in the war; and the lack of perfected standards at the time this country became a belligerent helped to impede our program.

Some of the bombs first designed and put into production were later rejected by our forces in France, as they had become obsolete before being shipped overseas. We managed to manufacture a great quantity of unloaded bombs by the time the armistice was signed, enough, in fact, to provide for the Army's needs during another year of warfare. These had to be loaded with explosives before they were ready for use. We lacked adequate facilities for loading bombs with explosives, although these facilities were being provided rapidly when the war ended. The result was that the thousands of completed American bombs remained unloaded, while practically all the bombs used by our fliers in France were of foreign manufacture.

Military science had had some small experience with aerial bombing prior to the great war. Italian aviators had dropped bombs of an ineffective sort during Italy's war in Africa. When Mexico was having a civil war in 1914 American air-sailors of fortune on one side or the other dropped bombs on troops from their planes.

In the great war the first nation to attempt bombing on any systematic scale was Germany, who sent her Zeppelins over London and Paris early in the conflict and released bombs upon the heads of the helpless civilians. Yet this early and impressive effort was, in its difficulties, out of all proportion to the actual damage done to the city of London, largely due to the fact that Germany had not yet produced effective aerial bombs. The frightful scenes and noises of a bomb raid probably did more to reduce the morale in these early days than the destruction caused by the exploding missiles.

It is an exceedingly difficult trick to drop a bomb from any considerable altitude and hit what you are aiming at. The speed of the airplane, its height above the ground, the shape of the bomb itself, and the currents of air acting on the falling missile influence its line of flight. The aviator approaching an enemy target drops the bomb long before his airplane is directly above the object aimed at.

The line of the bomb's flight is a parabolic curve. The speed at which the airplane travels at first propels the bomb forward, almost as if it had been shot from a stationary gun. As the downward velocity of the bomb increases very rapidly, it soon becomes so great in proportion to velocity forward that the course of the missile bends sharply downward until, as it nears the ground, it is falling nearly in a vertical line. Hence, it becomes evident that accurate bomb dropping is an art attained only by much practice on the part of the aviator.

The latest bombing machines were equipped with sights which enabled the birdman to drop these deadly objects with greater accuracy than had been possible earlier in the war. While some of the expert European bombers scorned the new inventions in sights and preferred to continue the use of makeshift sights which they themselves had invented and installed on their planes, the average accuracy of bomb dropping was considerably greater after bomb sights came into general use.

These sights were adjusted to height, air speed, and strength of wind. When these adjustments had been made, the two sighting points were in such position that, if the bomb were dropped when the target was in line with them, an accurate hit would be registered.

We adopted a British sight, tested and found satisfactory by the Royal Flying Corps, and known as the High Altitude Wimperis, and in the United States as the Bomb Sight Mark I–A. On November 11, 1918, American factories, working on contracts placed by the Ordnance Department, had produced 8,500 of them. The job of turning out this intricate mechanism was turned over to Frederick Pearce & Co., of New York City, in January, 1918. Later in the year additional contracts were given to the Edison Phonograph Works and to the Gorham Manufacturing Co. These contracts called for 15,000 sights. By December 12, 1918, these concerns had completed a total of 12,700 of them.

Airplane bombs are shaped so as to offer the least possible resistance to the air. They have fins on their tails to steady them lest they tumble over and over. On the smaller types of bombing planes, such as the De Haviland–4, the bombs were usually carried underneath the lower wings or under the fuselage, hanging horizontally by hooks or fastened by bands around the bodies of the bombs, according to their type. The bombs were dropped by a quick-release mechanism operated by a small lever within the fuselage. The production of these release mechanisms, of which several types were made, was one of the troublesome jobs in connection with the airplane bombing.

All bombs are carried on the planes either suspended under the wings or fuselage of the plane or in a compartment in the fuselage. The manner of carrying and the design of the release mechanism is determined by the type of plane used. Since the weight-carrying capacity of the planes is limited, release mechanisms must be designed with a view to lightness as well as safety. These mechanisms are so designed that the observer can release any desired number of bombs either as a salvo or in a "trail fire," and the order of releasing must be so arranged that the balance of the plane will be disturbed as little as possible; that is, if bombs are carried under the wings they should be released alternately from each wing. All bombs are fitted with a safety mechanism which enables the observer to drop them either "armed" or "safe," i. e., so that they will explode or not as desired. An occasion might develop where the aviator would have to get rid of his bombs over his own lines. These various points are all taken care of in the design of the release mechanism and are controlled by the observer with an operating-control handle placed in the observer's cockpit.

All of the bombs used by our fliers and by the fliers of the other nations at war were of three distinctive types—demolition bombs, fragmentation bombs, and incendiary bombs.

Our Ordnance Department built demolition bombs in five different weights: 50 pounds, 100 pounds, 250 pounds, 500 pounds, and, finally, the enormous bomb weighing 1,000 pounds—half a ton. The most frequently used demolition bombs, however, were those of the 100-pound and 250-pound sizes. The demolition bombs were for use against ammunition dumps, railways, roads, buildings, and all sorts of heavy structures where a high-explosive charge is desired. These bombs had a shell of light steel which was filled with trinitrotoluol—T. N. T., as it is more commonly known—or some other explosive of great destructive power. The charge was set off by a detonator held apart from the dangerous contents of the bomb by a pin. As the bomb was released by the mechanism the pin was automatically drawn out, and the detonator slid down into position so as to explode the bomb the instant it struck its object.

The first contract let for drop bombs of any type was given to the Marlin-Rockwell Corporation of Philadelphia in June, 1917. This contract was for the construction of 5,000 heavy drop bombs of the design known as the Barlow, and also for 250 sets of release mechanisms for this bomb. We were able to go ahead with the production of this bomb at this early date since it was the only one of which we had completed designs and working drawings when we entered the war. In November, 1917, this order was increased to 13,000, and in April, 1918, to 28,000.

The Barlow bomb, however, was destined never to cut any figure in our fighting in France. The production was slow, due to the necessity of constant experimentation to simplify a firing mechanism which was regarded as too complicated by the experts of the War Department. Finally, in June, 1918, when 9,000 of these bombs and 250 sets of release mechanisms had been produced, a cablegram came from the American Expeditionary Forces canceling the entire contract.

Meanwhile, the final type of demolition bomb, known variously as the Mark I, II, III, IV, V, or VI, depending upon its size, had been developed here. In December, 1917, a contract for 70,000 of the size known as Mark II, weighing 25 pounds, was given to the Marlin-Rockwell Corporation. But in June the American Expeditionary Forces informed us that this bomb would not be of value to the Air Service abroad because of its small explosive charge, and the contract was cut down to 40,000 bombs, which number the Army could use in training its aviators. By the end of November, 1918, bomb bodies of the Mark II size to the number of 36,840 had been completed.

By the end of March, 1918, we had developed here a series of demolition bombs that promised to meet every need of our Air Service abroad in projectiles of their class. We let contracts for the manufacture of 300,000 of the 50-pound Mark III size, these contracts being reduced later to a total of 220,000. The manufacturers were the A. O. Smith Corporation, an automobile parts concern of Milwaukee, Wis.; the Edward G. Budd Manufacturing Co. of Philadelphia; and Hale & Kilburn of Philadelphia. Six months later the A. O. Smith Corporation had reached a production of 1,200 of these bombs a day, and completed their contract in October. Both the other concerns also completed their contracts in the autumn of 1918.

The A. O. Smith Corporation had tooled up their factory so as to become one of our largest producers of airplane bombs. In addition to the contract already mentioned, during 1918 this concern received orders for approximately 300,000 demolition bombs of the

TWO OF THE LARGEST DEMOLITION DROP BOMBS.

The larger of these two bombs weighs 1,000 pounds and carries 570 pounds of explosive. The smaller weighs 550 pounds and carries 280 pounds of explosive. They are both made with a heavy cast-steel nose and pressed metal rear body.

MARK II BOMB RELEASE MECHANISM FOR HANDLEY-PAGE MACHINE, SHOWING MARK I AND MARK IV BOMBS IN PLACE.

MARK IX-A RELEASE MECHANISM AS ATTACHED TO MARK II RELEASE FOR HANDLEY-PAGE PLANE.

100-pound (Mark I) size. By November 11, 1918, they had turned out 153,000 of these and had developed a capacity for building 7,000 drop bombs daily. Another large manufacturer of drop bombs was McCord & Co., of Chicago, a concern which in 1918 received orders for nearly 100,000 bombs of the 250-pound, 550-pound, and 1000-pound sizes. By the day the armistice was signed this concern had produced 39,400 completed bombs. These bombs were the heaviest and largest ones intended for use by our service abroad.

The fragmentation bombs differ from the demolition bombs in that they have thick metal walls and consequently smaller charges of explosive. They throw showers of fragments like those of high-explosive artillery shell. The demolition bombs contain, on the other hand, the maximum possible amount of explosive and produce destruction by the force of explosion. Fragmentation bombs always have instantaneous firing mechanisms, while demolition bombs are usually provided with delayed fuses, allowing them to penetrate the target before explosion.

The fragmentation bombs produced by the Ordnance Bureau were smaller than the demolition type, the size most commonly used weighing 24 pounds. These bombs had thick cases and were constructed so that they would explode a few inches above the ground. As the bombs reach a velocity downward of over 500 feet per second, the mechanism had to operate to an accuracy of less than one-thousandth of a second. They were designed for use against bodies of troops.

The fragmentation bombs were a late development in this class of work. The timing device to explode the bomb at the proper distance from the ground was undertaken by three concerns. The contracts for approximately 600,000 of these devices were let in July, 1918. The John Thomson Press Co. of New York City completed its contract for 100,000 mechanisms by the end of October, 1918. The National Tool & Manufacturing Co. of St. Louis completed its contract for 100,000 shortly after the armistice was signed. The Yale & Towne Manufacturing Co., Stamford, Conn., which had contracted to build approximately 400,000 of these devices, had turned out 150,000 by the end of November, 1918. Other concerns which manufactured various parts for the fragmentation bombs were the American Seating Co. of Grand Rapids, Mich., makers of school desks and seats, and the Dail Steel Products Co. of Lansing, Mich.

Some idea of the quantity of fragmentation bombs in our program may be gained from the fact that the contract for the Cordeau-Bickford fuse used in the fragmentation bomb, let to the Ensign-Bickford Co. of Simsbury, Conn., called for the manufacture of 550,000 linear feet of fuse, or more than 100 miles of it. The contracts for fuse were placed in August and September, 1918, and the Ensign-Bickford Co. finished up the job on November 7, four days before the armistice was signed.

The Government discovered that 3-inch shell rejected for various reasons could be remachined and used to make these airplane fragmentation bombs. The various arsenals had a large supply of them in storage. In August and September, 1918, contracts were let to large numbers of concerns to convert over 500,000 of these shell into fragmentation bombs, and by November 30, nearly 21,000 of the new bombs had been delivered.

These bombs, made from the 3-inch shell, as far as the machining of the bodies is concerned were turned out in various quantities by the following firms:

Vermont Farm Machinery Co., Bellows Falls, Vt.
Richmond Forgings Corporation, Richmond, Va.
Bethlehem Steel Co., Bethlehem, Pa.
Consolidated Car Heating Co., Albany, N. Y.
S. A. Woods Machine Co., South Boston, Mass.
Westfield Manufacturing Co., Westfield, Mass.
Wheeling Mold & Foundry Co., Wheeling, W. Va.
A. P. Smith Manufacturing Co., East Orange, N. J.
Watervliet Arsenal, Watervliet, N. Y.
Keystone Machine Co., York, Pa.
McKiernan Terry Drill Co., Dover, N. J.

The nose-firing mechanism for these bombs was produced by the Yale & Towne Manufacturing Co., Stamford, Conn.; the National Tool & Manufacturing Co., St. Louis, Mo.; and the John Thomson Press Co., New York City; while the rear cap stabilizer assemblies were produced by the Dail Steel Products Co., Lansing, Mich., and the American Seating Co., Grand Rapids, Mich.

The last item on the bomb program to come into production was the fragmentation bomb Mark II–B, which was an exact copy of the British Cooper bomb, the most effective bomb of this type in use by the allied nations. Contracts for this bomb were not let until August 17, 1918, to the Lycoming Foundry & Machine Co., of Williamsport, Pa., and the Paige-Detroit Motor Car Co., of Detroit, Mich. The former company by December 1 was producing these bombs at the rate of 500 per day and the latter was just coming into quantity production the first week in December.

When the United States entered the war no satisfactory incendiary bombs had yet been produced by any country, and consequently a long period had to be given over to experimentation before quantity production could be attained. We produced two types of incendiary bombs, the first being of the scatter type, designed for use against light structures, grain fields, and the like, and the second of the intensive type, for use against large structures. Later on in

TWO VIEWS, MARK V RELEASE TRAP (RIGHT HAND) WITH UNIVERSAL NOSE AND TAIL BEAM, MOUNTED ON T-RAILS UNDER RIGHT WING OF DH-4 PLANE.

Upper—Front view, showing operating tube connected to alternating cam in fusilage. Two Mark III demolition drop bombs (150 pounds) held by supporting straps; one bomb released, showing free supporting strap. Lower—Rear view, showing method or retaining stabilizer by tail clip with three Mark III demolition drop bombs.

TWO VIEWS OF MARK X RELEASE TRAP ON PLANES.

Shows Mark X release trap (Cooper) mounted upon T-rails under wing of DH-4 plane. Bowden control wire and casing connected to fuselage. Two Mark II-B fragmentation bombs suspended—one arming vane retained, the other free.

our program we abandoned the manufacture of the scatter type incendiary bombs on cable instructions from abroad, as it was found that the wet climate made a bomb of this type of little value. The American intensive bomb, while it had not yet come up to our ideal and was in process of evolution during its manufacture, nevertheless was regarded by our officers as more effective than any other bomb of its type in existence, since it produced a larger and hotter flame.

Our intensive incendiary bombs weighed about 40 pounds each and contained charges of oil emulsion, thermit, and metallic sodium, a combination of chemicals that burns with intense heat. These bombs were used against ammunition depots or any structures of an inflammable nature. The sodium in the charge was designed to have a discouraging effect upon anyone who attempted to put out the fire of the burning charge, since metallic sodium explodes with great violence if water is poured upon it.

Of the scatter bombs we built 45,000 before abandoning the manufacture, an action taken in September, 1918. When hostilities ceased we had out contracts for 122,886 of the intensive bombs and about 86,000 of them had been delivered ready for loading.

One of the large manufacturers of incendiary bombs was the Conron-McNeal Co., of Kokomo, Ind., manufacturers of skates. The company had to equip its plant with new machinery especially for handling this novel manufacturing enterprise. In all, they produced 50,000 bombs and were turning them out at the rate of 400 per day when the armistice was signed. This concern was the pioneer in the manufacture, the subsequent contractors profiting by the experience of the Conron-McNeal Co., and consequently being able to obtain quantity production more quickly than the Kokomo plant had been able to reach it. The Globe Machine & Stamping Co., of Cleveland, Ohio, built 30,000 bombs and 36,400 firing mechanisms before hostilities closed, and eventually reached a production rate of 500 bombs and 1,000 firing mechanisms per day. Parrish & Bingham, also of Cleveland, produced 13,000, and were turning them out at the rate of 400 daily when the production was stopped. The C. R. Wilson Body Co., of Detroit, built 42,562 of the intensive bombs and reached a daily production of 500. The New Home Sewing Machine Co., of Orange, Mass., manufactured 20,000 firing mechanisms for the scatter-type bombs.

One of the interesting phases of the bomb manufacturing program grew out of the necessity for target practice for our aviators. For this work we built dummy bombs of terra cotta, costing about a dollar apiece. Instead of loading these bombs with explosive, we placed in each a small charge of phosphorus and a loaded paper shotgun shell, so that the bomb would eject a puff of smoke when it hit its object. The aviators could see the smoke puffs and thereby determine the accuracy of their aim.

The Gathmann Ammunition Co. of Texas, Md., was the first contractor for dummy bombs, building 10,000, which were delivered in the spring of 1918. In the spring and summer of 1918, the Atlantic Terra Cotta Co., the New Jersey Terra Cotta Co., both of Perth Amboy, N. J., and the Federal Terra Cotta Co. of Woodbridge, N. J., each built 25,000 of these bombs. In September additional contracts for 50,000 dummy bombs were given to each of these three concerns, while another contract for 25,000 went to the Northwestern Terra Cotta Co. of Chicago. By the end of November these concerns had delivered nearly 34,000 of the 175,000 bombs contracted for, and were turning them out at the rate of 1,300 per day.

The Essex Specialty Co. manufactured 10,000 phosphorus rolls for dummy bombs, and the Remington Arms–U. M. C. Co. supplied 10,000 shotgun shells for the first bombs produced. Later the Remington Arms Co. produced 100,000 shotgun shells for dummy bombs.

AIRPLANE PHOTOGRAPHIC SUPPLIES.

In four days of the final drive of the Yankee troops in the Argonne district the American photographic sections of the Air Service made and delivered 100,000 prints from negatives freshly taken from the air above the battle lines. This circumstance is indicative of the progress made by military intelligence from the days when a commander secured information of the enemy's positions only by sending out patrols, or from spies. The coming of the airplane destroyed practically all possibility for the concealment by day of moving bodies of men or of military works. Mere observation by the unaided eye of the airmen, however, soon proved inadequate to utilize properly the vantage point of the plane. The insufficient and often crude and inaccurate drawings brought in by the airplane observer were early succeeded by the almost daily photographing of the entire enemy terrain by cameras, which recorded each minute feature far more accurately than the human eye could possibly do. The airplane, to quote the common saying, had become the eye of the Army, but the camera was the eye of the airplane.

This development in military information-getting from start to finish was entirely the product and an evolution of the great war. When the war broke out in 1914 there were no precedents for the military photographer to go by, nor had any specialized apparatus ever been designed by either side for this purpose. As a result the first crude makeshifts were rapidly succeeded by more and more highly developed equipment.

At the outset of the war, before antiaircraft guns were brought to efficiency, it was possible for the observation planes of the British, the French, and the Germans to fly at low altitudes and take satisfactory pictures with such photographic appliances as were then in common use. But as the "Archies" forced the planes to go higher

TYPE DR-4, DE RAM CAMERA.

TYPE A-3, HAND-HELD AIRPLANE CAMERA.

TYPE L, 4 x 5 PLATE CAMERA.

MOBILE FIELD PHOTOGRAPHIC OUTFIT, USED FOR AIR SERVICE.

It includes a dark room, printing lantern, and light-generating plant.

in the air, special equipment had to be designed for longer distance work under the adverse conditions of vibration and speed, such as exist on airplanes. It is a tribute to the photographic technicians of the world that they were able to produce at all times equipment to meet these increasing demands.

As the airplanes moved into higher altitudes, longer focus lenses had to be employed, special dry plates developed, and special color filters provided to overcome the haze created by humidity in the long spaces between the cameras and the ground. When the war ended, cameras were in common use taking photographs at an altitude of 4 miles with such microscopic fidelity as to show even where a single soldier had recently walked across a field.

The American Army came into the war almost innocent of any information at all on the subject of war photography. Such technical information as the allied nations had developed during the war had been most carefully guarded from us and all other neutral countries, with the result that what information we had was of a meager and conflicting sort.

Although in the early months of our participation in the war the Signal Corps, which then had charge of all phases of aerial warfare, made large purchases of motion-picture cameras, hand cameras, and view cameras, it was not until the end of 1917 that our officers were able to begin their real development of aerial photography. By this time we had received much valuable information from the foreign high commissions and samples of their earlier apparatus. Aerial photography had become one of the leading activities of the air service. Thus in April, 1917, the British service made 280,000 pictures at the front, and a great part of all flying was done to secure photographs. Moreover, the art was advancing at such a pace that practices in approved use one week at the front appeared likely to become obsolete the next, as new methods and new equipment superseded the old.

For years America had been second to none as a photographic country, and it was to be expected that this country would make notable contributions to the new science. It may indeed be wondered why, with the experimental laboratories and the skilled technicians at our command, we did not start at once to develop our own aerial designs and equipment. Our officers, however, felt that such a course would be likely to duplicate much of the work already done by the allied countries, who stood ready then to furnish to us the results of their experiences. While original research work might result in the invention here of certain equipment of superlative merit, yet we would be sure, in the course of such an undertaking, to adopt methods which had been tried and discarded by the allies and which we ourselves would have to discard when experience had proven them to be without value.

The information in our hands in December, 1917, showed that the British system of air photography differed radically from that of the French. The French cameras made a relatively large negative, 18 by 24 centimeters in dimension, on a glass plate. The magazines of the French cameras held 12 plates, and extra magazines were carried in the plane. These cameras were fitted with lenses of relatively long focus—20 inches. Three operations were necessary to make an exposure. The photographer must change the plate, set the focal plane shutter, and press the release. When the negatives were developed, fixed, washed, and dried, prints were made by contact.

The British used a smaller-sized plate, 4 by 5 inches in size. Their cameras were equipped with the only lenses available in England in the early part of the war—lenses of relatively short focus, ranging from 8 to 12 inches in this respect. Instead of making contact prints from these plates, the British made enlargements, measuring 6½ by 8½ inches. In the earlier period of our development of aerial photographic apparatus, we were in the same position as the British as regards lenses. We had no adequate supply of long-focus lenses. Consequently we followed the British designs of cameras and adopted the British system almost explicitly in the training of aerial photographers.

It had been our first thought to use films to a great extent on the front, since America was the country which had perfected the photographic film, and was therefore, presumably, best equipped in skill to adapt it to war uses. But plates had been used practically exclusively by the British, the French, and the Italians; and it appeared wisest to follow their experience at first, though all agreed that film, with its small bulk and weight, would be greatly superior for airplane use.

The Photographic Experimental Department of the Air Service, which was organized in January, 1918, had as its major problems the design and test of aerial cameras and all their parts and accessories. Equally important with this problem was that of sensitive plates, papers, color filters, and photographic chemicals. The corps of photographic and optical experts, into whose hands these matters were placed, early secured the active cooperation of the chief manufacturers of photographic apparatus and materials in this country. In the laboratories in Washington, D. C., Langley Field, Va., and Rochester, N. Y., comprehensive development work was inaugurated, leading ultimately to perfection of new designs of cameras and the development of plates and other photographic materials equal or superior to any available abroad.

The first airplane camera which it was decided to put into production in America was a close copy of the British type "L," which use had proven to be one of the best mechanisms employed at the front. The operation of this camera was semiautomatic, the operator having nothing to do except to press the shutter-release to keep the camera at work. The operating power was derived from a small windmill or air propeller driven by the rush of air past the plane. The automatic mechanism changed the plate and set the shutter after each exposure. Because of the situation with respect to lenses these cameras were constructed to use lenses of 8-inch to 12-inch focus, and the English 4 by 5 plate. Some 750 of these cameras were constructed. They played an indispensable part in the training of nearly 3,000 aerial photographers in this country. They were also used by our bombing squadrons at the front.

At the same time it was generally agreed that we should plan to follow the French practice as soon as lenses of greater focal length could be manufactured in this country. Increase in focal length was becoming imperative, because aerial photographers were being compelled to make exposures from much greater heights than in the earlier part of the war. For the sake of those unacquainted with photography it may be stated here that lenses of short focal lengths will not record the details of objects a great distance away from the camera, the longer-focus, rarer, and more expensive lenses being required for distance work.

As a basis for the design of cameras of longer focus a sample of the 20-inch focus camera used by the French had been sent to this country by the American Expeditionary Forces. The first camera authorized of this focal length was similar in general character to this French camera. It was constructed on the unit system, each part—shutter, camera body, lens cone, and magazine—being of standardized dimensions. It was understood that these standard dimensions were to be followed in all subsequent cameras both in this country and in the countries of the allies.

The idea constantly put before all designers of aerial cameras has been that of the automatic type, in the use of which the observer or pilot will have a minimum of work. Late in 1917 the Photographic Section of the Air Service, American Expeditionary Forces, secured the rights for the manufacture of an ingenious design of automatic plate camera invented by Lieut. DeRam, of the French Army, and requested that this be put in production. In this camera the magazine, which carries 50 plates, 18 by 24 centimeters in size, rotates between each exposure, while the exposed plate is removed from the front of the pile and carried to the back. After some study here of the incomplete model, this camera was redesigned in such form as to fit it for methods of American manufacture. It was made semiautomatic in operation; that is, the work of the observer or pilot consisted merely in releasing the shutter at will, a fresh plate always being in place. At the time of the armistice 200 of these cameras were rapidly approaching completion.

Meanwhile experiments were actively pushed in the matter of the utilization of film. Various difficulties and problems had to be solved before film could be considered practical. Considerable time was consumed in overcoming the peculiar static electrical discharges which occur on film in cold, dry regions, such as in high mountains or the upper atmosphere, and fog the sensitive surface by their light. The film camera finally decided upon was based on a fundamental design by the Folmer & Schwing organization of the Eastman Kodak Co.

This camera, known as the "K" type, carries a film on which 100 exposures, 18 by 24 centimeters in dimension, can be made at one loading. The film is held flat by an ingenious device. The film strip passes over a flat perforated sheet behind which a partial vacuum is set up by a suction, or "Venturi," tube extending outside the body of the airplane. The camera is entirely automatic, and is driven either by a wind turbine of adjustable aperture or, in war planes, by electric current from the heating and lighting circuit. The observer in the airplane needs only to start the camera and regulate its speed according to the speed with which the airplane is passing over the ground below, and the camera thereafter will, of itself, take pictures at such intervals as to map completely the terrain under observation.

In conjunction with the use of film in cameras came the question of handling the film in the dark-room; that is, the ordinary manipulations of developing, fixing, washing, and drying—a serious problem when the large dimensions of the film, its length, and difficult characteristics in handling are taken into consideration. This problem was attacked and a film developing, handling, and drying machine was produced.

Some 200 of these automatic film cameras were on order at the close of the war. Altogether over 1,100 airplane cameras of all types had been and were about to be delivered when the armistice came. These were built by the Eastman Kodak Co., Rochester; the Burke & James Co., Chicago; the G. E. M. Engineering Co., of Philadelphia; and Arthur Brock, jr., of Philadelphia.

One of the most serious problems in aerial photography is the proper mounting of the camera in the plane. Not only does the plane travel at great speed, which makes necessary exceedingly short exposures and therefore highly sensitive photographic materials, but the motor causes a continuous vibration which, communicated to the camera itself, would be fatal to obtaining sharp pictures.

The experimenters of the Air Service carried out a long, extensive, and most interesting investigation at Langley Field to make clear the whole question of preventing the vibration of the airplane camera.

The scientists worked out a method of making the camera itself record the vibrations communicated to it by the plane when the box was not held by a proper vibration-neutralizing suspension.

The plan adopted was to send up a camera thus mounted on an airplane, focus it on a light on the ground below, open the shutter, and take a time exposure from the swiftly-flying plane. The result, of course, was a streak, or trail, written on the plate by the point of light below, the jagged or wavy character of this trail indicating the vibrations of the camera and determining the proper principles of a suitable mounting.

The first thought was to do this work at night, as the British had done, when the light below would pierce the darkness distinctly. But night flying is hazardous, and a better plan was called for. Nor would the proposal to use an extremely strong light in broad daylight do, because, while the light would indeed be photographed continuously across the plate, so also would the surrounding ground, and the general result would be a fogging or blurring of the outlines of the streak.

Finally the problem was solved by conducting the aerial experimental work over woodland in the late afternoon. A strong, reddish light was placed in the woods so as to be visible from above. The surrounding green foliage supplied a frame of sufficient contrast to the light to make its impression distinct on the plate. To emphasize the contrast, the camera lens was covered with a reddish colored ray filter, and this brought out sharply the outline of the streak.

These tests resulted in the design and production of new and unique camera mountings which successfully stopped all vibrations of the camera.

A problem on which it was necessary to have the closest cooperation of the plane designers was that of installing the large 20-inch focus cameràs in the airplane. There is little room at best in a plane, and the demands for armament, wireless, and bombing space all had to receive attention. In the American service a distinct advance was made in the design of a special plane intended primarily for photographic reconnaissance. Several of these planes, which were the most completely equipped for photographic purposes of any designed during the war, were built and would have been put into quantity production in the late fall of 1918.

Parallel with this development of apparatus went studies of the sensitive materials and methods of photography from the air. Because of the swift motion of the plane extremely short exposures are imperative. Consequently, the most advanced technique of instantaneous photography had to be applied. The cooperation of various plate manufacturers was obtained, who brought out especially for the Government several new plates which showed on test to be superior to any which had appeared in the war on either side.

As an airplane rises higher and higher in the sky, the moisture of the intervening atmosphere between the machine and the ground creates a haze which makes aerial photography above a certain height unsatisfactory and even impossible with the naked lenses as used on the ground. The problem of finding the best means for piercing aerial haze occupied the attention of a corps of experts working both in the laboratory and in the field. The solution lay in the use of special color filters of general yellow hue which obscured the bluish light characteristic of haze. Filters of new materials specially adapted to airplane use were made available as a result of this study.

Field equipment of quite new and special design for performing photographic operations had to be designed and built. Among the most interesting of these developments was the photographic truck or mobile photographic laboratory. This consisted of a specially designed truck and trailer containing all the equipment necessary for the rapid production of prints in the field. The truck body was equipped with a dynamo for furnishing the electrical current required for lights and drying fans, while each unit was provided with an acetylene generator for emergency use, if the electrical apparatus should break down. The mobile dark room carried on the trailer of each unit was equipped with tanks, enlarging camera, printing boxes, and other necessary apparatus. In all, some 75 of these field laboratories were constructed.

While the development of apparatus and new materials was from a popular standpoint in many ways the most interesting phase of the work of the photographic scientists, nevertheless it should be remembered that the great problem in this, as in all other fields of American endeavor, was to produce the supplies in tremendous quantities. In October, 1918, we shipped overseas 1,500,000 sheets of photographic printing paper, 300,000 dry plates and 20,000 rolls of film. We also sent 20 tons of photographic chemicals. These were merely the principal items in the consignment. Besides paper, plates and chemicals, the field force required developing tents, trays, printing machines, stereoscopes, and travelling dark rooms, to name only some of the principal items. Much of the material already on the market was not suitable for the purpose, a fact requiring the production of specially manufactured supplies.

THE FIREWORKS OF FLYING.

It is interesting to consider that without fireworks, and particularly some of the familiar forms of them used to celebrate the Fourth of July, war flying would have lost much of its efficiency. Night flying would have been well-nigh impossible, while day flying would

MARLIN MACHINE GUN WITH FIXED MOUNTING, ON A JN-4 FUSELAGE.

TWO LEWIS MACHINE GUNS WITH MOVABLE MOUNTING, IN THE OBSERVER'S COCKPIT OF A DE HAVILAND-4.

AIRPLANE FLARE.

HOLT WING-TIP FLARE HOLDER.

have had to invent substitutes for fireworks had the latter not been available.

The squadron fields near the front were kept as dark as possible at night for obvious reasons. The first inkling that a squadron commander might have of the approach of one of his aviators at night would be the sudden appearance high in the air of a green or red or white Roman-candle ball. This would be the signal inquiring if the landing field were clear. A pyrotechnic star of a predetermined color, shot from the ground, would answer the homing birdman; and, if the signal were in the affirmative, he would descend through the sheer blackness, unable to see clearly, yet confident that he would make his landing safely.

As the plane neared the ground suddenly under one of the wings a flare of dazzling power would commence to burn, for a few seconds flooding the field with light. In that brief space of time the plane would have made its landing, and soon field and quarters would again be obscured under the protecting blanket of darkness.

Every service airplane at the front was equipped with one or more signaling pistols. In appearance these weapons were more murderous than the "gat" carried by a desperado of the movies, but, like the prize bulldog with the undershot jaw, they were more deadly in looks than in deeds. Their formidable-appearing cartridges were larger than the shells used in shotguns, resembling the latter almost identically in appearance; but every one of these shells contained only a Roman-candle ball and a sufficient charge of powder to eject the star a good distance into the air. The sound of the discharge was a mere whisper of the shattering roar that might be expected from such a redoubtable piece of ordnance. These aviation pistols were similar to the Very signal pistols used in the trenches.

The stars shot were three colors, red, green and white, and the color of a cartridge's star was painted on the end of the shell. This base was also ridged with a different pattern for each color, so that the aviator at night could feel with his fingers and tell the color of the cartridge without seeing it.

Codes of numerous messages were worked out in different combinations of these three colors. The stars were quite visible in broad daylight, too, and were used for many signaling purposes. They indicated the position of enemy troops or the presence of hostile aircraft, they called for help from other airplanes, and they signaled squadron orders when the machines were flying in formation.

But the signal pistol had a more sinister use. If the pilot were driven down in enemy territory, it became his duty to destroy his machine. In some cases the signal pistol was used effectively to set airplanes on fire under such conditions. The pilot had only to open his gasoline tank and fire a Roman candle ball into the escaping fluid. In other cases when the aviator landed amid enemy troops he was able to hold them at bay with his signal pistol until his plane was burned beyond the possibility of salvage.

While we manufactured Very pistols in this country, all of those actually used by our fliers in France were purchased abroad.

Night-flying is one of the most hazardous duties of the aviator, the chief danger being in landing. The fields well back of the front were usually brightly illuminated by flood lights at night, but those nearer the enemy were left in darkness, as a rule, to protect them from the attacks of hostile aircraft. The aviator at night can usually see the ground faintly, but he is unable to make an accurate judgment of the distance of his machine above the ground. This danger was greatly alleviated when the wing-tip flares were invented. The wing-tip flare consisted of a small cylinder of magnesium material in a metallic holder, one flare being fitted under each lower wing of the plane. Each flare was controlled by a push button in the pilot's cockpit. Pressure on the button sent an electric spark into the magnesium and touched it off.

When the descending pilot at night judged that he was near the ground he pushed one of the buttons. Immediately the flare ignited and burned for about 50 seconds with the brilliant light of 20,000 candle power. Being hidden by the wing, this light did not dazzle the eyes of the aviator, but the reflection from the under surface of the wing lighted up the field for an adequate distance in all directions.

Another important use of pyrotechnics occured in those enterprises known as night-bombing raids. Since both sides kept their vulnerable ammunition dumps and their important buildings completely unlighted at night, even though the night raider knew he was in the general vicinity of his objective, hits from bombs dropped from aloft were almost accidental. To enable the night bomber to see his target the interesting piece of pyrotechnics known as the airplane flare was invented. This was a great charge of magnesium light held in a cylindrical sheet-iron case nearly four feet long and half a foot in diameter, the exact dimensions being 46 inches by 5 inches. The flare weighed 32 pounds. Within the cylinder was not only the magnesium stick but also a silk parachute 20 feet in diameter. The entire cartridge was attached to the airplane by a release mechanism similar to those holding the drop-bombs.

When over his objective at night the pilot or observer touched a button and the entire cartridge, iron case and all, dropped from the plane. A pin wheel on the lower end of the case was instantly spun by the rush of air, and the resulting power not only ignited the magnesium but at the same time detonated a charge of black powder sufficient in force to eject from the case the flare and its tightly rolled parachute. The parachute immediately opened; and the burning flare descended slowly, flooding a large area of the ground below with a light of 320,000 candlepower, this light burning for about 10 minutes.

Such a light not only enabled the bomber to drop his destructive missiles accurately, but it was found by experience that it dazzled the eyes of antiaircraft gunners below and made their aim inaccurate. The light of this flare was so strong that it was possible for the airplane above to obtain photographs of good detail on the darkest of nights.

We were just starting to produce these flares when the war ended. In fact the actual production of pyrotechnic supplies in this country was small, the American Expeditionary Forces depending almost exclusively for these supplies upon French and British sources.

KEEPING OUR FLIERS WARM.

When the commander of an airplane squadron sends an aviator into the high altitudes, he sends him into climate that much of the year is colder and more severe than any known on earth, even at the North Pole. Not only is the temperature of the air likely to be many degrees below zero at the heights which war planes attained, but the flier must face this bitter cold in the gale of wind that is never blowing less than 100 miles per hour.

Consequently when we trained a corps of aviators to fly at altitudes of 18,000 to 20,000 feet above the western front, it was necessary for us to design and manufacture for them the warmest clothing ever made. They were dressed more warmly than any Polar exploration party that ever set forth, more warmly in fact than any other class of men in the world. For we not only gave them the protection of all the fine wool, leather, and fur that they could wear without hindering their movements, but in addition we literally wrapped them in flexible electric heaters.

The first purchases of aviators' flying clothes were made by the coordinated action of the Council of National Defense and the Quartermaster's Department. It was soon apparent that the design of such clothing was a special matter which the aviation authorities themselves should control, and purchases thereafter were all made by the Bureau of Aircraft Production. There were no standard styles at the time, so it became necessary for us to develop our own equipment. This development resulted in an output for the flier that became standard.

In moderate weather the flier wore upon his head a woolen hood, or helmet, extending well down over the forehead to the eyes, and around the neck to the shoulders. In cold weather, or for high-flight work, this headgear was augmented by a silk helmet of double thickness, having between its layers an electrically heated pad connected by copper wire to the electric generator on the plane's engine. Outside of this was worn a soft leather helmet lined with fur, extending down over the back of the head, covering the ears and cheeks, and fastening under the chin. Then the face was entirely covered with a leather face mask lined with wool and having an opening for the eyes, over which were worn a pair of goggles. When the pilot was also required to operate the radio system, in place of the fur-lined helmet he wore the radio helmet. This was of leather and resembled the other in appearance, but it contained the receiver of the wireless telephone, enabling the flier to hear what was spoken to him in an ordinary tone of voice several miles away.

In addition to this equipment the aviator who went up to the great heights wore the oxygen mask. This was of rubber, and, besides supplying oxygen, it contained a transmitter, allowing him to speak as well as to hear by wireless.

Over the body was worn a one-piece flying suit extending from the feet to the throat, belted and buttoned tight at the ankles and wrists. The outer material of this suit was waterproof, and when it was buttoned on there were no gaps through which the air might penetrate. This suit was lined throughout with fur.

It was a considerable problem to find a fur of extreme warmth with a pelt strong enough to withstand rough usage and still not be too great in bulk, and purchasable at a price not too extravagant. After the furs of many beasts had been examined and tested, it was determined that the hide and fur of a Chinese Nuchwang dog met these requirements better than any other. We were making so many of these suits that we required all of the dogskins we could get, not only in this country, but in China. Merely the final purchase of these pelts before the armistice was signed was for nearly 500,000 of them, and that many dogs in an interior Chinese province gave up their lives that the American aviation warfare might succeed.

With its waterproof outer surface and its furry lining, it might seem that such a garment would be warm enough for any work. But the aircraft authorities of the United States were not content until they had installed between the fur and the outer covering thin, flexible, electric-heat units connected by silk-covered wire with the dynamo on the engine. Similar heating pads were placed in the gloves and moccasins of the fliers.

On their hands, besides the electrically heated gloves, the fliers wore gauntlets of muskrat fur, these extending well up the arms and being of special design which allowed the fingers of each glove to remain in a fur-lined pocket or to be withdrawn from the pocket without removing the gloves from the hand. Over the electrically heated moccasins were worn leather moccasins extending well up the calf of the leg and lined with heavy sheep wool. These were fastened with straps and buckles. Thus clad, our aviators were acknowledged generally to be the most warmly and efficiently equipped of any at the front.

Besides these special garments for warmth, the fliers required many other items of clothing, such as sweaters, leather coats, fur-lined coats, helmets, and many styles of goggles.

The total cost of air clothing, provided or in course of manufacture on November 11, 1918, was over $5,000,000. Some of the major items in round numbers were 50,000 fur-lined flying suits (at $36.25), 100,000 leather helmets, an equal number of leather coats, costing anywhere from $10 to $30 each, and over 80,000 goggles at $3.50 apiece.

PROTECTION IN HIGH ALTITUDE FLYING.

Even to-day the veteran of the air squadron scoffs at the new-fangled outfits of oxygen masks and tanks carried in an experimental way on some of the high-flying planes at the western front when hostilities ceased. Nevertheless, had the war continued a few months longer, it is probably true that the oxygen apparatus would have been included in the indispensable equipment of every airplane in the front areas. Such a development, had it occurred, would have been due largely to the efforts of the American Aircraft Service.

Many aviators who have gone into high altitudes, fought there, and lived to tell about it, doubt the necessity of oxygen-supplying apparatus, since they themselves returned safely without it. Nevertheless the experiments conducted by the Bureau of Aircraft Production demonstrated conclusively that the flyer artificially supplied with oxygen in the high altitudes is much more efficient than one who is without it. These experiments were conducted in a room which duplicated the conditions at high altitudes. At 19,000 feet the pressure of the atmosphere is one-half the atmospheric pressure at sea level. The lack of pressure in itself causes no appreciable physical or mental reaction; but the reduced pressure at 19,000 feet means that in a given amount of air there is only one-half the oxygen that there is in a similar amount at sea level. The lack of oxygen is serious.

Experienced aviators were placed in an air-tight chamber under the observation of Government scientists. The air in this chamber was then exhausted until it corresponded to the atmosphere at the 19,000 feet level. The subjects were then set at small mechanical tests, such as the pushing of certain buttons when different colored lights were turned on, these tasks requiring a degree of mental concentration. In this and similar tests it was discovered that not only do the subjects lose accuracy in the attenuated air, but their movements become conspicuously slower. In the parlance of the pilot they become "dopey." More than one returning aviator has confessed to this feeling when at a high altitude.

When the British analyzed their air casualties during the first year of the war they found that 2 of each 100 fliers in the casualty list were killed or hurt by the enemy, 8 of them owed their misfortune to defects in the planes, while the other 90 came to the hospital or the grave because of themselves, their carelessness or recklessness, their physical failings, and all other things which may be summed up in the human equation. A thorough study on the part of the British disclosed the fact that practically all of the flying personnel was suffering from what became known as oxygen fatigue, caused by flying so many hours each day in altitudes where there was not enough oxygen to feed the body properly.

Before the war broke out the aviation record was 26,246 feet above sea level. In January, 1919, this record had been lifted nearly a mile, the high point being an altitude of 30,500 feet. Early in the war pilots at the 7,000 feet level could laugh at antiaircraft fire, and few machines ever went above 10,000 feet. Thus with the first equipment the "ceiling"—that is, the average high level to which every day flying goes—was about 12,000 feet.

When the war closed, a pilot was not safe under the 15,000 feet level, due to the development of antiaircraft guns, and the safest machine had become that which could fly highest. The aviators were demanding a working ceiling of 18,000 feet, and were obtaining it, too, from the latest type of planes. It was evident that the reduced oxygen at this ceiling was responsible for casualties among the fliers, and we could expect the ceiling to be pushed even higher as antiaircraft guns became more powerful. The need of oxygen equipment was plainly indicated. Even at 18,000 feet the aviator relying upon the normal oxygen supply at that altitude, while he may feel perfectly fit, is actually slow to judge distances, to aim his guns, to fire them, and to maneuver his plane.

The first oxygen apparatus was designed for the British Air Service and was made at the plant of de Lestang in Paris. The demand for the apparatus was so great that an automobile was constantly kept waiting at the factory that as soon as each set was finished it could be rushed straight to the front. The first British squadron which used oxygen equipment reported that its men gave six times the service of any other British squadron.

Our Air Service adopted the Dreyer oxygen apparatus, which was the original device produced by the British. We found it to be a hand-made appliance, but under our direction we adapted it to American methods of manufacture. The British apparatus was built to supply oxygen to one man only. We changed it to take care of two men. The model received was too heavy; we reduced the weight. Finally we added improvements to make it more efficient and reliable and redesigned it to meet American factory methods.

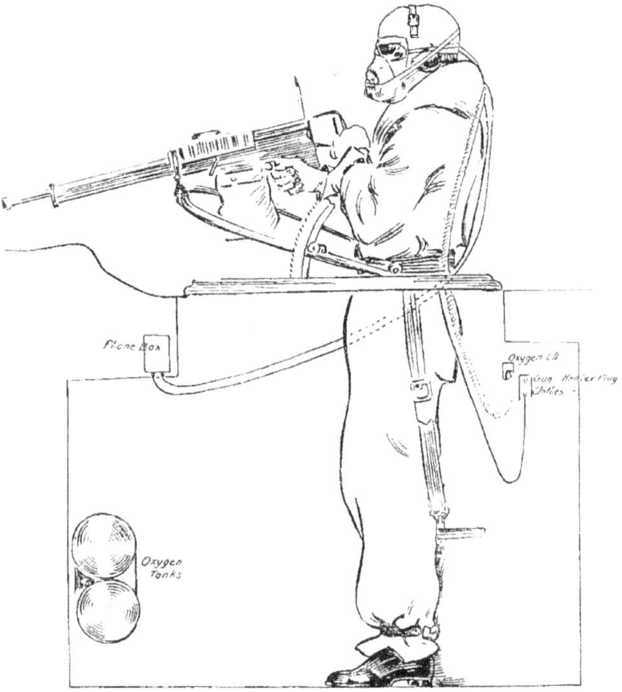

GUNNER IN COCKPIT EQUIPPED WITH OXYGEN HELMET AND TELEPHONE RECEIVER, OPERATING MOVABLE MACHINE GUN.

AVIATOR'S OXYGEN HELMET EQUIPPED WITH TELEPHONE RECEIVER.

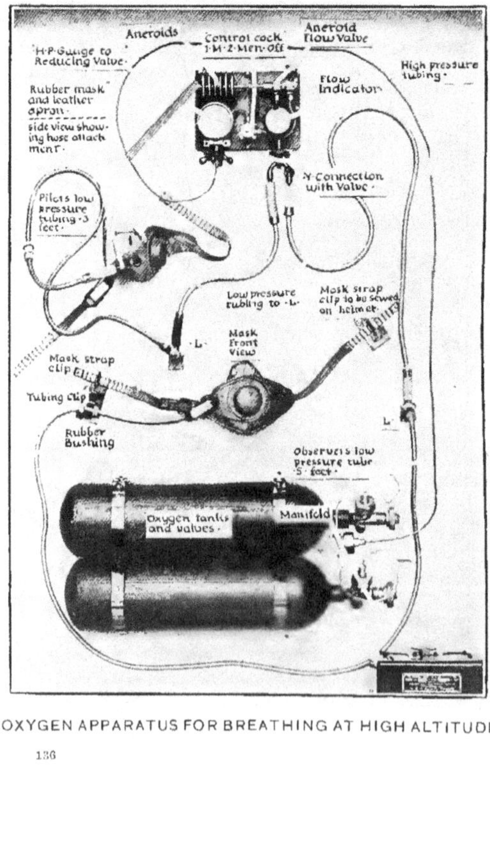

OXYGEN APPARATUS FOR BREATHING AT HIGH ALTITUDES.

Such an equipment has to be entirely automatic in its operation and as reliable as human ingenuity can make it. The Dreyer device embodies several instruments all of which must work perfectly under widely varying conditions. In use its tanks will contain oxygen under pressure ranging from 100 pounds to 2,250 pounds per square inch, yet the mechanism must deliver the oxygen to the aviator at a constant rate regardless of its tank pressure. Then the whole apparatus is subjected to temperatures that may be as high as 80° above zero or as low as 30° below. It must function evenly in the atmospheric pressure at any altitude up to 30,000 feet, delivering more oxygen as the atmosphere thins. Such was the problem of manufacture. Yet, taking up the work in January, 1918, we turned out six complete equipments by May 3, 1918, sending them overseas by special messenger for actual test on the front. Twenty-eight days later we shipped 200 sets. By the end of the war we had built 5,000 complete oxygen equipments. Of this number 3,600 had been sent to ports of embarkation awaiting shipment, and over 2,300 of these had been shipped overseas. In October we had reached a production rate of 1,000 sets per month.

Some of the difficulties of this production may be read in the description of the complicated character of the apparatus. The equipment consists of a small tank or tanks, the pressure apparatus, the tube leading from the reservoir, and finally the face mask covering the mouth and nose. The mask has combined with it either the interphone, a mechanism which cuts off the roar of the engine from the ears of the passengers and allows the pilot and observer to talk freely with each other, or in certain cases the receiver of the radio telephone or telegraph.

The flow-regulating apparatus consists of five parts. In front of the pilot is a high-pressure gauge to indicate the supply of oxygen in the tank. In the tank there is a high-pressure valve with an upper chamber which compensates for the temperature. There is also a shut-off valve, hand operated, which can be set to provide a flow of oxygen to one man, to two men, or to none at all. Then there is a regulating valve operated by an aneroid barometer which adjusts the oxygen flow to the altitude, the flow increasing as the machine goes higher. Finally in the pilot's view there is a flow indicator consisting of a small fan wheel which tells the aviator that the oxygen is actually flowing.

The mask presented a difficult problem, as it must be big enough to contain the radio receivers and still enable the aviator to see and work. Yet the mask must keep its adjustment in a gale of wind at least 100 miles per hour in velocity.

The actual use of the equipment on the front was just starting when the armistice was signed. We sent across to France a special division of experts to take charge of the installation of these equipments on the planes. At the close of hostilities we required all military planes flying above an altitude of 10,000 feet to be equipped with oxygen apparatus. This class included day bombing, pursuit, and chassé planes, and a certain number of night bombing planes, and Army and corps observation planes.

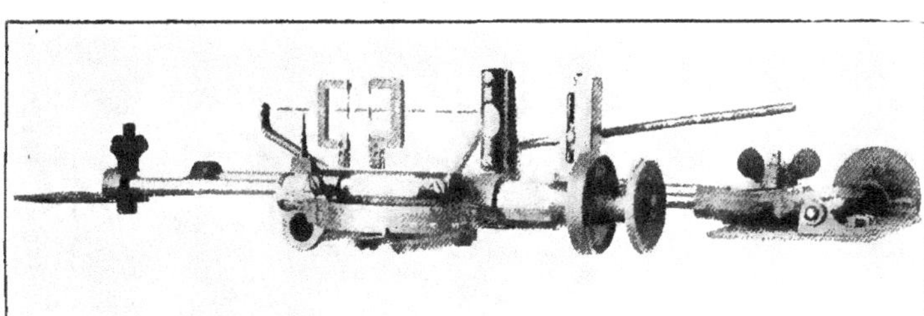

TWO VIEWS OF BOMB SIGHTS USED ON AIRPLANES.

Upper picture shows bomb sight on De Haviland 4. Lower picture shows high-altitude bomb sight. Set from readings of instruments showing altitude and air speed, it indicates to the bomber the precise instant for release of the bomb in order to reach the target.

AVIATORS EQUIPPED WITH TELEPHONE TRANSMITTERS AND HEAD SETS TO COMMUNICATE WITH EACH OTHER.

CHAPTER VI.

THE AIRPLANE RADIO TELEPHONE.

Electrical science was called upon to furnish marvels and prodigies indeed during the recent war as aids to the American arms, but in no respect did it respond in more successful and spectacular fashion than it did when asked to produce a wireless telephone system that would make possible the transmission of human speech to and from moving airplanes. It is doubtful if any other branch of science enlisted for war work produced any instrument or mechanism so far in advance of what was known before the war as the airplane wireless phone was in its class.

To be sure, we had the radio telephone some time before America entered the war or even before the war broke out in Europe in 1914. Ever since the scientists began experimenting with wireless electricity it has been axiomatic that, at least theoretically, whatever you can do with wires you can do without wires. And so following the development of the wireless telegraph came the production of the wireless telephone, and the invention had been so perfected in 1915 and 1916 that in the United States Navy's official test at the Arlington Station, across the Potomac River from Washington, human speech sent out by the transmitters there was heard simultaneously at the Eiffel Tower in Paris and at the Government's own wireless station in Hawaii.

But there is a vast difference between using the wireless telephone in the quiet of the radio rooms aboard ship or in the shore stations and using it amid the roar of the powerful engine propelling an airplane. The equipment, too, that had been used on the ground was altogether too cumbersome to go into the fuselage of an airplane.

As early as August, 1910, American genius had successfully accomplished wireless telegraph transmission from airplane to ground, and in October of the same year the idea of aerial fleet command by telephone was conceived and plans for its development discussed by Army officers on duty at the International Aviation Tournament at Belmont Park, Long Island. In 1911 a message was successfully transmitted from an Army airplane over a distance of 2 miles. In 1912 the Signal Corps had increased the distance to 50 miles. Two years later, in the Philippine Islands, a message had been successfully received on an airplane in flight over a distance of 6 miles.

In 1915 the Aviation Section entered upon a definite plan of development of aircraft wireless at the Signal Corps Aviation School, San Diego, Calif. This plan was based upon the Belmont Park idea and discussions, with the voice-commanded tactical air fleet as the ultimate goal. The airplane had changed from the pusher to the tractor type, with the noise of the motor of the latter driven back by the blast of the propeller into the face of the aviator. The airplane wireless problem was thus quite completely changed. Under these new conditions, however, the development was entered upon, and it has since been continuous. In October a spring-driven dictaphone was taken into the air and a record of speech made in the noise of the motor. This was contemporaneous with the successful long-range experiments in radio telephony at Arlington, referred to above. A study of this dictaphone record convinced the aviation officers that the idea of the radio telephone for airplanes was entirely practicable. Experiments during the fall and winter with various means of driving the wireless power plant resulted in a decision to develop the air fan as a source of power rather than the gear or belt system.

This development continuing through 1916, transmission by telegraph from airplane was accomplished up to 140 miles, means for receiving in the noise of the motor were worked out, and a message successfully telegraphed between airplanes in flight. The radio telephone was under construction, and in February, 1917, the voice was first transmitted by telephone from airplane to ground. Like Alexander Graham Bell's first wire telephone, the apparatus was crude. But the door was unlocked and ready to be opened upon the new field of development.

When on May 22, 1917, Gen. Squier, the Chief Signal Officer of the Army, called upon the scientists to develop at once an airplane telephone, he was not only introducing them into what was to many of them a new field, but he was asking them to produce what the science of Europe had been unable to create in nearly three full years of acquaintance with the successful ground system, although the needs of airplane fighting demanded this invention as they demanded almost nothing else.

It will thus be seen that when we began this development as a war measure we had a considerable basis of experience to work upon. The Army had established the foundation of operation on the airplane, made a study of the tactical requirements, and knew what it wanted. The Western Electric Co. in 1914 and 1915 had conducted extensive experiments with the radio wireless telephone at a ground station at Montauk, Long Island, and had played an important part in the long-range experiments at the Arlington station. There had been wireless voice communication before this time, but the apparatus and systems perfected at Montauk set the standard on which all subsequent development was built. The French Scientific Mission and other officers of the allies had arrived and enabled us to check up what had been done abroad and to confirm or modify our ideas of the tactical requirements.

At the conference with Gen. Squier in May was Col. Rees of the Royal Air Force of Great Britain; Col. C. C. Culver, United States Army, then a captain; and F. B. Jewett and E. B. Craft, respectively the chief engineer and the assistant chief engineer of the Western Electric Co.

At this meeting Gen. Squier outlined the future of the part the airplane was to play in the war, and pointed out how invaluable would be a successful means of communication between battle planes when flying in squadron formation. Mr. Jewett had received his commission as a major in the Signal Corps, and he was ordered to take charge of the work of developing radio communication for aircraft.

Capt. Culver had taken part in the 1910 experiments and discussions, and since 1915 had been conducting the Army development of airplane wireless at the aviation school at San Diego, Calif. He was detailed to work with Maj. Jewett and his engineers, bringing to their assistance the result of his experience and the point of view of the trained military man and the aviator.

The first development was carried on in the laboratories of the Western Electric Co. on West Street, in New York. Men and materials were drafted from every department of the company, and the laboratories were soon seething with activity. In a few weeks the first makeshift apparatus was assembled, and the first practical test of a radio phone on an airplane was made at Langley Field at Hampton, Va., less than six weeks after the Signal Corps had given the go-ahead. Three employees of the Western Electric Co. on that day established telephone communication between an airplane in flight and the ground. A few days later the first apparatus produced successful communication between planes in the air.

It is not possible here to go into a technical description of the wireless telephone. The most vital part of the apparatus, however, and the essential factor in airplane wireless telephone communication is a vacuum tube containing an incandescent filament, a wire mesh or grid, and a metal plate. By means of electrical current the wire filament is heated to incandescence. The tube has the property of receiving the energy of the direct current of a dynamo and, through the medium of the wireless antennæ, of throwing it out into space as a high-frequency alternating current. Such is the sending tube. A modification of the same tube picks up from the antennæ the high-frequency alternating vibrations from another sending apparatus and transforms them into direct current, carrying the sound waves of the human voice along with them.

The design of the radio apparatus itself was relatively simple for the experts who had undertaken the work, for the company had already developed some highly successful forms of vacuum tubes, and it was an easy matter for these technicians to assemble tubes with the necessary coils, condensers, and other apparatus of the transmitting and receiving elements and produce a system of such small compass that it could be carried on an airplane. But working this apparatus under ordinary conditions in the quiet laboratories and in a swift-moving and tremendously noisy airplane were two different propositions.

One of the first problems was to design a comfortable head set which would exclude all undesirable noises and admit only the telephone talk. A form of helmet was finally devised with telephone receivers inserted to fit the ears of the pilot or observer. Cushions and pads adjusted the receiver to the ears, and the helmet fitted close to the face so as to prevent as far as possible the transmission of undesirable sounds either through the ear passages or through the bony structure of the head, these bones acting as a sort of sounding board. The designers finally developed a helmet that solved this portion of the problem.

Not only was it necessary to exclude the roar of the engine and the rattle of the machine gun from the ears of the men receiving the radio communication, but it was also necessary to filter out these sounds from the telephone transmitter. Every person who has ever shouted into a telephone knows how sensitive the ordinary telephone transmitter is to extraneous noises. It requires no wide stretch of the imagination to hear in fancy how an ordinary transmitter would behave when beside the exhaust of a 400-horsepower Liberty engine. A brilliant line of experimentation conducted by one of the scientists at the laboratory resulted in a telephone transmitter or microphone which possessed the extraordinary quality of being insensitive to engine and wind noises and at the same time highly responsive to the tones of the human voice.

With the receiver and the transmitter perfected the scientists thought that the problem of airplane telephoning was solved; but nevertheless three months of hard work were required before the entire system could be adjusted and put in such shape that it might be considered a practical device for everyday use.

The question of weight was of utmost importance, and a structure that would adequately house and protect the delicate parts of the mechanism from the vibration and jars of flying and landing and at the same time not be too heavy for practical use on the plane was a difficult problem in mechanical design. Day after day the inventors took the mechanism up in flying machines and brought it back night after night for more work in the laboratory.

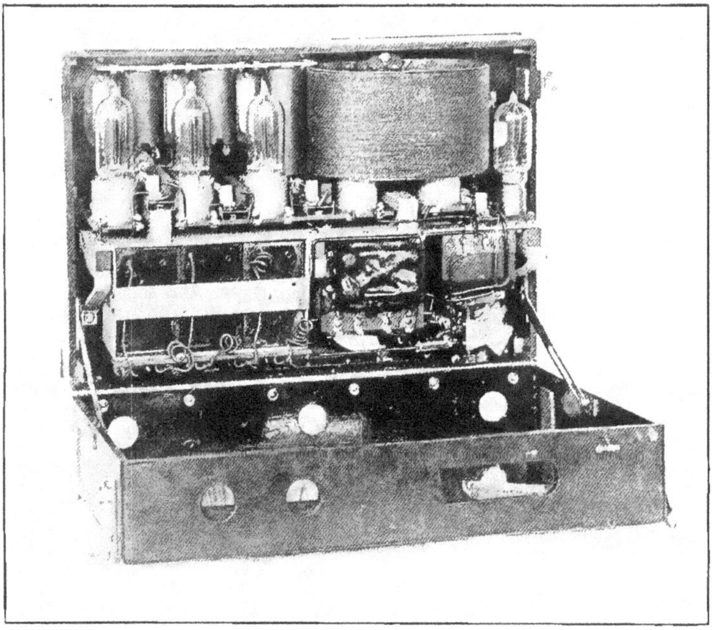

INTERIOR VIEW. AIRPLANE RADIO TELEPHONE SET BOX.

EXTERIOR VIEW OF SAME.

This was a period, however, of rapid progress. Officials appearing on Langley Field from time to time witnessed informal demonstrations of this development. In August Mr. Baker, Secretary of War, and Gen. Scott, Chief of Staff, listened to a conversation being carried on in the air, and some six weeks later Brig. Gen. Foulois witnessed a similar demonstration and from the ground directed the movements of the airplane in flight. The experimental apparatus had reached such a state of efficiency that on October 16, at Langley Field, communication by voice was carried on between airplanes in flight 25 miles apart and from airplane to ground over a distance of 45 miles. By September cables had been sent abroad telling of the progress made in this country on the development of this apparatus. Our officers abroad were skeptical and could not believe that this country could outdistance the scientists of the allies who had had three years of war experience to draw upon. By October the designers had brought the system to a perfection where they were willing to risk its use in actual war flying, and Col. Culver took to the American Expeditionary Forces in France several trunkloads of the apparatus to acquaint those abroad with what had been done and to test the apparatus under service conditions. Meanwhile the development work continued in this country. Early in December the operation of the apparatus was exhibited in an official test at the Morraine Flying Field at Dayton, Ohio.

A large number of military and civilian officials not only of our own country but of the allies had been invited to witness this test. It must be remembered that at this time even those who had heard about the progress being made were skeptical of the possibilities of the successful adaptation of the radio telephone to airplane work. The designers of aircraft never look with favor upon additional equipment which may clutter up the machine with trailing wires and the like and possibly compel alterations in standard lines. The pilots, also, do not usually give a friendly reception to new equipment for their planes.

The exhibitors at Dayton planned to have two planes in the air at once, so that the officials might listen in on their conversation at a ground station located on the top of a hill near the flying field. By hard work the inventors got their equipment installed, and just at dark on the evening before the day of the trial one machine equipped with wireless went up into the air and held successful communication with the ground.

The next morning when the official party arrived the members viewed the apparatus in the planes while the inventors explained what it was expected to do. The visitors were then conducted to the station on the hill, where those who were putting on the show had rigged up a megaphone attached to the wireless receiver so that everyone could hear without putting on a head set.

The attitude of some of the officials, particularly those from the foreign nations who had had experience in war flying, was skeptical, if not bored. The planes left the ground, and when the machines had gone up so high that they were but specks in the sky the receiver began emitting the premonitory noises that indicated that the men in the planes were getting ready to perform. Suddenly out of the horn of the loud-speaking receiver came the words: "Hello, ground station! This is plane No. 1 speaking. Do you get me all right?"

Looks of amazement came over the faces of all those who had never heard the wireless phone in operation before. Soon came the signal from plane No. 2, and then the demonstration was on. Under command from the ground the planes were maneuvered over much of that part of the country. They were sent on scouting expeditions and reported what they saw as they traveled through the air. Continuous conversation was carried on, and finally, upon command, the planes came back out of space and landed as directed.

From that moment there was nothing but enthusiasm in all quarters for the radiophone upon airplanes. It was no longer a question whether the device would work or was any good, but a question of how soon the company could start manufacture and in what quantities the device could be produced.

The demonstrations Col. Culver had been conducting in France began, too, to bear fruit. Both the British and the French had developed experimental apparatus by this time and this was examined and tested. Then cablegrams began to arrive from abroad requisitioning the American apparatus in large quantities—convincing evidence that it had greater promise than any other.

But still difficulties were ahead, for at this stage the wireless telephone consisted of a few experimental parts built by hand. It remained a heavy task to standardize the equipment and perfect the multitude of designs and drawings that must be in existence before quantity manufacture could begin. All sorts of mechanical details slighted in the experimenting and taken care of by makeshift devices had to be worked out as practical manufacturing undertakings. It was another case of day-and-night work to put the mechanism into condition for production. The factory of the Western Electric Co. is in Chicago but its drafting rooms and laboratories are in New York. As soon as any detail was finally worked out the drawings were taken by messengers and rushed to Chicago where the work of producing the manufacturing tools had begun. Only the fastest passenger trains between New York and Chicago were patronized in this part of the development.

As every detail was perfected it had to be checked by actual test in the field, so that the company's engineers were almost constantly in the air. One of these experts made 302 flights himself; and a total of 690 flights, of a combined duration of 484 hours, was required in the experimental stage of the mechanism.

Immediately after the official trial in December the Government ordered thousands of sets of the radio telephone. In spite of the enormous detail involved in making ready for production, the first systems were turned out early in 1918, well ahead of the delivery of the airplanes in which they were to be used.

All through this development the designers had to confine their activities within limits set by the producers of the aircraft. This in itself created some puzzling problems. For instance, a constant current of electricity must be supplied to heat the filaments of the vacuum tubes and to operate the transmitter. A simple way to provide this current would seem to be to connect a dynamo with the driving shaft of the airplane engine, but the airplane constructors would not allow any such connection with the engine. Current could be supplied from storage batteries, but the planes were already loaded down with all the gear they could carry, and the use of heavy batteries was out of the question. Therefore it was the task of the phone designers to supply a dynamo plant that would not add appreciably to the weight of the plane. This was done by installing on the outside of the plane a wind propeller, which was driven by the rushing air and had power enough to turn the dynamo.

The dynamo must deliver a constant and unvarying voltage to the radio phone, if its operation is to be possible, yet a wind propeller on the airplane would be driven by air rushing by at speeds varying from 90 or 100 to 160 miles per hour, the latter figure being the speed of a diving plane. This meant that the wind propeller, and hence the armature of the dynamo, would revolve at a speed varying from 4,000 to 14,000 revolutions per minute. It would seem to be impossible to procure current at a constant rate from a dynamo varying so widely in its speed of operation; yet one of the experts engaged in this enterprise solved the problem, and the dynamo thereafter performed always in a most steady going and dependable manner.

Incidentally as a sort of by-product of the undertaking the special transmitter and helmet may be employed as a means of communication between the pilot and observer in a two-seated machine. When the helmet is used for this purpose, the wireless is not employed at all, but the head sets are connected by wires so that notwithstanding the fact that one can not hear himself talk because of the noise on the plane the pilot and observer can converse over the telephone with ease. Then at any time by throwing a switch they can connect themselves with the radio apparatus and talk with the men in another plane 3 or 4 miles away or to their squadron headquarters on the ground.

One good result of the airplane telephone was to speed up the training of aviators in this country and to make that training safer. But the primary object of the wireless phone was to make it possible for the leader of an air squadron at the front to control the movements of his men in the air. For this purpose extra-long range was not required, and the distance over which the machines could talk was purposely limited to 2 or 3 miles so that the enemy could not overhear the conversation except when the planes were actually engaged in fighting each other.

The Navy made use of the wireless telephone sets in the seaplanes, and here the range of the equipment was made greater. The Navy also adopted a modified form of the set for the 110-foot submarine chasers. The subchasers hunted the submarines in packs, and by means of the radio telephone the commanders of the boats kept in constant touch with each other, thereby greatly increasing the effectiveness of their operations.

Altogether there were produced for the Army airplanes about 3,000 combined transmitting and receiving sets of the radio telephone and about 6,500 receiving sets alone.

CHAPTER VII.

BALLOONS.

When Stephen Montgolfier and his brother Joseph, in November, 1782, sent a sheep, a rooster, and a duck into the sky, lifted by a paper bag inflated with hot air, these Columbuses of ballooning could scarcely foresee the importance that their invention was to have in the great war 135 years later. To the humble observation balloon in France rather than to his dashing hero of a cousin, the airplane, must go the chief credit for that marvelous accuracy which long-range artillery attained during the great struggle.

The balloon itself was spectacular enough once its true character was known. The fact that the American production of observation balloons during our 19 months as a belligerent was a complete and unqualified success makes the story of ballooning in France of particular interest to the American reader.

After the animals of the Montgolfier barnyard had made their ascent, two friends of the brothers, M. Pilatre de Rozier and Girond de Villette, essayed to be the first human beings to take an aerial flight, ascending to a height of 300 feet and returning to earth sound of limb and body. Thereafter and until the great war in Europe the balloon remained the awe of the circus and country fair grounds and the delight of the handful of sportsmen who took up the adventurous pursuit; but, except for a limited use of captive balloons in our Civil War and in the siege of Paris, in 1870 and 1871, the balloon had no important military use.

The hot-air balloon never could have become of great value to armies. In the first place, it would descend when the balloon cooled off. This defect was overcome by the use of lighter-than-air gas. Moreover, the free balloon was subject to the whims of the breezes. To overcome this characteristic the balloon must be fastened by a cable or propelled by a portable engine. It was obvious, however, to military experts that a stationary observation post anchored thousands of feet in the air would be ideal in war operations; yet for all of this obvious need, until the great war military science had perfected nothing better than the spherical balloon. The spherical, anchored to a cable, bobbed aloft in the gales and zephyrs as a cork does on the ocean waves. Although there had been some experimentation with kite balloons before 1914, it was not until the great war had beeu in progress for some months that the principles of stream-line shape were applied to the captive balloon; and the kite balloon, the well-known "sausage," made its appearance, to be the target for enemy aerial operations and the chief dependence of its own Artillery.

The term "kite balloon" effectively describes the captive observation balloon as we knew it in the war. It rides the air on the end of its cable much in the manner of an ordinary kite, and some of the early "sausages" even flaunted steadying tails such as kites carry. These principles applied to the captive balloon gave to its observation basket a stability unknown by the pioneer aeronauts under their spherical bags.

In the first stages of the war the Artillery relied principally upon airplanes for firing directions. But, while the airplane observers could locate the targets fairly well, they frequently lost touch with their batteries because of the difficulty of sending and receiving wireless or visual signals upon their swiftly moving craft. This disadvantage brought the captive balloon into use, gradually at first, but before the end of the war on a scale which had practically displaced the airplane as a director of gun fire. The balloon came to be the very eye of the Artillery, which, thanks to the development of this apparatus, reciprocated with an efficiency beyond anything known before in the history of warfare.

Sitting comfortably aloft, the observer in the kite balloon basket had the whole panorama of his particular station spread before him. His powerful glasses could note accurately everything transpiring in a radius of 10 miles or more. He was constantly in touch with his batteries by telephone and not only could give by coordinated maps the exact location of the target and the effect of the bursting shell, but could and often did supply most valuable information of enemy troop movements, airplane attacks, and the like. He was a sentinel of the sky with the keen, long-range vision of the hawk. He played a part less spectacular than the scout airplane with its free and dazzling flights, but his duties were not less important.

Nor did he suffer from ennui during his period aloft. When a kite balloon went up it became the subject of alert attention by the enemy, because it was up there on hostile and damaging business. Long-range high-velocity guns turned their muzzles on it, and planes swooped down upon it from dizzy heights, seeking to pass through the barrier of shell from antiaircraft guns and get an incendiary bullet through the fabric of the gas bag, an eventuality which meant the ignition of the highly inflammable hydrogen gas, the quick destruction of the balloon and perhaps of the luckless occupants of the basket as well, unless they could get away in their parachutes.

Only quick work could save the men in the basket in such a case. From the time the gas leaped into flame until the explosion and fall of the balloon there was an interval of rarely over 15 or 20 seconds.

The pilot of the airplane could dodge and slip away from the guns, but not so the pilot of the kite balloon anchored to a windlass from 2 to 5 miles behind his own lines. He had to take what was coming to him without means of defense. He must carry on his scientific calculations unconcernedly and in his spare moments experience the questionable pleasure of watching on some distant hill the flash of an enemy gun trained upon him and then of waiting the 20 or 30 seconds for the whizzing messenger to reach him, the while he pondered on the accuracy of the enemy gunner's aim.

While the artillery on both sides paid considerable attention to the observation balloons, the fact was that few of them were brought down by direct shell hits. The diving airplane with its incendiary bullets was a far more deadly enemy to the ballon than the ground artillery. Certain pilots in all the air services made a specialty of hunting sausages, the nickname given to kite balloons because of their shape. In the 17 days between September 26 and November 11, 1918, our Army lost 21 balloons, of which 15 were destroyed by enemy planes and 6 by enemy shell. But it may be noted that our aviators and artillery exacted a toll of 50 German balloons in the same period and on the same front. Of 100 balloons lost at the front, an average of 65 were destroyed by enemy attacks and 35 by natural wear and tear.

The German general staff so strongly appreciated the work of the allied kite balloons that in its system of rating aviators it ranked a balloon brought down as the equal of one and one-half planes.

The average life of a kite balloon on an active sector of the western front was estimated to be about 15 days. Some of them lived only a few minutes. One American balloon passed unscathed through the whole period of American activity on a busy sector. While ordinarily five or six months of nonwar service will deteriorate the balloon fabric, there are many cases of useful service longer than this.

When the war broke out Germany is said to have had about 100 balloons of the kite type. France and England had few of them. The German balloon was known as the Dracham. Its gas cylinder of rubberized cotton cloth was approximately 65 feet long and 27 feet in diameter, the ends being rounded. To give it a kite-like stability in the air a lobe, which was a tube of rubberized fabric, of a diameter approximately one-third of the diameter of the main balloon, was attached to the underbody of the ggas bag as a sort of rudder, which curved up around the end of the balloon. This lobe was not filled with gas, but the forward end of it was open so that when the balloon rose the breeze filled the lobe with air. The inflated rudder then held the Drachen in line. The lobe automatically met the emergency. In calm, windless weather the balloon needed no steadying and the lobe was limp. Let the gale blow, and the lobe inflated and held the nose of the Drachen into the wind. As a further stabilizer three tailcups, with mouths open to the breeze, were attached 10 feet apart on a line descending from the rear of the balloon. In a strong wind these helped to keep the contrivance from swinging.

The tailcup was made of rubberized fabric, circular in shape, about 4 feet in diameter, and about 2 feet deep when inflated by the breeze. It looked like an inverted umbrella, and was attached to the tail end of the balloon for exactly the same purpose and with the same effect as the tail attached to a kite.

The Drachen type of balloon was still in the experimental stage here and in France and England when the Germans swept over Belgium. The Drachen balloon was clumsy and relatively unstable in high winds, yet its importance to the Artillery could not be ignored by the allies. The results of its work daily became more apparent. The first effort of the allies was to improve the Drachen to give it greater stability and permit it to go to higher altitudes. While this work was going on, Capt. Caquot, of the French Army, produced a kite balloon so superior that it quickly superseded what had been in use. Germany clung to the Drachen for a time, but finally abandoned it for the Caquot principles of design.

The earlier balloons of the sausage type had been merely cylinders with hemispherical ends. Now for the first time, in the Caquot model, appeared a captive that was sharply stream lined. Stream lines are lines so curved as to offer the least possible resistance to the medium through which a mobile object, such as a yacht, an automobile, or an airship, moves. The Caquot gas bag was 93 feet long, as compared with the Drachen's 65 feet of length, yet its largest diameter was only 28 feet, being but a foot thicker than the pioneer German type. The Caquot, as all balloons developed in the war, was made of rubberized cotton cloth. Its capacity of 37,500 cubic feet of hydrogen gas lifted the mooring cable, the basket, two observers, and the mass of necessary equipment, and in good weather the balloon could ascend to a maximum altitude of over 5,000 feet.

The principal innovation in the design of the Caquot balloon was the location of the balloonette or air chamber within the main body of the gas envelope. This chamber was in the forward instead of the rear part of the bag and along the bottom of the envelope. It was separated from the gas chamber by a diaphragm of rubberized cotton cloth, which was sewn, cemented, and taped to the inner envelope somewhat below the "equator" or median line from the nose to the tail of the gas bag.

When a balloon of the Caquot type is fully inflated, the diaphragm rests upon the underbody of the gas envelope, and there is no air in the balloonette. Then, as the balloon begins to ascend, at the higher levels the surrounding air pressure is reduced and the gas in

CAQUOT, TYPE R, CAPTIVE OBSERVATION BALLOON.

This balloon is 93 feet long and 28 feet in diameter. Its gross lifting power is 2,600 pounds.

BALLOON CONTROLLED BY WINDLASS ON A MOTOR TRUCK.

WINDLASS FOR CAQUOT BALLOON MADE BY JAMES CUNNINGHAM & SONS.

the balloon expands. This expansion would normally burst the envelope when the balloon is at a high altitude, except for a safety valve which pops at the danger point and relieves the pressure. Also, when the balloon is anchored it gradually loses gas, since no fabric can be made entirely gas-tight. A flabby balloon in a gale of wind is dangerous to the men in the basket. This flabbiness might be expected to increase, too, as the balloon was hauled down into the heavier air pressures.

It was to overcome this flabbiness that the interior balloonette was first invented, but the new location not only accomplished this end but increased the stability, lessened the tension on the cable and allowed an almost horizontal position of the balloon itself. As the balloon rises the wind blows into the balloonette through a simple scoop placed under the nose of the balloon. This forces up the diaphragm and compensates for any loss of gas from the envelope above. If the day is calm and no air is driven into the balloonette, there is no danger from a flabby balloon anyhow, and hence no need for the air chamber. The thing is automatic.

The Caquot was equipped with lobes of rubberized fabric to act as rudders. These lobes, which were spaced equidistantly around the circumference of the rear third of the balloon, filled with wind when wind was blowing and there was need of rudders. In calm weather the lobes, particularly the two upper ones, hung loosely, resembling elephant ears. On account of this characteristic the Caquots were nicknamed "elephants" by the soldiers.

The Caquot maintained its stability without tailcups, and its construction caused it to ride nearly horizontally and directly above its mooring, regardless of winds. In this position it put much less strain on the anchoring cable than the old-fashioned sausage. This balloon has been operated successfully in winds as high as 70 miles an hour, so that apparently no gale could keep it on the ground.

When we went into the war both our Army and Navy were practically without observation balloons, and we knew little about their construction, although we had been watching the developments in Europe. One local National Guard organization had taken to the Mexican border a locally designed captive balloon, the gift of the Goodyear Tire & Rubber Co., of Akron, Ohio.

In April, 1917, the total production capacity of the United States was for only two or three military observation balloons in a month. But when the emergency came the various concerns whose plants were adaptable to this class of manufacture—the list including the Goodyear and Goodrich organizations at Akron, the United States Rubber Co., the Firestone Tire & Rubber Co., the Connecticut Aircraft Co., and the Knabenshue Manufacturing Co.—all joined wholeheartedly with the Signal Corps to solve our balloon problems.

One of these problems was the production of balloon cloth, for which there had never been any commercial call in this country. Such cloth obviously must be of cotton, for in cotton we had our largest supply of textile raw material. The cloth must be closely woven, smooth, and strong, to serve as a base for the rubberizing process. The standard balloon cloth should have a weave of approximately 140 threads to the inch both ways. In our vast cotton industry only a few mills had ever made such a cloth, and then only in small quantities. In fact we found only a few looms in existence capable of weaving such cloth, which must be from 38 to 45 inches wide. A single loom could turn out only an average of ten yards of this cloth in a day. Our balloon program was to call for millions of yards of high-count cloth, and this meant the construction of thousands of new looms, as well as the training of hundreds of weavers.

Naturally our cotton manufacturers were reluctant to undertake such a production, and their fears were justified when we found that the earliest deliveries of balloon cloth were frequently as high as 67 per cent imperfect. By the middle of 1918, however, the mills had so perfected their methods that the wastage amounted to only 10 per cent of the cloth woven. This wastage was largely caused by "slubs," knots, and other imperfections which prevented an even surface for rubberizing. Because of the lives which depended upon having perfect balloon cloth, the fabric was literally inspected inch by inch, and hundreds of men and women had to be educated especially in this inspection work.

The development of the new art of weaving balloon cloth was an achievement of no mean degree. In April, 1917, all of our cotton mills put together could produce only enough cloth to build two balloons a week. In November, 1918, our looms were turning out cloth sufficient for 10 balloons a day, an expansion in the industry amounting to 3,000 per cent in 19 months. This expansion proceeded at a rate that always kept us a little ahead of the military schedule. To produce 10 balloons a day the cotton mills had to turn out 600,000 yards of special cloth a month. In addition to the small army of weavers, this production called into service 3,200 looms.

Had the war continued another year, we would have reached our goal of 15 complete new kite balloons produced every day. Our complete project of balloons and dirigibles of all types called for a total output of 20,000,000 yards of balloon cloth. Had we reached the quantity production planned, we would have been able to supply not only our own needs but also all of the balloon needs of the allies in Europe. America had the raw materials necessary for the whole anti-German balloon program.

CUTTING AND CEMENTING BALLOON PANELS IN THE GOODRICH PLANT AT AKRON.

SPREADER ROOM AT THE U. S. RUBBER CO. FACTORY, SHOWING MACHINES THAT RUBBERIZE THE CLOTH.

FINAL BALLOON ASSEMBLY ROOM AT GOODRICH FACTORY.

As it was, we supplied to France and England a considerable number of balloons when the materials shortage in those countries was becoming acute. The foreign users of this American made equipment reported that it was equal to the best European product. It should have been. No war material was ever manufactured more conscientiously than this. In addition to the painstaking care of the producers, from start to finish a large force of inspectors watched every step in the construction of each balloon, and when America sent a balloon to the front it was right for the work it had to perform.

The weaving of the cloth was but the first step in the production of the balloon fabric. The fabric of the balloon envelope resembles a sandwich in its construction, there being a thin film of specially compounded rubber between two plies of the cotton cloth. The outer ply of the cloth is cut on the bias. This method prevents any long straight tear down the grain of the fabric. The threads of the inner ply are set at an angle of 45° to those of the outer ply, thus distributing strain sufficiently to stop a "snag" practically where it starts.

The cotton cloth alone can not resist the seepage of gas, and, therefore, it is necessary to rubberize it, the rubber film being really the gas-resisting envelope. In this rubberizing process the cloth must be run through the spreading machine 30 to 35 times in order to build up the thin rubber film without a flaw in it of any kind. The outside ply of the balloon fabric is "spread," that is, painted with a rubber compound containing a coloring matter. This compound makes the fabric waterproof; it gives also protective coloring to the balloon when in the air, making it less visible to the enemy; and, finally and most important, this coloring absorbs the actinic rays of the sun which are so fatal to the life of rubber. In some of the fabric the rubber film itself was colored to withstand both the heat and ultra-violet rays, thus both protecting the rubber and reflecting the heat which would otherwise expand the gas in the balloon.

While in general we adopted the European standards of construction, we had to develop our own rubber compounds and cures as well as our various fabrication processes. The latest reports we received from the front stated that the American fabric not only was successful, but that it had an added characteristic which was a direct means of saving life. It was discovered that the American fabric burned more slowly than the European balloon fabric, giving the men in the observation basket more time to get away in the parachutes when the balloons were destroyed by hostile attack.

When we went into the war we had never built a windlass for a kite balloon. The ability of the American manufacturer solved this problem as it did almost every other problem in the development of war instruments. Steam was the motive power first used for windlasses, but before the fighting came to an end America had developed both gas and electric windlasses which were thoroughly efficient.

The best known type of gasoline windlass was that having two motors, one to turn the cable drum controlling the balloon's ascent and descent, and one for moving the windlass itself along the road. A record pull-down speed of 1,600 feet a minute, or more than three times the speed of the fastest passenger elevator, has been attained by the gasoline windlass.

The electric windlass, while pulling down the balloon at the slower rate of 1,200 feet per minute, was smoother in operation. The mobile windlass would move on a road under its own power at 20 miles an hour and could tow the balloon in the air at the rate of 5 miles an hour, or even faster if necessity demanded.

To play on the safe side at the start, we adopted a satisfactory windlass that had been developed in France. It was difficult to manufacture this entirely French machine with American materials and methods; yet James Cunningham, Sons & Co., of Rochester, N. Y., succeeded in obtaining a delivery of four complete windlasses per week.

In addition to this windlass we designed two of our own. One of these was the product of the United States Army Balloon School and was manufactured by the McKeen Motor Car Co., of Omaha, Nebr.; the other windlass was designed and manufactured by the N. C. L. Engineering Co., of Providence, R. I. Both were put into quantity production, assuring us a sufficient number of the best windlasses ever manufactured.

The first cable used to hold the balloon captive was approximately a quarter-inch in diameter, weighed 1 pound for each 8 feet of length had a breaking strength of 6,900 pounds, and was made of seven twisted strands of plow-steel wire, containing in all 133 separate wires. This cable, while it accomplished the original purpose, was early seen to have fine possibilities of development. The observers in the basket must be kept in constant communication with the Artillery and their own windlass and this communication could best and most efficiently be obtained by means of the telephone. The balloon telephone, as first used, was an entirely individual unit with its own separate cable from the basket to the ground. In this way communication was indeed established, but only at the cost of a decrease in possible altitude, increased cable resistance, and the necessity of an extra windlass for winding and unwinding the telephone cable.

Previous to the entrance of the United States in the war, preliminary experiments in France were being made with the view of putting the telephone wires in the center of the main cable, thus doing away entirely with the second cable and windlass. But there

BALLOONISTS READY TO ASCEND.

The picture shows balloonist with telephone equipment, also a parachute on side of basket.

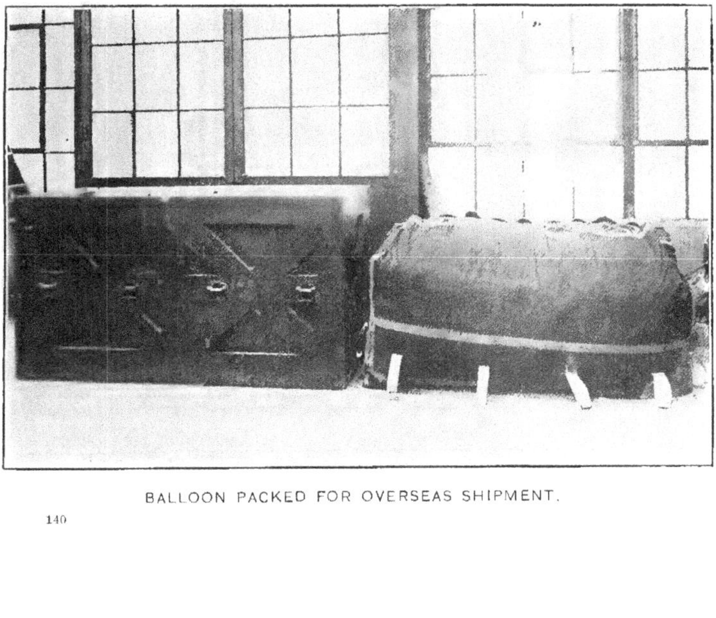

BALLOON PACKED FOR OVERSEAS SHIPMENT.

BALLOON IN GOODRICH FACTORY INFLATED TO BE SUBJECTED TO AIR TEST.

NURSE BALLOON CONTAINING 5,000 CUBIC FEET OF GAS.

It is used in the field to replenish kite balloons with hydrogen.

had never been developed a satisfactory cable of this construction. American inventiveness at the John A. Roebling Sons Co. and the American Steel & Wire Co. was set to work on this problem with the result that not only was a satisfactory cable developed but a steady production was attained, 50,000 feet per week being delivered regularly by the John A. Roebling Sons Co. alone. This new cable consisted of 114 separate wires of special steel besides the telephone center of 3 copper wires properly insulated and armored. The specifications demanded a breaking strength of 7,200 pounds while the actual test of the finished Roebling cable showed 8,250 pounds.

Another of the balloon problems was the supply of hydrogen gas. Before the war only a little hydrogen was used in this country, the element being a by-product in the manufacture of commercial oxygen. We met the additional demand for millions of cubic feet of hydrogen for our balloons by establishing Government gas plants and expanding privately owned plants already in existence. There were two methods of supplying hydrogen to our balloon units at home and abroad. One of these was by furnishing portable plants which would generate hydrogen at the place where it was to be used. The other was to take the hydrogen from the stationary plants, condense it by pressure in steel cylinders, and ship it to points of demand. By far the greater part of the gas which we used not only in this country but in France was produced at the permanent supply stations and shipped in cylinders. Each cylinder held about 191 cubic feet of gas under a pressure of 2,000 pounds per square inch at 68° F. temperature. When the war ended we had placed orders for 172,800 of these cylinders, of which 89,225 had been delivered and were in use. We developed a manifold filler which would take the gas from 12 to 24 cylinders at the same time and quickly inflate a kite balloon, a speed of 23 minutes for a complete inflation having been reported from one training camp.

In the production of portable hydrogen generators we had to produce not only the machine but the chemicals required in the process. We adopted the ferrosilicon and caustic soda process by which it was possible to produce 10,000 cubic feet of hydrogen per hour in a field generator. There was plenty of caustic soda to be had, but high grade ferrosilicon, a production of large electrolytic furnaces, was scarce, because of its heavy consumption in the steel industry. We procured, however, 2,482 tons of it for our generators, of which over 2,360 tons were supplied by the Electro-Metallurgical Sales Corporation alone.

An interesting feature of the gas supply in the field was the use of "nurse balloons." The nurse balloon was simply a large rubberized-fabric bag with a capacity of 5,000 cubic feet of gas. It was used for storage of gas, and the observation balloons were fed from it. We have not received the exact figures of the quantity of gas used by the entire Balloon Service; but, as one item alone, private manufacturers previous to the signing of the armistice produced and delivered 17,634,353 cubic feet of hydrogen and were in position to meet practically any demand for the gas. This figure is only a small part of the total, since it does not include the hydrogen produced in the permanent Government stations or by the field generators.

Hydrogen itself, while the lightest of cheap gases, and therefore the one universally used in balloons, has the grave fault of being dangerous to the balloonist. When mixed with the air it is highly explosive, if touched off by a spark of fire or electricity. For years balloonists have dreamed of a gas light enough to have great lifting power, but which would not burn nor explode. There was such a gas known to chemistry, and this was helium, discovered first in spectroscope examinations of the corona of the sun, but later found by chemists to exist rather freely in the atmospheric envelope of the earth. Although one of every 100 parts of air is pure helium, it was not until comparatively recent years that this light nonexplosive gas was discovered in our atmosphere.

Now helium was rare and expensive, and until the United States entered the war no one had considered its production as a commercial possibility. Up to two years ago the total world production of helium since its discovery had not been more than 100 cubic feet in all, and the gas cost about $1,700 per cubic foot.

It had been discovered that certain natural gases issuing from the ground in the United States contained limited quantities of helium. The question was whether we could extract this helium in sufficient quantities to make its use practical. The Signal Corps, the Navy, and the Bureau of Mines combined in a cooperative plan to develop a practical helium production. By adopting a method of obtaining the helium from liquefied gas produced in the processes of the Linde Air Products Co. and the Air Reduction Co., and also by the Norton process, we attained astonishing success in this enterprise.

On the day the armistice was signed we had at the docks ready for loading on board ships 147,000 cubic feet of helium. At its prewar value this gas would have been worth about $250,000,000. On November 11, 1918, we were building plants which would produce helium at the rate of 50,000 cubic feet per day, and the cost of obtaining it had dropped from $1,700 per cubic foot to approximately 10 cents.

None of this gas was actually used in the war, but its production by our chemists was hailed as the greatest step ever taken in the development of ballooning. It now seems to have opened a new era in lighter-than-air navigation. In war helium will nullify the incendiary bullet which destroyed so many balloons and airships. In peace it brings the possibilities of new types of construction of dirigible airships, since its use eliminates entirely all of the frightful dangers from lightning, static electricity, or sparks and flames from gasoline engines or any other source.

The Army and Navy cooperated in the production of balloons. The Army furnished the balloon cloth to the Navy. Navy balloons had two automatic safety valves for the expanding gas, one on each side of the balloon a third of the way back from the nose and just above the equator; while the Army held to the French and British idea of a single valve in the nose itself. The Navy adopted a Caquot-type balloon which rides at an angle of about 25° to the horizontal and is somewhat smaller than the Army model. The Navy used these balloons as spotters for submarines and mines. They were towed on cables from the decks of war ships, and were connected with the ships by telephone.

The use of parachutes with balloons is a comparatively recent development, the man who first successfully descended to earth in a parachute being not only still active and enthusiastic over aerial development, but being in fact the chief inspector of all United States Army balloons and parachutes. This is Maj. Thomas S. Baldwin, known the world over as Capt. "Tom" Baldwin, hero of innumerable aerial exploits of all kinds under all conditions and in all parts of the world, and at present chief of the United States Army balloon inspection. The Yankee balloon observer in France went up to his observation post in the security of knowing that the equipment on which his life depended had been O. K.'d by men who knew the business from beginning to end.

The parachute as it is known at the county fair and the parachute used in the recent war were far apart in type, the latter embodying all the improvements that the world's aeronautical experts could add to it. The need for parachutes developed when hostile aviators began shooting down the sausages. At first the one-man parachute was used exclusively, the men in the basket leaping overboard the instant their balloon was fired over their heads. Any delay on their part would be fatal, since the entire bag would be consumed in 15 or 20 seconds and the observer would then be unable to leap out of the falling basket. When the individual parachutes were used, the maps and records in the balloon basket were usually lost.

To overcome these difficulties, the designers invented the basket parachute. This was considerably larger than the individual parachute, and to operate it the balloonists pulled a cord which cut the basket away from the balloon entirely. The spreading parachute overhead then floated the basket, with the men themselves and all else it contained, safely and quickly to the ground.

Although hundreds and even thousands of parachute jumps occurred during the war, there were few fatalities from this cause.

During all the time our forces were at the front only one of our men was killed as the direct result of a parachute drop. In that particular instance the burning balloon fell on top of the open parachute, setting it on fire and allowing the observer to fall unprotected the rest of the distance to the ground. One of our observers was known to make four jumps from his balloon on the same busy day, and another leaped thrice in four hours. In the Argonne offensive 30 balloon jumps were made by our men.

As to the safety of our parachute equipment, the only complaint from the Yankee balloonists at the front was that they were too safe. The man who is escaping from a German airplane nose-diving at him with a machine gun spitting fire is in a hurry and does not wish to be detained by a parachute which floats him too slowly to the earth.

In the rigging of each kite balloon there are about 2,000 feet of rope of different sorts. There was a shortage of proper cordage in the United States at first, and the French thought they could furnish this rigging to us. But this attempt proved to be unsuccessful, and we were forced to develop a cordage manufacture in this country of high quality and great quantity. We did this so swiftly that there was no serious delay to the balloon production.

Up to November 11, 1918, we produced over 1,000 balloons of all kinds, 642 of these being of the final Caquot type which we adopted. This production included many propaganda balloons for carrying printed matter over the lines into the enemy's country. We supplied several target balloons for gun practice on our aviation fields. We developed new types of parachutes and built acres of canvas hangars for balloons. We produced 1,221,582 feet of steel mooring cable. These are only the major items in the balloon enterprise, and do not include hundreds of others of less importance.

The balloon production was one of the most important and successful of all our war projects. Although we had a limited knowledge of the subject in the beginning, our balloons stood the hard test of actual service and could bear comparison in every way with the best balloons of Europe, where the art of balloon building had been in existence for many years. Once our production actually started, we never had any shortage of balloons for our own Army; and soon we would have been in a position to produce the observation balloons for all of the armies fighting Germany, if called upon to do so.

Balloon production figures.

BALLOONS AND PARACHUTES.

Firm.	Observation balloons produced to—		Supply balloons produced to—		Miscellaneous balloons produced to—		Parachutes produced to—	
	Nov. 11, 1918.	Mar. 1, 1919.	Nov. 11, 1918.	Mar. 1, 1919.	Nov. 11, 1918.	Mar. 1, 1919.	Nov. 11, 1918.	Mar. 1, 1919.
Goodyear Tire & Rubber Co., Akron, Ohio	458	565	39	39	[1] 3 [2] 5 [3] 53	[1] 3 [2] 7 [3] 53	20	20
B. F. Goodrich Rubber Co., Akron, Ohio	169	279	10	10	[1] 3 [2] 1	[1] 3 [2] 1	15	15
Connecticut Aircraft Corporation, New Haven, Conn	22	37	72	72	[2] 1	[3] 3		
United States Rubber Co., East Cambridge, Mass	23	37						
Knabenshue Manufacturing Co., East Northport, Long Island	3	25	1	1	[2] 1	[2] 1		
French-American Balloon Co., St. Louis, Mo	1	1			[2] 2	[2] 2		
Scott-Omaba Tent & Awning Co., Omaha, Nebr					[3] 161	[3] 161	53	228
New York Tent & Awning Co., New York							63	
Follcmer-Clogg & Co., Lancaster, Pa							104	131
Bickford Bros., Rochester, N. Y							1	1
Firestone Tire & Rubber Co., Akron, Ohio			7	7				
Columbia Mills (Inc.), Wilkesbarre, Pa					[3] 1	[3] 1		
Total manufactured	676	944	129	129	[1] 6 [2] 10 [3] 215	[1] 6 [2] 12 [3] 215	256	458
Total shipped to ports	481		6				4	
Total shipped to camps, etc		463		123		[1] 6 [2] 12 [3] 215		448

[1] Target. [2] Spherical. [3] Propaganda.

WINDLASSES.

Firm.	Produced to—	
	Nov. 11, 1918.	Mar. 1, 1919.
McKeen Motor Car Co., Omaha, Nebr	23	35
Chris D. Schramm & Son, Philadelphia, Pa	20	20
Jas. Cunningham & Sons Co., Rochester, N. Y	5	35
N. C. L. Engineering Corporation, Providence, R. I	1	37
Deloies Hoisting Co., New York	1	1
Total manufactured	50	128
Total shipped to camps and depots		124
Total shipped to ports		

CABLE.

	Feet.	*Feet.*
John A. Roeblings Sons Co., Trenton, N. J	476,700	860,700
American Steel & Wire Co., Worcester, Mass	744,882	
Total manufactured	1,221,582	1,605,582
Total shipped to ports	486,000	
Total shipped to camps and depots		1,119,582

Balloon production figures—Continued.

HYDROGEN GAS.

Firm	Produced to—	
	Nov. 11, 1918.	Mar. 1, 1919.
	Cubic feet.	*Cubic feet.*
Oxygen Gas Co., Kansas City, Mo	2,715,845	3,568,600
Southern Oxygen Co., South Washington, Va	945,412	1,608,778
International Oxygen Co., New York	2,166,668	2,166,668
Walker Refining Co., Austin, Tex	496,000	595,783
Tarrifville Oxygen & Chemical Co., Tarrifville, Con	1,072,590	1,072,590
Louisiana Oxygen Co., New Orleans	477,500	904,576
Kentucky Oxygen Co., Louisville, Ky	276,779	540,752
Burdett Oxygen Co., Chicago	9,532,559	11,347,999
Total produced	17,634,353	21,805,746
Total shipped overseas	7,319,578	....
Total shipped to camps	....	14,456,168

HYDROGEN CYLINDERS.

National Tube Co., McKeesport, Pa	35,800	43,300
Harrisburg Pipe & Pipe Bending Co., Harrisburg, Pa	38,425	65,500
Tindel-Morris Co., North Eddystone, Pa	15,000	15,000
Total manufactured	89,225	125,800
Total to ports	30,000	....
Total to camps and warehouses	....	95,800

BOOK III.

THE ENGINEER CORPS.

CHAPTER I.

THE ENGINEERS IN FRANCE.

In describing the activities of the Engineers, we are carried to the front itself, into the zone beaten by enemy fire, where machine-gun bullet, bursting shell, and deadly gases have brought sudden death and painful wounds to many members of the technical services. A large proportion of the Engineers are combatant troops, constituting in the American Expeditionary Forces about 8 per cent of the total combatant troops engaged. These troops, trained and equipped to march and fight as Infantry, demonstrated their fighting qualities during the war on numerous occasions, both when used as Infantry to increase the rifle strength of that arm and when fighting as Engineers to obtain possession of terrain as a preliminary to the exercise of their technical art in its organization.

From the day the first sector was taken over by American troops in November, 1917, until the Meuse River was passed and the enemy, in flight, sought an armistice to save his armies from destruction, the combatant Engineers—the "sappeurs" of French soldier lore and song—fought and bled in a manner never to be forgotten. Railroad engineers, nominally considered noncombatant, at Cambrai dropped their tools to take arms and stand stubbornly shoulder to shoulder with their British brothers with whom they were learning to work under the special conditions of the front. From Cantigny to Chateau Thierry, Engineer troops fought as well as worked, and often not only advanced with the Infantry under or through the barrage, but actually led the first wave, to demolish or remove the obstacles placed in its path. Through the days when from March 21, 1918, until July 18, 1918, the German army made its rapid plunges toward Paris until checked and thrown back across the Marne at Chateau Thierry, the sapper troops fought and worked with the Infantry of their divisions, enduring the same dangers, privations, and hardships, and winning equal honors and commendation.

In the drive at St. Mihiel and through the Argonne, the combatant Engineers played a conspicuous part. Advancing with the tanks, they made possible the passage of many difficult points for these lumbering monsters, against which was directed a particularly destructive fire. Using elongated torpedoes of high explosive, known as Bangalore torpedoes, they prepared passages for the Infantry through the broad barbed-wire entanglements, echeloned in depth by numerous separate lines, each to be breached and passed before the objective could be gained. In this work the Engineers reduced the machine-gun nests that hindered their operation, cleaned up the strong points that delayed the advance of the tanks they were assisting, and threw extemporized footbridges across the streams which barred the further advance of the Infantry.

The combatant Engineers did their part in the winning of the reconquered ground as well as the lion's share of its organization for the defense and the maintenance of the communications behind it. In this last respect alone, the Engineers, as combatant troops, opened across No Man's Land the first communications practicable for the light field artillery, which pressed forward immediately behind the Infantry troops to their support and protection.

Filling in trenches, removing wire entanglements, building trestles across wide mine craters, searching for and rendering inoperative treacherous mines and traps of extreme ingenuity and destructiveness, the sapper found a wide field for the exercise of his functions. Shattered and obliterated by four years of shelling and mining, trenching, and countermining, the "roads" across No Man's Land existed only on the map; and as they retreated the Germans demolished and obstructed the highways behind the old front from which they had been driven, with the thoroughness and attention to detail for which they are noted. As our Infantry advanced, upon their heels, literally speaking, came our Engineers, to attack the problem of providing for the Artillery and supply trains a means of following. From the standpoint of the road builder in civil life, their methods were crude in the extreme, but for the military purpose and the pressing immediate needs, their road building achievement was adequate. The Engineers sometimes reopened abandoned quarries, and sometimes started them where none had existed before, to obtain a supply of road metal, which supply was sometimes supplemented and in some cases replaced by the use of débris from ruined villages and shattered farmhouses. From demolished structures many useful materials were extracted and adapted to the military purpose by the Engineers. Where bridge and trestle timbers were lacking, deserted buildings—in one case the tower of a ruined church—filled the need. Where shell hole or crater yawned a remnant of a stable wall might be pulled down by ropes and man power, and broken up to fill the void.

Through the dense woods the soft forest floor offered no support even to the light artillery, and miles of corduroy and brush path were built to permit the guns to advance to the reinforcement of the attack. In many places the tactical situation admitted of insufficient time to build even the crudest paths, and then the Engineers fell to and assisted artillery and supply wagons to get through and over the bad spots, replacing guns on the road where they had run into the ditch, righting and reloading combat wagons when they had turned over in shell holes or deep ruts.

While thus engaged the sapper troops were subjected to the fire of enemy artillery seeking to prevent the advance of the supporting guns, and, further, they were working within the zone of combat enemy aviators, the rattle of whose deadly machine guns, as they plunged at low altitude toward a busy working party, was as much to be dreaded as the high-explosive bombs which they dropped.

Behind the combatant Engineer troops, extending through the service of supply to the base ports and across the ocean to the United States was an organization of technical noncombatant supply and administration.

The work of these production, construction, and supply departments of the Engineer service in France was organized in the American Expeditionary Forces under the administration of three divisions of the office of the Chief Engineer. These were the division of military engineering and engineering supplies, the division of construction and forestry, and the division of light railways and roads.

ENGINEERING SUPPLIES.

The division of military engineering and engineering supplies was charged with the procurement, standardization, and distribution of all classes of supplies used by Engineer troops. During the 19 months of warfare this division handled 3,225,121 tons of supplies, storing them and distributing them from immense depots aggregating 25 acres of covered storage and over 756 acres of open storage. This service was further charged with the current investigations into new developments of the art of military engineering, and with the development, operation, and administration of certain technical branches of the American Expeditionary Forces, such as electrical and mechanical troops, water-supply troops, searchlight regiments, etc.

At its seven storage depots in the base, intermediate, and advance sections, this division had in service 23 locomotive cranes, mostly of 15 tons capacity and capable of handling an enormous amount of freight and material at warehouses and cars. The following table of principal items of engineer material shows the kinds and quantities of supplies which were received in France for issue through this division up to December 15, 1918:

Table of engineer supplies received in France.

FROM THE UNITED STATES

Item.	Tonnage.	Item.	Tonnage.
General machinery	45,454	Liquids	7,067
Iron and steel products	242,226	Explosives and accessories	952
Hardware and hand tools	26,780	Unit accountability	7
Railway rolling stock	343,888	Engineer supplies	52,106
Railway motive power	144,066	Miscellaneous office supplies	2,239
Lumber	39,086	Floating equipment and accessories	10,093
Track materials and fastenings	488,793	Material and tools for locomotive and car repair and erection shops	10,407
Automotive transportation, etc	22,127		
Horse-drawn transportation	7,967		
Building materials and supplies	98,671	Total United States	1,541,929

FROM EUROPEAN SOURCES.

Item.	Tonnage.	Item.	Tonnage.
Tools and equipment	462,027	General engineer supplies	581,149
Machinery	13,146	Cement from American Expeditionary Forces mills	54,860
Office supplies and equipment	1,781	Barracks	263,590
Auto and truck supplies	464	Bunks	9,892
Track and ties	115,438	Latrines	2,143
Locomotives and cars	8,649	Miscellaneous	49,200
Water supply, machinery	6,210		
Water supply, supplies	48,416		
Electric service, machinery	2,342	Total European sources	1,818,182
Electric service, supplies	4,083		
Construction materials	188,830	Total United States and European sources	3,360,111
Boats and barges	5,940		
Motorcycles and bicycles	12		

To facilitate the procurement of supplies in the existing world markets, this division established in Paris a purchasing board, having branches in England, Switzerland, and Spain. When the war ended this board had accomplished the tremendous task of buying over 1,800,000 tons of engineer supplies, with a total value of $205,242,728. In addition to this material, our own country furnished over 1,500,000 tons, with a value of $248,993,322. France sold to us through this board 1,234,968 tons, valued at $134,393,870, and England 396,000 tons, valued at $56,145,818. In Switzerland, purchases consisting principally of sectional barracks and technical equipment, totaled 96,867 tons, with a value of $14,643,410. Purchases from Spain amounted to only 797 tons, with a value of $59,630.

Much work was done in standardizing supplies of all classes, so that quantity-production methods could be used in their fabrication, thus promoting economy and stimulating the rapidity of supply.

In the procurement of cement for the use of the American Expeditionary Forces, the Engineers dealt successfully with a problem of large magnitude and importance. By contract with English and French mills, by direct purchase for specific jobs from local mills, and by their own manufacturing operations, the Engineers secured enough cement to supply the demands for construction both at the front and in the S. O. S., as the service of supply was generally known. Three large cement mills were leased from the French

ENGINEERS, ASSISTED BY INFANTRY, BUILDING A ROAD OVER WHAT WAS NO MAN'S LAND ONE WEEK BEFORE.

The stones from tottering walls of buildings are broken into small pieces and laid on the road bed, making a good military road. Fay on Haye, France.

WAREHOUSES AND DOCKS FROM THE RIVER. BASSENS DOCKS, BORDEAUX, FRANCE.

NORTHERN VIEW OF THE AMERICAN DOCKS AT OLD BASSENS, BORDEAUX, FRANCE.

owners and operated by special troops organized in the United States. To certain other French mills the Engineers furnished labor and materials in return for a certain proportion of their output. It is estimated that about 215,000 tons of cement were thus procured, representing a total cost of about $7,000,000.

The Engineers operated shops at various points near the front in which were manufactured standard material for dugout, trench, and emplacement construction, such as concrete beams, concrete slabs for overhead protection against high-angle shell fire, trench frames, revetment material, trench duck boards, mine and gallery timbers, knock-down bunk sets, etc.

CONSTRUCTION AND FORESTRY.

The division of construction and forestry was charged with all construction work in the service of supply, and also with the procuring of forest products for the American Expeditionary Forces. At the signing of the armistice its organization totaled 150,823 men, of whom about 127,000 were constantly engaged in production work. Using standardized building plans, this force performed a huge amount of construction work in France.

It was assumed that one-third of the American troops in France would have to be housed in new buildings erected specially for the purpose. Thus accommodations for about 750,000 men had to be built at the rate of 16 barracks, each 20 by 100 feet in size, for every 1,000 men. Contracts were let to British and French contractors for 23,000 demountable barracks, this order being based on the ultimate probable size of the Expeditionary Forces. During August, September, and October, 1918, these barracks were being received at the rate of 1,000 per month. To supplement a supply of even such magnitude, our own type of barrack was developed to be built with lumber furnished by the American forestry forces in France. One cantonment project involved the construction of 500 barracks, accommodating 55,000 men. A total of 11,862 barracks were erected for the American Expeditionary Forces in France, representing 225 miles of length, if all the barracks were placed singly end to end.

It was the policy of the American Expeditionary Forces to provide hospital room sufficient to give beds, if necessary, to 15 of every 100 American soldiers in France. On this basis the Engineers set out to provide hospitals with a total of 280,000 beds. Of these, 139,000 beds were in hospitals taken over from the French, 25,000 beds being added to this capacity by new construction. In entirely new base, camp, evacuation, and convalescent hospitals, 116,000 beds were ultimately made available for the casualties of the American Expeditionary Forces, requiring the erection of 7,700 hospital barracks of special type, all of which would have totaled 127 miles in length if placed singly end to end. As to the progress of this construction, on November 14, 1918, there were 190,356 beds occupied in American hospitals in France, but all 280,000 beds originally specified were ready and available.

The base hospital plants were complete municipalities in themselves, and had capacities varying from 1,000 to 6,000 beds. These units were built where nothing had existed before but little French rural communities, devoid of the improvements and modern conveniences with which we in this country are so familiar. To establish a modern military hospital, capable of caring for the varied casualties and illness arising from action and abnormal living conditions, it required the construction of roads, sidings, unloading platforms, sorting and classification buildings, operating rooms, surgical and medical wards, dormitories, morgues, cemeteries, complete water supplies, fire protection systems, sewage and garbage disposal plants, recreation buildings, electric light plants, and all that goes to make complete a modern installation for the care of the wounded and sick. Many of the camp and evacuation hospitals required construction of the same character, but differing in magnitude.

The Engineers developed port facilities at St. Nazaire, Bordeaux, La Pallice, Marseilles, Brest, and at less important harbors. In general, at these places the existing facilities were expanded to meet the needs for the debarkation of troops and the unloading and shipment of supplies. Originally 23 ship berths were placed at our disposal by the French. The Engineers expanded this equipment to a total of 89 berths, with authorized projects for 160 berths by June, 1919. Our overseas shipments grew from 20,000 tons in July, 1917, to 1,000,000 tons in October, 1918, but the port expansion kept abreast of this development. Fifty-eight 300-ton lighters were built by Engineer troops with French timber, and twenty-six 500-ton lighters with American timber. The Engineers constructed seven derrick barges with lifting capacities ranging from 30 to 100 tons.

The existing French railroads running from the base ports to the advanced zone were quite inadequate, so that it was necessary to supplement their facilities with many miles of new track and other construction, including important storage, classification, arrival, and departure yards, warehouse tracks, engine terminals, water points, and repair shops. At Bassens, St. Sulpice, Miramas, and Montoir, enormous storage depots were constructed to handle the supplies entering France for our forces. The American-built railroad yards at these points were comparable in magnitude and completeness to the important yard developments undertaken in this country in recent years by the large railroad systems, the yards at St. Sulpice having a trackage totaling 147 miles of single track. Those at Bassens and St. Sulpice were virtually completed during the war, while the con-

INTERIOR OF SUBSISTENCE WAREHOUSE AT NEVERS, FRANCE, MARCH, 1918, SHOWING FRENCH WOMEN TRUCKING RATIONS.

INTERIOR VIEW OF ONE OF THE LARGE WAREHOUSES AT BASSENS DOCKS, BORDEAUX, FRANCE.

Picture taken in April, 1918.

ICE PLANT, BUILT FOR THE AMERICAN EXPEDITIONARY FORCES AT GIEVRES, FRANCE. IT IS THE THIRD LARGEST PLANT IN THE WORLD.

INTERIOR OF FREEZING ROOM OF ICE PLANT AT GIEVRES.

struction at Miramas was well under way at the signing of the armistice. At St. Sulpice the project was designed on the basis of receiving, storing, and forwarding the supplies for 1,000,000 men for 30 days. The others were of like magnitude.

At Nevers, in the intermediate section, a condition existed requiring the construction of six miles of new double-track line, with a bridge over the Loire River 2,190 feet long. This piece of construction is known as the "Nevers Cut Off." It relieved the railroad congestion at this important point.

At Is-sur-Tille, in the advance section, was built a regulating station at which train loads of supplies and troops were dispatched to points where needed. Still farther toward the front, at Liffol-le-Grand, was another and smaller regulating station, controlling troop movements and the distribution of munitions and subsistence. Both of these projects were entirely new and were in useful operation when the war terminated.

In addition to the above projects, many storage yards, hospital tracks, ordnance depot yards, aviation center tracks, and construction tracks were laid out and built. In all 937 miles of single track were laid, thus fulfilling in the equivalent the prediction that to supply an American Army at the front we should have to build a double-track railroad from the French coast to the trenches.

Storage depots, remount depots, and veterinary hospitals erected by the Engineers proved entirely adequate for the needs of the American Expeditionary Forces at all times. A grand total of 535 acres of covered storage was built or acquired, of which about 482 acres was new construction. Space was provided in remount depots for 29,000 animals, and it was projected to accommodate 48,700 animals had it been necessary. Veterinary hospital space was provided for 17,250 sick animals. Each veterinary hospital required much special construction, such as concrete dipping tanks for the treatment of mange, operating rooms, exercising paddocks, hay sheds, living quarters for attendants and veterinary surgeons, and administration buildings.

At Gievres, in connection with the important storage depot built there, was constructed the third largest refrigerating plant in the world. This plant, built by the Engineers from plans prepared by experts, was capable of caring for 5,200 tons of meat at once, and of producing 250 tons ofice per day. Another similar plant at Bassens had a capacity for 4,000 tons of meat.

Miscellaneous construction work in France covered many fields of activity. The question of adequate water supply was ever present, and in most places where hospitals, depots, shops, or warehousing plants were built, a water supply development was incidentally necessary. Many systems were installed complete from the collection of the water at its source to its distribution to the points of consumption, while in some other cases only extensions and ameliorations of existing systems were undertaken. Water supply in the service of supply was placed under as fine and complete a system of bacteriological inspection and examination as is customary under more normal conditions. At Tours, Vierzon, St. Nazaire, and Dijon, where unfavorable bacteriological conditions existed, arrangements were entered into with these municipalities whereby the existing water supplies were chlorinated by the American water supply service.

At Is-sur-Tille was built a mechanical bakery at which 500,000 pounds of bread, fresh for immediate shipment to the troops at the front, could be produced in one day. Another such plant was built and put into service at Neufchateau, and at Liffol-le-Grand it was proposed, and plans had been prepared, to construct a third plant for 400,000 pounds of bread per day, but this project was canceled just after the armistice. In addition to these plants, bakery capacity for 240,000 pounds per day was provided at the base ports.

Oil storage was provided for 175,000 barrels of oil and gasoline. The large plants, with tanks having a capacity of 25,000 barrels each, built with enduring concrete foundations and equipped with connections and pumping plant for the loading of tank cars destined for the front, rivaled in size the installations at large refineries of this country.

For the operation of these many plants numerous power developments were undertaken, and a total of 5,000 kilowatts of new power, being provided for at the time of the armistice, was canceled. Plants of the capacity of 750 kilowatts each, providing 3,500 kilowatts of electric power in all, were in operation when the armistice was signed, not to mention numerous smaller units installed at various points where needed.

Ordnance repair shops were erected, as were also assembling plants for ordnance material, and heavy gun-mounting plants. Repair shops of enormous extent were established near the front, equipped with machine-tool equipment for the repair and maintenance of tank and motor transport material. Schools for the line and staff were constructed, the first and largest being at Gondrecourt and Longres. Laundry plants, salvage depots, aviation assembly plants, sewage disposal plants, refuse incinerators, mechanical repair shops, locomotive assembly plants and locomotive round-houses were placed at convenient points. At Chalmdray and at Colombey-les-Belles, both within a short day's automobile ride of the front, were the tank and air-service repair depots, each one covering many acres of ground and each provided with full equipment for any job of manufacture or repair in their respective fields.

AMERICAN ENGINEERS QUARRYING STONE TO REPAIR MILITARY ROADS DESTROYED BY GERMAN SHELL NEAR MENIL-LA-TOUR, FRANCE.

SAWMILL AT LARGEST LUMBER CAMP IN FRANCE. TWENTIETH ENGINEERS NEAR ECLARON, FRANCE.

HAULING LOGS BY MOTOR TRUCK, CAMP KELLOG, BORDEAUX, FRANCE.

STORAGE DAM AT SAVENAY, FRANCE, AND THE LAND THAT WILL BE INUNDATED WHEN IT IS COMPLETED.

The forestry work of the American Expeditionary Forces was developed to meet the heavy demands of our armies for forest products of all kinds. The first move in this direction was the dispatch to France of the Tenth Engineers, a foresty regiment of two battalions. This was in September, 1917. By the spring of 1918 we had recruited and trained the Twentieth Engineers, a forestry regiment of 10 battalions. Later additional forestry troops were sent across. Shortly before hostilities ceased all these troops were consolidated into a single regiment of 13,000 men, known as the Twentieth Engineers. To this force were added negro service troops to the number of 9,000, making 22,000 men engaged exclusively in the work of cutting down French forests and turning them into lumber required by our forces.

At first we had difficulty in supplying the necessary machinery. Until the sawmills came the forestry troops were engaged in building camps and hewing out railroad ties. In January, 1918, the machine equipment began to arrive. In February our troops cut about 3,500,000 feet of lumber; while in October the cut for the single month had reached the enormous figure of 50,000,000 feet. When the war ended we were expanding our forestry operations in France to produce 1,000,000,000 feet of lumber in a year.

The lumber produced by our sawmills in France up to November 30, 1918, would build completely enough barrack buildings 20 feet wide to stretch out to a distance of 600 miles if placed end to end, quarters enough for 3,107,600 men. In addition to this output the railroad ties produced would build 1,091 miles of standard-gauge railway and the small ties for the 24-inch track would build a double-track railroad behind 185 miles of trenches.

Just the posts and poles produced, if all cut into 6-foot posts, would be sufficient to support a wire fence, with posts one rod apart, reaching one-third of the distance around the earth. The piling, if stood end to end, would make a flagpole 362 miles high. The cord-wood produced would make a rack 1 yard wide, 1 yard high, and 600 miles long.

The sawmill machinery installed to accomplish such a production comprised 30 mills of 20,000 feet per day capacity, 56 mills of 10,000 feet per day capacity, and 92 smaller mills capable of producing ties and rough timber.

In the base and intermediate sections a large amount of work was necessary in the maintenance of the existing roads and highways, and in the construction of new roads in the vicinity of the various new projects. Experienced road engineers, drawn from civil life and commissioned as officers of the Army, were put in charge of this work, and specialist engineer troops and labor battalions were assigned to them. Quarrying the rock, grading the road, surfacing it, and maintaining it in good condition thereafter—all these duties fell within the province of the engineers.

LIGHT RAILWAYS AND ROADS.

The light railway and road regiments of engineers attached to the armies at the front, while their duties did not carry them so far or so much into the zone of enemy fire, may be considered as combatant units, since they operated with and in support of combatant troops in the field. To the light railway regiments were assigned the construction, operation, and maintenance of the light railroads of 60-centimeter gauge (about 24-inch gauge). A great quantity of such trackage was used during the war. These narrow-gauge railroads, capable of being operated under extreme conditions of grade and curvature, and powered with light steam and gasoline locomotives, were essential to the proper supply of a stabilized sector. They were the lines of communication between the railheads of the broad-gauge system and the dumps and depots within the front sectors. At the very front, sometimes within a few hundred meters of the German lines, these light railroads were operated by hand or animal traction, while further back the gasoline locomotive, less conspicuous than the steam engine, came well within range of the enemy's light field pieces. In periods of activity and during an advance these railroads did a tremendous service, not only in transporting troops, munitions, materials, and subsistence stores, but in affording a means of bringing up rapidly a certain class of railway artillery adapted for use upon 60-centimeter gauge trucks. Built of light rail and steel ties assembled in portable sections, this track was easily destroyed by shell fire, and such was often its fate, yet it was but short work for the engineers to replace broken sections with new material, a work frequently done under heavy fire. Engineer troops suffered many casualties in this service.

In cooperation with the Engineer Department in the United States, a practical, efficient, and standard type of narrow-gauge motive power and rolling stock was developed by American manufacturers. This material was shipped to France knocked down, and was assembled and set upon the rails at Gondrecourt, where a plant for this purpose had been established. Up to November 30, 1918, there had been built and placed in operation 538 miles of 60-centimeter track, with 347 steam and gasoline locomotives furnishing motive power for the operation of 3,281 cars of different types.

The road-building regiments in the zone of the armies built and maintained the roads immediately behind the front. Equipped with modern road-building machinery and motor trucks, these regiments maintained the roads in shape to handle the abnormally dense and heavy traffic incidental to operations at the front. The Army road troops were recruited from among men accustomed in civil life to road building, quarrying, and construction operations. They usually worked well within sound of the enemy guns, and frequently

TYPE-PRINTING, BASE PRINTING PLANT, 29TH ENGINEERS, A. E. F.

LITHOGRAPHIC PRESSROOM, BASE PRINTING PLANT, 29TH ENGINEERS, A. E.

BINDING DEPARTMENT OF THE PRINTING PLANT OF THE SOCIÉTÉ PUBLICATIONS PERIODIQUES.

The new pay books were manufactured at this plant and were turned out at the rate of 100,000 a day, using 36 tons of print paper, 16 tons of parchment cover, 15 tons of paper for envelopes, 6 tons of paste, 10,000 rolls of moleskin. Paris, France.

under their direct fire. During the advances made from the stabilized line of June, 1918, these regiments improved and perfected the hasty roads thrown across No Man's Land by the sapper regiments of the fighting divisions, so that transport of supplies and troops could be maintained to the advancing armies. To furnish materials for this construction many quarries were opened or taken over from the French road service. A total of 42,000 cubic meters of rock was quarried and prepared for use in quarries operated exclusively by American engineers, while in quarries jointly operated with French forces 75,000 cubic meters were produced.

MAP MAKING AND PRINTING.

A vitally important part of the work of Engineer troops was the making and reproduction of the many maps required for the conduct of tactical and strategic operations by the American Expeditionary Forces. A highly specialized regiment was organized to conduct the topographic surveying operations, map reproduction, and printing work in France. Many of the officers of this regiment had been formerly connected with the American Coast and Geodetic Survey and the Geological Survey, and they were well qualified for the work of war-map making. At Chateau Thierry a portion of this organization rapidly mapped to a large scale the new region in which the theater of operations suddenly found itself, thus supplementing the excellent small-scale map which was in existence for the whole of France, but which was not sufficiently precise for the conduct of our artillery fire. This work was done under pressure, but it contributed its share to the later American successes in that locality. These troops also were charged with furnishing to the Artillery the mathematical azimuths and coordinates, on the basis of which artillery indirect fire was executed.

The maps in use even on stabilized fronts were in a constant process of revision and change. The data and information on which these changes and revisions were based were constantly pouring in from the photographic branch of the air service, from the intelligence service, the Artillery, and from the sapper regiments at the front. Consequently new maps had to be prepared continually and furnished to all the organizations and officers concerned with their use. Then, too, an Army as large as ours required an impressive amount of field printing in order to distribute its orders and information.

As soon as our forces reached France it was apparent that the French map-production plant could not take care of our needs. The Chief of Engineers in the United States thereupon ordered the purchase of equipment for a base printing plant large enough to take care of all the map printing for an army of 1,000,000 men.

The special machinery ordered in the United States for this plant did not arrive in France during 1917, and so the American Expeditionary Forces purchased abroad five large rotary lithographic presses, several type presses, and a number of linotype machines and other printing equipment.

The base printing plant was established at Langres, France. In the spring of 1918 the American equipment arrived, and thereafter the base printing plant was able to print not only the current maps required but also the base maps which the French had been supplying. In addition, during the heavy fighting in July and August, 1918, our printing plant supplied to the Seventh and Eighth French Armies the base maps of their fronts.

The demands for maps and printing steadily increased until the base printing plant grew to have a working force of 35 officers and 750 men. From July 15 until September 15, 1918, the plant worked continuously 24 hours a day to turn out the work required. By this time the shop had 10 rotary lithographic presses, 4 linotype machines, and several job presses, printing each month an average of over 1,200,000 lithographic impressions and 500,000 sheets of printed matter. In November the plant turned out 1,900,000 lithographic impressions and over 1,000,000 sheets of type work.

To supplement the base printing plant we had at each army headquarters an advanced printing shop to supply maps when they were needed within a few hours. At the base printing plant we had a department for making relief maps, which work had been done for us previously by the French Government.

The equipment for military map making was enriched during this period by an invention of Maj. James N. Bagley, United States Engineers, called the aerial cartograph, or map camera. The Bagley camera's three lenses at the height of 5,000 feet could photograph a strip of territory 3½ miles wide.

MILITARY BRIDGES.

The science of building military bridges is an old one. When war with Germany was declared the United States had developed its heavy ponton equipment, which was standard in design and yet which had changed but little since the Civil War. As soon as we formally took the step to send troops against Germany the Engineers ordered great quantities of this equipment and by the latter part of 1917 had plenty of it ready to go overseas. Our deliveries to France, however, were hindered by the shortage in ocean tonnage, particularly after we had begun to use every available ship for the transport of men.

Meanwhile the efforts of the Engineers were being directed to the development of standard ponton equipment strong enough to carry

TRUCKS LOADED WITH PILING EN ROUTE TO FRENCH RAILWAY YARDS NEAR BRUYERS, FRANCE.

PILE DRIVING FOR FOOTBRIDGE BY ENGINEERS IN FRANCE.

FOOTBRIDGE BUILT BY ENGINEERS ACROSS CANAL DE L'EST NEAR THE VILLAGE OF BRIEULLES, FRANCE.

This bridge was constructed under heavy fire from enemy guns.

tanks and the ponderous artillery of the present day. The old ponton bridge was first strengthened to carry loads of 5 tons on each of two axles spaced 10 feet or more apart. The standard prewar equipment would support only 3 tons similarly spaced.

The next step was to develop a bridge that would hold up axle loads of 10 tons with a distance of 12 feet or more between axles, although in actual use this bridge showed itself capable of supporting a load of 15 tons concentrated on one axle. As soon as these developments were made, the plans were mailed to the American Expeditionary Forces, so that the Engineer Corps abroad could provide the beams and metal parts at its own mills and shops in France. When the fighting ceased, the Engineers were designing a raft capable of transporting the heaviest portable ordnance then under manufacture in the United States.

In 1917 the Engineer Department made designs for a standard sectional steel bridge, consisting of short latticed steel truss sections capable of being assembled to form trusses varying by increments of 11 feet up to a maximum span of about 90 feet. Two of these trusses with the span mentioned were capable of supporting a load of 30 tons, and they could be erected in a matter of hours over abutments prepared in advance or extemporized from the ruins of a demolished structure. These bridges had been manufactured in quantity in this country and were ready for shipment when the armistice was signed.

In the Argonne push Army bridge troops repaired and replaced the bridges destroyed by the retreating enemy as fast as material and labor could be provided at the points needed. For this work much heavy timber was utilized, and, in general, trestle structures were erected as best meeting the conditions of relatively soft crossings and soft river bottoms.

The fighting in the French terrain with its numerous narrow but deep streams and canals indicated to us the desirability of a portable floating footbridge. Such a bridge was designed and produced by the Engineers in France. Many of the crossings of the Meuse River and near-by canals under machine-gun and artillery fire from the high hills on the eastern side were made possible by the use of these bridges.

CAMOUFLAGE.

While camouflage has existed in nature since the beginning of time, its application to warfare on a grand, scientific scale was almost solely a development of the great war. Camouflage, due to the great developments of aerial observation and aerophotography, as well as of air bombing and indirect artillery fire, became a vital necessity for every branch of the service, far back in the rear as well as at the front. Any matériel or personnel the position of which was observed was at the mercy of the enemy, but, further, such observation might betray strategic plans. The need for camouflage became universal.

Camouflage organization was carefully developed for both the field and the factory, while one of the most important duties, that of instruction, was carried out in Army and corps schools and artillery camps, where thousands of officers and men were taught both the necessity and the methods of camouflage.

Our undertakings in this direction were based largely upon the methods developed by the French and British. In one respect camouflage was a matter of quantity production. This was in the manufacture of material used for concealing guns, roads, and other strategic locations which fall under the eyes of enemy observers on the ground or in the air.

In this work the British did nothing without the most careful scientific investigation, including the aerial photography of all their materials used, while the French relied more on their innate artistic sense of color and form. The camouflage material produced in quantity by the British consisted principally of burlap cut in strips about 1 inch wide and 12 inches long, colored in the desired hues with oil-emulsion paint. For artillery cover this was knotted in fish nets and chicken wire. The French for this purpose used raffia, a common product of Madagascar, whose natives use it largely for making their fantastic garments. The raffia was dyed and then knotted on nets and wire in the French camouflage factories.

After a careful study we adopted the British system and used burlap. Our Engineers made this decision because of the impossibility of finding permanent dyes for raffia and because raffia is more inflammable than burlap and scarcer and higher in price.

The first demand for camouflage material which we received embraced coverings for guns, sniper's suits, dummy heads, silhouettes, and some airplane hangar covers. In order to supply this material at the outset the engineering forces abroad leased a factory building in Paris and turned out a sufficient quantity with a working force of 30 enlisted soldiers and 100 French women.

But, as the American troops at the front increased in number, the demands for camouflage material became rapidly heavier. Battery positions of some types required about 4,000 square yards of camouflage cover. Aviation hangar covers were demanded in large numbers, and each one was a special order due to the varying conditions of terrain encountered. It became evident that we needed a vast increase in our camouflage-factory space.

In January, 1918, the Engineers secured about 20 acres of ground in Dijon, Haute Marne, a city on the main supply line north through the regulating station at Is-sur-Tille. They started to erect buildings immediately, and within 20 days this plant began turning out material. By November the Dijon factory numbered about 40 buildings, including blacksmith and machine shop, a sewing shop, a paint shop, laboratory, and a toy shop where dummies and silhouettes were made. The factory turned out artillery cover at the rate of 50,000 square yards per day.

The total output of camouflage cover for all purposes required about 3,000,000 square yards of burlap per month. When the fighting stopped the American Expeditionary Forces were using camouflage materials to the value of $1,500,000 monthly.

By new methods of manufacture, we succeeded in reducing the weight of fish-net covers. We designed two important field devices, one being an improved frame and set for mobile artillery protection, this equipment being later adopted by the British, and the other an umbrella machine-gun cover having special advantages. The central camouflage works of the American Expeditionary Forces at Dijon was declared by unprejudiced observers to be the best equipped and most efficient of any on the western front.

When the Dijon camouflage plant was projected it was expected that the American forces would require great quantities of camouflaged observation posts, silhouettes, dummy heads, snipers' suits, and other concealing devices. It was for this production that the toy shop at Dijon was erected, this shop being a kind of studio for the painters and sculptors connected with the Fortieth Engineers, which was the camouflage regiment. These various devices for deceiving the enemy, however, were used principally in the stagnant action of the trench warfare deadlock. By the time American forces came into the war in large numbers the struggle had become one of movement in which the trenches were left far behind. Also, American troops found themselves largely in sectors which were well wooded and therefore provided plenty of secure observation. The result was that there was never a great use on the part of American troops of these clever and interesting exploits in camouflage with which the public is familiar.

One of the best observation posts was the imitation of a tree trunk made of armor plate and set up in advanced positions during the night. Both the British and the French made considerable use of these. The British tree consisted of an oval shell of manganese steel. This was covered with tin, crimped in imitation of bark, and further camouflaged with paint and plaster and natural bark.

When it was desired to set up such a post a camouflage artist would surreptitiously make a faithful sketch of the tree trunk to be duplicated in armor plate. This sketch was then taken back to the workshop, where the spurious tree was built in exact duplication. The metal tree was built to rest on a base with hinges holding it down on one side. During the night two saps, or trenches, would be dug to the natural tree selected. Workers in one of these trenches would fell the branchless stub and carry it back out of the way. The armor plate tree would be drawn up in the other trench. The base would be set in place, and then the whole tree was raised on its hinges by means of ropes until it was upright and as life-like as artistry could make it. Inside the tree was a flight of iron steps leading to a seat in the upper part of the trunk. At this seat were peepholes and a stand for the phone which was connected up with the exchange at the adjoining trench. The sap, covered over, served as the gallery leading back to the trench.

The American camouflage force built only a single one of these trees, using it as a training device. Such objects were useless in an advancing movement, since, under such circumstances, they would play an important part only a short while and would then be left far in the rear.

The Dijon factory, however, turned out a number of small observation posts for use at the edges of shell holes. These were known to our troops as beehives and to the English soldiers as domes. Each one was built of light metal and covered with chicken wire and plaster. It was camouflaged with paint and bits of grass to simulate the appearance of the surrounding terrain, often being studded with tin cans or old shoes to make it appear to be an accumulation of rubbish. The favorite way of making the peephole for a beehive was to cover with gauze a hole cut in the bottom of an old shoe, which was then fastened to the observation post.

Another device built by the Dijon factory was the trench periscope. This was built and set up to look like an ordinary stick, thrown down casually upon the ground. For periscopes, too, we also used imitation stakes placed naturally in the barbed-wire entanglements. The British on occasions used imitation trench telephone poles to mask their periscopes.

The Dijon shop turned out large numbers of silhouettes and dummies. They were drawn from life by artists at Dijon and then cut out from ordinary wall board. Soldiers of the Fortieth Regiment posed as models for these silhouettes. All sorts of postures were employed, but nearly all of them represented soldiers in the act of climbing out of a trench or running, gun in hand, towards the enemy. The uniforms were painted in neutral shades, but the faces and hands were highly colored to be visible at considerable distances during the gray and mist of dawn, when silhouettes were usually employed.

The object of these dummy heads and silhouettes was to draw the fire of the enemy so as to make him reveal his strength and positions. The usual method of use was to place a number of silbouttes, possibly several dozen of them, in shell holes out in front of the trenches. The silhouettes were mounted so that they could be made to stand erect instantly whenever the ropes were pulled from the trenches. At the appointed moment the ropes would all be pulled at once, and the appearance to the enemy would be that of a raiding party starting out at top speed.

The British troops called this operation the Chinese attack. The Germans made no extensive employment of it. The silhouettes nearly always fooled the enemy, as indeed they would deceive anybody in such light and under such circumstances. The British were often amused to read in the German communiques that these Chinese attacks were regarded by the enemy as the real thing. More than one such "repulse" of silhouettes has gone down into the German records as a local success. On one occasion the Germans took a Chinese attack so seriously that they concentrated troops against it with the result that the British were able to gain considerable ground at the points weakened on both sides of the pseudo attack.

The Dijon factory made a thousand or so silhouettes, as well as a large number of dummy heads, these latter devices to draw the fire of snipers. These simulacra, however, had their principal use at the training schools in France, since they were peculiarly adapted to trench warfare, and by the time the American forces reached the front in strength the war of movement was in progress.

Several thousand sniper suits were turned out at Dijon. These suits were made of burlap, resembling in appearance the teddy-bear pajamas which little children wear. They were colored to match the terrain, either painted to resemble rocks or fitted with a grasslike covering. An adjunct to this suit was the cloth cover for the sniper's rifle.

Sniper suits were so deceptive that they would protect a man from observation even at short distances, and if exceptional care were used in the making of one, a man could conceal himself so effectually that an observer might step on him before seeing him. An American camouflage officer upon his return from France brought a sniper's suit with him and found a novel but practical use for it when he was invited to go duck hunting with a party of sportsmen. The other hunters stayed in their blinds, but the officer in the sniper suit went out in the open and shot more ducks than all the other gunners together were able to bring down.

The Dijon camouflage factory also turned out a large number of covers for the so-called Bessenaux hangars for airplanes. These hangars were large tents set up at aviation fields near the front. It was soon found to be impracticable to attempt to camouflage such tents by day, as they gave plenty of indication of their position in spite of the best efforts at concealment on the American and allied sides. However, the great danger at aviation fields came at night when the German bombing planes were abroad. Even on a dark night a white tent proved to be a good mark for the hostile airmen. Consequently the attempt was made to camouflage Bessenaux hangars at night only. It was found impracticable to paint the tents themselves, since the waterproof canvas would not take the paint readily. The solution was a large cover of burlap. This was painted in broken patches of color, much as artillery was painted in camouflage. At the factory such covers were spread out on the ground and painted with floor brushes dipped into water color.

All machinery at Dijon, with the exception of two lathes, two drill presses, and a shearing machine, was designed and built at the plant itself. The work of providing camouflage cover required enormous quantities of burlap to be cut up into strips. The English camouflage shops used stationary knives and a machine operated by a crank. American Engineers at Dijon designed a power-driven cutting machine with a large number of circular, whirling knife blades. The invention of this machine increases the production of burlap strips 900 per cent with the same force. The engineers at the plant also designed paint tanks and special machinery that would mix 4,000 gallons of color in a day.

There were about 1,000 French women employed in this plant. The executives paid great attention to their welfare. A special nursery, the "Creche," was built for their children. American Red Cross nurses cared for the babies during the time their mothers were at work. Many of the women employed were refugees driven from once comfortable homes. Their children fattened up with the good food provided by the army mess, and the mothers were correspondingly happy. Entertainments were frequently provided for the operators of the factories. The artists at the shop worked during their leisure moments and eventually produced the scenery and equipment for a genuine Yankee circus, animals and all, the menagerie, however, being made principally of papier mâché with human operatives inside the beasts. The first performance of the circus was given on Thanksgiving Day, 1918, and the audience was so delighted that it demanded a repetition. After three encores of this sort it was suggested that performances be given in Dijon, a city of upward of 50,000 population, with admittance charged. This advice was followed, and the circus made such a hit that the Engineers were able to turn over to the French orphan fund a considerable sum of money.

CONCLUSION.

The foregoing account gives in a broad way an idea of the scope of activities and the achievements of the Engineers during the 19 months of actual warfare in France. To furnish the organization of technical troops and specialists which made all this possible, the original Engineer Arm of the United States Army was increased to 131.5 times its prewar strength, and the proportion of Engineer troops relative to the total forces was increased from 1.6 per cent to 10.8 per cent. To accomplish this, a heavy demand was made upon the technical professions and upon the industries of this country. In filling this demand most necessary assistance was given by the engineer societies and the engineering journals, whose patriotic work demands the highest praise.

In situations requiring special knowledge almost always there could be found some specialist capable of adapting himself and his work to the military needs. Engineer officers for the combatant regiments were younger members of the technical professions, who were sent to the training camps provided for the purpose and there given the essentials of strictly military knowledge. This training was later supplemented by courses in Engineer and line schools located in France. The training officers of the regiments were supplied from the Corps of Engineers, these men having both the military and technical knowledge fitting them for the command. The diversity of education and experience necessary in all branches of the Engineer service may be understood by a consideration of the duties of the different units sent to France during the war—specialist units, in addition to the strictly combatant divisional regiments, who also numbered among their commissioned and enlisted personnel many technical specialists of high attainment.

We had, for instanoe, seven railway construction regiments, two railway construction battalions, one regiment and five battalions for railway maintenance of way, two battalions for maintenance of railway equipment, four regiments and one battalion to operate our main military railways in France, three regiments to operate the light railways in France and their repair shops, two regiments for operating the regular railway shops, two regiments and six battalions for constructing buildings and other general construction work, two regiments for storing and transporting Engineer supplies, a forestry regiment, a light railway construction regiment, a regiment for building roads, a water supply regiment, a mining regiment, a quarrying regiment, a technical regiment for handling surveying, sound ranging, and location of enemy positions by means of special apparatus, three survey and printing battalions, two railway transportation battalions, an electrical and mechanical regiment, several companies to operate cranes, a camouflage service, five inland waterway companies, five ponton trains, a ponton park, a railway transportation and stores battalion, and a searchlight regiment.

Utilizing and applying the new knowledge and scientific achievements of recent years, drawing upon the fund of experience acquired by the Regular Army in its theoretical studies and past wars, making available the vast amount of technical skill which has assisted this Nation to its present commercial and industrial status, the Engineers of the United States Army worked and fought, planned, and accomplished in France a work which in magnitude exceeds any similar undertaking recorded in American history. From base port to first waves of an assault upon the enemy's position, Engineer troops have been constantly in action first to last and have "carried on" always with the high ideals of the profession and with the motto of the Corps of Engineers, "*Essayons,*" before them.

CHAPTER II.

MILITARY RAILWAYS.

In establishing contact between our great bases of supply on the French coast and interior points, as well as with the fighters in the various fields of operations, the Department of Military Railways of the Engineer Corps found it necessary to provide thousands of miles of railway track ranging from the standard gauge down to the narrow 60-centimeter type built right up to the border of No Man's Land, to construct and ship across seas thousands of almost every kind of freight cars, to build hundreds of locomotives and transport them to Europe, to provide in addition fabricated track that could be laid under heavy shell fire, and hospital trains that could care for our wounded.

It was on July 10, 1917, that Gen. Pershing cabled stating that the French had asked for 300 locomotives and 2,000 kilometers of track, in addition to numerous items of accessories that go with an order of this size. Delivery of the locomotives was requested by October 15, 1917, and of the track by December 31, 1917.

It was ascertained that the American Locomotive Works had built consolidation engines for France of an entirely satisfactory type, and that similar locomotives for the use of British forces on French soil had been turned out by the Baldwin Locomotive Works. After the decision to adopt the consolidation type of locomotive, which is generally used in freight service in the United States, arrangements were made at once with these two concerns to build 150 locomotives each.

The consolidation locomotive weighs 166,400 pounds, and is about the heaviest that can be used in France. It has one pair of engine truck wheels and four pairs of drivers. The engine is just as large as it is possible to use within the French tunnel and platform clearances. The type sent to France was, however, not nearly so large nor so heavy as the general run of freight engines used here.

The order for 150 engines was placed with the Baldwin concern on July 19, 1917, and the first locomotive of this order was ready for shipment on August 10, 1917, just 20 working days elapsing between the date of the placing of the order and the day when the first engine was completed and all set up ready for shipment.

This is believed to have established a new record for locomotive construction in the United States and probably in the world for an engine of this size. All the other locomotives in this order were delivered promptly—36 of the Baldwin engines being freighted from the factory in August, 71 in September, and the final 43 in October, Of the locomotives ordered from the American Locomotive Works. 133 were freighted in October and the remaining 17 in November.

On account of differences in the details of construction the original price fixed for the locomotives turned out by the American Locomotive Works was $51,000 each and for those of the Baldwin Works $46,000 apiece. Advance payments on these engines reduced the price by $1,000 each.

Changes in the painting and other small details resulted in a saving of $60 additional on each locomotive built by the Baldwin Works and $400 on each engine turned out by the American Works, so that the net cost of each Baldwin locomotive was $44,940, and of each American locomotive $49,600.

After much consideration, and after this initial order had been disposed of, it was determined that the Baldwin type of engine should be made the standard, and all subsequent orders for engines went to the Baldwin Works.

As orders were placed from time to time with the Baldwin people, reductions were made in price, so that the last engines of the total of 3,340 ordered from this concern were obtained for $37,000 each. Orders for 1,500 of these engines eventually were canceled without cost to the United States Government. The saving effected by the reduction in price on the engines ordered, using the original price as a basis of comparison, was $22,989,385.

There were shipped in all to the American Expeditionary Forces 1,303 locomotives, of which 908 had been put into service by November 11, 1918.

During the severe winter weather of 1917–18 and the simultaneous shortage of motive power on American railways, 142 of these consolidation engines built for the American Expeditionary Forces were turned over to the American railways to help out a critical situation in this country. It was possible to use these engines here by making changes in the couplers and some other slight additions to meet the requirements of our safety appliance laws.

At the time these engines were turned over to the Railway Administration we were producing locomotives for France much more rapidly than it was possible to provide tonnage to transport them overseas. These locomotives were in service helping out the transportation facilities in this country an average of 6 months and 28 days each before being recalled for shipment to France. They earned profits for the Government while in service for the Railroad Administration at the rate of 32.3 per cent a year.

STANDARD GAUGE 10-WHEEL CONSOLIDATION (BALDWIN) LOCOMOTIVE. CYLINDER, 21 INCHES X 28 INCHES.

Driving wheels, 56 inches; wheel base, engine, 23 feet 8 inches; wheel base, engine and tender, 57 feet 4½ inches; weight in working order, engine 166,400 pounds, tender 112,000 pounds; tractive power 35,600 pounds; capacity, water 5,400 gallons, fuel 9 tons.

RATION TRAIN NEAR MENIL-LA-TOUR, FRANCE.

It might also be noted that the Director General of Military Railways was appointed custodian of undelivered locomotives ordered by the Russian government from the Baldwin and American Works. In January, 1918, a total of 200 of these Russian locomotives was purchased, and the engines were converted to meet American requirements by a change in the gauge from 5 feet to 4 feet, 8½ inches, and a change in the coupling system to meet our standard. The price of these was $55,000 each. The Baldwin Works turned its 100 over to the Railroad Administration between February 3 and May 20, 1918, and the American Works made its deliveries between February 19 and May 30, 1918.

The combined cost of these locomotives was $11,000,000 and their total rental revenue from the railroads was $2,585,475 up to December 31, 1918, or 23.5 per cent of the cost price, or at the rental rate of 29.8 per cent per annum.

Orders for 90,103 freight cars to be used by the American Expeditionary Forces were also placed with American contractors. Of these the orders for nearly half—40,915 cars, in exact figures—had been placed just before the armistice, and these contracts were canceled at slight cost to the Government. Up to the end of the year 1918 a total of 18,313 freight cars had been shipped overseas, nearly all of these cars being of the 60,000-pound size. Close bargaining in the purchase of these cars resulted in a saving of $15,737,633 under the prices originally quoted.

For the first time in history American locomotives were shipped across the Atlantic stacked in ships on their own wheels. In our normal foreign trade, and even in the early locomotive shipments to the American Expeditionary Forces, both engines and cars had been disassembled at the seaboard and their parts put up in packages for convenient and economical loading on ships. Each of the first locomotives sent to France was crated in 19 packages, while the parts for an ordinary box car were put up in 26 packages.

On October 29, 1917, however, Gen. Atterbury called attention to the fact that the English were shipping locomotives across the Channel on their own wheels and stated that it would result in very great economies of time, money, and man power if such an arrangement could be made for shipments from the United States. Manufacturers of the locomotives, however, advised against this. So did our own embarkation people and the Shipping Control Committee. Efforts were unsuccessful to get car ferries from the Key West and Habana line and from Quebec for the transport of locomotives on their own wheels over the ocean.

Finally, however, after numerous efforts to get ships with large hatches the ore steamer *Feltore* was loaded with 33 locomotives on their own wheels, packed in baled hay. This steamer sailed May 18, 1918, and its arrival in France resulted in the following cable from Gen. Pershing:

Shipment of erected locomotives transmitted on the *Feltore* very satisfactory. Boat completely discharged of locomotives and cargoes in 13 days with saving of 15 ship's days in unloading the 33 locomotives erected as compared with same number of locomotives not erected and further saving of 14 days of erecting forces. Observations of Capt. Byron, who came with these locomotives, show that by loading locomotives in double tiers, placing cab parts and tools, now in separate packages, within tender space and fire boxes, 40 to 45 locomotives can be loaded.

Subsequently the steamers *Cubore, Firmore,* and *Santore* were assigned to the task of carrying these engines over on their own wheels. The total number of locomotives that went abroad in this manner was 533. After the signing of the armistice we sold the French Government 485 locomotives, of which 142 had been shipped up to January 1, 1919.

Efforts were likewise made to ship over freight cars already set up but this was also met with much objection. Finally, 1,000 cars were built to go over complete but the signing of the armistice stopped the shipment. The saving in the cost of shipping locomotives on their own wheels amounted to $775 for each one, and an average of $20 a car would have been saved by sending the cars over on their own wheels. But, in addition to this, the cost of erection on the other side, amounting to $800 for each locomotive, should also be added to the saving.

The number of cars actually shipped overseas for the American Expeditionary Forces, if made into one solid train, would be 140 miles long.

In August, 1918, there came a call from abroad to produce locomotives at the rate of 300 a month and freight cars at the rate of 8,200 monthly. Machinery for getting this production was started at once and was so effectual that during the months of September and October and up to the signing of the armistice engines were actually being produced and shipped from the Baldwin Locomotive Works at this rate. This company was turning out the greatest number of locomotives ever produced by any one company in the same length of time.

Arrangements for increasing production of freight cars to meet every possibility of tonnage facilities on the ocean were also made, and had the armistice not been signed we had planned to produce during the month of December 11,000 complete freight cars and to maintain this production rate until we had filled all orders from Gen. Pershing.

On our first purchase of rails, amounting to 102,000 tons, the price paid was $38 a ton for Bessemer steel and $40 a ton for open-hearth steel, as against a price of $59 a ton that the Russians were paying and

LOADING RAILWAY LOCOMOTIVES, COMPLETE, ABOARD SHIP.

60-CENTIMETER GAUGE TANK CAR.

Capacity in gallons 2,500, in pounds 22,000. Length over end sills 22 feet 1¼ inches; width over side sills, 5 feet 7 inches; weight, 12,000 pounds.

60-CENTIMETER GAUGE V-SHAPED DUMP CARS.

Capacity, 27 cubic feet. Length over couplers, 6 feet 9 inches. Width of body, 48⅜ inches.

RAILROAD LOCOMOTIVES PACKED WITH BALED HAY IN THE HOLD OF A SHIP.

prices between $54.36 and $61.87 that were being paid by the French. There was a saving in this item of approximately $2,040,000 as compared with the prices paid by the Russians and of $1,938,000 compared with the prices paid by the French.

In connection with our first purchase of this steel rail, it should be stated that the Lackawanna Steel Co. and the United States Steel Products Co. agreed to sell us rail on this basis. Orders were placed with these companies, but not with two other companies—the Cambria Steel Co. and the Bethlehem Steel Co.—who declined to accept the price offered.

All subsequent orders for steel rail were on the basis of $55 and $57 a ton for Bessemer and open hearth, respectively, which figure was established by the War Industries Board pursuant to the Government policy to stabilize industry by establishing fixed prices alike for all purchasers—the Government itself, the allies, and the public.

A total of 937 miles of standard-gauge railway track was laid in France with material shipped from this country.

A big money saving was effected by changing the design of the freight cars asked for by our overseas forces. Their original call was for 17,000 four-wheel cars of the French type, these varying from 10 to 20 tons capacity per car. Our investigations here convinced us that the American type of car with 30-ton capacity could be used on the French railroads. Consequently we recommended that 6,000 of the 30-ton American-type cars be sent abroad instead of smaller-capacity French cars. Our recommendation was approved by officers abroad, and as a result there was a saving of $12,640,000 in the cost of this initial order of cars. From that time all cars shipped from the United States were of the American 8-wheel type, a fact which resulted in a saving of approximately $189,600,000 over what it would have cost to build and ship the lighter French cars.

Had the light French type of cars, as originally suggested, been adopted, 270,309 cars would have been required instead of 90,103 cars, and probably twice as much tonnage would have been necessary to transport these cars overseas.

Most of the steel rails were purchased from the Cambria Steel Co., the Lackawanna Steel Co., the Bethlehem Steel Co., the United States Steel Products Co., and the Sweets Steel Co. Raised pier, gantry, and locomotive cranes were turned out by the several crane builders in proportion to their ability to produce. The Standard Steel Car Co. made millions of dollars' worth of metallic parts for freight cars, and the Colorado Fuel & Iron Co. produced rails and bars. As previously mentioned, the Baldwin Locomotive Works got the bulk of the orders for locomotives, although the American Locomotive Co., the Vulcan Co., the H. K. Porter Co., and the Davenport Locomotive Works also made locomotives for our Expeditionary Forces.

HOSPITAL TRAINS.

Ambulance trains were called for by Gen. Pershing in his cablegram of July 15, 1917. It was stated in this message that plans for these ambulance trains would be furnished by the Surgeon General of the Army.

To build these ambulance trains, with their complicated designs and specialized equipment, in this country would have entailed lengthy delay and very heavy expense, as after they had been constructed it would have been necessary to knock them down for shipment. With this fact in mind our officers here took up the question with Sir Francis Dent, of the British railway commission, who was in this country at the time. He stated that ambulance trains built by the London & North Western Railway, which had proved wholly satisfactory in three years of service, could be turned out by that same concern there quickly if the English design were adopted for our Army.

After considerable discussion and consideration the English design was followed, and orders for our ambulance trains were placed abroad. Up to December 7, 1918, there had been completed for our Army 19 of these trains, with a total of 304 cars, and there were in the course of completion at that time or under order 29 additional ambulance trains.

Information from England shows that it was indeed the part of wisdom to order these ambulance trains abroad, as figures from England stated that the first 14 of these trains were produced at a cost to us of £3,845 per car, including repair parts. This means that at the present rate of exchange the cost of each coach was $18,302.20, while to have built these cars in this country, knocked them down, and shipped them overseas would have cost $40,000 each.

NARROW-GAUGE RAILWAY EQUIPMENT—FABRICATED STEEL TRACK.

The urgent necessity for narrow-gauge railway equipment for our armed forces in Europe was first brought home to us when Gen. Pershing cabled on July 15, 1917. In this message he asked for large quantities of 60-centimeter locomotives, cars, and track. The types requested were entirely new in this country.

Specifically, there were required 195 60-centimeter strain locomotives with a low center of gravity and with a maximum of 3½ tons axle load; 126 40-horsepower gasoline locomotives; 63 20-horsepower gasoline locomotives; and 3,332 freight cars of various types, including box cars and flat cars of 10-ton capacity, tank cars, and dump cars. To aid in the building of this new equipment many photographs and designs brought over from France were used. It was decided to build the 10-ton cars fitted with small 4-wheel

60-CENTIMETER GAUGE STEAM LOCOMOTIVE; TRACTIVE POWER, 6,225 POUNDS. CYLINDERS, 9 x 12, DRIVING WHEELS 23½ INCHES, WHEEL BASE 5 FEET 10 INCHES; WEIGHT IN WORKING ORDER 34,500 POUNDS; CAPACITY: WATER 476 GALLONS, FUEL 1,700 POUNDS.

60-CENTIMETER GAUGE GASOLINE LOCOMOTIVE; 50 HORSEPOWER. CYLINDERS 5¼ x 7, DRIVING WHEELS 30 INCHES, WHEEL BASE 4 FEET; WEIGHT IN WORKING ORDER 14,000 POUNDS; FUEL CAPACITY, 30 GALLONS.

ARMORED RAILWAY MOTOR CAR. HALL-SCOTT GASOLINE ENGINE; LENGTH 62 FEET 9 INCHES, WIDTH 9 FEET 11 INCHES, TRUCK CENTERS 46 FEET.

ARMORED MOTOR CAR, OIL-ELECTRIC ENGINE.

ARMORED CAR EQUIPPED WITH 3-INCH GUN AND SEARCHLIGHT ON CAR ATTACHED.

trucks at each end, rather than to make them of the 4-wheeled type, as with this construction they would be better adapted for the rounding of short curves.

In turning out the different kinds of locomotives for the 60-centimeter railways new designs were made in order to produce locomotives that would run with equal facility in either direction. For the gasoline locomotives, designs of types similar to standard-gauge engines, a few of which had been in the service in this country, were made, and orders were placed with the Baldwin Locomotive Works for the first lot.

The first steam locomotives were delivered by the builders on October 3, 1917, and the first gas locomotives on November 7, 1917.

Orders for the freight cars for these narrow-gauge railways were placed with a number of the larger car-building companies of the country. The first of these cars were delivered November 24, 1917.

When the armistice was signed a total of 1,841 locomotives and 11,229 cars of the narrow-gauge type had been ordered and 427 locomotives and 6,134 cars completed. Up to the 11th of November 361 of the locomotives and 5,691 of the cars had been shipped overseas.

Of the 361 locomotives sent to France, 191 were steam engines, 108 had 50-horsepower gasoline engines, and 62 had 35-horsepower gasoline engines. Of the 5,691 cars that went to the Expeditionary Forces prior to the signing of the armistice, 600 were box cars, 166 were tank cars, 500 were flat cars, 1,555 were 8-wheeled gondola cars, 330 were dump cars, 100 were artillery truck cars, 970 were motor cars, 180 were inspection cars, 300 were hand cars, and 990 were push cars.

For the construction of the narrow-gauge railroad used in the combat areas behind the front line trenches a special type of fabricated track was designed. This consisted of short sections of rail bolted to steel crossties. The American narrow-gauge railway was so arranged that it could be packed in knockdown shape to save shipping space. Most of this track was in 5-meter lengths, although many shorter sections were used. All, however, were in multiples of $1\frac{1}{4}$ meters, accurately sawed so as to insure absolute fit of intermediate sections when shell fire made replacement necessary. Vast quantities of curved track, as well as innumerable switches and turnouts, also were built.

In all about 605 miles of fabricated narrow-gauge steel track were purchased and 460 miles shipped to France. All but 192 miles of the fabricated track was built at the Lakewood Engineering Co., near Cleveland. The balance was obtained through the United States Steel Products Co. The cost of the straight track was about $7,400 a mile, while the cost of the curved sections was $8,000 a mile.

Much of this narrow-gauge track that went to France was manufactured at the rate of between 5 and 6 miles of completed track a day.

Great quantities of the fabricated track produced by the Lakewood Engineering Co. were loaded upon camouflaged steamers in Cleveland in May, 1918, and sent direct to France, via Lake Erie, the Welland Canal, and the St. Lawrence River.

CHAPTER III.

ENGINEER ACTIVITIES AT HOME.

A vast quantity of motorized or portable equipment was required by the Engineer units of the American Expeditionary Forces and this had in the main to be furnished under the supervision of the Engineers in this country. The extent to which this material was produced is shown by such items as 6,923 trucks of all kinds, 2,082 portable buildings, 124 portable shop and material trucks, 51 portable pile drivers, 90 electric storage trucks, 6,006 boilers, and 3,504 dump cars. Two-thirds of this equipment was shipped overseas before the armistice was signed.

The development of mobile shops was one of the most interesting phases of this branch of engineering work. Quite early in the war, when we began the construction of the great base shops in France, we developed these portable machine shops, blacksmith shops, carpenter shops, and storeroom shops in demountable truck bodies, to be used for general service in the field. The shops were so constructed that they could be entirely closed up when the unit was in motion; but when the shop was ready for use the sides and ends of the inclosing structure were lowered, forming work tables when the shop was left on the truck chassis. If the shop were entirely demounted, these sides and ends, let down, formed extensions of the floor. With this arrangement a wide range of general repair and construction work could be handled on the spot on short notice. If it were necessary for the shop to stay in one place for several days or weeks, the body could be demounted, and the truck chassis was then used for transporting materials to and from the shop.

Each portable shop contained about 800 different items of tools and equipment. Each was mounted on a 5½-ton truck. The portable machine shop contained a workbench, a drill press, a portable electric drill, a grinder, and a 14-inch lathe, these being operated by an electric power plant carried on the truck; and it also had an equipment of necessary small tools and supplies, including an oxy-acetylene welding outfit.

The portable blacksmith, plumbing, and tin shops each contained a workbench, forges, hoists, pipe-fitting machines, a shear and punch, vises, and a welding and cutting outfit, together with a power plant and switchboard and the necessary small tools and supplies. The portable carpenter shop contained boring machines, a drill press, a bench grinder, a workbench, a saw bench, a winch, power plant and switchboard, small tools and supplies.

A complete machine shop on wheels cost the Government about $8,500. The carpenter shop cost $7,600. As supply units for the portable shops, the Government built 30 material trucks, each containing about 600 items of tools and supplies. These material trucks cost $6,100 apiece.

Another successful development of this nature was the portable photolithographic press truck, already referred to in the account of the American Expeditionary Forces' lithographic equipment. These automobile presses, which were at our front soon after our troops went into the trenches, were able to print and distribute lithographic sketches and maps within 12 hours after the original sketches were submitted for reproduction. The French and British armies also had mobile photolithographic units which were much less portable than ours and much slower in operation. The best time made by the French and British outfits was four days for the same work.

We also supplied to the Engineering forces abroad special water sterilizers and water tanks, mounted on trucks. The Engineers put small job-printing shops on trucks and photographic dark rooms on trucks for use in the field.

They equipped trucks with derricks, capstans, and wrecking machinery. They furnished automotive road sprinklers and oilers, as well as trucks with special dump bodies for highway work.

They developed a light, portable pile driver unlike anything used theretofore in commercial work. This machine was constructed of structural steel and had a total weight of 4 tons. It was mounted on a truck drawn by horses or mules, and the pile driver itself was operated by a 25-horsepower gasoline engine. The pile driver could be used within 16 minutes after its arrival at any point.

One development of this sort, the mobile clam-shell derrick, is worth noting. This unique piece of machinery was built by the Winther Motor Truck Co., of Kenosha, Wis. When the American Expeditionary Forces issued a requisition for 120 clam-shell derricks mounted on motor trucks, no such piece of equipment was in existence anywhere on earth. The Winther Co. volunteered to attempt to produce the machine. By giving a wider tread of rear axle to the Winther motor truck, the company could provide a suitable vehicle, but, search as they might, they could not find a derrick of sufficient power to operate a half-yard clam shell and also light enough to mount on a 7-ton truck. No such derrick existed. The company, therefore, without knowing anything about the manufacture of derricks, put its engineering force to work to produce a design. This design was developed in two weeks, and the derrick built from it was less than half the weight of any derrick of equal capacity. After being perfected, the mobile derrick in tests showed that it could move

ENGINEERS' TOOL WAGON.

ENGINEERS' BLACKSMITH SHOP, CLOSED.

ONE OF THE ENGINEERS' PORTABLE MACHINE SHOPS.

350 cubic yards of sand or gravel per day or 500 or 600 tons of coal. One man could operate it and the motive power was a 4-cylinder gasoline engine.

The Engineering Department approved this design and ordered 32 such clam-shell units. Nine of these were delivered before the armistice was signed. The company has continued production of these derricks with a view of selling them to road builders and excavators in civil life.

For use of the various Engineer units we manufactured 1,610 tool wagons and shipped most of them to France. Because of the rough nature of the shell-torn ground over which these wagons must be used, we designed each to be uncoupled and operated as two 2-wheeled carts.

The development of mobile industrial units mounted on motor trucks is likely to have a profound effect on American industry in the future. For instance, the special derrick or crane trucks which we built are almost certain to be adopted in commercial use. The locomotive crane has always been a useful machine, but its chief use has been in handling heavy materials which were being loaded on or off railway cars. A crane which can be moved rapidly to places where railway tracks are not located should be of almost equal importance. An accompanying illustration shows in operation one of the derrick trucks which we built for overseas use.

In the same way the mobile pile drivers designed by the Engineer Corps should be of great future service in road building in this country.

The various machine shops which were built for war purposes will, in their duplications and adaptations, undoubtedly serve a useful purpose in future commercial activities in this country. The increased use of motive power on farms has created a demand for machine repairs. The day may come when the traveling machine shop will be a familiar sight on our rural highways.

The Engineer troops required a great quantity of hoisting machinery. Our purchases in this respect amounted to 700 cranes, mostly of the locomotive type, and 886 hoisting engines, at a total cost of $4,996,000. About two-thirds of this equipment was sent to France and installed at the ports of debarkation and at depots. The rest was used at the shipping points in this country. This machinery was of great aid in the rapid handling of materials at tidewater.

A vast amount of small tools and construction material was required. Some 21,000 tons of barbed wire, shipped abroad to be used principally in the construction of entanglements in front of American battle positions, were manufactured principally by the United States Steel Products Co., Jones & Laughlin, the Gulf States Steel Co., and the Colorado Fuel & Iron Co., although several other firms also supplied barbed wire.

The Engineering Department ordered in the United States, during the fighting, equipment and supplies which cost approximately $754,201,407.

We furnished in all 85,120 steel shelters of various sizes, of which 38,320 were of the individual type which could be carried by one man. The steel used in these individual shelters was about one-eighth of an inch thick.

There may be expected to be great incidental benefit to future American industry from improvements and inventions brought out by American military engineering in 1917 and 1918.

One important work, for instance, which the Engineer Department undertook was that of standardizing the requirements for paints and varnishes. At the outset our Army needs ran into 29 shades of color in 315 different paint and varnish mixtures. Without affecting any of our camouflage projects or other important undertakings, we reduced the number of shades required from 29 to 16 and brought the total number of commodities down from 315 to 99. This reduction in the range of commodities will be of great use to the paint and varnish industry in the future.

At the beginning of the war the mechanical rubber industry had but few standard specifications. The Engineers, after considerable research, developed 30 standard specifications for mechanical rubber goods, which class included such materials as hose, packing, and sleeves. The representatives of the rubber industry verbally stated that the Engineer Department in this short time did more good to the trade than it had been able to accomplish for itself in the previous three or four years of effort. Immediately after hostilities stopped rubber concerns began asking the Engineer Department for its standard specifications.

In the manufacture of hardware and kitchen utensils there was also considerable standardization done, and changes in manufacturing methods were recommended which were put into effect by the producers. All spun goods were eliminated, and the industry confined itself to straight stamping, which meant a reduction in labor. A standard cobalt coating for enamelware was developed by which the industry conserved about 30 tons of nitre per month and made a more durable and satisfactory enamel coating, with the result that to-day the Army is purchasing its vast quantities of enamelware subject to certain tests, whereas, in the past, practically all this material was bought purely upon the manufacturers' statements. The shortage of tin was of considerable importance. Upon the recommendation of an Engineer officer enormous quantities of cafeteria trays were coated with zinc and large amounts of tin thereby conserved. The finished tray was entirely satisfactory and gave essentially the same service as that plated with tin. Horseshoe nails, formerly a variable product, were standardized and tested, and methods were devised by which the Army was enabled to control their quality.

Before the war there was no standard rating for internal-combustion engines, each manufacturer rating his motors according to his own ideas. Our studies of small engines of the type used for driving pumps or operating wood-working and metal-working machines resulted in many improvements, which have been adopted by the manufacturers of internal-combustion engines. Out of these studies came the so-called army rating, a standard which is bound to result in the more careful rating of commercial engines.

The Engineer Department brought out a modification of the design of the existing line of gasoline-driven shovels by applying caterpillar traction to the larger sizes, thus doing away with the labor required to plank up and block shovels that move on wheels.

When we entered the war, the explosive trinitrotoluol was standard for our Army for mining and demolition purposes. The Bureau of Mines, in cooperation with the Engineer Department, developed an explosive which is cheaper than T. N. T. and promises to replace it for engineering operations.

We also improved the devices commercially used in electrical detonation of distributed charges, our improved detonators being more certain and reliable than anything in use.

Commercial machines for detonating as many as 250 standard No. 8 caps were developed for the Panama Canal, but the machines in common use had seen little improvement for 25 years. As a result of the development by the Engineer Corps, a machine capable of detonating 120 caps was obtained, weighing no more than the 30-cap commercial blasting machine and costing slightly less.

A second machine was developed, capable of exploding 500 caps, at a price not greatly above the price of a 30-cap commercial machine. Mining engineers who saw this development stated that it would have a high commercial value, as these improved machines would make electric blasting more positive and dependable than any other form of detonation, as well as making it possible to set off a large series of charges simultaneously. The Panama Canal machine weighed 35 pounds and cost $126. Our 500-cap machine weighed 30 pounds and cost $35. The du Pont 30-cap machine weighed 25 pounds and cost $25. Our small machine weighed 20 pounds, cost $22.50, and would fire 120 caps.

In addition to this there might be mentioned other projects developed primarily for war purposes but which will be available for the industrial uses of peace. These included portable well-drilling outfits of a new type, alcohol stills of a small size for the utilization of waste products in small units, sound reducers on the exhaust pipes of gasoline engines, air strainers to minimize the chances of dust and grit entering gasoline engines. When the war ended we were working on the problem of hastening the setting of concrete and were also studying the production in this country of photographic colors and tone chemicals formerly secured only from Germany.

In general, mention should be made of the exhaustive tests in many industries conducted by the Engineer depot and by special detachments of Engineers. Tests were made of hundreds of pieces of apparatus, and these tests led to many improvements in American manufacture. Illustrating how these tests were regarded by individual concerns, the Cleveland Tractor Co., after a test of its equipment conducted by Army Engineers, stated: "Our people consider this test to be the most valuable ever undertaken by this company." This is indicative of benefits scattered throughout American industry by the engineering war tests.

While practically all of the research work which resulted in the developments and improvements noted was conducted by Engineer officers while on duty at the General Engineer Depot in Washington, since the transfer of the functions of the General Engineer Depot to the Division of Purchase, Storage and Traffic, November 1, 1918, much of this research work has been and still is being carried on by the latter division.

For handling Engineer materials there were established the General Engineer Depot at Washington, D. C., embarkation depots at South Kearney, N. J., and Norfolk, Va., and shipping depots at Baltimore, Md., Philadelphia, Pa., Jacksonville, Fla., New Orleans, La., and Mobile, Ala. In addition, subdepots were organized at all of the divisional camps and cantonments.

The war demanded the production in America of quantities of precision instruments. These were required not only by the Ordnance Department for the equipment of artillery with sights and indirect fire-control apparatus but also by the Engineer Corps, the Signal Corps, the Bureau of Aircraft Production, and the Medical Department. These instruments were such things as aneroid barometers, pocket compasses, measuring tapes, surveyors' equipment generally, map-drawing outfits, draftsmen's supplies, and so on. For a large period of the war the procurement of precision instruments was in the hands of the General Engineer Depot. Later, when the War Department's supply activities were being consolidated, the purchasing of precision instruments, except the highly technical sound-ranging devices, was taken over by the Director of Purchase and Storage, the organization of the General Engineer Depot going along in the transfer. The development and the production of searchlights and sound-ranging apparatus remained in the hands of the Engineer Corps.

In April, 1917, there were probably not more than a dozen recognized American manufacturers of high-grade precision instruments. As an indication of the expansion of manufacturing capacity required by the war, one concern, the Taylor Instrument Cos., of Rochester, N. Y., which had manufactured in peace times watch-pocket compasses at the rate of 15,000 a year, were called upon to turn them out at the rate of 10,000 weekly to fill an order for 200,000 of them. In order to handle this contract the Taylor Instrument Cos. built a new factory building in 20 days. A certain type of aneroid barometer required by the exigencies had never before been produced in America. The Taylor Instrument Cos. succeeded in producing 1,240 of these barometers.

The Lufkin Rule Co., of Saginaw, Mich., was called upon to manufacture 700 band chain measuring tapes for surveying, graduated throughout according to the metric system, and also 1,240 special outfits for repairing such tape. These band tapes when broken are fastened together by tiny rivets, which are produced by special machinery. Because of the inability of the machine-tool industry, swamped as it was with war demands, to produce the special rivet-making machines, it was necessary to reduce in the specifications for repair outfits the quantity of metal rivets for each kit from 4 ounces of rivets to 2 ounces.

Field artillery required a precision instrument known as the miniature telescopic alidade of the Gale type. It is unlikely that 150 of these instruments had been made in the United States during 10 years, yet the Artillery demands called for the production of 1,110 of them. The W. & L. E. Gurley Co., of Troy, N. Y., not only manufactured half of this order, but, in order that the Government might obtain a sufficient supply of these instruments, it turned over to a competing firm, the Eugene Dietzgen Co., of Chicago, the lenses, prisms, hermetically-sealed bubbles, and other parts for 555 instruments.

The Army required large numbers of hand tally registers to be used by checkers and observers. The Benton Manufacturing Co., of New York, had been making less than 15,000 registers of this sort in a year, yet it increased its facilities and turned out 62,000 of them for the Army within two months.

The Army required 35,000 complete sketching outfits for the use of military observers. The contents of these outfits were manufactured by a dozen different concerns.

Drawing instrument sets were produced by the Eugene Dietzgen Co. Each set included a pair of proportional dividers. Our draftsmen had always obtained their dividers from Europe. The divider, which nearly everyone has seen, appears to be a simple device, yet it must be made with the utmost precision, or else it is valueless. In manufacture it goes through more than 100 distinct factory operations.

Marching compasses for troops were made by the Sperry Gyroscope Co., of Brooklyn, N. Y., the quantity in manufacture being over 200,000 instruments.

Many other delicate instruments of most difficult manufacture, whose description is too technical to be set down here, were produced successfully in America during the war period.

DERRICK TRUCK FOR OVERSEAS USE.

LIGHT MICROPHONE SET.

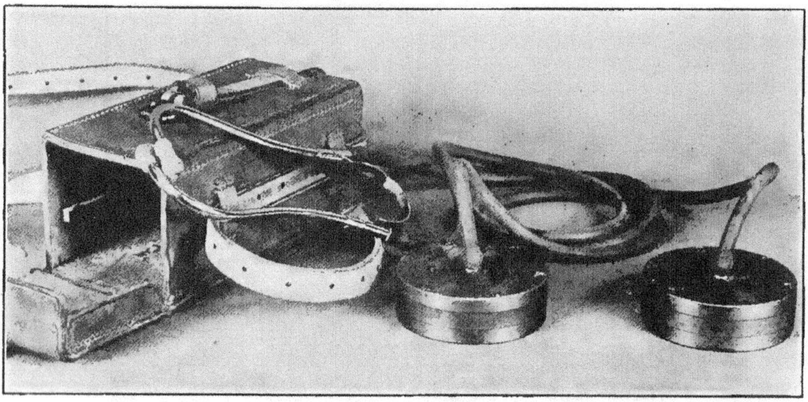

GEOPHONE SOUND-RANGING SET.

AMERICAN T-M SURFACE SOUND-RANGING SET.

CHAPTER IV.

SOUND AND FLASH RANGING AND SEARCHLIGHTS.

In childhood we were enthralled by the tales of those magic persons whose keen hearing could detect even the whisper of the growing grass. As camouflage developed, modern warfare yearned for such supernatural gifts of sense that troops might detect the unseen presence of the enemy. Accordingly Science, the fairy godmother of to-day's soldiers, raised her wand, and lo, the Army was equipped with the wonderful ears of the fairy tale, uncanny no longer, but a concrete manufacturing proposition.

Artillery practice nowadays abhors the wasted shot. The time when cannon fired in the general direction of the enemy, and hoped to hit something, passed when the long-range rifles and howitzers, with their marvelously accurate sighting instruments, came into existence. Whole books have been written on the subject of pointing a modern cannon in the modern way. A great proportion of our industrial effort in the recent conflict was devoted to the sole end that we might aim our artillery accurately.

For instance, to this end almost exclusively was devoted the enormous production of aircraft material. The observer in the airplane or balloon trusted not to his eyes but to the finer sight of the photographic camera; and this again occasioned a large war industry—the production of cameras and their operation in the field, which included the production of finished photography in the field dark rooms. But, as the airplane and aerial camera were perfected, camouflage was undertaken as a protection from discovery from aloft; and so might be brought in another chapter—the production of camouflage material and the work of camouflage experts in the field. Presently camouflage succeeded in baffling the camera to a great extent, and this made necessary the development of instruments that could detect the location of the enemy by sound. Since the unaided ear was not keen enough to supply the desired information, applied science came to the rescue with the various devices embraced in the general classification of sound-ranging equipment. The production of this equipment was under the direction of the Engineer Department of the Army.

In three classes of military work we needed hearing refined to the razor-edge. With keen enough ears we could detect those subterranean operations of the enemy known as mining; with ears of that sort we could detect and locate the positions of hostile cannon; and still again we could employ such sensitiveness of hearing to find in the darkest sky at night the hostile raiding airplane.

One of these long-distance ear drums which man invented for himself as an aid to his military operations was known as the geophone. The first geophone used by the western powers in the war was invented by the French. It was a simple mechanism. The device or drum which received the sound waves and magnified them consisted of a small closed box with a confined air space. This box was weighted with a leaden disk to give it the required inertia. The geophone was placed upon the ground and the vibrations of the earth were communicated through the medium of the confined air space. The sounds then reached the listener's ears via a rubber tube and an ordinary stethoscope horn. By means of this device the slightest vibrations of the ground were rendered audible.

The geophone was used to detect enemy mining operations. The listener placed the weighted box on the floor of an underground gallery or on solid earth or rock. If the enemy were burrowing in the ground anywhere within a distance of 75 yards the geophone would tell about it. In order to enable the listener to know in what direction the sounds came, two geophone boxes were provided, one connected with each ear. By placing the boxes a small distance apart from each other and them moving them until the vibrations in both ear horns were equalized, the listener could tell approximately in what direction the enemy mining operation was located.

Geophones were used by both sides, and so effective did they prove to be that it is reported that they were largely instrumental in stopping mining operations altogether. If an enemy mine were located by one of these devices, a counter mine could be started at once and carried through, usually with disastrous results to the hostile miners.

As our first step in the production of geophones, we adopted the French device; but later on we developed an instrument with nearly one-third greater range than the French geophone had. This improvement was developed by the Engineers and bureau specialists at the Bureau of Standards in Washington with money provided by the Engineering Department. We produced the improved model in sufficient quantities to meet the requirements of the American Expeditionary Forces.

We also developed an electromechanical geophone that could be connected up by wire to a central listening station some distance back from an exposed location. The sound-receiving boxes or microphones were placed out in No Man's Land and hidden under trash or earth. They were so sensitive that they would not only record any subterranean activities of the enemy within their range, but at night would betray enemy raiding parties attempting to cross to our positions, the sensitive boxes picking up the vibrations of their speech or footsteps. The central listener could locate approximately the position of hostile operations by observing which boxes were receiving the sounds in greatest intensity. The boxes could also pick up and send to the central listening stations conversations carried on by the enemy parties even in low tones, the apparatus thus acting as the dictagraph of the war.

But by far the most important work done by listening instruments was in locating the positions of enemy gun batteries. This was one scientific instrument, at any rate, which the Germans were never able to produce successfully for themselves. During the final months of the war more enemy guns were located by listening instruments than by any other means. An American instrument with the Army spotted 117 German gun positions in a single day by surface sound ranging. This was the high American record set in the war, but at all times our sound-detecting equipment had an uncanny accuracy. Up to the end of the fighting, no way had been discovered to conceal the location of a gun from sound-ranging instruments suitably placed and properly operated.

The instruments used for locating gun positions were of such a highly complicated and technical nature that no one but designers and mechanics skilled in the production of complex electrical equipment could build them at all. The recording instruments, or microphones, were of a sort so delicate that their use theretofore had never been considered outside of laboratories. Yet they were required to operate successfully amid the din and concussion of heavy bombardments. All useless sounds and jars were filtered out so that only the sought-for vibrations could come to the central recording mechanism.

Studies of gunfire showed that when a cannon fires an explosive shell of high velocity there are three distinct concussions. One of these is the sharp crack produced in the air when the shell, dragging a short vacuum trail behind it, passes over the head of the observer. As the air rushes into this vacuum and collides with itself, it produces a crack that is similar in origin to ordinary thunder. The second concussion to be heard is that produced at the muzzle of the gun by the expanding gases that propel the shell. There is still a third, the break, or explosion. In order to locate a battery or gun exactly only one of these concussions—the explosion at the muzzle of the gun—must be picked up by the microphone. The first and third shocks, and all other sounds not useful to the work should be damped out and excluded.

A number of these microphones would be placed in scattered positions, usually in a trench, and then connected with the central recording mechanism. When a microphone picked up a hostile gun explosion the disturbance was instantly transmitted through several miles of wire. An ingenious and complicated mechanism actuated an electromagnetic needle, which instantly recorded this disturbance on a tape of photographic paper, calibrated to show fifths of seconds in time. Each microphone on outpost duty was represented on this tape by a parallel line; and, as six microphones were usually used, the tape was striped with six parallel lines. As the other microphones at the front picked up the concussion of the gun, their records were made on their respective lines; and the observers at the central station, by noting the difference in time between the reports of the various microphones, and by making calculations based on the rate at which sound travels, could locate the gun that set up the disturbances by means of ordinary surveyor's triangulations. So accurately would this mechanism do the work that a gun position could be determined within 50 or 60 feet.

Incidentally, it is interesting to note that the practice of our Army was to secure in advance, by means of surface sound ranging and other methods, the positions of all the enemy's guns that could be learned. Then, often after intervals of hours or even days, the fire began simultaneously upon all these gun positions just as our attack started.

In this country we had two experimental stations for the development of sound-ranging apparatus. We began experiments in this work in June, 1917. Before we had perfected any satisfactory instruments, the British had met with great success with the Bull-Tucker system; and we adopted that type for the use of the American Expeditionary Forces. From plans and models sent to this country we produced an American Bull-Tucker machine, utilizing standard American electrical equipment wherever we could. At the close of the war we had in operation along the American front 12 complete American outfits. The six microphones of each recording machine in action were set about 5,000 feet apart along the front, so that each sound-ranging section covered a frontage of approximately 5 miles. The 12 outfits in use were sufficient to locate the guns of the enemy on a 60-mile front.

About a month before the fighting stopped we sent to France a new model sound-ranging set which had been developed with the cooperation of the Bureau of Standards. The reports from the American Expeditionary Forces indicated that this American development was superior in several important particulars to anything else in use when the war came to an end. The American instruments were lighter, easier to carry about, easier to install, and much cheaper than those of the British type, and would operate under more adverse weather conditions. The impulses received by the microphone in this equipment were recorded on a running tape smoked by an acetylene flame.

Sound ranging for the detection of airplanes at night requires an equipment which consists fundamentally of a sound-gathering device and a listening mechanism, the combination enabling the observer to tell the direction from which the sound is coming. When a bombing plane approaches at night the hum of the motor can be heard at a distance from 1 to 3 miles, or even more, depending upon conditions. But the direction of this sound is elusive to the unaided ear, as anyone can testify who has heard an airplane in broad daylight but could not locate it with his eyes. Before the invention of aerial sound ranging the searchlights hunting for the hostile airplane were obliged to sweep the sky aimlessly in an endeavor to locate it; and the pilot of the plane could often maneuver to keep out of the light. But by use of the sound detectors not only can the approach of the airplane be detected at a distance beyond the hearing range of the unaided ear, but, what is more important, its direction can be determined within an angle of 3°. The use of these sound detectors greatly increased the chances of locating airplanes at night by searchlight.

The Engineer Department conducted extensive experiments in the development of aerial sound detectors. One form developed consisted of a set of long horns with listening tubes attached to the small ends and leading to receivers on the observer's head set. These horns were mounted on a turntable which the observer could revolve, so that the horns could be turned in the general direction of the sound. Four horns were used in this mechanism—two to indicate the direction of the airplane on a horizontal circle (in azimuth), and the other pair to indicate the direction on the vertical arc (in elevation). Under favorable conditions the sensitiveness of this device was three times that of the unaided ear, and the airplane could be located within an angle of 1°. The horn detector, however, was large and cumbersome and not satisfactory for a mobile unit.

For field sound ranging, when the listener may wish to move from place to place, the parabloid sound reflector was developed. This hemispherical object, like a huge fountain basin in shape, was made of material similar to building board and shaped in parabolic lines. Such a sound collector echoed or reflected the sound from every point of its surface to a focal point where the listening instrument was located. The observer turned the parabloid on its universal mount until the sound was equalized in his ears, and then the exact direction of the airplane would be indicated by the azimuth and elevation pointers on the machine. The parabloids developed by our Engineering Department had a sensitiveness three times that of the unaided ear and could locate sound within 3° of arc.

We were not pioneers in developing the parabloid, however, the French having built them ahead of us; but our apparatus possessed marked advantages over that of the French. In the first place, the French collecting device weighed 3½ tons and was so heavy and cumbersome that it could scarcely be moved at all. The total weight of the American collecting device was only 1,300 pounds. The American instrument was thus much lighter and more portable. It was so simple that it could be set up in about one-sixth the time that it took to erect the French device. The cost of our machine was only about two-fifths that of the French mechanism.

Although valuable work in detecting gun positions was done by sound ranging, yet both sides located guns by watohing their flashes. We improved flash-ranging sets of the allies. These were simple in principle. A number of observers at posts commanding good views were equipped with observation telescopes mounted on tripods to watch for the flashes of enemy guns. Whenever two or more of them observed the same flash and reported its direction, the position of the gun could be determined by ordinary triangulation.

However, in operation the system was not so simple, because of the fact that the observers reporting might not have turned their instruments upon the same flash. This difficulty was met by furnishing each observer with an outpost switch set. As soon as he observed the flash through his telescope he closed the switch, and that action turned on a small electric light at the headquarters station, which might be miles away. Then, as soon as he could, he telephoned in the direction of the flash observed. If the operator at the switchboard saw two or three of the lights flash simultaneously, he knew the observers at the front had probably caught the same flash. Lights that came on a little ahead or a little behind the simultaneous lights were disregarded when the observers telephoned reports.

In developing the telescope for this system considerable difficulty was experienced on account of the shortage of the proper optical glass in this country. We were, therefore, obliged to buy our telescopes in France until our supply would be available. These telescopes were expensive mechanisms, and in some of the work of the flash-ranging sections two of them were originally required at each observing station—one to determine the position of a shell burst in elevation and the other its position on the horizontal circle in azimuth. Since the declaration of the armistice an American Engineer officer has designed a telescope eyepiece which enables this work to be done by observing through a single instrument, thus effecting a marked saving in the number of telescopes which might be required in the future.

AMERICAN PARABLOID TYPE ACOUSTIC DETECTOR.

60-INCH OPEN TYPE PORTABLE SEARCHLIGHT.

60-INCH HIGH INTENSITY SEACOAST TYPE SEARCHLIGHT.

When the fighting stopped our military scientists and others cooperating with them were developing a type of ground sound-ranging apparatus which it was hoped could be utilized to give troops warning of the firing of heavy artillery shells in their general direction. Preliminary experiments show that at a distance of 4.1 miles this mechanism could record the firing of a gun some 19 seconds before the arrival of the shell. Under proper circumstances this elapsed time would enable troops properly warned to seek cover from the explosion of the projectile. This development of sound-ranging apparatus and its application to the protection of personnel was made possible by the far greater speed with which shock vibrations travel through a dense medium like the earth than through the usual sound-conveying medium, the atmosphere.

SEARCHLIGHTS.

The searchlight equipment of the United States Army prior to 1914 consisted chiefly of lights located at our coast defenses. In 1916 we began the development of mobile searchlight-and-power units for field-army work, four horse-drawn equipments, with 36-inch lights, being ordered first, and later eight other sets, with extensible towers and gasoline electric generators. When the war was approaching we ordered 85 sets of the limber-and-caisson type. The caissons of these sets carried 24-inch lights on extensible towers. In January, 1917, we ordered 50 high-intensity lights to replace as many low-intensity lamps at our seacoast fortifications. The first war order was placed in April, 1917, and consisted of 20 additional searchlights of the 60-inch dimension, the largest light ordered by the War Department.

After the entrance of America in the war the Engineer Department began studying the requirements abroad for searchlights used in defense against hostile aircraft; and in September, 1917, this investigation resulted in orders for 360 high-intensity searchlights, 693 high-intensity arc mechanisms, and 1,000 glass mirrors of standard design.

About this time we began looking to the improvement of existing searchlight equipment. The cooperation of leading scientists, manufacturers, and Government bureaus was obtained, and the product of exhaustive experiments was 18 different new kinds of searchlights either partially or wholly developed.

The first of these were produced, shipped, and were in operation with the Second Field Army in France on October 1, 1918. This was a new form of searchlight more powerful than any that had been produced before that time. It weighed one-eighth as much as lights of former design, cost only one-third as much, was about one-fourth as large in bulk, and threw a light 10 per cent stronger than any other portable projector in existence.

Without going into the details of this mechanism its most striking innovation, from the standpoint of the nontechnical observer, was the absence of the front glass through which the beams of the older type lamps are sent. The absence of the glass, while reducing the weight and cost of a light, also increased the intensity of the beam of the searchlight, since any glass, no matter how conducive to rays, absorbs light to a considerable extent.

In the first part of the war we took the 36-inch lights which the Government had on hand and mounted them on motor trucks. For generating power for the lights, motor trucks were equipped with electric generators operated by the crankshaft of the truck engine. In moving about each truck carried not only the light and power unit and accessories, but provided space for the crew and their equipment.

When we went into the war there was only one firm in the United States that could make the large searchlight mirrors, but two other concerns developed the art and the facilities during the hostilities. These mirrors were of glass and cost about $1,000 at prewar prices. The maximum output in the United States before the war was three 60-inch mirrors per week. As the result of governmental encouragement the production of the 60-inch mirrors increased until it reached the stage of 15 a week in November, 1918; and the price was reduced to about $900 per mirror, even under wartime conditions with respect to labor and material. This was equivalent to a price of about $700 per mirror under normal conditions, or a saving of 30 per cent.

A remarkable contribution of the United States to searchlight science was the production of a satisfactory metal mirror for projecting the beam. The metal mirror not only weighed a little less than the glass mirror, but it cost only one-third as much as the glass one, could be produced in one-fifth the time, was much less fragile, and extended the possibility of manufacture to a wide number of industries. The metal mirror possessed 97 per cent of the reflectivity of the glass mirror. This slight dullness is inappreciable in searchlight work and more than compensated for by the other qualities of the metal reflector. This type of mirror, however, had not yet been put in production when the war ended.

Our inventors during the 19 months of hostilities succeeded in reducing the size of carbons used in 200-ampere lamps from 2 inches in diameter to 1⅛ inches. This cut the cost of carbons in two, but the improvements tripled the amount of light developed.

In November, 1918, we were working with assurance of success to develop a simple system whereby field searchlights could be pointed and controlled from a distance. Such controls had been used in experimental work prior to 1917, but the mechanisms were complicated and not suitable for field service.

The searchlight section of the Corps of Engineers also developed optical finding devices, which doubled the range of all searchlights without requiring any modification of the lights themselves. Neither the ordinary telescope nor night glass is suitable for target finding by searchlight. The result of our investigation was the development of a combined observer's chair, eye protector, and searchlight target finder, the new equipment adding only 10 per cent to the cost of the searchlight unit.

The range of our modern high-power searchlight, whose target is a ship at sea, is about 15,000 yards; the range of this searchlight when its target is an airplane is about 15,000 feet.

BOOK IV.

CHEMICAL WARFARE.

CHAPTER I.

TOXIC GASES.

The first recorded use of suffocating gases in warfare was about 431 B. C., sulphur fumes having been used in besieging the cities of Platea and Belium in the war between the Athenians and the Spartans. Similar uses of toxic substances are recorded during the Middle Ages. In August, 1855, the English Admiral Lord Dundonald, having observed the deadly character of the fumes of sulphur in Sicily, proposed to reduce Sebastopol by sulphur fumes, and worked out the details of the proposition. The English Government disapproved the proposition on the ground that "the effects were so horrible that no honorable combatant could use the means required to produce them."

That the probable use of poison gases was still in the minds of military men is evidenced by the fact that at The Hague conference in 1899 several of the more prominent nations of Europe and Asia pledged themselves not to use any projectiles whose only object was to give out suffocating or poisonous gases. Many of the Powers did not sign this declaration until later. Germany signed and ratified it on September 4, 1900, but the United States never signed it. Further, this declaration was not to be binding in case of a war in which a non-signatory was or became a belligerent. Admiral Mahan, a United States delegate, stated his position in regard to the use of gas in shell, at that time an untried theory, as follows:

> The reproach of cruelty and perfidy addressed against these supposed shells was equally uttered previously against firearms and torpedoes, although both are now employed without scruple. It is illogical and not demonstrably humane to be tender about asphyxiating men with gas, when all are prepared to admit that it is allowable to blow the bottom out of an ironclad at midnight, throwing four or five hundred men into the sea to be choked by the water, with scarcely the remotest chance to escape.

The Second Hague Peace Congress in 1907 adopted rules for land warfare, and among them was article 23 which read as follows: "It is expressly forbidden to employ poisons or poisonous weapons."

The use of toxic gas in the great war dates back to April 22, 1915, on which day the Germans employed chlorine, a common and well-known gas, in an attack against the French and British lines in the northeastern part of the upper Ypres salient.

The methods of manufacturing toxic gases, the use of such gases, and the tactics connected with their use were new developments of this war; yet during the year 1918 from 20 to 30 per cent of all American battle casualties were due to gas, showing that toxic gas is one of the most powerful implements of war. The records show, however, that when armies were supplied with masks and other defensive appliances, only about 3 or 4 per cent of the gas casualties were fatal. This indicates that gas can be made not only one of the most effective implements of war, but one of the most humane. It will, of course, be necessary to remove the noncombatant population from a greater depth of country immediately in the rear of the fighting lines than formerly, in order that women and children may not be gassed. This additional sacrifice of territory for war uses is another element of effectiveness in the weapon.

Since Germany had chosen to utilize toxic gas in warfare, the allied nations were compelled to adopt like tactics; accordingly England and France, faced with the desperate situation resulting from advantages secured by the Germans through the employment of these new weapons, immediately turned their attention not only to devising methods for protecting their own troops, but also to securing supplies and equipment necessary for the utilization of toxic gas as an agent of warfare against the German Army.

Germany originated thereafter the use of most of the new forms of gas, but the allied nations and America were actually producing, at the time of the armistice, gases on a much greater scale than Germany was ever able to attain. In fact, America itself was producing gases at a rate several times as great as was possible in Germany.

Prior to the entry of America in the war our overseas observers had been collecting information bearing upon gas warfare, referring the facts so obtained to the Ordnance Department in Washington, where the information was turned over to Lieut. Col. E. J. W. Ragsdale, who was then in charge of the Trench Warfare Section.

In the early days of our belligerency it was seen that we should need a plant for filling artillery shell with toxic gases. The Government in the fall of 1917 bought a large tract of land near Aberdeen, Md., to be an artillery proving ground. Approximately 3,400 acres of this reservation, about one-tenth of it in area, was set aside as the site for the gas shell-filling plant. This reservation was known as Edgewood, and the plant erected on the site was called the Edgewood Arsenal. Work started on the construction of the arsenal on November 1, 1917.

None of the toxic gases in use in Europe, except chlorine and small amounts of phosgene, had ever been commercially prepared in the United States. It was the original intention to interest existing chemical firms in the manufacture of these gases; but there were many difficulties in the way of such a project, not the least of which was the ruling of the Director General of Railways that such products as poison gas be transported only on special trains.

Also we discovered that the private chemical companies were loath to undertake such manufacture. The exhaustive investigations necessary before quantity methods of manufacture could be devised would be uncertain and expensive. There would be great danger to the lives of those employed in such work. Many of the private concerns were already crowded with war work. Finally, the new plant equipment which must be set up would be worth nothing when the war ended, since the manufacture of such gases would be limited to the period of hostilities. These and other considerations explain the reluctance of the commercial chemical industry to undertake the production of war gases.

Consequently the Government was forced to adopt the plan of building various chemical plants at the Edgewood Arsenal in connection with the filling plant. By December 1, 1917, it had been decided to build at Edgewood a chlorpicrin plant and a phosgene plant. The contracts were immediately let, and the work was pushed through the rigorous winter of 1917–18.

In March, 1918, the Edgewood project was taken from the Trench Warfare Section of the Ordnance Department and made an independent division under the command of Col. Wm. H. Walker. In June, 1918, the Chemical Warfare Service was organized, and the Edgewood Arsenal was transferred to it. Gen. W. L. Sibert, Director of the Gas Service, took charge of the activities of the arsenal in May prior to the official transfer.

Chlorine, the raw material for the manufacture of which is common salt, was one of the principal materials required in the gas-production program. Although chlorine was a standard product in the United States prior to the war, it was soon seen that we had an inadequate commercial supply to meet the requirements of our proposed gas offensive. Chlorine was used not only by itself, but it was also the active agent in the manufacture of nearly all the other toxic gases which we required. Consequently we decided to build a Government chlorine plant with two 50-ton units, giving a daily capacity of 100 tons of liquid chlorine. The ground for this plant at Edgewood was broken on May 11, 1918, and the actual production of chlorine begun on September 1.

In July, 1917, the Germans introduced the so-called mustard gas. It was immediately realized that for certain purposes of fighting this chemical was the most effective product so far employed, and a large number of Government experts here at once concentrated their energies in developing methods for its manufacture on a large scale. Not only were the uniformed experimenters busy at the Gas Service's American University Camp, at Washington, D. C., but experimental units were established at the plant of the Dow Chemical Co., at Midland, Mich., at the plant of Zinsser & Co., Hastings-on-Hudson, N. Y., and at the Government plant which had been started by the Trench Warfare Section, at Cleveland, Ohio.

Eventually it was decided to erect a large plant at Edgewood for the manufacture of mustard gas. Not until April, 1918, however, did we feel that we possessed sufficient knowledge and information to justify the construction of a mustard-gas plant on a large scale. France and England also were long in working out satisfactory methods of mustard-gas production. We began to make mustard in June, and continued with rapidly increasing output until the signing of the armistice.

It soon became evident that we could not depend upon civilian labor in the operation of the various chemical plants at Edgewood because of the danger involved. It was decided, therefore, to utilize enlisted men in the working crews. As the projects at Edgewood increased in size and number, the forces at the arsenal grew, until at one time there were 7,400 troops at this point.

Meanwhile the Government had at last been able to persuade a number of private chemical firms to manufacture toxic gases. The Government agreed to finance all new construction, but the operation was to be in the hands of the contracting companies. At each plant the Government stationed one of its representatives with necessary assistants. In the spring of 1918, these scattered factories by official order were made part of the Edgewood Arsenal, each plant being designated by the name of the city or town where it was located. Thereafter in Army usage the term "Edgewood Arsenal" embraced not only the group of factories on the Edgewood reservation, but also included the following projects:

Niagara Falls plant, operated by the Oldbury Electro-Chemical Co. Project—the manufacture of phosgene.

Midland, Mich., plant, operated by the Dow Chemical Co. Project—the sinking of 17 brine wells for the purpose of securing adequate supplies of bromine.

Charleston, W. Va., plant, operated by the Charleston Chemical Co. Project—the manufacture of sulphur chloride.

Bound Brook, N. J., plant, operated by Frank Hemingway (Inc.). Project—the manufacture of phosgene.

Buffalo plant, operated by the National Aniline & Chemical Co. Project—the manufacture of mustard gas.

In addition to these, the Edgewood Arsenal built at points advantageous to supplies of raw materials four other plants, and operated them as well. These were as follows:

Stamford, Conn., plant. Project—the manufacture of chlorpicrin.

Hastings-on-Hudson, N. Y., plant. Project—the manufacture of mustard gas.

Kingsport, Tenn., plant. Project—the manufacture of brombenzylcyanide.

Croyland, Pa., plant. Project—the manufacture of diphenylchlorarsine.

In constructing and equipping the Edgewood Arsenal we laid 21 miles of standard-gauge railway and 15 miles of narrow-gauge railway, built nearly 15 miles of improved roadway, and set up two water systems, one with a capacity of 1,500,000 gallons per day for the manufacturing purposes of the chemical plants, and the other providing a fresh-water supply pumped 4 miles with a daily capacity of 2,000,000 gallons. In all 558 buildings were put up on the grounds of the arsenal. There were 86 cantonment buildings, with a capacity of 8,400 men, as well as adequate quarters for officers and civilian employees. Three field hospitals, a complete base hospital, and separate buildings for Y. M. C. A. and K. of C. activities indicated the extent of the building equipment. Three power houses were provided, with a total capacity of 26,500 kilowatts.

In the construction of buildings every precaution was taken to avoid accidents from the handling of toxic gas, the ventilating systems being as near perfection as human science could make them. It is notable that out of the thousands of men employed only four met their death by gas poisoning. Three of these casualties were due to phosgene and one to mustard gas.

To show that all of the danger of the war was not confined to the front, the following table of casualties in 1918 at the Edgewood Arsenal proper is here given:

Toxic agent.	June.	July.	August.	September.	October.	November.	December.	Total.
Mustard gas	14	41	190	153	227	47	2	674
Stannic chloride		3	8	15	21	3		50
Phosgene			3	7	22	17	1	50
Chlorpicrin		14	18	9	3			44
Bleach chlorine		2	39	2	1			44
Liquid chlorine		1	3	2	7	5		18
Sulphur chloride			2	1	6			9
Phosphorus		2	7	5	1			15
Caustic soda			3		3	4		10
Sulphuric acid			4	3	1			8
Picric acid			2					2
Carbon monoxide					1			1
Totals	14	63	279	197	293	76	3	925

As has been noted, chlorine was the only war gas produced on a commercial scale in America prior to the war. At the ordinary temperatures chlorine is a greenish-yellow gas of strong, suffocating odor. Through the combined effects of cold and pressure it is readily condensed to a liquid and is ordinarily shipped in this form, stored in strong cylinders.

Chlorine is prepared commercially by the electrolytic process. A current of electricity is passed through a solution of common salt. The greenish gas at once arises, leaving behind it a residue of caustic soda. The apparatus in which the salt is decomposed by the electric current is known as a cell. The Government plant used Nelson cells, each with a capacity of 60 pounds of chlorine and 65 pounds of caustic soda per 24 hours.

The Government chlorine plant at Edgewood was ready for operation in August, 1918, but was not actually started until September 1. The plant consisted of (1) a cell house, which had a total capacity of 100 tons of chlorine per 24 hours; (2) an electric substation for supplying the current; (3) a brine building, where the salt was mixed with water and the resulting brine purified; (4) a boiler and evaporation building, for concentrating the caustic soda from the cells; (5) a caustic fusion building, for drying the caustic soda and fusing it into solid form for shipment; and (6) a liquefying plant to condense and liquefy 50 tons of chlorine per day.

With the exception of chlorine, chlorpicrin was the first war gas to be manufactured on a large scale in this country. When pure, chlorpicrin is a colorless liquid which boils at a temperature approximately of 112° C. The compound has been known since 1848. While not so poisonous as some of the other products used in gas warfare, it is, nevertheless, an active poison, and has the additional advantage of being a fair lachrymator, or tear producer.

Chlorpicrin is made by the reaction between picric acid and chlorine. The chlorine is best supplied in the form of so-called bleaching powder, which is ordinary chloride of lime. In the manufacturing process as originally carried out, free picric acid was mixed with bleaching powder held in suspension with water. Later it was found advantageous to use calcium picrate instead of picric acid.

Accordingly, the final process was as follows: The bleaching powder was creamed with water and mixed with a solution of calcium picrate in large stills holding 5,000 gallons or more. A jet of live steam was then introduced at the bottom of the still, and the reaction began at once, the rapidity depending upon the amount of steam introduced. The resulting chlorpicrin, together with a certain quantity of steam, passed out of the still and was liquefied in the condenser. The resulting mixture of chlorpicrin and water was run into tanks, where the chlorpicrin, being insoluble in water, gradually settled to the bottom and was run off and used directly in gas shell.

In developing this process the Government was assisted by the Dow Chemical Co., the Semet-Solvay Co., and the American Synthetic Color Co., of Stamford, Conn., the principal work being done by representatives of the Bureau of Mines at the Stamford plant.

AIRPLANE VIEW OF CHLORINE PLANT, EDGEWOOD ARSENAL.

CHLORINE PLANT, EDGEWOOD ARSENAL.

This is the largest single chlorine and caustic soda plant in the country. Its capacity, when entirely completed, is 100 tons of chlorine and 112 tons of fused caustic soda per day.

CHLORPICRIN PLANT AT EDGEWOOD ARSENAL.

Bleaching powder, lime, and picric acid are received by rail. In the mixers appearing in the right foreground lime, picric acid, and water are mixed to form a solution of calcium picric, and bleach and water are mixed to form a cream. These solutions are pumped together into any of the several stills, where they react to form chlorpicrin. This plant was rated at 12½ tons of chlorpicrin a day, but reached a production as high as 31 tons on one day.

MIXER BUILDING OF PHOSGENE PLANT AT EDGEWOOD ARSENAL.

The capacity of this building is 20 tons of liquid phosgene per day. Dry 98 per per cent gaseous chlorine, as obtained directly from the cells of the chlorine plant, and pure carbon monoxide obtained from the producers, are mixed in approximate equal volumes and the mixture passed through catalyzers, where the two gases combine to form phosgene. The resultant gas is liquefied in the condensers, appearing in the left.

CHLORINE PLANT, EDGEWOOD ARSENAL.

One of eight cell rooms, capacity 12½ tons gaseous chlorine per day. Each cell room consists of six circuits—74 cells per circuit, or a total of 444 cells per room.

IN THE FOREGROUND THE CHLORINE PIPE LINE FROM CHLORINE PLANT PASSING TO CHEMICAL PLANTS IN RIGHT CENTER OF THE PICTURE. ON THE EXTREME RIGHT THE MUSTARD GAS PLANT. IN UPPER LEFT AND CENTER, VIEW OF FILLING PLANTS AND SHELL DUMPS.

FILLING 1-TON CONTAINERS WITH PHOSGENE.

Each empty cylinder weighs 1,300 pounds and will contain 1,650 pounds of liquid. The plant shown fills 25 cylinders per day.

MACHINE FILLING 75-MILLIMETER SHELL WITH MUSTARD GAS.

FILLING MUSTARD GAS SHELL AT EDGEWOOD ARSENAL.

Empty grenades are first completely immersed in a shallow hot-water bath, shown on extreme left in photo. In a tank that is not shown in picture white phosphorus is melted under water, and this molten phosphorus is pumped by a small centrifugal pump into a system of distributing pipes. Through a flexible tube and by hand, each grenade is completely filled with molten phosphorus, displacing the water in them. While the grenades are still immersed in the water bath, a suction tube is inserted in each grenade to remove the molten phosphorus to a certain depth below the top of the grenade, this molten phosphorus being displaced by water in the bath. The operation shown in the photo depicts the grenades thus filled with molten phosphorus to a definite heighth and with the remaining heighth filled with water, having the water removed from the top of the phosphorus by suction, after being taken out of the bath.

FILLING HAND GRENADES WITH WHITE PHOSPHORUS.

Inspected empty shell, as shown inverted on the left in the foreground, are placed on small filling trucks, shown in the right middle ground, and run under filling machine. Filled shell with boosters screwed down leave the tunnel, as shown in center of picture, where any possible mustard gas liquid on them is vaporized by gasoline torch. A draft from this operation into the tunnel prevents the distribution of mustard gas vapor throughout the plant. The loaded shell are then placed on trucks, as shown in foreground of photo.

FILLING LIVENS DRUMS AT EDGEWOOD ARSENAL.

This photo shows the Livens drums being filled with phosgene. The range of this special type of projectile, known as the Livens drum, is about 1,500 yards. Its empty weight is about 30 pounds, and it contains a charge of about 30 pounds of gas.

PAINTING AND STRIPING FILLED SHELL AT EDGEWOOD ARSENAL.

After leaving the filling plants, shell are classified by weight into four groups and each group maintained separately. The shell then are stored in an inverted position to detect leaks. After testing for 24 hours, the shell are buffed, painted, and striped by spray painting, as shown on the endless conveyor, and then are ready for packing. In the left background will be noted Livens drums being similarly painted.

SHELL DUMP AT EDGEWOOD ARSENAL.

This picture shows filled shell being stored for leakage test before being painted.

FILLED CONTAINERS OF PHOSGENE, READY AT EDGEWOOD ARSENAL FOR OVERSEAS SHIPMENT.

Each container holds approximately 1 ton of liquid.

PHOSPHORUS CLOUDS FROM BURSTS OF 75-MILLIMETER SHELL AT LAKE-HURST, N. J., PROVING GROUNDS.

GAS CLOUD FROM 4.7-INCH GAS SHELL EXPLODING 8,533 YARDS AWAY FROM THE GUN AT LAKEHURST.

America's whole supply of chlorpicrin during the war came from the American Synthetic Color Co. and the Edgewood Arsenal. The Stamford plant was the first to reach large-scale production.

The contract with the American Synthetic Color Co. was dated December 13, 1917; and the company shipped over 111,853 pounds of the gas to Edgewood on March 11. This, when mixed with the necessary stannic chloride, supplies of which were already on the ground, was sufficient to fill approximately 100,000 75-millimeter shell. In the spring of 1918, due to certain internal troubles at the Stamford plant, it was agreed that the United States should lease this factory and operate it as a Government plant. Under Government operation the total production of chlorpicrin at the Stamford plant amounted to 3,226,000 pounds, of which 2,703,300 pounds were shipped overseas in 660-pound drums.

The chlorpicrin plant at Edgewood went into entire operation on June 14, 1918. Up to the signing of the armistice this plant had produced 2,320,000 pounds of chlorpicrin.

Phosgene was one of the deadliest gases employed in the war. Numerous other gases were used to annoy the enemy and force the wearing of masks, but phosgene was a killer employed to produce as many casualties as possible. The gas did not persist long in the air or on the ground after the shell had exploded, so that it was an ideal chemical for use in an attack. The gas would clear away by the time the troops following reached the place of gas concentration.

Phosgene at ordinary temperatures is a colorless gas, but it condenses to a liquid at 8° C. It is formed by the combination of two gases, chlorine and carbon monoxide, in the presence of a catalyzer. The reaction is best conducted in iron boxes lined with lead and filled with charcoal of proper quality, into which boxes a stream of the reacting gases, mixed in proper proportions, is introduced. The reaction creates heat, and means must usually be taken to keep the reaction boxes cooled. The resulting phosgene is condensed to a liquid by passing the gas through a condenser which is surrounded by brine kept cold by refrigeration. The liquid is then stored in strong steel containers or run directly into Livens drums or artillery shell.

Prior to 1917, the Oldbury Electro-Chemical Co., of Niagara Falls, N. Y., had set up a small experimental phosgene plant in the hope that the experiments might lead to the commercial utilization of carbon monoxide which was obtained by this company as a by-product in the manufacture of phosphorus. When we entered the war the company had developed its process to such efficiency as to warrant the construction of a large phosgene plant, and the Government entered into a contract with the company for the creation of facilities with a capacity of 10 tons of phosgene per day.

Also, because of the great importance of phosgene in warfare, it was decided at the same time to build a Government phosgene plant at Edgewood. A little later the Government financed a phosgene plant at the factory of Frank Hemingway (Inc.), at Bound Brook, N. J.

The total output of the original small experimental plant at Niagara Falls, which was later leased by the United States, was 83,070 pounds of phosgene, of which 24,800 pounds were shipped overseas. The contract with the Oldbury Chemical Co. for its main phosgene plant was signed on January 15, 1918. Production here began on August 5 and by August 20 had reached a daily average of 5 tons. On November 1 the average daily production was 7 tons. The total quantity produced at this plant was 435 tons. The plant loaded 18,768 Livens drums with phosgene, each drum holding about 30 pounds. This plant was operated by enlisted men.

The contract with Frank Hemingway (Inc.) called for a factory producing 5 tons of phosgene per day by a secret process controlled by the company. The construction of the plant was begun on February 2, 1918, and phosgene was first manufactured on May 17. This concern reached its maximum of 5 tons per day by August 1, and produced in all 205 tons of phosgene.

Construction of the phosgene plant at Edgewood was begun on March 1, 1918. The plant consisted of four catalyzer buildings, each building having four units, each unit possessing a projected capacity of 5 tons of phosgene per day. The total capacity, therefore, was designed to be 80 tons per day. The carbon monoxide used in the process was produced by passing a mixture of oxygen and carbon dioxide over heated coke in a gas producer, the oxygen being supplied by a Claude machine with a capacity of 100,000 cubic feet of oxygen every 24 hours. The chlorine used came partly from the Edgewood chlorine plant and partly from outside sources.

The actual production of phosgene at Edgewood began on July 5. 1918, and worked up to an output of 20 tons per day by the date of the armistice. The total production of phosgene at Edgewood was 935 tons. The total output of phosgene from all three plants, Edgewood and the Bound Brook and Niagara Falls operations, at the date of the armistice was 35 tons per day; and this was increasing to reach 95 tons per day by May 1, 1919. The total phosgene produced by all the plants before the armistice was 1,616 tons.

The Germans, in spite of their attainments in chemistry, were never able to improve their clumsy and expensive methods of producing mustard gas. The best reports we have show that at the time the fighting ended, all of Germany's chemical warfare facilities could not produce more than 6 tons of mustard per day. The

United States alone had ten times that capacity on the same date, while France and England both reached a heavy output. So concerned was the German high command because of the fact that Germany was being outdistanced in the production of mustard gas that the ablest spy of the German Empire was sent into France in October, 1918, to find out the French method of making mustard. One of the Chemical Warfare officers who accompanied our forces into German territory reported that the Germans had decided to adopt the American method of making mustard gas and to stop their former process.

Mustard gas was by no means a child of the great war, having been prepared in experimental quantities since 1886. It is a colorless, slightly oily liquid, boiling at 220° C. with some decomposition. When perfectly pure it freezes at 14° C.; but, since it usually contains small percentages of impurities, it usually remains liquid at 0° C., or even below that. In chemistry the substance is known as dichlorethyl sulphide.

The first commercial process proposed for the manufacture of mustard gas depended upon the use of ethylene chlorhydrin; and on April 13, 1918, a contract was made with the Commercial Research Co., Flushing, Long Island, for the manufacture of 10 tons per day by this process. In the spring and summer of 1918 a new process was developed both abroad and in the United States, one which used sulphur monochloride. Accordingly, the contract with the Commercial Research Co. was canceled, and efforts were concentrated on the later process.

This process consisted in blowing gaseous ethylene into liquid sulphur monochloride in large iron reaction vessels. The reaction developed much heat. Sulphur is set free by this reaction, and the temperature must be controlled in order to prevent the formation of solid sulphur in the reaction machine.

At the date of the armistice three mustard gas plants were either completed or nearing completion. The construction of the Edgewood plant was begun on May 18, 1918, and the first mustard was produced exactly a month later. The changing of processes, however, hampered production somewhat, but by September 20, the arsenal was producing 10 tons per day, and by November 11 had increased this to 30 tons per day. The total production of mustard gas at Edgewood during the war period was 711 tons, of which approximately 300 tons went into shell.

On July 8, 1918, the Government began the construction of a mustard gas plant at Hastings, N. Y. This factory was to have a capacity of 25 tons per day, afterwards increased to 50 tons per day. The first unit of this plant was ready to operate when the armistice was signed.

On July 6, 1918, the Government signed a contract with the National Aniline & Chemical Co., Buffalo, N. Y., calling for a mustard gas plant with a capacity of 50 tons daily. On November 11 this plant was 80 per cent complete. The cost of the plant was met by the Government, but the operation was to be in the hands of the Buffalo concern. The total daily capacity of all three plants when complete was estimated to be 200 tons.

To insure an adequate supply of sulphur monochloride for its mustard gas production the Government built a special plant at Edgewood with a capacity of 300 tons of sulphur monochloride per day.

As soon as toxic gas warfare had developed to a considerable extent, the perfection of gas-absorbing masks had given almost a complete protection against this new weapon, if the soldier put on his gas mask in time. But the mask, especially the earlier forms of it, was not easy upon the wearer, due to the difficulty of breathing through it and also because it restricted the soldier's vision. It was soon discovered that a force compelled to wear its gas masks for any considerable period lost in efficiency. The employment of gas by both sides for the purpose of forcing the opposite sides to wear masks continually was an important element in war at the close of hostilities.

For this purpose the so-called tear gases were produced. Gassing the enemy with tear gas was much cheaper than with poison gas, yet it forced him to remain masked. The tear gases were highly effective. Even a trace of tear gas in the air would in a few moments blind a man temporarily. A single tear-gas shell could force the wearing of masks over an area so wide that it would require from 500 to 1,000 phosgene shell to produce the same effect.

Most of the tear gases had bromine bases; so it was early determined that we should have to increase the American supply of bromine considerably if we were to meet our gas-warfare requirements. Bromine is a deep red liquid which boils at 63° C. The domestic source of bromine is principally in certain subterranean brines found in the United States, these solutions containing bromine in its compounds. The brines obtained in the vicinity of Midland, Mich., are especially rich in bromine, and by far the largest amount of bromine obtained in this country comes from that locality.

In December, 1917, at a conference with Mr. Dow, of the Dow Chemical Co., Midland, Mich., it was decided that the Government should finance the sinking of 17 brine wells near Midland, the Dow Chemical Co. to supervise the work and to produce the bromine from the brine. The work on this project was not begun until March, 1918, but the entire project was practically completed when the armistice was signed. This plant is a future war asset of the United States.

It is capable of yielding approximately 650,000 pounds of bromine per year.

The tear gas which we prepared to manufacture was brombenzyl cyanide. It is a brownish oily liquid which solidifies to white or brownish crystals at 29° C.

The production of brombenzyl cyanide involves a fairly intricate chemical process. The first step is to chlorinate ordinary toluol, one of the coal tar bases, to produce benzyl chloride. This chloride is then mixed with sodium cyanide in alcoholic solution and distilled, benzyl cyanide being the result. It is then only necessary to brominate the benzyl cyanide by treating it with bromine vapor.

The first manufacture of brombenzyl cyanide in the United States was conducted at an experimental plant at the American University Station at Washington. After this a large scale plant was authorized at the plant of the Federal Dye & Chemical Co., at Kingsport, Tenn. The construction of this factory began on July 8, 1918, and operations started on October 29, the total production of brombenzyl cyanide being a trifle over 5 tons. In November the plant reached a capacity of 3 tons per day.

The bromine gases were not poisonous in the sense of being killers, but were merely highly irritating to the membranes of the eye. The killing gases were phosgene, chlorpicrin, and chlorine. Mustard gas in sufficient amount was also fatal, its effect being identical to that of a deep burn. It attacked the lungs, the eyes, the skin, and even the intestines if food contaminated with mustard gas were swallowed. An insidious feature of mustard gas is the fact that its action is practically always delayed. It might be several hours after a man was gassed, even fatally, with mustard before he became aware of it, and then it was too late to administer the treatment that might save his life. Goggles alone would have been sufficient protection against tear gas, except for the fact that it was invariably mixed with the deadlier gases.

The various experiments preliminary to our production of gases were conducted in provisional laboratories at the Bureau of Standards, Washington, D. C., Bureau of Mines, Washington, D. C., the Geophysical Laboratory, Washington, D. C., the Ohio State University, Columbus, Ohio, and Johns Hopkins University, Baltimore, Md. A control laboratory for the solution of problems arising in manufacture was eventually established at Edgewood. A total of 167,092 single chemical determinations were made at these laboratories under the direction of 20 commissioned officers, 45 noncommissioned officers, and 204 privates.

The production of gases and other chemicals was only part of the work of the Edgewood Arsenal and its subsidiary plants. The other chief activity was that of filling artillery shell with the toxic substances. The description of the plant which filled shell with phosgene will indicate the scale upon which this operation was conducted.

The empty shell, after being inspected, were loaded on trucks, together with the proper number of loaded boosters. The booster was the device which exploded the shell and scattered the gas. Electric locomotives then pulled the shell trucks to the filling buildings. There were four of these to a single shell-filling plant, radiating at right angles from a common center. From the trucks the empty shell were lifted by hand to a belt conveyor and the conveyor carried the shell slowly through a room kept cold by artificial refrigeration. Although the shell moved only 70 feet through this room the conveyor traveled so slowly that they were 30 minutes in transit, and during this time they were cooled to a temperature of about 0° F. This chilling was necessary because phosgene has a low boiling point, and it was necessary to keep the temperature of the metal of the shell considerably below the boiling point of phosgene in order that the gas might remain in liquid form while the filling was going on.

The chilled shell cases were next transferred to small trucks, each carrying six of them. The loaded truck was then drawn through a filling tunnel by means of a chain haul. This tunnel was so ingeniously contrived that the human assistance to the filling and closing machinery could all be conducted from the outside. The phosgene, kept liquid by refrigeration, was run into the shell by an automatic filler.

The truck was then moved forward a few feet to a point where the boosters were inserted into the noses of the shell by the hands of the operator reaching in through an aperture in the tunnel. The final closing of the shell was then accomplished by motors. The air in the filling tunnel was constantly withdrawn by strong ventilation, the exhaust air being washed in stone towers by chemical agents to neutralize any gases that might be present. The filled, inclosed shell were next conveyed to a dump, where they were classified and then stood nose down for 24 hours to test them for leaks. Then they were painted, striped, and stenciled by air paint brushes. The final process was to pack them in boxes and store them for shipment. This was done in large storage magazines on the grounds of the Edgewood Arsenal.

A similar method was used for filling shell with chlorpicrin, except that refrigeration was unnecessary. Mustard gas required another sort of filling machine.

Several filling plants were designed and constructed for filling grenades with stannic chloride and with white phosphorus, and also one for filling incendiary drop bombs.

The capacity of each of these plants per day was as follows:

Stannic chloride plant, hand grenades, 25,000. White phosphorus grenade plant, 30,000. White phosphorus smoke-shell plant, 155 millimeter shell, 2,000; or 4.7-inch or 5-inch shell, 4,000; or 75-millimeter shell, 6,000. Incendiary drop-bomb plant, 2,000.

The following sentences summarize the production and expectations of the Edgewood Arsenal:

(1) The gas program as of March 1918. called for approximately 545 tons of toxic gas weekly.

(2) The Chemical Warfare Service program of August 12, 1918, called for a much larger amount, viz, about 4,525 tons per week.

(3) The approximate filling capacity of the Edgewood Arsenal plant from August to November, 1918, was nearly 1,000 tons per week.

(4) The toxic gas production during this same period increased from 450 to 675 tons per week.

(5) The capacity of all projectiles received, unlimited by boosters, varied during the same period from 125 to 450 tons per week.

(6) The maximum capacity corresponding to boosters received was less than 100 tons pei week.

In these facts it will be seen that the numbers of empty shell delivered to the plant was far less than the number required to accommodate the gas production. Many of the shell received were without boosters and therefore without value until boosters were provided, so that the limiting factor was really the supply of boosters. The booster supply was sufficient to take care of only a relatively small fraction of the toxic gas actually produced. The filling capacity of the plant was also in excess of the delivery of shell and boosters. The 75-millimeter shell-filling plant had a capacity of 1,200,000 shell per month, eventually double that, while delivery of shell was slightly over 300,000 per month and of boosters less than 200,000.

Because of the nature of toxic gas it is impossible to store it up in any large quantities. Early in the summer of 1918 large amounts were shipped in bulk overseas and there loaded into shell. Later we received instructions to stop all shipments in bulk except a limited amount of chlorine, and thereafter our production was limited to the number of shell and boosters available.

In June, 1918, we shipped in bulk 15 tons of mustard gas, 705 tons of chlorpicrin, and 48 tons of phosgene. This was to be exchanged for gas shell produced by the French. In late July the French had no more extra shell to be filled with American gas and this fact terminated the arrangement. However, we sold excess gas both to England and to France. England received 900 tons of our chlorpicrin and 368 tons of American phosgene. France took 300 tons of chlorpicrin and 1,408 tons of chlorine, equivalent to 1,226 tons of phosgene, since phosgene is 80 per cent chlorine including allowance for wastage in manufacture. France furnished phosgene shell to us in exchange for chlorine. In addition 200 tons of mustard gas were shipped to England and utilized by the English.

We therefore shipped to Europe in bulk 3,662 tons of gas or its equivalent, which gas was largely loaded in shell and used by the United States troops or those of the allies. This quantity was sufficient to load 1,600,000 shell, two-thirds of them being of the 75-millimeter caliber and the other one-third 155-millimeter, the total number being thought to be at least equal to the total number of gas shell fired by American troops in action. Thus while American gas was not actually fired in American shell against the Germans, American gas was used against the enemy and America furnished at least as much gas as she fired.

In addition to this we shipped 18,600 Livens drums loaded with phosgene. These contained 279 tons of gas, and some of them were fired at the enemy. We began producing loaded gas shell in the summer of 1918 and by August 9 had shipped 75,000 loaded 75-millimeter shell. These shell were unassembled for firing in the guns, the Ordnance Department having decided in June to assemble gas shell in their cartridge cases in France.

The Chemical Warfare production organization developed and manufactured a large number of special containers for the shipment of toxic gases. These were of special construction in order to guard against dangers that would result from leaks, and all had to stand the tests required by the Bureau of Explosives before they would be received for railroad shipment. The 1-ton containers, all of which would hold 1 ton of liquid chlorine, were designed by the Ordnance Department and would withstand a pressure of 500 pounds per square inch. The 300-pound phosgene cylinders, designed by the Ordnance Department, were made to withstand a 500-pound hydrostatic pressure and a 250-pound air test.

We purchased standard 55-gallon acid drums and standard-pattern cylinders for holding 75 pounds of chlorine.

We constructed chlorine tank cars, each tank with a capacity of 15 tons and a strength that could withstand a pressure of 500 pounds to the square inch. We also designed a tank car originally for the shipment of chlorpicrin and later used it for shipping sulphur monochloride.

CHEMICAL WARFARE PRODUCTION DATA.

Production of toxic materials.

[All figures are for tons of 2,000 pounds.]

1918.	Liquid chlorine.[1]	Gaseous chlorine.[2]	Chlorpicrin.	Phosgene.	Mustard oil.	Brombenzyl cyanide.	White phosphorus.	Tin tetrachloride.	Titanium tetrachloride.
January	....	....	10	....	....	....	....	....	....
February	....	....	27	....	....	....	34	....	....
March	40	....	59	....	....	....	74	38	....
April	176	....	33	15	....	....	59	116	....
May	378	....	130	18	....	....	70	51	50
June	546	....	263	23	6	....	60	95	....
July	512	....	499	100	21	....	80	112	27
August	243	....	646	314	36	....	162	94	53
September	438	191	564	327	144	....	125	96	26
October	242	649	445	664	361	....	265	75	25
November	148	264	100	155	143	5	77	18	....
Total	2,723	1,104	2,776	1,616	711	5	1,006	695	181
Amount shipped overseas	1,488	....	1,903	420	190	....	171	106	....
Total monthly producing capacity, Nov. 1, 1918	895	1,500	1,500	1,050	900	90	100	91	30
Estimated capacity, Jan. 1, 1919	1,100	2,250	1,500	1,650	4,000	90	100	91	30

[1] Procured from commercial agencies. [2] Manufactured at Edgewood.

Shell, grenades, Livens drums, and drop bombs filled.

1918.	75-mm. shell.			Grenades.		Livens drums.	Incendiary drop bombs.	
	Chloropicrin.	Phosgene.	Mustard oil.	White phosphorus.	Tin tetrachloride.	Phosgene.	Mark I.	Mark II.
July	62,866	....	....	8,696	1,639	....	....	....
August	125,951	....	....	170,160	56,763	1,738	350	....
September	110,358	1,988	75,529	51,421	127,319	6,355	....	1,998
October	109,704	12	79,272	110,928	147,669	12,026	184	100
November	15,892	9	224	98,948	30,386	5,570	8	6
Total	424,771	2,009	155,025	440,153	363,776	25,689	542	2,104
Total number shipped overseas	300,000	....	150,000	224,984	175,080	18,600	....	....

Total monthly capacity of filling plants on date of armistice (Stokes shell, drop bombs, and other special containers not included).

75-mm. shell (ultimate capacity)	2,400,000
4.7-inch shell	450,000
155-mm. shell	540,000
6-inch shell	180,000
Gas grenades	750,000
Smoke grenades	480,000
Livens drums	30,000

CHAPTER II.

GAS DEFENSE EQUIPMENT.

During the spring and summer of 1917 two marked tendencies were to be observed in the fighting in France. One of these was the greatly increased use by both sides of poisonous gases and chemicals, frightful in their effect; the other the almost complete censorship that hid the knowledge of this tendency not only from the people of Europe but particularly from those of the newest belligerent, America. The French and British Governments, who then controlled all news from the front, feared, and perhaps with reason, that if the picture of gas warfare, as it was then developing, should be placed before the American people, it would result in an unreasonable dread of gases on the part of the American Nation and its soldiers.

One year later, with tens of thousands of American troops facing the Germans, there was almost no censorship upon the details of fighting with chemicals. The mysterious gases of 1917 were then known to almost every reading individual in the civilized world. The once secret formulas were published in the technical journals. Noncensored photographs of defensive equipment were freely published, and masks and other paraphernalia were exhibited for the public interest. Except for secret plans for the future and the various surprises being prepared by one or more of the belligerents, the whole subject of chemical warfare had become an open book.

What occasioned this change in policy on the part of governing authorities? The reason was that the American troops brought with them to France the best and most protective gas masks the world had seen; and they brought these with them by the millions. Starting a mask-production effort in May, 1917, America turned out a total of 5,250,000 gas masks before the armistice was signed, and sent more than 4,000,000 of them overseas. As to the quality of these masks, it is only necessary to say that they gave twenty times the protection afforded by the best German gas masks. In other words, we protected our soldiers against the poisons which Germany had brought into warfare, and protected them completely. No American soldier was ever gassed due to the failure of an American gas mask, and such gas casualties as did occur were due to the fact that the masks were not quickly enough utilized when gas was thrown over, or because the soldier was unaware of the presence of gas. With such protection there was no longer reason to fear that the frightfulness of chemical warfare would reduce American morale.

The production of gas masks was one of the most picturesque and successful phases of our entire war preparation. It engaged the attention of some of the principal chemical engineers of the country, and millions of men, women, and children in the United States contributed something to the success of the undertaking, if only to obey the "Eat More Coconut" slogan or to save peach stones for the benefit of the production of the charcoal essential to efficient gas masks.

The problem of making masks in such quantity and under such supreme demands for perfection was one which might well stagger manufacturers accustomed to large-scale operations. We started in with practically no knowledge whatsoever of the fundamental principles of a perfect mask. Yet the apparatus was as difficult to build as a rifle. It must, perforce, be made of perishable materials, and this fact brought the question of durability to the fore at the very start. It was evident that no chemical substances known in our past commercial life would give protection against the new poisons which had been developed in Europe. With the exception of phosgene and chlorine, the various war gases which had been brought out prior to our entrance in the struggle were completely unknown in our trade or commerce and had existed only in our experimental laboratories. Then it was discovered that as these toxins increased in power they could penetrate the ordinary fabrics known in commerce, and this necessitated the creation of new types of materials to be used in the masks. Finally the increasing use of gases forced the soldiers to wear their masks for much longer periods than had been necessary at the beginning of gas warfare; so that the problem of comfort became one of great importance. All of these basic considerations indicate to some extent the difficulty of the undertaking.

The chlorine, which floated in a pale greenish-yellow cloud down upon the defenseless Canadian troops at Ypres, with such terrible effect upon the men, was, as has been said, the first gas used. Chlorine, though easy to obtain, the principal source of supply being common table salt, was, from the standpoint of strategy, far from being the ideal gas of warfare. Troops could be quickly and easily protected from it. But even as it was, only lack of faith in their new weapon prevented the Germans from winning the war with it then and there. Had they brought into the fighting a sufficient supply of this chlorine, they might have gassed their way to Paris in short order. In fact, they brought to the line an almost negligible supply and they themselves were insufficiently protected to go through their own gas and follow up the attack. By the time they were able to renew gas warfare the French and British had equipped themselves with masks which were sufficient to protect men against chlorine.

Thereafter the tendency was toward new and strange gases which were heavy in weight and highly toxic in their physiological action. This development led to new, slightly volatile liquids, the so-called mustard gas being the best example. Mustard gas (properly called dichlorethyl sulphide) is similar to lubricating oil in many of its physical characteristics but smells like ordinary mustard. Ground soaked with the mustard gas remains impregnated for days, the vapor rising continually.

A perfect mask is one which will remove completely every trace of gas or poisonous vapor before the air can reach the eyes, nose, or mouth of the soldier.

The first masks adopted by the allies were simply gauze pads saturated with neutralizing chemicals. These became unsuitable as soon as new varieties of powerful poisons were brought out. The mask development thereafter progressed to the box respirator type. This consisted of a mask or helmet connected to a box filled with absorbing and neutralizing chemicals which purified the air for the mask wearer. This was the type of respirator in use to the end of the fighting.

It is quite clear to us now that only such a mask could be efficient in chemical warfare, but in the early part of 1917 the matter was not clear either to us or to the allies. The first requisitions from the A. E. F. called for masks of two types, each soldier to be supplied with one of each. The reserve mask was to be of the gauze type and the regular mask of the box respirator type, affording protection from the more powerful poisons that were then just coming into use. We wasted considerable energy at the beginning in our attempt to produce both types. Eventually, however, when we were just ready to start manufacturing the gauze-type mask, orders came to abandon the effort, since it was even then apparent that our soldiers must be prepared at all times to withstand all gases.

The box respirator equipment, the general principle of which was finally adopted by all the nations at war, fell into two classes. In a single-protection mask the wearer breathed air from inside of the face piece, so that any leakage around the edges of the face piece would result in a casualty when the wearer was in a strong concentration of gas. The other sort, known as the double-protection mask, consisted of a gas-tight face piece, similar to that of the single-protection mask. In this type, to guard against any possible leakage around the edges between the mask and the wearer's skin, the breathing system was sealed away from the air inside the face piece by means of a rubber mouthpiece and a nose clip, the wearer inhaling through the mouthpiece.

VARIOUS TYPES OF GAS MASKS.

Top row, left to right.—First type U. S. Navy mask, now obsolete; U. S. Navy mask as finally developed; U. S. C. E. respirator (production started October, 1917); U. S. R. F. K. respirator (production started February, 1918); U. S. A. T. respirator (production started August, 1918); U. S. K. T. mask (production started August, 1918); U. S. model 1919 mask (ready for production when armistice was signed).

Middle row, left to right.—British black veil mask (first mask used after initial gas attack in April, 1915); British P. H. helmet (stops phosgene but not tear gases); standard British box respirator used by all British forces after 1916; French M-2 mask used by the French; French Tissot mask used by artillerymen; French A. R. S. mask.

Bottom row, left to right.—Late type of German mask; Experimental mask; Italian mask (similar to French M-2 mask); British Motor Corps respirator; U. S. rear area emergency respirator; U. S. Connell mask never passed the experimental stage).

AMERICAN C. E. TYPE OF BOX RESPIRATOR.

This side view shows face piece ,harness, hose, flutter valve, and knapsack. This is the mask most used by our troops.

The United States and English double-protection masks consisted of 11 principal parts as follows:

1. A knapsack slung from the shoulder or neck. This contained the canister and a pocket for storing away the mask when not in use.

2. A metal canister in which was contained the absorptive neutralizing chemicals.

3. A flexible hose reaching from the canister to the face piece.

4. A flutter, or exhalation, valve, which opened when the wearer exhaled his breath and closed when he inhaled, thus bringing the inhalation through the canister but allowing the exhalation from the lungs to pass out without polluting the chemicals of the canister.

5. The face piece, or hood, fitting snugly around the edges and covering the eyes, cheeks, lower forehead, nose, mouth, and chin.

6. The eyepieces, or lenses, through which vision was maintained.

7. An elastic harness for the head, to hold the face piece in place.

8. A body cord to tie around the chest and hold the knapsack firmly, so that the mask could be seized in both hands and pulled out of the knapsack.

9. A metal flange connection or angle tube which carried the hose through the face piece to the mouthpiece.

10. A rubber mouthpiece through which the wearer breathed and which helped to hold the mask in place.

11. A wire nose spring and rubber nose pad to hold the nostrils shut and force breathing through the mouth.

The first order for gas masks was issued on May 16, 1917, when the Chief of Staff asked the Surgeon General to supply 1,100,000 masks before June 30, 1918, or within about one year. Meanwhile 25,000 masks were needed at once in order to equip Gen. Pershing's first division, then about to sail overseas. There was but one man in the Army who knew anything at all about the subject and who could even attempt to produce this quantity in three weeks. This was Maj. (later colonel) L. P. Williamson, of the Surgeon General's Department, who had been spending some months at the Army War College at Washington studying as a side issue such papers on gas warfare as came from abroad. It was due to his knowledge and the volunteer staff of the Bureau of Mines that we were able to begin the actual manufacture of masks within a few days after the requirements were fixed, and actually to turn out 25,000 masks in but little more than three weeks' time.

Col. Williamson's first step was to consult with Dr. Van. H. Manning, the Director of the Bureau of Mines, and with his assistant, Mr. G. A. Burrell. Since February, 1917, the Bureau of Mines had been experimenting with gas masks and had built up a corps of scientists for this work. Within this organization was Mr. Bradley

Dewey, a chemical engineer, who, though then director of the research laboratory of the American Sheet & Tin Plate Co., of Pittsburgh, had been loaned to the Bureau of Mines. To Mr. Dewey was turned over the job of directing the production of the first 25,000 masks for the American troops then sailing.

To produce 25,000 gas masks in three weeks meant to compress England's two years of experience into 21 days. The military authorities of this country at that time could plead entire ignorance of the qualifications of an efficient gas mask. The prevailing idea seemed to be that you could go out into the market and buy them by the hundreds of thousands, as you might buy Halloween masks. But this was not any ordinary poison which we were to fight. These powerful chemicals attacked the human tissues as would acid. As the result of the effort, we did supply the first division going overseas in July. However, the masks were inferior to the British and were quickly replaced in France by British equipment. It was not until the following January that we developed an apparatus which we regarded as satisfactory to undergo the supreme test of battle.

To indicate some of the difficulties overcome between May and December, 1917, there are here set forth some of the features of an effective mask.

In the first place, the face piece must fit perfectly; it must not leak gas around the edges. It must fit into the hollows of the temples and must give the jaws a free space in which to work, and yet not slip back and press against one's Adam's apple. The pressure of the mask on the forehead must come above the supraorbital nerves which are just above the eyebrows, or else intense headaches will result from a few moments' wear. Moreover, to fit all faces and heads, several graduated sizes of masks are required. We first attained the gas-tight fit with a padded band around the edge of a flexible rubber-cloth face piece. Later we developed a thicker, stiffer face piece, but maintained a gas-tight fit by the elasticity of the face piece and the head harness.

Then the material of the face piece must be gas-tight in itself. At first we manufactured a fabric made by spreading rubber on cotton sailcloth; and, after testing it, we found that the smallest molecule known, that of hydrogen, would not pass through it in large amounts. This seemed to be a suitable fabric, until tested by the newer gases. Then we found that some of these gases were soluble in rubber compounds and could dissolve their way through thin rubber so quickly that the face piece cloth offered practically no protection at all. Another difficulty with the rubber fabric was that it was likely to absorb and hold certain of the poisons, so that a man might be gassed by the mask itself. The rubber companies, principally at Akron, Ohio, experimented continually until they discovered a coating that would not only withstand gas concentrations for a sufficient time, but would also aerate promptly and lose as much gas as it had absorbed.

The eyepieces or lenses offered another problem. Celluloid is strong but it is not so transparent as glass. It ignites easily and is easily scratched. Glass is ideal in transparency and will not burn, but is fragile. It was evident that we must provide eyepieces which would not break easily, since even so slight an accident as the breaking of a lens might cost a soldier his life by admitting concentrated gas to the mask. A material known as triplex glass had been experimentally made. This consisted of a thin celluloid strip sandwiched between two layers of glass, all three welded together. This glass would not splinter, and even if cracked or broken, would still be gas-tight. However, this had never been made in quantity and it was necessary to work out many kinks and to start a large plant to provide the necessary millions of lenses.

Then there was also to be overcome the tendency of the eyepieces to dim, particularly in cold weather, as the wearer breathed moist breath into the mask. The answer to this problem was a soapy compound which put a slippery surface on the glass and avoided the droplets of mist. The first masks were also equipped with deep plaits so that the wearer could wipe off the lens with the interior of the facepiece itself, though the final development (the invention of a Frenchman by the name of Tissot) was to bring the cold air into the mask so that it flowed directly against the lenses and evaporated any condensed moisture. This kept them clear under all ordinary circumstances.

It was evident that the metal tube passing through the face piece must not contain pinholes and must be able to stand rough handling without pulling loose. The harness must maintain a gas-tight connection between the wearer's face and the face piece, but not at the cost of pain or chafing of the face or head. The flutter valve must fit with absolute tightness and must work perfectly and instantaneously at all times.

The flexible hose leading from the canister to the face piece must be strong and without flaws or leaks, and yet flexible in the extreme. A stiff hose would be likely to swing and displace the face piece whenever the wearer moved. The mouthpiece must be comfortable and must be built along lines to prevent irritation to the gums or lips, yet it must be reinforced so that in his excitement the soldier can not bite down and shut off his air supply.

The canister must withstand corrosion and must be gas-tight. Smooth sided canisters can not be used, for the gas would slip up the sides without coming in contact with much of the chemical filling. The sides of the canisters were, therefore, ribbed so that the charcoal and other ingredients working into these ribs baffled the gas and threw it out into the body of the chemicals. The canister, moreover, must be equipped with a perfectly working check valve which will stop exhalation through the canister and force the air to pass out through the flutter valve.

The web sling of the knapsack must not curl and chafe the neck or shoulders of the wearer. The knapsack must be waterproof and must have easily and quickly workable fastenings.

The canisters were filled with charcoal and with cement granules. These were crushed into carefully sized small bits about the size of a pinhead and packed in layers in the canisters. The air could pass through them easily and the particles of both substances absorbed gas. The chief quality requirements for the carbon and the cement were that they must have long life and great activity.

Of the canister ingredients the charcoal offered the more difficult technical problem. It had long been known that charcoal was highly absorptive of certain gases, but except in rare instances no thorough study had ever been made of the subject. It was evident, however, that the more charcoal or carbon which could be packed into the canister and still allow the free passage of air the greater the amount of gas that would be absorbed. Consequently a search was made for carbon existing in the natural state in the most compact form. This search is described later.

Each canister also contained concrete granules in a definite proportion to the carbon pieces. These granules were made of cement mixed with strong alkalis and oxidizing agents to digest the poisons as they passed through the canister.

It will be seen that the manufacture of good gas masks was a highly technical undertaking, one calling for the best talents of eminent men of science. The mask was not something that could be improvised on the spur of the moment, but each part of it must be worked out after the most painstaking research. The Gas Defense Division of the Chemical Warfare Service never at any time approved a type of mask which its own officers or men did not themselves wear in the most deadly concentrations of gas.

To get back to the chronological order of development, on May 21, 1917, the making of the first 25,000 masks was started with frantic haste; though, as it developed later, there was no need for such an effort, since there were available in England and France plenty of masks for the first American troops. Working to produce in the shortest possible time some sort of protection for the first overseas division, the officers in charge were forced to adopt methods which, had they been followed throughout the manufacturing program, would have been extremely costly. There was no time then to stop and study the problem either here or abroad. Before the end of June 20,088 masks had been started overseas, and 5,000 more were ready a little later. The most that can be said for this effort was that it gave our officers the experience which was the groundwork of the solid development later on.

The production of these first 25,000 masks called upon the services of various manufacturers. The assembling of the masks was conducted by the American Can Co., at Brooklyn, N. Y. The B. F. Goodrich Co., of Akron, manufactured the face pieces with the eyepieces inserted, also the connecting hose, the check valve of the canister, the flutter valve, and the rubber mouthpiece. The American Can Co. produced the canisters. The Day Chemical Co., of Westline, Pa., gave the charcoal its first burning. The Ward Baking Co., of Brooklyn, patriotically baked the charcoal—to activate it—in their bread ovens free of charge. The General Chemical Co., of New York, supplied the soda-lime granules. The Doehler Die Casting Co., of Brooklyn, manufactured the angle tubes. The Simmons Hardware Co., of St. Louis, produced the waterproof knapsacks. The Seaver Howland Press, of Boston, printed the cards of instructions that went with the mask outfit; and the Beetle & MacLean Manufacturing Co., of Boston, printed the record tags.

Though Maj. (now colonel) Williamson was formally in charge of this emergency work, he requisitioned the masks from the Bureau of Mines, which took entire charge of the first contract. Following this, on August 31, 1917, the Gas Defense Service of the Surgeon General's Department was established by official order, and Mr. Dewey, who had been working as a volunteer in the Bureau of Mines, was commissioned major and put in charge.

The next step was to prepare for the permanent development and manufacture of gas masks. Contracts were let for the manufacture of 320,000 component parts of masks as we then knew them, and a price was fixed for the assembling of the entire original requirement of 1,100,000 masks. The assembling contract went to the Hero Manufacturing Co., of Philadelphia, which remained until the end of the war the sole private contractor assembling our gas masks.

The spirit of cooperation and desire to serve the Government was evident from the start. The B. F. Goodrich Co. had been the only producers of the rubber parts of the first 25,000 masks. In this original contract it had gained valuable technical and cost knowledge; but in order that the Government might not be limited to one source of supply for such parts, the Goodrich Co. voluntarily imparted to the Goodyear Tire & Rubber Co. and to the United States Rubber Co. the information that would enable them to bid intelligently for portions of the work. This was a distinct departure from the usual practice in competitive industry.

109287°—19——27

All during the fall of 1917 and early winter of 1917–18 the development of the mask continued, the Government experts working hand in hand with private contractors. Because of the newness of this sort of manufacture and because of the wide variety of unusual articles required, entailing in some instances the actual creation of hitherto unknown commodities, the Government at all times was required to act as the procurer of raw materials for the masks. In this period of development America designed her own typical mask—a gradual evolution, but one which, though based on the British design, arrived at a perfection which had been unknown in warfare before.

The triplex glass used in the eyepieces was a patented commodity produced only in one small factory in Philadelphia. It was necessary to expand the facilities for the production of this necessary material. Meanwhile some of the men engaged in the work had improved the eyepiece by providing it with an aluminum mounting. But this very improvement brought embarrassment to the work, since the Akron rubber contracts had provided for eyepieces inserted in the fabric itself, and to apply the aluminum frame brought about a radical change in the manufacturing methods at the rubber factories.

There were also many other problems that had to be solved before our authorities were satisfied to go ahead in quantity production. There was the matter of rubberizing the face-piece fabric, for instance. Two methods of rubberizing cloth were in use. The first method was to roll out a thin sheet of rubber and then press it into the cloth fabric by running the whole thing under heavy rollers. This was known as the calender method. The other method, called the spreader method, was more intricate. In this process the sailcloth, tightly stretched, was carried around a roller. Above the roller a few thousandths of an inch was a knife blade extending from edge to edge. The rubber compound in liquid form was then fed upon the roller in such manner that a thin film of it pressed under the knife blade and upon the cloth on the roller. The rubberizing method finally adopted was a combination of the calender and spreader methods. The rubber was applied green to the cloth. The curing process thereafter was highly important. If the curing process were too short, the rubber would be sticky and would pull off the sailcloth too easily. If the rubber were overcured, it would crack and split.

Nothing short of absolute perfection in every part would do, since the slightest imperfection anywhere was likely to cost a man his life. Consequently we installed at the various producing plants not only 100 per cent inspection, but we constructed laboratories for putting the materials through the most elaborate and exhaustive sorts of control tests, and then reinspected the parts at the assembly plants, both before and after the assembly.

CHEMICAL DEVELOPMENT DEPARTMENT OF LONG ISLAND LABORATORY, GAS DEFENSE DIVISION, SHOWING INTERMITTENT FLOW CANISTER TESTING MACHINE.

PHOTO TAKEN AT PHILADELPHIA CONTROL LABORATORY OF GAS DEFENSE DIVISION, SHOWING APPARATUS USED IN EXPERIMENTAL WORK ON THE EFFECT OF RESISTANCE TO INHALATION AND EXHALATION OF MASKS.

SIDE VIEW OF GAS CHAMBER AT CHEMICAL DEVELOPMENT LABORATORY, SHOWING SUBJECTS ON OUTSIDE BREATHING TESTS.

ANOTHER VIEW OF GAS CHAMBER AT CHEMICAL DEVELOPMENT LABORATORY, SHOWING SUBJECTS INSIDE.

All the rubber used was continually sampled and analyzed in the laboratories. The tensile strengths of all fabrics were determined by standard destructive tests. We also tested the adhesion of the rubber coating by standard chemical methods and worked out flexibility tests for the breathing tube.

After all of the factory inspection and material-control tests, the masks themselves were sampled and worn in highly toxic atmospheres. In this work thousands of our masks were worn by the officers and men of the Gas Defense Division in concentrated atmospheres of the most deadly gases. For such work we constructed testing rooms whose atmosphere could be completely exhausted and changed in 90 seconds. The efficiency of canisters was tested either by the lungs of the inspectors or by mechanical breathing into telltale solutions.

The story of the carbon (charcoal) which went into the American canister is one of the most interesting phases of the whole undertaking. Investigations carried on by the research staff of the National Carbon Co., aided by a clue from the University of Chicago, led to the selection of coconut shell as a raw material. Any carbon absorbs a definite number of times its weight of gas. Therefore the densest carbons will be most efficient, volume for volume, as gas absorbers in a given space. Coconut shells and other nut shells were found to be the most compact form in which carbon exists in nature in commercially practicable quantities, being considerably superior in this respect to anthracite coal and to such woods as ironwood and mahogany. Another essential for charcoal used in the canisters was that it must be so hard that it would not crumble easily and produce dust that would clog up the air passages and prevent easy breathing through the canister. Coconut shell fulfilled both of these conditions better than any other known material.

Further study by the National Carbon Co., backed up by wonderful large-scale development work, paid for and carried out by the National Electric Lamp Association under the direction of their Mr. F. N. Dorsey (who later became Col. Dorsey and chief of the Development Division of the Chemical Warfare Service), gave us the details of a new process for treating the charcoal to make it absorptive. After the original burning of the nut shells, or other carbon materials, the resulting carbon was given a second highly specialized heat treatment, and this activated it until it had a powerful affinity for gas. Such carbon, made from nutshell material, would absorb 150 times its own volume of chlorpicrin, one of the most deadly of the war gases, the action being approximately instantaneous.

It must not be supposed, however, that investigation of carbons stopped with these experiments. In the search for the ideal carbon we experimented with almost every hard vegetable substance known. Literally, hundreds of kinds of carbon were tested. Next to coconut shells, the fruit pits, several common varieties of nuts abundant in the United States, and several tropical nuts, were found to make the best carbon. Pecan nuts, and all woods ranging in hardness from ironwood down to ordinary pine and fir, were found to be in the second class of efficiency. Among other substances tested were almonds, Arabian acorns, grape seeds, Brazil-nut husks, balsa, osage oranges, Chinese velvet beans, synthetic carbons, cocoa bean shells, coffee grounds, flint corn, corn cobs, cottonseed husks, peanut shells and oil shale. While many of these substances might have been used in an emergency, none of them would produce carbon as efficient, volume for volume, as that of the coconut shells and other hard nuts.

Some idea of the scale of the American mask production may be seen in our requirements for coconut shells. In our survey of raw materials we included the entire coconut resources of the world. Such figures were relatively easy to obtain because the copra, or dried coconut meat, industry is an important one, particularly in southern Asia and the South Sea Islands of the Pacific. Ceylon was the greatest single source of coconuts, 2,300,000,000 nuts being gathered there annually. British India was next with 1,500,000,000 nuts. Our own Philippine Islands were third, with an annual production of 900,000,000 nuts. Then followed in order the Dutch East Indies, British Malaya, French Indo-China, Siam, and the Pacific archipelagos, the total production of the Orient being 7,450,200,000 nuts annually. This was a supply that would provide 4,000 tons of coconut shells every day. The total production of coconuts in Central America, the West Indies and the Carribean coast of South America amounted to 131,000,000 nuts annually, equal to a supply of 75 tons of shells daily.

When we first began to build masks our demands for carboniferous material ranged from 40 to 50 tons a day of raw material; but by the end of the war, due to vastly increased mask requirements, we were in need of a supply of 400 tons of coconut shells per day. This demand would absorb the entire coconut production of the tropical Americas five times over. It was equal to one-tenth of the total coconut production of the Orient. Since transportation from the oriental countries was out of the question on the scale demanded by our mask program, it was evident that we were likely to be seriously embarrassed by the lack of raw materials; and, indeed, at no time before September, 1918, did we have on hand a reserve supply of shells and other charcoal materials that would last for more than a few days, though at no time after the start was the actual output of masks retarded by lack of these materials.

In building up our supply of coconut shells we naturally turned first to the resources in the United States. America normally con-

AIRPLANE PICTURE OF CARBON PLANT AND CANTONMENT OF GAS DEFENSE DIVISION AT ASTORIA, LONG ISLAND.

GENERAL VIEW OF CARBON PLANT NO. 3 ON LEFT, AT GAS DEFENSE DIVISION PLANT, ASTORIA, L. I

Storage bins are in central background, with administration building and carbon plant No. 2 in the right foreground.

CARBON PLANT NO. 2 AT ASTORIA, L. I., SHOWING ALSO OFFICE AND LABORATORY.

sumes fresh coconuts at a rate sufficient to supply about 50 tons of shells daily. The war restrictions on the use of sugar had the effect of cutting down the consumption of coconuts, used largely in candy and cakes, and consequently one or our efforts was to increase by widespread propaganda the use of coconut. The "Eat-More-Coconut" campaign more than doubled the American consumption of coconut in a brief space of time; and the 50 tons of shells daily, which had been the original supply, grew in volume until in October, 1918, with the help of importations of shell, we averaged about 150 tons per day exclusive of the Orient.

The first heating of coconut shells to make charcoal reduces their weight 75 per cent. Therefore it was evident that we could most economically ship our oriental supply in the form of charcoal produced on the other side of the Pacific Ocean. For this purpose, in August, we established under the direction of an officer of the Chemical Warfare Service a charcoal plant in the Philippine Islands. From this plant agents were sent to Ceylon, India, Siam, and other oriental countries to purchase enormous supplies of nutshells. This work was only gaining momentum when the armistice was declared. As it was, the Philippine charcoal plant actually shipped over 300 tons of coconut shell carbon to the United States and had 1,000 tons on hand ready for shipment on November 11.

The method adopted in the Philippines was to burn the shells in long, shallow trenches. As soon as the smoke had disappeared and the flames came clear and lambent through the incandescent mass, the bed of coals was smothered by means of galvanized-iron lids thrown over the trenches. It is interesting to note that the coolies hired by the Chemical Warfare Service in the Philippines would not work at charcoal burning more than a few hours each day, because they declared that the heat from the pits would give them tuberculosis and other lung troubles.

Meanwhile agents and officers of the Gas Defense Division were searching the tropical regions of Central and South America for other nuts valuable for this purpose. The best of these was found to be the cohune or corozo nut. These nuts are the fruit of the Manaca palm tree. They grow in clusters, like bananas or dates, one to four clusters to a tree, each cluster yielding from 60 to 75 pounds of nuts. Cohune nuts grow principally on the west coast of Central America in low, swampy regions from Mexico to Panama, but are also found along the Caribbean coast. Before the war created a demand for cohune nuts none of them had ever been imported commercially in this country, although it is understood that France had a prewar commercial use for them.

The chief virtue of the cohune nut from our point of view was its extreme thickness of shell, the kernel of this large nut, which is 3 inches or more in length and nearly 2 in diameter, being relatively small. We were importing cohune nuts at the rate of 4,000 tons per month at the time of the armistice. A disadvantage in the use of cohune nuts was that their husks contained a considerable amount of acid which rotted the jute bags and also caused the heaps of nuts to heat in storage. The fire department at the Chemical Warfare Service nut storehouse at Astoria, N. Y., was kept busy putting out spontaneous blazes in the storage piles of cohune nuts. We also sent agents to West Africa and there arranged for the shipment of some hundred tons of palm nuts a month.

A third source of tropical material was in the ivory nuts used in considerable quantities in this country by the makers of buttons. In the button factories in this country there is considerable waste of this nut material, amounting to 400 or 500 tons a month, this waste including the nut dust which was useless to us and had to be screened out. The price of ivory-nut waste was high, because of the use of this material in the manufacture of lactic acid. Nevertheless, we used a considerable quantity of it.

Another great branch of activity in securing carbon supplies was undertaken in this country. In the search for fruit pits and for domestic nuts it was found that the quantity of apricot pits, peach pits, cherry pits (largely from the canning industry), and walnut shells on the Pacific coast amounted to 23,600 tons annually. We arranged for the whole Pacific coast supply of these commodities and converted a part of a San Francisco plant of the Pacific Gas & Electric Co. into a plant for the preliminary carbonization of 100 tons a day of these materials.

The next step was to turn to the consumers of the country and ask them to save their peach and apricot stones, their prune, plum, and olive pits, their date seeds, cherry pits, butternut shells, Brazil nut shells, and their walnut and hickory nut shells. The work of securing these and advertising the Government's need to the public was turned over to the American Red Cross. There was some question at the start as to whether the charter of the Red Cross would permit it to undertake such a war activity; but, since it was determined that this was purely a defensive operation, the legal forces of the Red Cross decided that the organization could go into a campaign of this kind.

"Help us to give him the best gas mask." That was the slogan which was carried on the posters, catching the attention of almost every person in the United States. More than 1,000,000 pieces of literature were distributed. The Red Cross established 163 collection points, and collection barrels appeared on the streets of practically every community in the United States. The Junior Red Cross, the Food Administration, and the Department of Agriculture

BAREFOOTED NEGROES IN SPANISH HONDURAS SHOVELING CORROZO NUTS INTO BASKETS TO BE LOADED INTO BOATS FOR SHIPMENT TO GAS DEFENSE DIVISION.

NUT SHELLS STORED ON DOCK OF EAST RIVER WHARF, ASTORIA, L. I., AFTER BEING UNLOADED FROM BARGES. SHELL-CRACKING TOWER ALSO SHOWN.

5,000 TONS OF PEACH PITS PILED UP AT SAN FRANCISCO.

This is enough to produce 1,600 tons of carbon for use in gas mask canisters.

1,200 TONS OF APRICOT PITS AT SAN FRANCISCO READY TO BE REDUCED TO CARBON FOR GAS MASKS.

TROOPS IN TRENCH AT LONG ISLAND CITY READY FOR A GAS ATTACK.

"GAS!" TROOPS HASTILY DON THEIR MASKS AT THE ALARM.

TROOPS WEARING GAS MASKS CHARGING IN OPEN ORDER AT LONG ISLAND CITY.

TROOPS WEARING GAS MASKS CUTTING BARBED-WIRE ENTANGLEMENTS IN TESTS AT LONG ISLAND CITY.

gave valuable assistance. The Boy Scouts organized nut gathering parties. The governor of Massachusetts proclaimed November 9, 1918, to be gas mask day for the collection of carbon material, and 28 other States fixed gas mask days in November. Two reels of motion pictures were shown through the country. Journalists aided the campaign in newspapers and magazines. Frederic J. Haskin sent out a valuable article which was published in many of the important newspapers of the United States. One Oklahoma town took a day off en masse and gathered a whole carload of nuts.

This campaign started September 13, 1918, but was abruptly cut short on the 11th of November. Thus it is impossible to give exactly the result of it, since many of the scheduled shipments of nuts and fruit pits were canceled and found their way into fuel bins. However, at one time there were on the rails, en route to the carbon plant at Astoria, 100 carloads of materials supplied by the patriotism of the American people. It was estimated that some 4,000 tons were collected in this brief period, exclusive of the material from the California canning industry.

The procurement of the nuts, however, was but the first step in the production of carbon for use in our mask canisters, for after charcoal is first burned its pores are still filled with various impurities which may be summed up by the word "tar." When the charcoal was given a second heating, under careful temperature regulation, this tar was burned out, with the result that the charcoal itself became much more active in its absorption of gas. In fact, properly activated charcoal is more than absorptive—it is catalytic in its action toward the gaseous poisons used in the war, not only absorbing them but hastening their breakdown (digestion) into less injurious substances.

The activating of charcoal offered at the start considerably more of a problem than the question of making the charcoal itself, since activating had never before been conducted on a commercial scale. Two months of experimentation showed us that the best distillation of shells and pits for charcoal was that conducted in illuminating-gas-making retorts. The activation thereafter had to be done in special equipment permitting of fine control of temperature. The Government eventually spent more than $1,000,000 in a charcoal activating plant, providing for America the best protection known to science against the poisons which Germany had introduced into warfare.

The cement granules, which also had to go into the canisters, supplied another problem. We originally used a special soda-lime for this material, but only obtained a satisfactory product after Maj. H. W. Dudley, R.E., came to America as our British advisor and brought to us the British granule formula. The basis of this cement was lime, to absorb gases of an acid nature. Portland cement was used, to give hardness and prevent disintegration and the formation of dust in the canister. Then infusorial earth was added, to make the compound porous in texture. A little sodium hydroxide was put in, to increase the alkalinity of the mixture. Finally there was an infusion of sodium permanganate, which is a powerful oxidizing agent. This latter chemical was added as a precaution against arsine. Arsine and arsenical compounds were difficult to use in warfare, but the Germans had introduced them to some extent, justifying us in adding this protection.

In making the granules the sodium permanganate solution was mixed with the cement. The mixture was roughed out into slabs, allowed to set for three days, dried, ground up, screened to the proper size, and packed in drums for future use.

As has been noted, the charcoal and cement were packed in the canister in alternate layers. The cement had the virtue of working while the carbon slept—that is, the carbon was active when there were gases present to be absorbed, but the cement kept on thereafter, digesting the gases which had been absorbed by the charcoal. The cement was not quick in action, but it had a remarkable capacity for consuming some poisons.

To return to the chronological development of manufacturing facilities, after we had placed the contracts for the first 1,000,000 masks in the early fall of 1917, we began looking around for facilities for producing carbon and cement in the quantities which we should need in the near future. We found at Astoria, the district near Hell Gate at the junction of the East River and Long Island Sound in New York, the large gas works of the Astoria Light, Heat & Power Co. perhaps the largest illuminating-gas plant in the world. This was a subsidiary of the Consolidated Gas Co. of New York, which concern readily agreed to turn over to the Government some of its retorts and to permit the construction of a Government-operated plant on its grounds. We might have been seriously delayed in the production of gas masks except for the extraordinary and continuing efforts of Mr. W. Cullen Morris, Chief Construction engineer of the Consolidated Gas Co., and Mr. Addicks, its vice president. It was due to Mr. Morris that a $150,000 granule plant was constructed, heavy complicated equipment installed, and operations started in the short space of 30 days.

Let us now go back to the history of actual mask production. At the start it was estimated that when the Hero Manufacturing Co. had reached full capacity it could assemble and turn out 6,000 masks a day. The fuel shortage and the railroad congestion of the late fall and early winter of 1917–18 hampered our supplying the Hero Manufacturing Co. with parts, until the mask production, averaging 2,430 a day as it had in November, dwindled to 1,500 a day in

Slabbing the doughlike mixture of carbon and spreading it on screen-bottomed trays at carbon plant No. 2.

Chemical laboratory, showing apparatus for testing the absorbent power of the carbon and gran powder.

Retort house. Discharging machine drawing out the nearer half and pushing out the far half of the hot carbon.

Carbon plant No. 3, showing treater room and west side batteries of Dorsite treaters.

FOUR VIEWS OF CARBON PLANT AT ASTORIA, L. I

Zigzagging and combing work on masks.

Assembly department.

Doping the masks.

Final inspection department.

FOUR INTERIOR VIEWS OF GAS MASK PLANT.

December. The Goodyear Co. at Akron had meanwhile established its Akron-Boston motor truck line. This was put at the service of the Gas Defense Division, hauling various supplies from both Akron and Boston to the assembling plant at Philadelphia. Sometimes in the mountains of Pennsylvania the trucks would be blocked in snow and the patriotic citizens of the community would get out with shovels and work until the supplies again started on their way.

All of the masks produced in the fall of 1917 were still regarded as experimental and not yet up to the standard of masks which we were willing to supply for actual service at the front. Consequently, not one of them was exported, but the entire 1917 production, after the first order of 25,000, was sent only to the training camps in this country. By January 8, 1918, we were producing masks which we were willing to put into actual service, and on that date the manufacture of masks for export was started.

In January we exported 54,000 masks, which was 16,000 less than the schedule which we had set for ourselves. But by February 20 we had wiped out this deficit with a little over, for our schedule by that date called for the production of 141,000 gas masks, and we had produced 142,000.

Late in the fall of 1917 the requirements of the Expeditionary Forces were reanalyzed in the light of information gathered abroad and in accordance with the new military program. Requirements were multiplied almost fourfold. Let us see how these requirements were met, and what difficulties were solved in the course of the effort.

Experience had already shown that for many reasons the Government needed its own mask factory, where improvements could be adopted as soon as made and where inspections and the storage of parts could be more centralized than in private plants.

With the necessary expansion then confronting us, any other policy would have meant making face pieces in half a dozen or more private plants, all starting at once with organizations untrained for this work. This would have been fatal, for even with the Goodyear and Goodrich companies manufacturing face pieces in Akron and the Kenyon Manufacturing Co. making them in Brooklyn, we found it most difficult to maintain uniform standards in all the plants. As new points came up, it was constantly necessary to interchange inspection personnel and to send men from one plant to another to teach manufacturing wrinkles. Such practices consumed more personnel than we could train in the time available. Moreover, it was impossible under the conditions that we were then facing to build up more than barely adequate supplies of gas mask parts and such raw materials as special fabrics. To have operated many more face-piece plants would have meant to divide these stocks of fabrics, elastic, tape, etc., still further. To have kept each of these plants properly stocked, under the existing traffic conditions, would have been impossible. A big central gas-defense plant was the only solution of our difficulties.

The order approving the establishment of the gas-defense plant was signed by Secretary Baker on November 20, 1917. The officers of the Gas Defense Division found in Long Island City, not far from the new chemical plant at Astoria, a group of modern concrete factory buildings which had been put up in this newly developed section by several different concerns, among them the Ford Motor Co., the Goodyear Tire & Rubber Co., and the National Casket Co. One of these buildings, known as the Stewart Building, was taken over by the Government and modern machinery was installed. Mr. R. R. Richardson, of Chicago, was appointed plant manager with a salary of $1 per year. He quickly set to work organizing the factory and its staff. On January 9, 1918, the first few factory operators were hired. Five days later the executive offices at the plant were ready for occupancy. The plant grew apace. One by one the other buildings were absorbed and added to the establishment—first the Goodyear Tire & Rubber building, then the National Casket building. Next a long storage building was built between the Stewart and the Goodyear buildings. Runways were built which connected up the various buildings, and, finally, in July, the Ford Motor building was taken over and connected up to complete the group.

Thus by the summer of 1918 we occupied five large buildings, with a total of over 1,000,000 square feet, or 20 acres, of floor space, connected up to make the gas-defense plant. Of the 12,000 employees in this plant, 8,500 were women. Endeavors were made as far as possible to hire those who had near relatives with the American Expeditionary Forces. The degree of care required in the manufacture of masks was beyond anything known in normal industry, and we rightly believed that this personal interest in the work would bring about greater care in manufacture and inspection. Since the factory was working at top speed a great deal of attention was paid to welfare work. Women employees were given 12-minute rest periods both in the morning and the afternoon, and completely furnished attractive rest and recreation rooms were set apart for women in the factory.

The plant was unique in more than one respect. At the very start it attempted the supposedly impossible, for it combined in its staff and in its working organization civilian and military personnel. The manager was a civilian, the assistant manager was Lieut. Col. Coonley. Below them on the next tier of the organization were Army officers in charge of several departments and civilians in charge of others. Throughout the plant were certain groups of women workers or

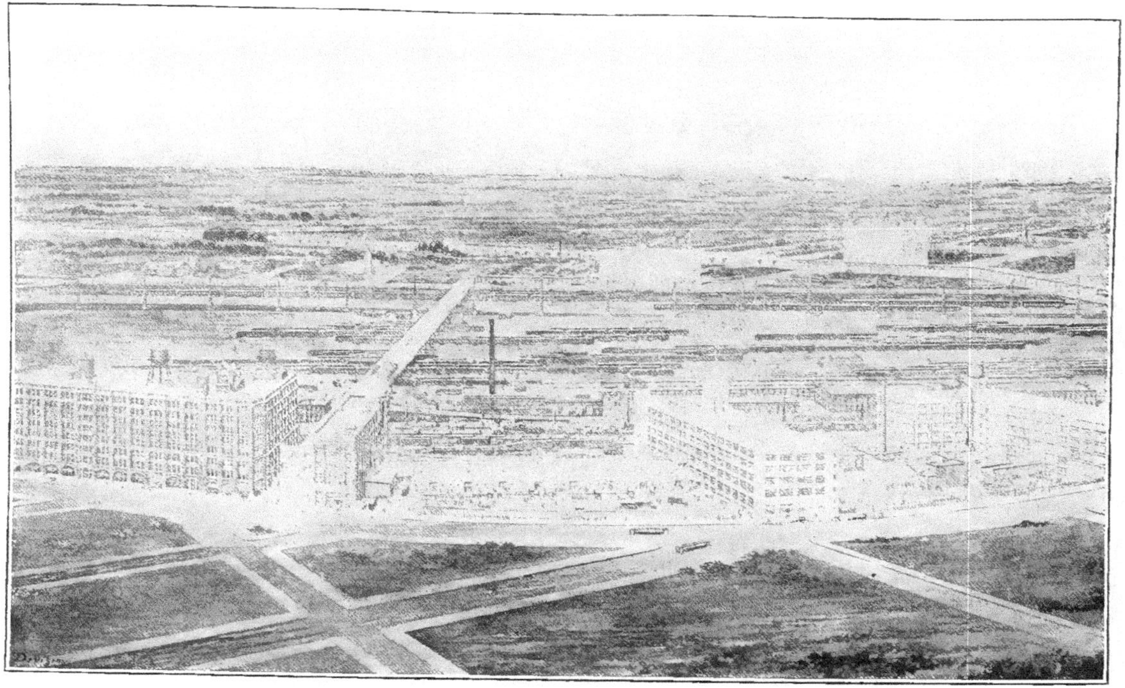

GAS DEFENSE PLANT AT LONG ISLAND CITY.

EMPLOYEES OF GAS DEFENSE PLANT, LONG ISLAND CITY, NOVEMBER 11, 1918.

inspectors in charge of civilians were others; in charge of sergeants or even privates. The arrangement worked out well and the whole organization pulled together as one team, without reference to civilian or military status. Again, at the start there was laid down a policy of inspection at every single stage of manufacture. The incoming parts, though already inspected at their source, were reinspected and retested. After every operation in the manufacture of the face piece there came an inspection by specially trained women set apart from the operators. Then again, there was a special control inspection. After the face piece was finished, and when assembly was complete, the entire mask went to a final inspection where it was looked over by several trained women, who worked in dark closets and inspected the face pieces over a bright light to make sure that no pin pricks had been made, either maliciously or otherwise. Furthermore, wherever there was an inspector there was a system of checking his or her accuracy, for 5 per cent of every inspector's work was periodically selected at random and checked over by other inspectors.

Hand in hand with this went many of the latest developments of factory operation. The best machinery was employed, conveyors were used wherever possible, and, when changes in the size of the operation or the design of the mask made it advisable, the factory was at once rearranged in order that the flow might always be orderly and continuous.

From all of this the reader might judge that the operation, lasting, as it did, for only a little more than eight months, was a costly one. Such, however, was not the case, for a well-ordered and accurate cost system, kept from the very start in accordance with the best practices of factory accounting, showed that after charging in all equipment changes and overhead, the plant made complete masks which cost the Government about 50 cents per mask less than it cost to get complete masks by purchasing parts and assembling them under private contracts.

Along with this manufacturing development went the building up of an elaborate procurement force charged with the responsibility of providing parts to be assembled at the gas-defense plant and the Hero Manufacturing Co. This section faced a hard and intricate task, but though there were instances where a shortage of parts temporarily slowed down production, these were remarkably few. Many were the difficulties of buying new parts; many of the parts were the product of elaborate die work; die makers in the country were overworked. Specifications had to be written, checked, and approved, and a field inspection first had to be organized and trained so that the product from all the different plants could be relied upon as satisfactory for the assembling plants. But this problem was still further complicated by ever-recurring changes in design, made necessary as improvement followed improvement. Officers had to be trained in a day and then sent out to train inspection corps in manufacturing plants in many parts of the country. Inspection and procurement detachments were maintained in most of the eastern industrial centers. There were over 100 enlisted men and 9 officers in Akron, 30 enlisted men with 6 officers in Boston, and men and officers in over 60 cities. Here again the civilian and Army officer worked hand in hand; for Mr. Robert Skemp, a volunteer civilian from Pittsburgh, was in charge of this procurement, reporting to Lieut. Col. Besse and directing an organization made up almost entirely of officers and enlisted men.

The March output of masks was 183,000; that of April, 363,000; May, slightly less than this figure; that of June, 504,000; that of July, 671,000. In all, between January 1 and November 11, 1918, we built more than 5,000,000 gas masks.

In February, 1918, shortly before the German drive commenced, we received requisitions for sample lots of oiled mittens and oiled union suits as protection against mustard gas and also for chloride of lime to neutralize poison-impregnated earth. In their March drive the Germans used gas in much more protracted concentrations than before. Originally the masks had been worn only during the sporadic gas alarms, and then only for a brief period at a time.

The double-protection mask which we had been building had been admirable in its day, but it was no longer adapted to the sort of use to which it was evident it must now be put. In long-continued wear the mouthpiece would irritate the gums and lips of the soldier, and the face-piece band would cause excruciating headaches after a few hours. It had now become frequently necessary for men to wear their masks for eight hours at a stretch. The word discomfort is a weak description of the feelings of a man wearing one of our masks for that period.

Our authorities in France decreed for a single-protection mask and more comfort, even at the expense of a little safety. The result of these new conditions together with the establishment of closer relationship with our Expeditionary Forces, through a visit of Col. Dewey to France, was the determination to build masks in this country which should give the protection of the masks which we had been turning out and at the same time be comparatively comfortable. There had been brought out in France a single-protection mask, that is, a mask in which the inlet tube entered directly into the space between the mask and the face, with the orifices so arranged that the fresh air was drawn across the eyepieces. This was known as the Tissot mask. The principle of the Tissot was correct as far as comfort was concerned, since it did away with the irritating mouthpiece, but the chief danger in this mask arose from the fact that it was made of thin, pure gum rubber. We took the Tissot and endeavored to produce a mask of this type which should be gas-tight and yet rugged. In this work we experimented on hundreds of subjects to determine face and head sizes and shapes. It is interesting to note in this connection that the size of a man's face has nothing to do with the size of his head, as large heads with small faces and small heads with large faces occur not infrequently.

We made two developments of the mask without mouthpiece or nose clip. Both were ready for field tests in August, 1918. The one produced in Akron and assembled at the Philadelphia contracting plant was known as the Akron Tissot, or Type A–T.

At the start of operations in Long Island City Mr. Waldemar Kops, of New York, a manufacturer of corsets, came to the Government, asking an opportunity to do his part in the war. He was assigned to the gas-defense plant, and later, with the commission of major, took charge of the gas-defense Long Island laboratories. Maj. Kops had no experience with gas masks until he came to the gas-defense plant, but his experiments soon led to an improvement in the design of the Tissot mask. It was called the Type KT mask—the Kops-Tissot. Only a few hundred thousand were produced, though the latest model was scheduled for enormous production beginning in December, 1918. It possessed much of the protective efficiency of the old uncomfortable mask, the cut of the face piece insured a gas-tight connection with the head, it was relatively comfortable, and it was durable.

The call of the allies in the spring of 1918 for American troops in as great numbers as the ships could carry them to France resulted in still further increases in our mask requirements. At the height of the drive we were making over 40,000 masks a day. Approximately 35,000 employees were engaged in the manufacture of various gas-mask parts. Our carbon requirements were expanding at a rate that would have needed 400 tons of raw materials a day by December, 1918. We built 336,919 KT masks and approximately 200,000 A–T masks. In exact figures the total production of masks of all types was 5,692,499. Of these 3,666,683 were built at the gas-defense plant and 2,025,816 were assembled by the Hero Manufacturing Co. In addition, we furnished 3,189,357 extra filled canisters for the replacement of those used up by 40 hours of field service.

Hand in hand with this procurement and manufacturing achievement went the development of the technical section of the Gas Defense Division. This was known as the Long Island laboratories, manned by a personnel of several hundred men and officers. Here in its laboratories were solved the knotty problems that bridged the gap between experimental work and production. Many new designs were worked out, only to be rejected when tested. Here there were workrooms that could make sample lots of 1,000 masks, and here were located the chemical laboratories and the gas chambers in which the product of the gas-defense plant was tested daily by control chemical analysis and by actual breathing and wearing tests.

In spite of this elaborate technical section, the testing of masks did not stop with it. There was a special field-testing section of the Gas Defense Division, composed of about 150 men who were trained to the minute in field maneuvers and did most of their work in gas masks. They were constantly in and out of gas with regular production and experimental masks, they played baseball in them, they dug trenches, laid out wire, cut wire, and fought sham battles at night, both with and without actual gas. This section was not organized until July, but it should have been one of the first of our units. It was there that we learned all the fine points of gas mask comfort and durability. The work of this section even went so far in the case of the later designs as to include a test where six men worked, played, and slept in the masks for an entire week, only taking them off for 30 minutes at each mealtime, and each day entering high concentrations of the most deadly gases, without any ill effects whatsoever to the wearers. When it is remembered that eight hours was the limit of time which a strong man could wear the old-type mask, something of the efficiency of the new mask may be realized.

We also built 377,881 horse masks. Investigation showed that a horse's eyes did not shed tears in the presence of even strong lachrymatory gases. Moreover a horse never breathes through his mouth; and it was, therefore, necessary only to cover his nostrils. Furthermore, horses proved to be more resistant to the toxic gases used in Europe than were men, and his mask, accordingly, needed to be only a bag of many layers of chemically treated gauze. The horse masks were all manufactured by the Fifth Avenue Uniform Co., of New York City, under the supervision of a detachment of the Gas Defense Division.

We furnished 191,338 dugout blankets to be used at the doors of dugouts to make them gas proof. These were specially woven all-cotton blankets which were treated in France with a special heavy oil, shipped from the United States.

Toward the end of the war we received large requisitions for protective suits and gloves to safeguard men against mustard gas burns. The suits were made of oiled fabric and the gloves were of cloth impregnated with chemicals. As a work just starting, we produced 2,450 suits and 1,773 pairs of gloves.

A total of 1,246 tons of a new ointment known as sag paste was made and shipped. This was an ointment to protect the skin against mustard-gas burns.

Gas warning signals were of several types, watchmen's rattles and Klaxon horns being the most commonly used to sound the gas alarms.

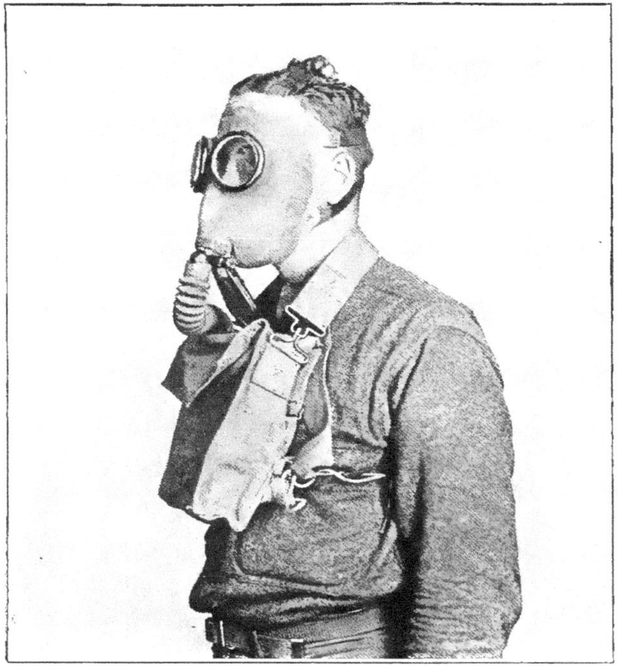

THE AKRON TISSOT MASK, AN IMPROVEMENT OVER THE MASK THAT WAS IN GENERAL USE AMONG OUR TROOPS.

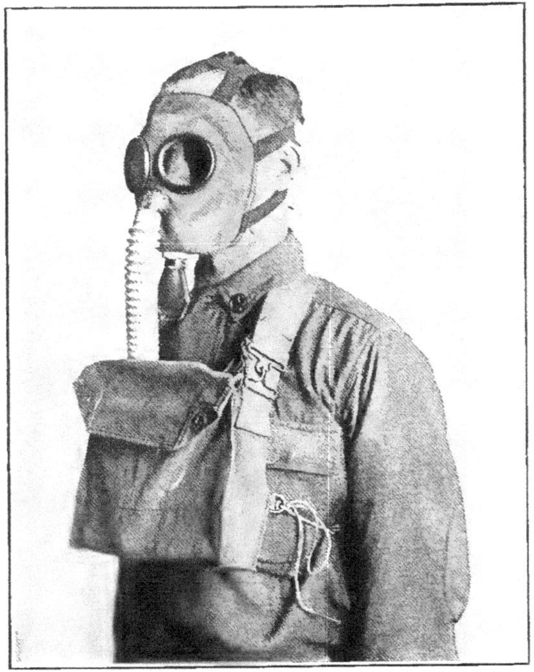

THE KOPS TISSOT MASK, OUR LATEST DEVELOPMENT AND CONSIDERED TO BE THE BEST OF ALL MASKS.

GAS MASK OF FLANNELETTE TYPE FOR HORSE.

We shipped 45,906 of these special hand horns. The rattles were secured in Europe.

Trench fans, for fanning out gas from trenches and dugouts, were produced, to the number of 50,549.

Production data of gas defense division of Chemical Warfare Service—Final statistical report.

Item.	Production.			Shipped overseas—	
	Up to July 1, 1918.	Up to Nov. 11, 1918.	Dec. 31, 1918 (total production).	July 1, 1918.	Nov. 11, 1918.
Respirators	1,719,424	5,276,515	5,692,499	1,196,787	3,938,808
Extra canisters	507,663	3,144,485	3,189,357	484,236	1,805,076
Horse masks	154,094	366,529	377,881	101,250	351,270
Bleaching powder (tons)	1,484	3,677	3,590	586	1,867
Extra antidimming (tubes)	....	2,855,776	2,855,776	....	2,855,776
Sag paste (tons)	20	1,136	1,246	....	915
Dugout blanket oil (gallons)	....	95,000	95,000	....	5,000
Protective suits	....	500	2,450	....	....
Protective gloves	....	1,773	1,773	....	....
Dugout blankets	....	159,127	191,338	....	36,221
Warning devices	....	33,202	45,906	....	19,620
Trench fans	11,343	29,977	50,549	9,600	27,690

BOOK V.

QUARTERMASTER ACTIVITIES.

CHAPTER I.

SUBSISTENCE.

When the American soldier went to war against Germany he took his appetite with him. The task of keeping that appetite satisfied with good food (and the soldier, therefore, contented and well) fell to the Quartermaster General. The average American soldier at the end of the fighting in 1918 is said to have weighed 12 pounds more than he did when the Selective-Service Act or his own enlistment brought him into the Army. This is the complete testimonial to the quality and quantity of the food served to the American troops in 1917 and 1918. Assuming 3,700,000 to have been the greatest number of Americans under arms, this average increase in weight means that the beans and bacon and fresh meat of the American Army ration were transmogrified into some 45,000,000 pounds of Yankee brawn to be the basis of untold resources of health and energy during the coming quarter of a century.

Consider these millions of soldiers as one composite, gigantic man in khaki; compress the war period into a single hour, the dinner hour; and it will be seen that the American fighter consumed what might be called a sizeable meal. Let us say that he started off with the main course. The roast of beef weighed over 800,000,000 pounds. It was flanked by a rasher of bacon weighing 150,000,000 pounds. Over 1,000,000,000 pounds of flour went into the loaf of bread, while to spread the bread was there a lump of butter weighing 17,500,000 pounds and another lump of oleomargarine weighing 11,000,000 pounds. As a side dish this giant had over 150,000,000 pounds of baked beans, half of these coming in cans ready baked and flavored with tomato sauce. The potatoes weighed 487,000,000 pounds. To add gusto to his appetite there were 40,000,000 pounds of onions. Then scattered over the table were such items as 150,000,000 cans of corn, peas, and string beans; while the salad contained 50,000,000 cans of salmon and 750,000 tins of sardines. Then there was a huge bowl of canned tomatoes, nearly 190,000,000 tins supplying its contents. For dessert he had 67,000,000 pounds of prunes and 40,000,000 pounds of evaporated peaches and apples. The sugar for sweetening various dishes weighed 350,000,000 pounds. He washed it all down with a draft made of 75,000,000 pounds of coffee thinned with 200,000,000 cans of evaporated milk. The bill for the meal, paid by the American public, amounted to $727,092,430.44, this figure to December 1, 1918.

In supplying such vast quantities of food, scientific attention was concentrated upon the details of the effort. At the time the armistice was signed the American troops in France were eating about 9,000,000 pounds of food every day. Never before in history had any nation been compelled to send subsistence so great a distance to so many men. It was not possible to ask France and England to divide their food supplies, as they were already rationing their civilian populations. We were required to purchase practically all food in America and transport it nearly 5,000 miles. Ships were relatively scarce. There was a strong bid for every inch of tonnage space. The tonnage allotted to subsistence must be filled with sufficient food not only to supply the immediate consumption, but to overcome losses due to the sinking of ships and the possible capture of base depots. These contingencies required two pounds of food to be shipped where one would ordinarily be sent; yet because of the shortage of ships the subsistence authorities were asked to pack these two pounds into almost the space of one. The result was foods in forms never before known by American soldiers and in some cases never before known at all—such forms as dehydrated vegetables, boneless beef, and the so-called shankless beef. Trench warfare made new demands for food. Calls came for such rare articles as soluble coffee or the wheat-and-meat cake of the emergency ration.

These problems were solved only by the assistance of the American food industry. In numerous instances new factories, or even whole new types of food manufacture, were built up as rapidly as three shifts of men could work and money accomplish results.

The cost of food rates high among the war costs of 1917 and 1918. Back in 1897 the average meal in the Army cost about 4 cents, and the daily three meals 13 cents. At the end of 1918 the cost of the ration was approximately 48 cents. The advance was not all due to the advance in living costs. Much of it was on account of the improved standards of the ration. In 1916 Congress appropriated $10,000,000 to feed the Army; the fiscal year beginning July 1, 1918, brought an appropriation of $830,000,000 for the same purpose.

The American fighting man of 1917–18 was a good feeder. He ate nearly three-quarters of a ton of food each year, or over ten times his own weight. Without counting any transportation costs or the expense of handling at all, each man's yearly supply of food cost more than $165. In spite of the most rigid and painstaking economies in the purchase of this subsistence the American people were paying at the peak of Army expansion more than $2,500,000 per day to feed the troops.

The distance of the American Expeditionary Forces from the source of their food supplies required that their food be largely purchased in nonperishable forms. That is, meats must be cured, meats and vegetables tinned, vegetables and fruits dried. We literally paved the way to Berlin with tin cans. The various foods put up in tins and purchased during the year 1918 totaled over 1,000,000,000 cans, or enough, standing on end, to make a road wide enough and long enough for a force of men marching in columns of four to go from the port of embarkation at Hoboken, N. J., to the heart of Germany. The largest closing machine can seal 240 tin cans per minute. If such a machine could be operated eight hours a day seven days a week, it would take it 23 years and 6 months to seal these tins.

During the spring of 1918, when the demand for men in France resulted in reducing the available tonnage for supplies, the cry came from France to cut every nonessential. As a result most of the canned vegetables and fruits, including peas, corn, sweet potatoes, asparagus, pineapple, pears, and apples were stricken from the list of food supplies for the American Expeditionary Forces.

From France came calls for tomatoes and men, men and tomatoes. This phrase did not mean that bread and bacon, beans and beef, should be eliminated; but it emphasized the importance of this one vegetable, the tomato. The total purchases of tomatoes exceeded those of all other vegetables combined. In addition to the many ways of serving tomatoes, they were used in the trenches to relieve thirst, being, perhaps, more effective than any other substitute for water. Because of its food value and slight acidity, a quart of tomato juice was worth several quarts of water to the thirsty men in the field. The Army took 45 per cent of the total 1918 American pack of tomatoes. These tomatoes were bought from 5,000 firms scattered throughout the rural districts of the United States.

The demands of the overseas forces for meat during the summer of 1918 were so heavy that they created a shortage of beef in the United States. Beef is the mainstay of the soldier's diet. The Army allows 456 pounds of beef per year for each soldier. This does not mean that the soldier actually eats that much beef, beef being simply the Army's meat standard. Pork, usually in the form of bacon, is substituted for 30 per cent of this quantity of beef, 12 ounces of bacon being considered the equivalent of 20 ounces of beef. The major portion of the American Expeditionary Forces' beef was fresh beef shipped frozen all the way from the packing plants in the United States to the company kitchens at the front, through an elaborate system of cold-storage warehouses and refrigerator cars and ships.

The Food Administration asked that the people substitute corn meal, rye flour, and other grain flour for 20 per cent of the wheat flour ordinarily used in making bread. The troops in the United States complied with this ruling and saved 1,000,000 barrels of flour. The use of substitutes in France was not insisted upon, as bread making in the field is more difficult. Field bakeries are not adapted to experimenting with doughs and yeasts, as is required when substitutes for flour are used. The Army allowance of flour for a year for one man is 410 pounds. Flour was usually issued in the form of bread, 1 pound of bread being allowed for each man each day. Other yearly allowances are 56 pounds of beans, 27 pounds of prunes, 27 pounds of coffee, 73 pounds of sugar, 11½ pounds of condensed milk, 3½ pounds of vinegar, and 13½ pounds of salt. For variety other items are specified which may be substituted for these foods.

Food was purchased by the Quartermaster's Department and furnished to the individual companies at cost of the food. In charge of the mess was a sergeant, who had had special instruction in schools as to methods of feeding the Army. The mess sergeant checked over his stocks daily and made up a list of what he would require for the coming day. This list, in turn, was given to the camp supply officer, under whose direction the order was made up and delivered to the kitchen on Army trucks.

This order was based on a ration allowance, as has been stated, a ration being the food required to subsist one man for one day. The general components of the overseas camp ration consisted of the following:

Component articles and quantities.		Substitutive articles and quantities.	
Beef, fresh...ounces..	20	Mutton, fresh...ounces..	20
		Beef, fresh, boneless...ounces..	16
		Bacon...ounces..	12
		Pork, fresh...ounces..	16
		Sausages, canned pork or Vienna...ounces..	16
		Canned roast beef or corned beef...ounces..	16
		Hash, corned beef...ounces..	16
		Fish, dried...ounces..	14
		Cheese, not exceeding 10 per cent of total issue...ounces..	10
		Fish, canned...ounces..	16
Bread, soft...ounces..	16	Flour, corn meal, oatmeal, or macaroni, in lieu of an equal quantity bread, but not exceeding 15 per cent of total issue.	
Baking powder (to be issued only with flour or corn meal, 1 ounce to 20 ounces)..ounce..	.08		
Beans, dry (not to exceed 4 issues in 10 days), ounces...	4	Beans, baked (not to exceed 4 issues in 10 days)...ounces..	8
Rice or hominy (not to exceed 6 issues in 10 days)...ounces..	2		
Potatoes, fresh...ounces..	20	Onions, fresh, in lieu of an equal quantity of potatoes, but not exceeding 20 per cent of total issue. Tomatoes, canned, in lieu of an an equal quantity of potatoes, but not exceeding 20 per cent of total issue.	
		Canned potatoes...ounces..	15
		Other fresh vegetables (not canned) when they can be obtained in the vicinity by purchase or from the U. S. Garden Service, or can be transported in a wholesome condition from a distance, in lieu of an equal quantity of potatoes.	
		Dehydrated vegetables to be issued only in case fresh vegetables are not available..ounces..	4
		Corn, canned...ounces..	12
		Peas, canned...ounces..	20
Jam...ounces..	3	Prunes, or evaporated apples, or peaches, or apricots, or figs, or dates, or raisins, in lieu of an equal quantity of jam.	
		Sirup...gill..	.64
Coffee, R. & G...ounces..	1.12	Tea, black or green...ounce..	.32
Sugar...ounces..	3.2		
Milk, evaporated, unsweetened...ounce..	1		

Component articles and quantities.			Substitutive articles and quantities.	
Vinegar	gill..	.16	Pickles, cucumber, in lieu of an equal quantity of vinegar.	
Salt	ounce..	.64		
Pepper, black	ounce..	.02		
Cinnamon	ounce..	.014	Cloves, or ginger, or nutmeg, or sage, thyme, or allspice....ounce..	.014
Butter	ounce..	.5	Oleomargarine or lard or lard substitute, ounces	.5
Flavoring extract, vanilla	ounce..	.014	Flavoring extract, lemon	.014
Candy (issued ½ pound once in 10 days)	ounce..	.8		
Tobacco, smoking (100 cigarette papers for each 4 ounces smoking tobacco)	ounce..	.4	Cigarettes....number..	.4
			Tobacco, chewing....ounce..	.4

The ration at home was practically the same. The home ration, however, did not include candy and tobacco. The commanding officer had authority to modify or change all rations to meet special conditions. For instance, in times of great cold and when the men were subject to great exposure, or after long and tedious campaigns or marches, or when the work required of the troops was abnormal, the ration might be increased. The ration also included soap, candles, matches, towels, and a few other items considered necessary in the daily life of a soldier. The value of a ration fluctuated with the market from month to month. Each day's food weighed about 4.6 pounds per man.

The men actually in the trenches sometimes made use of the emergency ration, the little flat can of compressed nourishment which every soldier carried in his pocket. This ration, however, was used only in severe straits, on the order of an officer, or on the enlisted man's own responsibility in the direst emergency, when the activity of the enemy made it impossible to get hot food to the men during daylight hours. Hot food was served in the trenches whenever possible. The hot food consisted principally of soups and soluble coffee. Specially constructed cans, made on the principle of thermos bottles, kept the food hot when it was being carried to the front. The chief quartermaster of the American Expeditionary Forces relates that on a tour of inspection made by him, during the Argonne-Meuse offensive, on November 1, 1918, he inspected the meals served at noon to the troops of the Fifth Corps actually engaged in battle on that day, and found in a number of cases that Artillery organizations were being served beefsteak, potatoes, onions, tomatoes, white bread and butter, rice pudding, and hot coffee, the men eating in reliefs in order that there might be no cessation of fire. The hot meals for the Infantry were prepared at their rolling kitchens a short distance in rear of the line, and sent forward to them in "marmite" cans.

The company was the unit on which the feeding of the men was based. Each month the company was given credit at the quartermaster's store equal to the number of men in the company multiplied by thirty times the ration allowance. On the basis of this credit the mess sergeant of the company made purchases to feed his men. He might be as economical as he desired, provided that he fed the men sufficiently. If the entire credit extended him at the camp quartermaster's office was not used up during the month, a check was given for the difference. This went into the company's funds, with which the mess sergeant might buy in the open market such extras and delicacies as the savings would permit, up to the quantity specified in the ration.

But this system was followed only in the United States. Savings were not allowed in France, all food there being issued on a straight ration basis. This was due to the fact that the shortage of tonnage made it imperative that no article not absolutely essential be shipped from the United States, while difficulties of transportation in France necessarily eliminated all except the most essential articles of food.

Under the procedure in vogue previous to the recent war, subsistence was purchased by depot quartermasters located in 13 principal cities throughout the United States. The plan gave the Army a large number of purchasing officers for subsistence, working without coordination and even in active competition with each other. This condition resulted in a wide range of prices and a lack of uniform quality; while under war conditions, with the enormous quantities to be procured, it would cause at times a congestion of buying orders, with consequent disturbance of market prices.

A plan of control was soon worked out whereby the Subsistence Division, with headquarters at Washington, received at regular intervals the estimates of the needs for subsistence for the Army, both at home and abroad. These estimates were compared and a budget made up. Bids were then asked through zone supply officers, who reported the bids to the control body in Washington. The lowest or most advantageous bid was accepted, and the purchase was completed by the zone supply officer in whose zone the seller was located. The plan eliminated one army zone bidding against another. At the same time it enabled every manufacturer or producer to bid on the needs of the Army. In this way active competition was secured and low prices obtained. A decided advantage of the plan was that purchases were made with a minimum of disturbance to prices paid by the civilian trade.

Not only was it necessary to coordinate army organizations, but it was also found that the independent buying of the Army, the Navy, and the Allied Provision Export Commission was having the effect of increasing prices of a number of food products. These buying agencies were unconsciously bidding against each other. In December, 1917, at the suggestion of the Food Administrator, with the consent and approval of the Secretary of War and of the Secretary of the Navy, the food purchase board was organized to coordinate all of the purchases of food products in this country intended for military purposes. The plan adopted was to allot through the Food Administration the required quantity to the industry producing the commodity in question, dividing the business among the various producers in proportion to their capacity. Products so controlled were those in which there was an actual or prospective shortage. The prices were determined by the food purchase board after studying and investigating the costs of production. The products so purchased included flour, sugar, all canned vegetables, canned and evaporated fruits, salmon, sardines, canned milk, rice, and, for a time, fresh beef. These products totaled about 40 per cent of all food requirements for the Army.

Practically all purchasing of meat was done by the Subsistence Division's packing-house branch, located in Chicago. Circular proposals were submitted by the various packers whose headquarters are located there. The Subsistence Division ordered the required purchases made, and the Chicago office at once allotted the amount needed among the packers. After the butchering and inspection of the meat, it was sent to the freezers and, after being frozen, was loaded in the cars and shipped to the embarkation points. The whole process from the time the animal was killed until it was loaded on the boat took about two weeks.

The Middle West produced practically all the beef which nourished our fighting men. Some of the cattle were bought in California, inspected at the packing-house plants along the Pacific coast, and sent to France via the Panama Canal.

The packers of Chicago and other cities found their plants, gigantic as they were, all too small to handle the demand of our troops for meat products packed in special forms; and extensive additions, both in buildings and machinery, were required by the Army's demands.

It was only by careful vigilance on the part of its inspection branch that the millions of men dependent on the Subsistence Division for their food were protected from deterioration of supplies and abuses by certain dealers and manufacturers. Such firms were in the minority, for the food industry backed the Army with great loyalty, giving honest and patriotic support. In a certain week the inspection service found oatmeal flour moldy and unfit for use, having been stored too long before using; large amounts of potatoes, shipped to Camp Devens, undersize and frostbitten; 3,000 pounds of butter at Camp Greene too old for use; and 12 carloads of tomatoes of poor quality. The system in vogue of demanding reinspection was responsible for discovering many such cases, and traveling inspectors also kept the products up to the highest standard. Any information from outside sources was immediately investigated.

Samples of all shipments of food stuffs were required to be sent to the inspection branch. In this way many violations of the food laws were found. One packer was found to be using pork which contained large numbers of skippers. Another tried, consciously or unconsciously, to pass off wormy dried fruits. Milk has in some cases been found to be much below standard. All of these supplies were promptly rejected as improper for Army use. In many cases the fault has been found to be the result of improper manufacturing conditions, and in this event the manufacturer has been compelled to make good the loss to the Army. The general result of this inspection was that manufacturers gave the Army their very best products.

One of the most important divisions of the inspection branch was the meat and meat-products section. Its function was the supervision of the reinspection, storage, and handling of meat and meat products, butter, and cheese. Special care was taken to see that there were no embalmed meats. Meat and meat products, butter, and cheese are all highly perishable articles; and, although they may be delivered in perfect condition, many imperfections may develop if diligent care is not exercised during shipment, handling, and storage. One of the first steps taken at the camps was the installing of complete cold-storage plants with adequate chill rooms, so that the proper preservation of fresh meats was assured after arrival at camps. From the first the most rigid inspection of meat and meat products was insisted on and no product allowed to pass which did not comply with Army specifications. The carcass might be from a perfectly healthy animal, yet be rejected, as lightweight carcasses were not approved for consumption in the Army. Instructions as to Army requirements were placed in the hands of every inspector, covering the inspection, storage, and handling of meat and dairy products. Supervisory traveling inspectors visited all stations at irregular intervals to insure these instructions being followed and to instruct quartermasters in posts which were too small to warrant a qualified meat inspector being stationed there.

One object of the Subsistence Division was to educate the proper officers throughout the Army to be inspectors. To accomplish this the inspection branch compiled a manual covering practically all the principal items of Army subsistence, the exact methods of inspection, and how to detect imperfections in foods. Complete Army specifications for all supplies were included. Gen. Pershing cabled for 250 copies to be used in France, and the University of California adopted the manual to be used in zymology classes. It placed exact knowledge in the hands of the men who received the food and who had the responsibility that it be up to specifications.

The overseas forces were the particular concern of the Subsistence Division. It was planned to have approximately three months' advance supply of food sent over each month for the number of troops actually sent to France during that month. This was called the initial supply. In addition to this, there was sent over a monthly automatic supply, equivalent to the amount of food the troops already in France would consume during that month. In this way a 90 days' reserve was usually maintained overseas.

The problems of the overseas forces demanded quick solution. The new modes of warfare gave rise to many needs unknown in peace times. The result was that calls came in for commodities which were not at the time being produced in adequate quantities. Factories had to be built, labor secured, and machinery manufactured; in instances entirely new industries had to be created.

The Service of Supply found it was impossible to secure sufficient fresh vegetables in Europe to take care of the requirements of our troops, and the Subsistence Division at home was called upon to supply dehydrated vegetables for overseas requirements. To send fresh vegetables from the United States was impossible, due to the great necessity for conserving ship tonnage, and a substitute was imperative. To supply dehydrated vegetables meant the development of an industry. Dehydration was practically unknown in the United States, there being but three small plants in existence. The Subsistence Division searched the country for advantageous locations where there were prospects of having such factories established. Within a few months the cooperation of companies was secured and factories were built whose combined output for the month of December, 1918, amounted to 6,000,000 pounds, there being 15 large plants in the United States at that time. Up to the date of the signing of the armistice 62,000,000 pounds of dehydrated vegetables had been ordered by Gen. Pershing.

The difficulty of supply was increased by the delicate process which is required to make dehydrated vegetables. The moisture of the fresh product must be removed without extracting the nutritious juices or destroying the food value or flavor. After the vegetables have been peeled and sliced or cubed, they are blanched, in order that they may retain their starch components. They are then placed on trays in huge kilns, through which heated air is blown until only the small required amount of moisture is retained. The product is then packed in hermetically sealed cans.

Dehydrated vegetables occupied a prominent place in the soldier's menu in France. Reports from overseas made by inspectors of the Subsistence Division indicate that dehydrated vegetables were quite satisfactory. The Surgeon General's Office has approved their use. However, when fresh vegetables could be purchased in foreign markets they were used in preference. The use of dehydrated vegetables saved two-thirds of the cargo space in ships over the amount required for fresh vegetables. Their use came at the time when the cargo space was as valuable as life itself, and it enabled men and munitions to be transported sooner than would otherwise have been possible. Dehydrated vegetables were also found especially adapted for use at the front when food was carried forward from the rail heads to the trench kitchens under shell fire.

The emergency ration and its production make another interesting story. Designed to be used only in dire extremity, primarily for No Man's Land fighting, the ration was packed in small cans to be carried in the soldier's pocket, usually the upper left-hand jacket pocket. This ration corresponded to the starvation ration of the allies. Its components were adopted after experiments at the battle front and after consultations with food experts. It represented the greatest amount of food that could be concentrated in the smallest compass.

The complete ration consisted of three cakes of a mixture of beef and ground cooked wheat, each cake weighing 3 ounces; three 1-ounce cakes of chocolate; three-quarters of an ounce of fine salt; and 1 dram of black pepper. From the beef the preparation process removed all fat, sinew, and white fibrous tissue. The meat was then heated, and all of its moisture was evaporated so skillfully that no flavor was lost. The wheat or bread component of the cake was prepared by removing the chaff from cooked wheat which had been kiln-dried, parched, and then ground to a coarse powder. The meat and bread were compounded together, about two parts of bread to each part of meat, making a perfectly homogeneous cake. The chocolate of the ration was prepared by combining equal weights of fine chocolate, containing not less than 20 per cent of cocoa butter, and pure sugar, and molding the product into cakes weighing 1 ounce each.

The several components were packed into oval tin cans, which were camouflaged to render them inconspicuous. These cans bore the legend:

> "U. S. Army Emergency Ration. Not to be opened
> except by order of an officer, or in extremity."

Many ways of preparing the emergency ration for eating in the field were found by experiments. The bread and meat cake could be eaten dry; or, when boiled in 3 pints of water, it made a palatable soup; boiled in 1 pint of water, it produced a thick porridge which could be eaten hot or cold; the cold porridge could be sliced and fried when circumstances permitted. The chocolate could be eaten as candy or made into a drink by placing the chocolate in a tin cup with hot water.

The gas attacks in the trenches made it necessary that the soldier's food be packed in containers impervious to mustard gas poison, mustard gas, when swallowed, attacking the intestines. The first call for such a ration came during October, 1917, and it called for the shipment of 100,000 sealed rations a month for 20 months. The food was to be packed in hermetically sealed galvanized iron containers, holding 25 rations each. The contents of each can consisted of 25 pounds of meat in 1-pound cans, 25 pounds of hard bread in 8-ounce cans, and 25 rations each of soluble coffee, sugar, and salt. Tobacco and cigarettes were added for the comfort of the men. The addition of tobacco and cigarettes was accidental. It was found necessary at first to fill the surplus space in the containers with excelsior. The office force of a large corporation learned of this fact and got permission to fill the empty space in some of the containers with tobacco. The Subsistence Division thought so well of the idea that orders were issued for the tobacco ration to be placed in all reserve ration containers.

One of the most difficult elements in supplying the reserve ration was the securing of tin cans for hard bread. These, because of their unusual size and shape, could only be manufactured after new can-making machines had been designed. The demand for such cans exceeded 10,000,000 in number. Within a comparatively short time, however, hard bread in cans for special reserve rations was being produced on a large scale, and the overseas requirements were filled.

Next the manufacture of the necessary galvanized containers and crates was contracted for. A packing plant was then designed to pack the components into the containers, which was an intricate operation in itself, the number of rations being so great. This plant was so contrived that the parts of the packing material came in at one end of the plant, and the hard bread, canned corned beef hash, canned roast beef, and canned corned beef, canned fish, coffee, sugar, salt, and can openers were packed into the galvanized containers as they traveled on a conveyor belt, until all the components were included.

Only the best of Army purchases were put in the reserve ration. A study was made of the best packers of the various commodities, and their products were used exclusively. Everyone connected with the packing knew the purpose of the ration. It was to be used only when the trenches were under the heaviest fire—when hot food could not be carried forward, and when the men were most in need of good food. The reserve ration became the quality ration of the Army as a result. After the packing was complete, the cans were hermetically sealed by solder and camouflaged with olive drab paint. The container of the ration, when packed, was so buoyant it would support two men upon it when thrown into the sea, thus being a potential life raft.

It was also necessary to feed our men in German prison camps. A ration for American prisoners was prepared by the Subsistence Division of the Quartermaster Corps, in conjunction with the Food and Nutrition Division of the Surgeon's General office. This ration was distributed by the American Red Cross from Denmark and Switzerland. Individual packages each containing sufficient food to supply one man were sent to the prison camps each week. The chief components of the package were corned beef and salmon (with an occasional substitution of corned beef hash and canned roast beef), hard dry bread, dry beans, rice, baked beans, and fresh potatoes (where possible). Prunes, jam, apples, peaches, coffee, sugar, evaporated milk, vinegar, salt, pepper, and pickles were also supplied. Potatoes and onions were procured when possible in Ireland, France, and Italy. Otherwise dehydrated potatoes and onions were used.

Special food was sent for the invalid prisoners, this ration containing potted chicken, crackers, concentrated soup, dehydrated spinach, creamed oat meal, cornstarch pudding, sweet chocolate, extract of beef, soluble coffee, etc. There were several substitutes for all items mentioned, among the substitutes being dried eggs, potted veal, cheese, peanut butter, dried apricots, honey, corn meal, gelatine, malted milk powder, bouillon cubes, apples, oranges, lemons, cocoa, and tea.

When the American troops entered the trenches it was found impracticable to use the ordinary roasted and ground coffee. Its preparation required too much fire, the smoke of which made a target for the enemy. Experiments were made with soluble coffee, looking toward guaranteeing a warm stimulant in the trenches. It was found necessary to give hot drinks to the men before they went over the top or after they had undergone periods of exposure. The British and French troops were supplied with brandy, wine, or rum on such occasions. But issues of intoxicants to soldiers were contrary to the American policy, and quantities of soluble coffee were substituted. Solidified alcohol was supplied so that the coffee could be served hot.

The soluble-coffee industry was in its infancy in the United States. So great was the demand for soluble coffee from the overseas forces that the calls were for over thirty times the prewar production. A cablegram was received in October informing us that after January 1, 1919, the troops would require 25,000 pounds of coffee each day in addition to the amounts packed in the trench rations, these latter quantities alone amounting to 12,000 pounds daily. Allowance was also made for possible sinkings of 5,000 pounds daily, making a total of 42,000 pounds necessary to meet the daily requirements of the American Expeditionary Forces.

The entire American output of soluble coffee was taken over for the Army, but this amounted to only 6,000 pounds daily. A number

MANUFACTURING BARRINGTON-HALL SOLUBLE COFFEE AT THE BAKER IMPORTING CO. PLANT, NEW YORK.
THREE BATTERIES OF 21 PERCOLATORS, CAPACITY 5 CASES OF COFFEE EACH.

MANUFACTURING SOLUBLE COFFEE FOR THE U. S. GOVERNMENT IN THE PLANT OF C. J. BLANKE TEA & COFFEE CO., ST. LOUIS, MO.

of manufacturers of other food products were induced to turn their entire plants into soluble-coffee factories. The greatest difficulty was incurred in the securing of the necessary equipment for these new plants. There was but one company in the entire United States which made the revolving bronze drums essential to the manufacturing process. This company ran its plant seven days a week, with three shifts daily, to produce the necessary materials. The metals which went into these drums were vital in the manufacture of other munitions, but it was even more important that men in the front lines be given hot drinks when tired and worn from long fighting and exposure.

The signing of the armistice saw the difficulties of supplying soluble coffee about overcome. The Subsistence Division had won one of its hardest fights. The cooperation of American manufacturers had made the achievement possible.

The problem of supplying good coffee to the troops was a difficult one. To make good coffee for a unit as large as a company is not easy for the average cook. To guarantee that good coffee would always be available, the Subsistence Division made one of its most radical changes in handling supplies. This change was so complete that whereas the Army formerly was served with coffee from three to six months out of the roasters, it came to be supplied with coffee freshly roasted every day.

At the beginning of the war coffee was purchased, ready roasted and ground, from competitive dealers. It was then held in New York for about 30 days before being shipped overseas, the transportation requiring 30 days more. Received in France, the coffee often was kept for 90 days before it was distributed to the troops. In addition, a 30 days' supply must be kept on hand, making the coffee 6 months old by the time it was used. The result was that when the coffee finally reached the men it had lost half of its value as a stimulant and was greatly deteriorated in flavor, often being in a crumbly condition. "Muddy" coffee on the mess tables resulted.

The only way for the troops to secure fresh coffee was for us to send over the green product for roasting as it was needed. Buildings were erected to house coffee-roasting machinery at home and abroad; men were trained as quickly as possible in the process of coffee-roasting, and sent out to take charge of the plants. In a relatively short length of time 16 plants were in full operation in France, and an increasing number at home. Eventually all the coffee used in France was shipped over green and roasted in the plants there. These plants were capable of roasting sufficient coffee to take care of 3,000,000 men at a considerably lower cost to the Government than under the old system.

The Expeditionary Forces, as is noted elsewhere, organized a purchasing office in Paris. Its purpose was to save tonnage space by securing as many products as possible in Europe. Its scope covered all classes of supplies, but a large section was devoted to subsistence. Candy, hard bread, and macaroni factories under the direction of the Quartermaster Corps were built or secured from the French Government. Large quantities of beans, fresh potatoes, onions, coffee, rice, salt, and vinegar were secured from European markets. Many thousands of tons of foodstuffs were purchased and manufactured in Europe for our Army, every ton representing space on ships saved for additional men and munitions. Overseas purchases were generally discontinued after the signing of the armistice, as the Director of Purchase and Storage and the Expeditionary Forces were firm for the policy of favoring American manufacturers wherever possible.

To reduce tonnage still further, extensive experiments were made in the packing of beef for overseas consumption. All bones, surplus fats, and waste portions were removed. The remainder, all edible, was pressed into 100-pound moulds and frozen. The initial shipment was composed of 16 carloads of boneless beef. The meat arrived in France in splendid condition, and was carefully watched from its arrival at the ports in France to its consumption in the front-line trenches. Officers, mess sergeants, and cooks were enthusiastic over the boneless beef, as it took much less time to prepare it and so conserved labor to a great extent. The men were gratified, as the inferior portions of the beef were not included, and much better meat resulted for the mess. After the success of this experimental shipment, as much boneless beef as possible was sent to France. Trouble was encountered in securing the skilled butchers to bone the great quantities needed, but this shortage was largely overcome.

No means was discovered so effective for reducing tonnage as boning beef, dehydrating vegetables, and purchasing foods in France, but in many of the smaller items there were stories just as interesting. Efforts to save tonnage brought about the reduction of moisture in soap. While the Subsistence Division was securing toilet paper it found that the entire supply for the Expeditionary Forces could be stored in the waste space of Army rolling field kitchens. A special formula for vinegar was devised, and double-strength vinegar was shipped. This, when mixed with an equal quantity of water in France, was a good product.

The saving of space in the transportation of subsistence stores makes a long story in itself. Just so much tonnage was allotted to food each month, and the ablest men in the food industry spent much time in working out how the maximum amount of essentials and luxuries in foodstuffs could be sent in the minimum amount of space.

COFFEE-ROASTING PLANT AS OPERATED AT A NUMBER OF CAMPS.

FURNISHING THE ARMY WITH TOBACCO.

Packing machines in a Winston-Salem, N. C., plant. These machine automatically weigh the tobacco, wrap and seal tins in which this particular brand of tobacco is marketed.

MAKING CIGARETTES FOR THE ARMY.

Leaf tobacco being blended for cigarette-making purposes in a Richmond, Va., tobacco plant.

The Subsistence Division not only looked after the working fighter but the playing fighter as well. The American soldier is fond of candy, tobacco, and chewing gum. The supply of these commodities brought much pleasure to the troops. Long lines of men waiting for free candy and tobacco in France, men who just came from the front, formed one of the interesting sights of the war. Tobacco has established its claim to a recognized place in the soldier's life. Probably 95 per cent of the soldiers of the American Expeditionary Forces used it in one form or another. In May of 1918 it was decided to adopt the practice of the allies, namely, to allow each soldier a certain amount of tobacco per day. This unusual innovation was the official recognition of tobacco as a necessity for men in active service. To men enduring physical hardships, obliged to live without the comforts and often even the necessities of life in times of battle, tobacco fills a need nothing else can satisfy. The daily ration of four-tenths of an ounce was given to every man overseas who desired it. The soldier had the choice of cigarettes, smoking tobacco, or chewing tobacco. If he chose smoking tobacco he received cigarette papers with it. In addition the men could buy at any Army or other canteen the most popular brands of cigars and cigarettes in unlimited quantities.

The Subsistence Division purchased for overseas shipment a monthly average of 20,000,000 cigars and 425,000,000 cigarettes. Abundant supplies of tobacco were on hand in the commissaries overseas, and the soldier could buy it at actual cost. There was no profit or tax added on any tobacco shipped to France, and it was sold at retail to the troops at a cost lower than the price paid by the biggest wholesalers in the United States. The plan for the purchase of cigars and cigarettes was to divide the contracts among the most popular brands in the same proportions as the latter are sold in this country.

Candy in the days of the old Army was considered a luxury. The war with Germany witnessed a change. The old popularity of chewing tobacco waned; that of candy increased. Approximately 300,000 pounds of candy represented the monthly purchases during the early period of the war. This amount included both the home and overseas consumption. Demands from overseas grew steadily. The soldier far from home and from his customary amusements could not be considered an ordinary individual living according to his own inclinations, and candy became more and more sought after. As the demand increased, the Quartermaster Department came to recognize the need of systematic selection and purchase.

The first purchases were made from offerings of manufacturers without any particular standard, 40 per cent being assorted chocolates, 30 per cent assorted stick candy, and 30 per cent lemon drops. A standard was developed through the steady work of confectionery experts. This standard offered no opportunities for deception, and it guaranteed candy made from pure sugar and the best of other materials. The specifications furnished all bidders covered raw materials, the methods of manufacture, packing, and casing. Specifications were adopted after many conferences with the leading manufacturers of the country. These men cooperated in the work by giving their best suggestions and often their trade secrets.

Huge purchases of candy were made during the days when sugar was scarcest in the United States. The Food Administration was convinced that the Army should have all the candy it desired, and sufficient quantities of sugar were allotted for the purpose. From 300,000 pounds monthly the candy purchases increased till they equaled 1,373,300 pounds in November, 1918, the highest amount purchased up to that time. In December, 1918, an innovation was adopted, consisting of giving the troops a regular monthly ration of candy. The candy which had been shipped every month for sale in the various canteens had always been quickly disposed of. Many men did not get the opportunity to make purchases. The ration plan, however, assured each man a pound and a half a month, without exception. It required 3,495,000 pounds the first month of the ration system to provide each soldier overseas with his allotted portion.

In December, 1918, the Subsistence Division took over the purchase of all candy for the various organizations conducting canteens for our troops. The purchase for that month totaled 10,137,000 pounds, all of which was shipped overseas. It was the largest exportation of candy on record. The candy purchased for the canteens, commissaries, and other agencies was manufactured by the best known candy firms in the country. A portion of the candy consumed overseas was manufactured in France. This French supply was discontinued January 15, 1919, and thereafter all requirements were shipped over from the United States. The candy was sold to the men at just half the price it would have cost individuals here. After December, 1918, 50,000 pounds were furnished each month for sales purposes for every 25,000 men in France. Up to February 1, 1919, 21,000,000 pounds of candy had been sent across. The demand for candy jumped skyward after the signing of the armistice, the men then having more time on their hands in which to enjoy luxuries. Tobacco demands likewise increased.

The suffering sweet tooth of the Yank was not appeased by candy alone. The third of a billion pounds of sugar bought for the Army represents a tremendous number of cakes, tarts, pies, and custards. An old soldier recently stated that the ice cream eaten by the Army during the war would start a new ocean. The serious shortage of sugar which at one time threatened to reduce sweets to an irreducible minimum on the civilian bill of fare did not interfere with the soldier's ration, which continued to be 6 pounds monthly in this country and about 9 pounds overseas. The ration for the civilian population was reduced to 2 pounds monthly. Army officers were placed on the same status as the civilian population and were allowed to purchase only the amount stipulated for civilians for use in their homes.

Up to the signing of the armistice the total amount of granulated, cut, and powdered sugar purchased by the Subsistence Division equaled 342,745,862 pounds and cost $28,465,050. Of this amount the greater portion was shipped to the troops in France.

A close companion in popularity to candy and tobacco was the typical American product, chewing gum. This confection was found of great value on the march as a substitute for water. Its importance is shown in the vast amount sent overseas. A total of 3,500,000 packages represents the overseas shipment in January, 1919. The shipment for February was 3,200,000 packages. The winter consumption of gum was heavier than that of summer, the average monthly supply being only 1,500,000 packages during the summer of 1918. Chewing gum came to be considered a necessity by the men in France and has been found to be an invaluable aid to keeping up their spirits in the midst of hardships.

Every complaint against meals served in the Army reaching the attention of the Subsistence Division was investigated. These investigations were made in conjunction with the Inspector General's Department of the Army. Where complaints were justified, remedial action was taken. A study of the complaints revealed that most dissatisfaction was among new troops who, when first separated from the luxuries of home, wrote of their adventures at the mess table, enlarging any lack of home comforts into stories of privation. The more solid food, however, soon became popular, as the hard work in training gave an appetite for sustaining rather than for the more fancy foods.

Subsistence to the value of $327,060,097 was shipped to our forces overseas from the United States from the start of the war to December 1, 1918. The following table gives the quantity, unit price, and total value of these subsistence items:

Item.	Quantity.	Unit price.	Total value.
	Pounds.	*Cents.*	
Ham	1,772,917	34.42	$610,238
Bacon	147,956,223	44.42	65,722,154
Beef, fresh, frozen	250,584,692	23.36	58,536,584
Beef, tinned	140,843,476	32.46	45,717,792
Fish, salmon	30,961,801	14.24	4,408,960
Cheese	314,203	27.75	87,191
Flour	542,874,797	5.25	28,500,927
Hard bread	27,978,830	12.92	3,614,865
Corn meal	16,074,687	4.58	736,221

Item.	Quantity.	Unit price.	Total value.
	Pounds.	*Cents.*	
Oatmeal	4,661,732	6.35	$296,020
Beans, dry	39,646,677	10.84	4,297,700
Beans, baked	54,731,786	9.55	5,226,886
Rice	25,466,547	7.97	2,029,684
Hominy	1,826,269	8.54	155,963
Tomatoes	100,081,789	6.02	6,024,924
Peas, green	4,689,425	5.60	262,608
Corn, sweet	7,639,786	5.65	431,648
Beans, stringless	2,148,759	5.92	127,207
Vegetables, dehydrated	12,971,935	30.25	3,924,000
Prunes	15,748,931	10.35	1,630,014
Fruit, evaporated	8,976,848	13.27	1,191,228
Jam	26,029,028	18.74	4,877,840
Apples, canned	1,831,096	6.39	117,007
Peaches, canned	2,415,182	10.56	255,043
Apricots, canned	863,415	9.12	78,743
Pears, canned	1,150,120	10.22	117,542
Cherries, canned	423,444	12.21	51,703
Pineapples, canned	899,258	9.12	82,012
Coffee	39,185,167	12.07	4,729,650
Sugar	106,169,345	7.43	7,888,382
Milk, evaporated	42,922,743	10.48	4,498,303
Lard and substitutes	15,781,228	24.47	3,861,666
Butter and substitutes	16,200,799	39.71	6,433,337
Candy	7,895,053	27.76	2,191,667
Tobacco	27,449,645	67.06	18,407,732
Salt	13,707,276	.88	120,624
Vinegar ... gallons	1,319,877	27.85	367,586
Pickles ... do	1,333,210	46.94	625,809
Sirup ... do	6,171,808	59.22	3,654,945
Cigars ... each	160,180,225	4.85	7,768,741
Cigarettes ... do	2,439,260,097	.62	15,123,412
Special reserve rations ... do	15,623,150	76.00	11,873,594
Emergency rations ... do	765,400	52.50	401,835
Total			327,060,097

MEAT FOR THE U. S. ARMY RECEIVES THOROUGH INSPECTION. FEDERAL INSPECTORS ARE EXAMINING SHEEP AS THEY PASS ALONG THE ENDLESS TROLLEY IN THE PROCESS OF DRESSING. ARMOUR & CO.

MANUFACTURING RAZORS FOR SOLDIERS: BLADE HONING DEPARTMENT. DURHAM DUPLEX RAZOR CO.

MAKING UNIFORMS FOR THE ARMY.

View of the cutting department of a plant at Red Bank, N. J.

MANUFACTURING SPIRAL PUTTEES FOR THE ARMY.

This photo shows yarn being wound in a Cleveland, Ohio, factory.

CHAPTER II.

CLOTHING AND EQUIPAGE.

The Army raised against Germany had to have stout shoes for its feet. It required warm uniforms and overcoats and good socks and underwear. It had to have heavy blankets for its beds. The men needed raincoats and rubber boots for wet and muddy weather. Tentage was required, pup tents for the front and large tents and flies at the camps. Belts and bandoleers of cotton webbing added to the soldier's efficiency as a rifleman or machine gunner.

To procure these and other supplies for an American Army that eventually reached the strength of 3,750,000 men required the best brains in the textile, rubber fabric, and leather goods industries. From the counting rooms, from the laboratories, and from the American factories the needs of the Government called to Washington several hundred men, experts in a thousand lines, and put them into American officers' uniforms. Eventually the various agencies of the War Department purchasing these supplies were centralized in a single division known as the Clothing and Equipage Division of the Office of the Director of Purchase and Storage, which in turn was part of the Division of Purchase, Storage, and Traffic.

The total cost of this necessary equipment of textiles and leather and rubber goods was approximately $2,100,000,000. Of the enormous sum of money appropriated for the so-called quartermaster activities, a full one-quarter went for clothing and equipage of this sort.

The group who handled this enormous manufacturing effort not only conducted one of the biggest undertakings of the war but did it in a way to command the admiration of those who knew the story of what was going on. The division turned scientific attention, and that means the attention of real scientists, to the proper construction of all sorts of articles. It designed new styles of soldiers' clothing adapted in every curve and line to the service in France. It standardized dyes and made studies of protective coloring. It produced highly specialized shoes. It saved millions of dollars by the scientific study of specifications of various articles. It educated manufacturers in the production of articles strange to their experience, and in some cases developed entirely new industries. At one time it constituted the entire wool trade of the United States, since it had optioned every pound of wool in sight and had its agents out gathering up the excess wool of the earth. It was a shipmaster, an employer of men, a reformer in labor conditions, and an inventor and originator of new products.

The organization was important not only for the size of its business but because it dealt more intimately with the individual soldier than perhaps any other production branch of the Government, with the possible exception of the branch which fed him. It might seem to be a fairly easy proposition to buy clothing for a soldier, his tent, and the bed clothing that kept him warm in active service or when he was a patient in a military hospital. But it was not a simple task. None of these articles was standard for civilian use, either in material, color, or pattern. Everything had to be made to order. The ordinary factory could not begin work on contracts for these supplies on a minute's notice, but usually only after special and sometimes costly preparation.

And as the Army grew in size it had to have large quantities of special clothing. Cooks needed cotton aprons, and blacksmiths leather ones. Linemen had to have special gloves; hospital orderlies and waiters at messes required white duck suits; motorcyclists needed hoods; laborers, overalls; and firemen, helmets. There were special garments for aviators. We began capturing prisoners and they had to have special uniforms. Convalescents at hospitals needed special suits. The women nurses of the Army were supplied with uniforms, something entirely outside of previous Army experience.

The Government was something more than the designer and manufacturer of these goods, drawing the specifications, placing the orders, and then teaching the processes of manufacture in the thousands of factories which had virtually become Government plants. The clothing and equipage organization had to go further back and become the actual procurer of the raw materials; and this phase of its work eventually became one of the largest and most spectacular and romantic elements of the whole undertaking. In addition to procuring the raw cotton and the raw wool and the hides, the Government had to go into the manufacture of cloth and the tanning of leather to supply these commodities to the manufacturers of the finished articles. The Government went into a raw materials market which was already glutted with orders from the allied governments and from domestic consumption. It went into this market at first without money, since funds on the scale demanded were not available between March 4, 1917, and June 15 of the same year; and it had to buy on credit and secure the commodities in the face of cash bidding for them.

Nevertheless the whole enormous undertaking was successfully carried through. Except in rare instances, the American soldier never lacked for necessary supplies of this character. The organization which handled the work originally consisted of 6 officers and

25 clerks. When the armistice was signed this great purchasing and manufacturing agency had an enrollment of 1,693 people.

Wool was the most important of the raw materials to be procured, since wool entered into the composition of more items than any other material. Uniforms, overcoats, underwear, socks, breeches, shirts, and many other articles had to be made entirely or partially of wool. The purchases of woolen breeches alone during the war period amounted to 13,176,000 pairs. On September 10, 1918, the wool experts of the army estimated the Nation's total needs for wool up to June 30, 1919. The War Department, it was found, during this time would require 246,000,000 pounds of clean wool; the allotment to civilian needs was but 15,000,000 pounds. In other words, the war demands were to absorb practically the entire supply of wool; civilians were to be forced to do without it almost entirely.

Soon after the declaration of war the Quartermaster Corps estimated that it would require about 100,000,000 pounds of scoured wool to meet the initial demands of the Army in 1917. A meeting was called of the principal wool dealers of the United States, most of them from Boston, and a quick inventory was taken of the available wool supplies, not only in the United States, but on order from foreign countries. It was found that there was in sight 78,000,000 pounds of greasy wool, which, after being scoured, would produce 35,000,000 pounds of wool of the quality needed. This was barely one-third of the Army's demand alone. It should be noted, however, that this inventory was taken just before the annual American clip, which would be finished by the end of July.

To insure that the Government would secure every pound of wool in sight, options were promptly obtained on all wool in American warehouses or on the sea, and speculation in the prices of the domestic clip for 1917 was thus headed off by the entry of the Government itself in the raw wool business. The prices were fixed for the 1917 clip as of July 31. A year later the clothing and equipage division had become the entire wool trade of the United States. There was no wool market again and no public sale of wool until after the armistice was signed.

To handle this enormous undertaking the division appointed a wool administrator to buy wool, a wool purchasing quartermaster to pay for it, and a wool distributor to sell it to the Government contractors. The Government's wool headquarters was in Boston, with branches at Philadelphia, Chicago, St. Louis, San Francisco, and Seattle. This organization arranged to procure the whole 1917 clip, if needed, took over all wool destined for the United States under import licenses, and sent its agents to foreign markets.

The largest of the foreign markets practically available from the standpoint of distance was the Argentine in South America. Australia and New Zealand were, of course, enormous markets, but the dearth of shipping made it impossible to spare many bottoms for the long voyage into the Antipodes. As a matter of fact, when the fighting ceased, the whole world was suffering for wool, except Australia and New Zealand. America was short of wool, France had practically none, there was a little in England, but Australia and New Zealand had the staggering surplus of 1,000,000,000 pounds. This was due to the fact that there had been no shipping available to bring this wool to America or Europe.

The Government's wool administrator secured such Australian and New Zealand wool as he could; but he had to rely principally on sailing vessels, which could not, under the most favorable conditions, go to Australia and back again in less than seven months, while nine or ten months were more often required. A quick sailing voyage to Argentina and back required five months.

Nevertheless, and this was particularly true in the early fall of 1918, when preparations were being made for the equipment of the Army in 1919, every effort was made to secure foreign wool. A South American wool-buying commission was formed and sent to Buenos Aires, arriving there October 30, 1918. By that time, however, the end of the war was in sight, and the commission never opened up its Argentine headquarters.

The Government conducted its raw-wool business on the lines of a great department store. Headquarters were established in Boston, where the wool distributors kept samples of almost every kind of wool produced on earth, these samples representing stocks on hand in the various Government warehouses in Boston and elsewhere. Charles J. Nichols, a member of a large Boston wool firm, was the wool administrator and E. W. Brigham was wool distributor. Prices were fixed, and the manufacturers bought from the samples. Carpet wool was sold at an office in Philadelphia. The wool administrator did a business that averaged $2,500,000 per day during his incumbency, his total purchases amounting to about 722,000,000 pounds of wool.

At first the supply of the better grades of wool seemed to be adequate to meet the Army's demands. Later, however, changes were made in the specifications for various cloths, uniform cloth being increased from 16 to 20 ounces in weight, overcoating from 30 to 32 ounces, shirting flannel from 8½ to 9½ ounces, and blankets from 3 to 4 pounds. These increases made it necessary for the Army to use grades of wool previously made only into coarse materials like carpet. The lower grades of wool were blended with the finer grades to provide the necessary weight and warmth, even at the expense of fineness of texture and appearance. This action explains why at the end of the period of hostilities some of the American soldiers' uniforms looked rough and uneven in color. But the necessary cloth was provided, and it was warm.

The Government saved every ounce of wool that it possibly could save. More economical patterns and layouts for the cutting of uniforms were designed in Washington and furnished to the manufacturers. The American soldier's uniform did not meet the approval of officers of the American Expeditionary Forces as to style, after the latter had become used to seeing the smartly dressed troops of Europe. Accordingly, after Gen. Pershing had recommended a better-appearing uniform, a new one was designed, incidentally with an eye to saving cloth. The coat of the uniform—formerly called the blouse, a designation which is now obsolete—was cut with new lines, making it slimmer without sacrifice of warmth or comfort. The patch pockets of the original blouse were usually unsightly bulges when the soldiers filled them with articles. On the new coat the patch pocket was retained only in appearance, the pocket actually being on the inside.

It is not known to most Americans that the breeches, which have been typical of the American service uniform for many years, were abandoned late in the war in favor of long trousers. This change was also due to studies made by the army clothing experts. The soldiers themselves were not enamored of breeches, since they had to be either laced or buttoned below the knee, a process which took time always, but seemed to take more when a man was in a hurry. The laces sometimes chafed the leg under the leggins. Then, too, it was often impossible to remove the breeches from soldiers wounded in their legs without cutting the cloth. Long trousers did away with all these objections and had the added virtue of being warmer than the breeches.

The overcoat, too, was redesigned, following Gen. Pershing's recommendations, the stock overcoat being too long to be worn in the trenches. A knee-length garment was provided which was much smarter than the older coat.

The redesigning of the overcoat and the uniform (although the new uniform never appeared in the field) accomplished numerous economies. Merely by the elimination of lacings, eyelets, tape, and stays, the new trousers cost 95.25 cents less than a pair of army breeches. By July 1, 1919, this change in design would have saved the Government $16,988,440 in orders for trousers already placed or in sight. The change in overcoat styles saved 62 cents per garment, or a total saving to July 1, 1919, estimated at $897,140. The service coat, made by redesigning the blouse, saved the Government $1.598 on each garment, or an estimated saving of $4,977,770 to July 1, 1919.

This was not only financial saving, but what was more important, it was saving the consumption of the raw material, wool. The

Government could always raise more money; but if the wool supply were exhausted, all the money on earth could not buy any more of it.

A more economical cutting pattern saved twenty-three one-hundredths of a yard of cloth in the manufacture of every pair of trousers. This resulted in the total saving of 2,300,000 yards of woolen cloth. Part of the facings of the service coats and overcoats were eliminated without sacrificing warmth or serviceability, and cheaper cotton linings were substituted. Another important cloth economy came when the Army designers cut off the right-hand pocket of the O. D. shirt, on the ground that this pocket was seldom used. The designers also substituted an oblong elbow patch on the Army shirt for the circular patch formerly specified. This substitution was not economy in cloth, but the original circular patch, put on the sleeve to reinforce it at the point of greatest wear, actually resulted in reducing or shortening the life of the garment by tearing loose at the stitches, a fault which the oblong patch overcame.

In the earlier contracts the garment makers were stimulated to save wool by being allowed a percentage of the cost of yardage saved. Each contractor, too, was permitted to sell his own clippings. But as the Government obtained a more scientific grasp of the clothing problem and produced pattern layouts which utilized the maximum percentages of the cloth, the issues of cloth to the garment makers were calculated more closely. Thereafter the contractors received no reimbursement for cloth savings, and the Government itself took all the clippings.

These clippings were shipped to a base sorting plant at New York, where they were baled and shipped out to mills to be used as reworked wool in blankets and other articles. The clippings were sorted at a cost of 1.7 cents per pound and sold at an average price of 23 cents per pound, the total sales bringing in to the Government $5,500,000.

The history of the Government's wool enterprise during the war illustrates how hard it was to check the momentum of the whole production undertaking against Germany once it had attained full speed. A week before the armistice was signed the wool stocks looked small, and shortages plainly existed to cause anxiety for the executives in Washington. That was because we were thinking in terms of consumption made familiar by the terrific destruction of war. A week later the same stocks looked overwhelming in size, and the shortages had become enormous surpluses. It had been a constant worry to procure a sufficient quantity of blankets, yet as soon as the armistice was signed, we had on hand a 47-months' supply of blankets for 1,000,000 men in the United States and 2,400,000 men overseas. As soon as the German plenipotentiaries affixed their signatures to the armistice agreement at Spa an apparently small stock of marching shoes turned into a 4-year supply for 3,400,000 soldiers at home and abroad. On November 1, 1918, the Clothing and Equipage Division had on hand a reserve stock of goods valued at $811,000,000.

The entire woolen industry, from the handlers of raw wool to the textile mills, worked splendidly with the Government. At all times there was plenty of available machinery to make all the cloth for which wool could be furnished. Mills which found no Government use for their regular business output went heartily to work to make something else that the Government would need. The Government's uses for carpet, for instance, were practically negligible; so that the carpet mills, many of them, swung their entire production to Army blankets and Army duck.

Blankets, in fact, were one of the largest items. The total purchases brought to the Government warehouses about 22,000,000 blankets, at a total cost of over $145,000,000. Melton cloth for overcoats and uniforms consumed an enormous quantity of wool. The total purchases of melton amounted to more than 100,000,000 yards, or enough to stretch twice around the world at the Equator, with a strip left over long enough to reach from New York across Germany and Russia and into Siberia. The total quantity of raw wool bought by the Government up to December 14, 1918, cost over $504,000,000.

After the Government had secured the wool and various types of cloth, there still remained the task of making this cloth into uniforms. The usual method was for the Government to furnish the materials and to pay the contractor his cost of manufacturing.

All Army clothing was made up according to the so-called tariff sizes. The average coat for a man is a 38 or 40, and experience shows how many men in a given number will need this average. But there were always exceptions. One camp sent in a special order for 46 overcoats for "fats."

Through a scientific study of the problem, notable reforms in the matter of fitting soldiers were brought about. When the men were coming in greatest numbers from civilian life to the training camps they were often put to great inconvenience in securing proper clothing. Each man would ask for such sizes as he thought were correct, but it often happened that the garments supplied to him did not fit him, and he thereafter spent some hours or even days swapping garments with other recruits until he eventually acquired an outfit somewhere near his size. Then, too, there was confusion in the way the articles were supplied to the men, who sometimes had to stand in line all day long, awaiting their turn at the issue windows.

The matter of fitting was satisfactorily solved by adopting the so-called foolproof size labels. The labels originally used were merely paper tags pinned to the garments, and in the handling of garments by men unfamiliar with the fitting of ready-made clothing mistakes often resulted. As in the case of civilian clothing, all Army clothing was divided into four classes, known as "longs," "shorts," "stouts," and "regulars." A garment of any size would come in these four classes. The labels were marked with diagonal, colored stripes to indicate the general characteristics of the garment to which it was attached. Thus green meant a "short," red indicated a "long," and yellow showed the garment to be a "stout." The soldier was pretty sure to remember the color of the stripe attached to the garment that fitted him. If he were a green striper, he would refuse to accept anything that did not bear a green stripe on its ticket.

Before hostilities ceased a system providing a more scientific issue of clothing to recruits had been introduced. Under this system the recruit would enter the supply building at one end and there, in a special room, strip himself of his civilian clothing. He would thereupon enter the mill as naked as the Lord made him. He would stop first at the underwear counter, where he would procure garments that fitted him, would don them, and then pass on to the hosiery counter. Thus he would progress down the line, eventually emerging from the other end of the building a fully dressed American soldier, the process reminding one of the progress of an automobile through the Ford factory.

It required the services of some 4,000 inspectors to supervise the garment-making in thousands of shops scattered throughout the country. This inspection also looked at the character of the shops taking contracts, and the Government was sometimes hard put to it to prevent child labor and sweat shop production in the work.

At one time there came a rush order from France to supply several hundred thousand mackinaws. An officer who was familiar with mackinaws was sent out from Washington to buy them from goods in stock. He accomplished his mission in 10 days, literally baring the shelves of the United States of these garments, his purchases including the extensive quantities of mackinaws held by mail-order houses in Chicago.

It was always a problem in clothing the Army to find olive-drab dyes that were fast in color. The first dyes used were apt to fade quickly. A certain dye was of the proper color, yet it was found on test to have the peculiar characteristic of being visible at a distance. As the new American synthetic dye industry expanded and processes were perfected, the officers of the Clothing and Equipage Division were able to cooperate with the American dye makers to produce satisfactory dyes.

Yet while the olive-drab dye used in dyeing coats and trousers seemed to withstand the sun and rain, that used in coloring the leggins proved to be fugitive to a remarkable degree. It seemed to be impossible to produce a dye that would hold its shade in leggins.

The experts working on the dye problem had expended a good deal of valuable energy in worry and had grown a few gray hairs in their heads over the failure of leggin dyes when they discovered the true cause of the fading. The men were deliberately bleaching out their leggins, usually by using salt solutions on them, since anything but a faded leggin indicated that the soldier who wore it was a rookie and a greenhorn.

The materials which went into the manufacture of clothing came from various sections of the country, since the several garment industries had grown up around centers. For instance, the melton cloth came generally from the Boston district. Linings were supplied from Atlanta, buttons from Philadelphia, and duck from Chicago. This geographic distribution of supplies simplified the Government's problem of supplying materials to the various contractors. It was possible to supply materials on short notice to any garment-making district.

At one time Chicago wired that unless 500,000 yards of flannel shirting were supplied immediately hundreds of shirt factories in Chicago and the Chicago district would have to close down. Accordingly, a special freight train was loaded with shirting in the East and started for Chicago on a special movement in charge of a "live tracer"—that is, an officer who saw that the train was put through to its destination. The train arrived in Chicago on the second day after the order was received, so rapidly had the goods been procured and loaded.

In addition to the regular uniforms for the men, almost half a million articles of clothing for officers were also bought by the Government.

The Quartermaster Department went into an entirely new field when it bought uniforms for the women nurses of the Army. There was a Norfolk suit which cost about $30 and a cotton uniform that cost about $3, an overcoat costing nearly $28, and then there were waists made from navy blue silk and from white cotton, and hats.

Before leaving the subject of clothing, it is interesting to refer again to the clothing furnished for interned prisoners. This was not manufactured for the purpose. Uniforms discarded by our own men were reclaimed and dyed a special shade of green. Over 50,000 of these garments were prepared at an average cost of less than 30 cents per garment. It had been the original intention to make a special prisoner's uniform striped in resemblance to the prison suits worn in American penitentiaries.

Another interesting development in the manufacture of Army clothing was the production of a special uniform for expeditionary troops sent to Russia. The uniforms were so warm that they could well serve as the equipment for an Arctic exploration party. The determination to send an expedition to Russia was made suddenly by the Government, and the decision brought with it the problem of producing in a jiffy an equipment of garments not only expensive in themselves, but of a character unknown to the American garment trade. An agent for the division in New York at once bought on the New York market large quantities of muskrat, wolf, and marmot fur. Other agents were sent into our own Northwest and to Canada to pick up such suitable garments as these markets afforded. The Siberian equipment as specified by the commanders of the expedition called for fur caps, fur mittens and fur overcoats, mucklucks, moccasins, felt shoes, fur parkas, and underwear for 15,000 men or more. The order for the equipment came in the latter part of August, 1918, so that only the fastest kind of work would produce the garments in time to catch the last steamer that could get into the northern Russian and Siberian ports before the ice closed navigation for the season. The result was that whenever the articles specified could not be procured on time, suitable substitutes were provided.

The specifications called for 80 per cent wool underwear. Underwear with that percentage of wool could not be provided, but underwear of equal weight was substituted. Where fur-lined garments were unobtainable, fur-trimmed ones were procured. The specifications called for Buffalo coats. The division sent a man to the north woods country of Minnesota and Wisconsin, and there in the supply cities he bought sheep-lined coats with moleskin or duck shells as a substitute. These coats were the sort used by woodsmen and Alaskan miners and explorers. There was no time to procure mucklucks, moccasins, and felt shoes, so an agent of the division was sent into Canada to buy shoe pacs (or lumbermen's boots) and lumbermen's knee-length socks. The total cost of the whole outfit was more than $100 per man.

It was impossible to find any substitute for the Alaskan parka. A parka is a sort of overshirt, wind proof and waterproof and hooded, to be worn over the overcoat and cap of the uniform. Consequently it was necessary to produce the parkas in this country, although our garment makers were entirely unfamiliar with such manufacture. The work was undertaken by the International Duplex Coat Co., at 114 Fifth Avenue, New York. It was necessary from the start in turning out this order for the employees of this plant to work overtime. In order to speed the production the principal member of this firm himself took his place at the bench and worked almost day and night in cutting out garments.

The day approached closer and closer when the shipment would have to start across the country if it were to catch the last boat from San Francisco. On the home stretch of the race the entire working force of the plant went 36 hours, stopping only for meals. The last

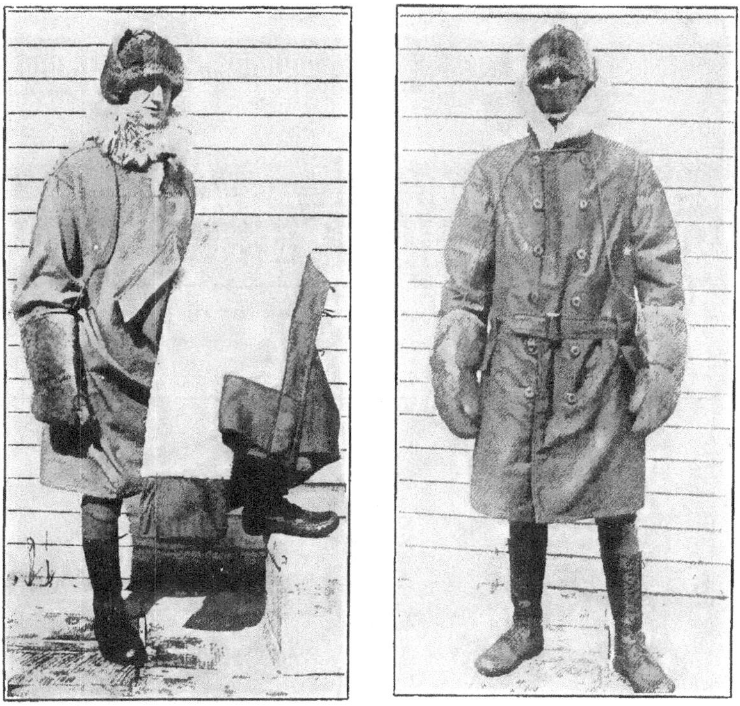

CLOTHING WORN BY OUR SIBERIAN EXPEDITIONARY FORCES.

MANUFACTURING WOOLEN UNDERWEAR FOR THE ARMY.

The weaving department of a plant at Cohoes, N. Y.

MANUFACTURING HOSIERY FOR THE ARMY.

The knitting room of a hosiery mill at Durham, N. C.

stitch was taken at 1.30 o'clock in the morning. The garments were then piled upon auto trucks to be rushed to the baling plant in Brooklyn. One of the loaded trucks developed engine trouble and stopped in the middle of a bridge across the East River. The officer in charge thereupon commandeered every automobile that came along, piled them all full of parkas and sent them to the baling plant. The entire shipment was aboard the train less than one hour before its starting time.

It was not only necessary for the Government to furnish cloth for the uniforms, shirts, and other articles, but it had to supply the fittings and findings as well, such needs as linings, tape, buttons, and hooks and eyes. In the calendar year 1918 the purchases amounted to over 46,000,000 yards of cotton lining and 2,500,000 yards of felt lining, worth over $18,000,000. The Government spent over $100,000 for hooks and eyes, $150,000 for tape, $1,250,000 for thread, and practically $3,000,000 for buttons.

When it was found that the standard specifications for Army uniform buttons favored a certain class of manufacturers and excluded many others, new specifications were drawn so as to make it possible for every button manufacturer in the country to compete for contracts. An exclusive study was made of new materials for buttons. They had been made of brass or bronze, but due to other war necessities for metals an effort was made to provide a substitute. It was found, too, that metal buttons sometimes resulted in infection of wounds received on the battlefield.

Substitution of vegetable ivory for metal in buttons was attempted. The Bureau of Standards in Washington tested the taqua, or ivory, nuts from which buttons are made and found them suitable. A vegetable ivory button with a shank was developed, although no such ivory button had been known before, and the Government's insignia was stamped on this button. Gen. Pershing approved the use of ivory buttons, and thereafter many manufacturers produced millions of gross of them. Every manufacturer who took button contracts agreed to turn over the ivory nut waste to the Chemical Warfare Service to be used in making charcoal for the gas-absorbing canisters of the gas masks. Most of the buttons were produced by firms in Rochester and Philadelphia. Many concerns made them who had never made buttons before. Manufacturers of electric goods, hardware, billiard balls, celluloid, pearl buttons, and phonograph records turned their plants into ivory-button factories. Enormous quantities of buttons were required. For the Army shirts alone the Government needed 216,000,000 buttons in 1918.

Flags constituted another class of goods requiring wool. In all, the division produced 40,000 flags during the war period, most of these being made at the Government's own shop at Philadelphia. It is a grim fact that many of these flags were used to wrap around the bodies of soldiers who died at sea. Thirty million chevrons for non-commissioned officers were also turned out by the Government.

The production of overseas caps for the American Expeditionary Forces was likewise an extensive undertaking. When the requisition for overseas caps came from France, it was not possible to design one here because of lack of knowledge of what was required. Later a courier bearing a sample cap came to the United States from Gen. Pershing. As soon as this sample was received a meeting of cap makers was called in New York, and 100 manufacturers attended. One and all agreed to turn over their factories to the exclusive production of overseas caps until the requirements were met. It took these cap makers only two weeks to turn out the first order. In all 4,972,000 caps were delivered.

Our experts on this side of the water were not satisfied with the overseas cap. It shrank after being wet, it quickly lost shape, it absorbed much water and did not dry out quickly, and it was unattractive in appearance. Also it did not shade the eyes, and the experience in France showed that the soldiers usually improvised peaks to their caps by sticking their girls' letters between their caps and their foreheads. Then, moreover, the standard cap was made of 20-ounce melton, which was a fabric hard to get. But there was plenty of rabbit fur available to make felt caps for an army of 6,000,000 or 7,000,000 men. Accordingly a new cap was designed, made of felt and doing away with the bad features of the melton cap; but this cap improvement came at the end of the war and was never used.

Wool was required not only for the outer clothing of the Army—for the uniforms, overcoats, and caps—but there was also a tremendous war demand for it for the manufacture of such knit goods as undershirts, drawers, stockings, gloves, and puttees. The matter of providing the Army with these necessary articles offered a problem of peculiar difficulty, since, in addition to the ever-threatening shortage of raw wool, there was an actual shortage of machinery in the knitting industry. When it was found that the regular mills could not turn out all the woolen knit goods the Army required, numerous mills which had been turning out specialties exclusively, such as women's underwear or men's union suits, were converted into factories to knit garments according to the Army specifications. Some idea of the extent of the Army's demand for this class of goods may be read in the fact that toward the close of hostilities every machine in the United States that could make hosiery at all was knitting socks for the Government.

At one time there was an acute shortage of needles. Germany had previously supplied America with knitting needles. When this

SEWING OVERSEAS CAPS IN A ST. PAUL, MINN., FACTORY.

HEAPS OF OVERSEAS CAPS READY FOR SEWING IN A ST. PAUL FACTORY.

UNDERWEAR FOR THE ARMY AT A MILL IN ST. JOHNSVILLE, N. Y.

KNITTING ARMY UNDERWEAR IN A SYRACUSE, N. Y., MILL.

source was cut off, we turned to Japan. The Japanese needles proved disappointing; they were not correctly tempered and frequent breakage caused great loss. At one time it was rumored that there were 10,000,000 knitting needles in Sweden, and the need here was so urgent that several buyers were sent to that country. Their effort was well worth while, for they actually secured a million needles to help relieve the situation here. Meanwhile, American needles were improved and American needle makers were pushed to the limit; but until the close of the war there was always an acute shortage of needles for the knitting industry.

It was soon discovered that there was not enough machinery in America to knit one-tenth of the seamless woolen gloves that the soldiers required. Consequently it was necessary to adopt a substitute—a glove of knit fabric cut to pattern and sewed up with seams. In actual service this glove did not stand up to the hard usage required of it. Consequently there was designed an overglove of canton flannel with the palm cased in leather, this to be worn outside the seamed woolen glove. In the effort to produce gloves which would give longer wear the so-called ambidextrous glove was designed so cut that it could be worn comfortably upon either hand.

Puttees, the spirally wound leggins that had long been used by the British Army, were unknown articles to American manufacture when the American Expeditionary Forces adopted them as standard articles of equipment. A puttee of knitted wool was designed and 6,000,000 of them were ordered in the spring of 1918, these to be preliminary to future orders for 8,000,000. The work required the installation of much new machinery in the textile plants. On November 1, 1918, we had produced all the puttees required by the troops then in France and had a surplus of 1,500,000 of them.

In the production of knit goods, economies in the use of material were constantly effected. An original article of equipment for the overseas troops had been a knitted woolen toque, which was a sort of stocking-cap. The toques had cost the Government $1 apiece, and some 1,500,000 of them had been piled up in the quartermaster warehouses before the toque was abandoned as a piece of standard equipment. Later a requisition was received for 400,000 woolen mufflers to be used by drivers of automobiles and motor trucks. According to the specifications these would cost about $3 apiece. Then it was discovered here that the abandoned toques might be sewed together to make mufflers. With this stock in hand it cost the Government only 20 cents each for the mufflers instead of $3, a clear saving of over $1,000,000.

The Quartermaster Department was the Mecca of inventors during the war period, who came bringing real or fancied improvements in many lines of apparel and personal equipment. One brought in a trench shower bath, consisting of a hot-water bag and a hose. He was much chagrined when informed that if this apparatus were set up in the trench there would be no room for soldiers to pass it. In no respect did the inventors have more novel ideas than they had in the manufacture of underwear. One of them brought in a patented vacuum suit of underwear which acted on the principle of a fireless cooker or thermos bottle to exclude the cold from the wearer's body. However, he had failed to take into consideration the fact that not only must cold be kept out, but perspiration must be given a chance to escape. The vacuum underwear would never dry out, after a man had become sweaty in it. For that reason it was not adopted.

A woman of Iowa invented cootie-proof underclothing by impregnating underwear with vermin-destroying chemicals. The State of Iowa was so interested in her invention that there was a public movement to clothe all Iowa troops in this underwear, should the Government fail to adopt it. The underwear was submitted to the experts of the Bureau of Entomology (the Government agency that deals with bugs), whose experts tested the invention. They found that the underwear was indeed death to the cootie. However, if the chemicals were applied in weak strength they soon evaporated and left the underwear harmless to the insect; if applied in great strength, the poisonous chemicals irritated the skin of the wearer.

During the first winter the men were in camp, the winter of 1917–18, there was no time to provide the troops with standard Army underwear. Consequently Government agents went into the underwear market and bought outright whatever was in sight. As a result, the soldiers that first winter wore underwear of almost every description and grade of merit. This gave the Army's underwear experts a fine opportunity to study the qualities of underwear of various types as proved by actual use. These studies contained hints of use to the civilian. For instance, the warning is plainly given to wear no fleece-lined underwear. A study was made of the causes of colds, and it was discovered that soldiers wearing fleece-lined underwear caught cold more easily than those wearing any other sort. The fleece of the lining absorbed perspiration and retained it, staying damp. Since many of the soldiers slept in their underclothing, they were thus encased in damp clothes 24 hours a day. Sick reports plainly showed the result of it.

When it comes to the production of cotton cloth for the Army's uses, the figures are so large as to appear almost fantastic. In all we procured over 800,000,000 square yards of cotton textiles. This was enough to carpet an area nearly four times as large as the District of Columbia. In a strip 3 feet wide there was enough of it to wrap 18 layers of cloth around the equator. Spread this strip out on some cosmic floor, and you could place upon it side by side 55 globes as large as the earth.

In addition to the cotton khaki required for uniforms and other purposes, the principal other cotton items were duck, denim, webbing, gauze, venetian, sheets, pillowcases, and towels.

The purchases made by the Army were beyond anything that had been known in the textile industry. In March, 1918, the supplies of cotton khaki on hand seemed to indicate a surplus of 21,000,000 yards beyond the needs of the immediate future. Then came the start of the German drive, and by the middle of April this great surplus of khaki cloth was not sufficient to the need. In other words, there was a shortage of khaki, since the Army needed at once 25,000,000 yards and thereafter would require a monthly supply of 10,000,000 yards. This was looking toward the great increase in the number of men soon to be called to the colors. It was planned to draft 300,000 in June alone, and subsequent drafts would be on a like scale.

In order to supply summer uniforms for these men it was necessary for Army officers to get every yard of khaki goods in the country. All stocks of goods in the hands of dealers and manufacturers were inventoried, and the positive order went out of Washington forbidding the use of khaki in articles for civilians. In spite of the Government's tremendous demand upon a limited supply, these stocks of khaki were acquired at a price 20 per cent lower than the prevailing market.

The requirements for cotton duck and cotton webbing also leaped upward as soon as the United States began to avalanche soldiers upon France. The demands were greater than could be supplied by the output of mills regularly producing these materials, and consequently the Clothing and Equipage Division called upon manufacturers of similar materials to adapt their plants to the production of duck and webbing. This they did, in many cases at considerable inconvenience and expense. Among the concerns which assisted in supplying these materials were manufacturers of carpets, automobile tire fabric, and even lace.

Owing to the scarcity and the high cost of leather a great deal of cotton webbing was substituted in the manufacture of such equipment as cartridge belts, suspenders, gun slings, and horse bridles. Here was additional demand, and to meet it factories which had been making such things as asbestos brake linings, hose, lamp wicks, suspenders, garters, cotton belting, and other similar fabrics, became webbing mills. All these plants thus adapted to the emergency manufacture of webbing were dependent on purchased yarns, which they had to secure in the open market from yarn manufacturers.

In the South particularly, where most of this yarn was purchased, the securing of power was a serious question. Many of the mills depended upon electricity generated by water power. These power plants did not always have good railway connections and many of them had no steam power equipment even if fuel could have been furnished. In the late summer of 1918 the rivers of the South ran nearly dry, and in order to operate many of the southern mills it was necessary for the Government to allocate according to most pressing needs the available power among the mills working on contracts. Also, for a long time when transportation facilities were seriously overtaxed, it was hard to secure a steady flow of materials from the South to the northern mills.

With regard to labor, employees in the cotton and webbing mills had to be educated in the manufacture of the new types of work to which these plants had been shifted. In the South, more especially, there was a question of child labor and of hours of labor for women and minors; for the Government inserted clauses in the later contracts requiring certain standards for the benefit and protection of labor. In some instances contracts were returned because of the child-labor clause. In such cases compulsory orders were often issued, practically compelling the mills to produce the goods called for.

Considerable burlap used for packing, as well as burlap bags, silk for flags, hat bands, and badges were also purchased in quantity.

The United States was never forced to turn to the use of paper in the manufacture of clothing, as the central powers were compelled to do; nevertheless preparation was made for the time when the cotton supply of the United States might become unequal to the demand. Garments made of paper cloth captured from the Germans were shipped to the United States and carefully studied by the Clothing and Equipage Division to learn the possibilities of paper fabrics should the need for them develop.

Over 100,000,000 yards of denim were bought. Denim was used particularly in making working clothes for the soldiers. At one time the factories were consuming denim at the rate of 13,000,000 yards a month. Brown denim which was required by regulations was a material hard to get, blue denim being the standard fabric for American overalls, and consequently heavy gray goods and drills were dyed olive-drab and put into use.

As to gauze, about 140,000,000 yards of it were purchased. Sheets and pillowcases were required in such quantities that at one time every mill in the country whose normal business was the production of sheeting was working for the Government. There were over 120,000,000 yards of webbing purchased, and nearly 300,000,000 yards of the various kinds of duck.

The duck and webbing just mentioned went into the manufacture of a numerous class of articles, known as textile equipment, including such articles as belts, tool bags, tool kits, flasks, canteen covers, and the like. The procurement of the webbing for these articles was in itself a manufacturing achievement. Before the war there were only a half dozen plants in the United States which could make webbing of the grade demanded by the Army. When the armistice came there there were 150 such plants. At the beginning of the war an order for 5,000,000 yards of webbing fairly staggered the industry, but that industry was to witness the day when an order for 50,000,000 yards would be absorbed as a matter of course.

But even after the webbing was secured there were practically no factories in the United States that had machinery heavy enough to make the Army's textile equipment. This work for the standing Army had been done exclusively by the Rock Island Arsenal. In order to increase the manufacturing capacity of the country it was necessary to get the Singer Sewing Machine Co. to build special machines adapted to this heavy work; and we also had to send experts from the Rock Island Arsenal to teach all new contractors how to make the articles. Many of the factory workers were women.

In spite of all difficulties production was wonderfully increased. Along in January, 1918, about 100,000 pistol belts a month were being made; while at the time of the signing of the armistice 560,000 were being manufactured monthly. Of cartridge belts in the same period the production was increased from 85,000 to about 410,000 monthly, and of haversacks from 290,000 to about 850,000 monthly.

No soldier could be sent overseas without a haversack, a cartridge belt, and a canteen cover; yet during the period of active hostilities no movement of troops was delayed one day on account of the lack of textile equipment. Up to December 1, 1918, the production of haversacks was over 2,500,000 in number, costing over $8,000,000; of canteen covers, about 3,750,000, costing $2,250,000; of cartridge belts, about 1,500,000, costing over $4,000,000. Another large item was bandoleers, which were procured to the number of over 31,000,000 at a cost of $5,500,000. These are only a few of the major items, but they serve to illustrate the extent of the purchases of textile equipment.

At the end of hostilities the Government was buying textile equipment at the rate of $22,000,000 a month, and was working toward the goal of being able to supply 750,000 men a month with all articles of textile equipment.

When the Army began to expand in size at an unexpected rate in the spring of 1918, it created a great shortage in cotton underwear. Government agents went out over the country and bought all cotton underwear stocks. In order to provide a sufficient manufacturing capacity for cotton underwear, women's underwear factories were enlisted for war work, and so were even corset factories.

The Army experts in cotton textiles also effected many economics. A standard pattern layout was drawn for the overall makers with consequent large savings of cloth in the manufacture of brown denim fatigue clothing, or soldiers' working clothes. At one time practically every overall factory in the United States was making fatigue clothing for the Army, after Gen. Pershing had cabled an order for 3,000,000 garments to be delivered in 90 days.

In making the soldiers' barrack bags, in which they pack their clothing and personal effects, the manufacturers in cutting out the pattern left a 3-inch strip of cloth. Army officers discovered these 3-inch strips and also noted the fact that every barrack bag must be provided with a draw-string. The specifications were thereupon changed so that these 3-inch strips could be used as draw-strings in the barrack bags, a trifling economy apparently, yet amounting to a saving of 6 cents in the cost of each one of millions of these bags.

A vast amount of tentage was required, not only for tents themselves, but also for such articles as paulins, tent covers, bed rolls and clothing rolls, canvas basins and buckets, bags for stakes, tool bags, coal bags, and mail bags, cargo covers, wagon covers, horse covers, and many similar articles.

Valuable work was done in substituting cotton thread for linen. Linen thread became so scarce that the Ordnance Department commandeered the whole supply. This worked havoc in the shoe industry, and as a result the Council of National Defense secured from the Ordnance Department enough linen thread to take care of the Army shoe contracts. Nevertheless it was discovered that cotton thread might be substituted for linen in many industries. In fact, it often proved to be better than linen.

Valuable standard tests for waterproof cloth were also worked out. These tests were developed at the Bureau of Chemistry, a branch of the Department of Agriculture in Washington. In these tests cloth was required to withstand a deluge of water equivalent in intensity to a tropical rain, and also to undergo a dry temperature of 120° Fahrenheit. There were also tests to determine under what conditions the cloth would mildew. These tests are expected to have a use in the waterproof-goods industry in normal times.

Another important contribution of the Army to peace-time industry was the design of the oversuit for the use of truck drivers. This was a waterproof garment, air-tight and cold-proof. It is expected that this new garment will continue in commercial use.

The principal items of rubber goods bought by the Army were rubber boots and overshoes, raincoats, and slickers. The production of rubber boots for the Army took practically the entire capacity of all mills in the United States, the rubber boot manufacturers having pledged themselves to discontinue their civilian business until the needs of the Government were taken care of. Of different types of rubber boots, the purchases were considerably over 4,000,000 pairs, at the cost of $20,500,000.

Incidentally there was worked out an improvement in rubber boots to prevent them from blistering the heels of wearers. It was discovered that a rubber boot blisters the heel because it rubs slightly as the wearer walks, no matter how well fitted to the foot the boot may be. To the specifications for the Army's rubber boots was added the requirement that straps be incorporated in the article to be buckled both around the ankle and around the instep, thus holding the boot so that it can not slip.

Raincoats caused a good deal of trouble, as there was not a sufficient manufacturing capacity in this country to meet the requirements. Practically all stocks of commercial raincoats were purchased, on the theory that even a poor cover was better than none. As these garments were made for civilian use, they were not built according to Army specifications, and considerable criticism was made as to their quality.

When the manufacture of raincoats commenced on a large scale, many new concerns went into the business, and some of them, either through lack of experience or through carelessness or intent, made garments that were not properly cemented. This led to investigations and indictments. The total purchases of ponchos, raincoats, and slickers amounted to about 10,000,000 garments, costing over $46,000,000.

In all 7,000,000 service hats of felt were manufactured on orders placed by the War Department. The felt for these hats was made of rabbit fur imported from Australia, New Zealand, and Russia and produced in the United States. Hats were made principally at Danbury, Conn., and Fall River, Mass., with smaller sources of supply at Yonkers and Peekskill, N. Y., and Reading, Pa.

The numerous requirements of the Army for pillows created a shortage in feathers. In all there were manufactured on Government order 500,000 pillows weighing 2½ pounds each. It had been the original intention to fill these pillows with duck feathers; but when the American duck-feather supply was exhausted and thousands of the ducks of China had given up their plumage for the comfort of the American soldiers, and still there were not enough feathers for the pillows, adulterations with goose feathers and other light plumage were permitted.

The procurement of leather for the Army, both the raw material and the finished products of leather, was one of the most important undertakings, the principal war uses for leather being in shoes for the soldiers and in harness for the horses and mules.

When the Government entered the leather market it found a high level of prices, due to the large quantities of leather and leather equipment which America had been exporting to the European nations at war. The tanners were called together, and they came to an agreement with the Government as to the prices of all grades of equipment which the Army expected to buy. The packers next agreed on a maximum price for hides suitable for Army leathers. The Government took an option on 750,000 hides then in the hands of the packers.

By consulting with the industry at all times the Government officers were able to stabilize prices of leather. The price of harness leather, which was originally fixed at 66 cents per pound, was advanced only 4 cents during the 18 months of the war period, while russet leather never advanced more than 4 cents per pound above the $1.03 fixed at the beginning of the war.

As the stocks of leather on hand diminished it became necessary to stimulate the production of leather goods, and there was formed a hide and leather control board, with a representative on it from each branch of the trade, one for harness, one for sole leather, one for upper leather, and one for the sheepskin trade. This board also inspected leather at all the tanneries and the finished leather in the various factories, a course of action which resulted in great improvement in the quality of leather, particularly in leather used in shoemaking.

At the outset the Quartermaster Corps, the Ordnance Department, the Signal Corps, the Engineering Department, the Medical Department, the Navy, and the Marine Corps were all buying leather or leather equipment, and the Y. M. C. A. and the Red Cross were also in the market for large amounts of leather materials. These activities, except those of the Navy and Marine Corps, were all eventually brought under the administration of the Clothing and Equipage Division, thus virtually eliminating competition in the leather market.

At the signing of the armistice it is safe to say there was enough leather equipment, either in the United States or in France, or in process of manufacture here, to meet the needs of 5,000,000 men. Leather equipment was available at all times. The principal items of leather were harnesses, shoes, jerkins, gloves, and mittens.

In all, $75,000,000 was spent for harness and leather equipment. The procurement of saddles in itself was a hard problem, since there were only three or four makers of saddletrees in the United States, and only one of these could get the ash or basswood required. The division induced various furniture factories to install the special lathes required for turning saddletrees, and in this way built up eight factories, which gave us sufficient capacity. Belting manufacturers and manufacturers of shoes were educated in the art

Dieing Out Uppers with Clicking Machine.

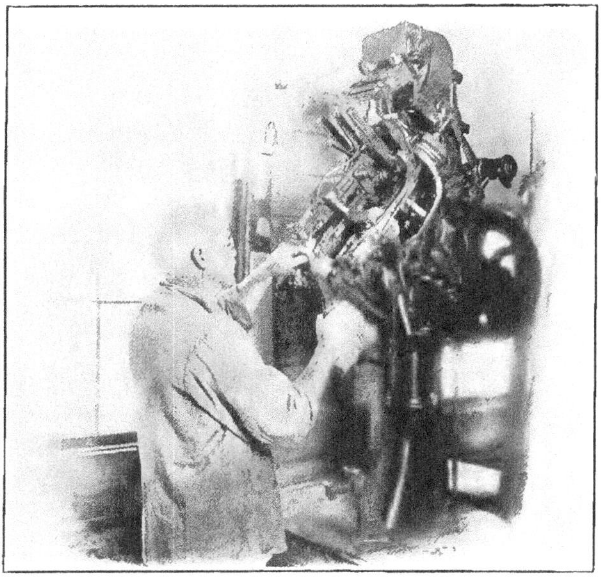

Pulling Upper Over Last.

MANUFACTURING ARMY SHOES.

Stitching Vamp to Quarter.

Attaching Heels.

MANUFACTURING ARMY SHOES.

of producing the leather for the saddles. The Army harness is of russet leather, a product for which there is no commercial demand. The result is that surpluses of Army harness can not be disposed of to advantage.

The former American Army shoe built on the Munson last and known as the russet marching shoe was machine sewed, had an upper of calfskin with the rough side turned in, and was lined with duck. This shoe proved to be short lived when subjected to the severe service in France. At the beginning of the war a new shoe was designed for trench service. This was a much heavier shoe and the calfskin of the upper was turned rough side out. There was no lining in the shoe. The shoe had two heavy soles, the outer one being hobnailed. Yet this shoe, too, proved to be unsatisfactory for the service. The uppers wore fairly well, but the soles could not stand the constant submerging in mud and water. The demands of trench service eventually led to the design of what was called the Pershing shoe. This was a shoe with three heavy soles, stitched, screwed, and nailed together. It had steel reinforcements on toe and heel. The outer sole was studded with hobnails.

The original requisitions from France for this shoe called for leather tanned with bark. As bark tanning is practically obsolete in the United States to-day, it was necessary to go into the tanneries and build up what was virtually a new industry. It may be mentioned that the design for the Pershing shoe was completed in 30 days.

The culmination of the shoe development was the model known as the Victory shoe. This model corrected certain defects in the Pershing shoe. The Pershing shoe was prone to rip along the back stays, and the upper did not fit snugly. In the Victory shoe the entire back of the upper was one piece.

At one time 52 shoe factories in 13 States were working on Army shoe contracts. A scheme of packing shoes for overseas shipment in burlap bags instead of in boxes was worked out, and it resulted in saving a great deal of space on board ship.

Machinery and tools for the shoe repair shops of the salvage division were purchased by the Clothing and Equipage Division. This was the first time that Uncle Sam had ever acted as cobbler for his soldiers. About 2,000 machines for repairing shoes were bought, besides some 28,000 repair kits, each one of which cost $135. Among the items of supplies for the Army shoe repair shops may be noted 20,000,000 pairs of half soles.

A shoe waterproofing grease, or dubbin, as it is called, which had no odor and which would not turn rancid, was developed. The experts worked closely with officers in the field in training soldiers in the care of shoes to make them last as long as possible. Every man who received a new pair of shoes was required to break the pair in by standing in them in water for a certain period and then walking for an hour until the shoes dried on his feet. The men were cautioned not to dry their shoes by placing them too closely to any heating apparatus, as this shortens the life of the leather. Good care of the soldier's feet has long been standard Army practice with us. No soldier in 1917 and 1918 was permitted to wear darned socks, unless he wore two pairs at once. At regular intervals officers inspected their men's feet, treated any blisters or sores that might exist, and dusted the feet with powder.

Bad shoe fitting means foot troubles, leg troubles, and sometimes even spinal and mental troubles. E. J. Bliss, a Boston manufacturer of shoes, developed a shoe fitting system which was adopted as being unexcelled. The fitter was an implement about like a roller skate, with movable wings on the sides and a movable plunger in front of the toes. The soldier to be fitted equipped himself with rifle and loaded pack. With this weight on his shoulders he stepped both feet upon the skate-like devices and then raised on the balls of his feet, until the weight and movement pressed out the wings as far as they would go and advanced the front plungers. With the size thus automatically determined, the next step was to check the accuracy of it. This was done by inserting a pair of implements with knob-like ends in the toes of the shoes, the implements just filling the space in front of the soldier's toes. Wearing shoes and implements, the soldier then walked about the room, stepped upon a platform, climbed a cleated ramp, and otherwise simulated actual service demanded of shoes in the field. If the checking implements in the shoes did not hurt his toes the fit was regarded as correct.

Clothing and equipage produced and shipped to the American Expeditionary Forces Apr. 6, 1917, to Nov. 11, 1918.

	Produced.	Shipped overseas.
Blankets	19,419,000	3,127,000
Coats, denim	10,238,000	3,423,000
Coats, wool	12,365,000	3,871,000
Drawers, summer	38,118,000	3,889,000
Drawers, winter	33,766,000	10,812,000
Overcoats	7,748,000	1,780,000
Shirts, flannel	22,198,000	6,401,000
Shoes, marching and field	26,423,000	9,136,000
Stockings, wool, light and heavy	89,871,000	29,733,000
Trousers and breeches, wool	17,342,000	6,191,000
Undershirts, summer	40,895,000	4,567,000
Undershirts, winter	28,869,000	11,126,000

CHAPTER III.

MISCELLANEOUS QUARTERMASTER UNDERTAKINGS.

Sergt. Irving Berlin, one of the fountain sources of American jazz music, found a special job cut out for him when he was drafted into the military service. The needs of the war machine called upon a wide range of individual talents, and this range did not exclude the artists. The painters engaged in camouflage work and made sketches and pictures of such things as unusual surgical operations for the permanent records of the Government, the poets fired the zeal of the country, and the musicians inspired the soldiers by providing them with music.

The American Expeditionary Forces as they grew in size found themselves possessed of some 390 regimental bands. These bands organized themselves, gathered such music as they could get, practiced, and presently regaled the soldiers of units to which they were attached; and then the inevitable happened—they played and played the same old pieces until their audiences yearned for something new. One day a cry of distress trickled through the cables, and then the plight of the hapless lover of band music in France became the problem of the quartermaster organization in the United States, resulting in the largest purchase of band music ever made, 200,000 sheets of it, costing nearly $50,000.

The music problem of the American Expeditionary Forces was put into the hands of a special committee of three well-known authorities in the musical world. Sergt. Berlin was the authority on popular numbers; Lieut. R. C. Deming, the bandmaster at Camp Meigs, Washington, D. C., was the member in charge of the ceremonial numbers; while Mr. Ward Stephens, the well-known composer, organist, and accompanist, was in charge of the concerted numbers.

This committee picked out a repertoire of 333 selections, consisting of 172 concert pieces, 43 ceremonial numbers, and 118 popular numbers. Four hundred complete sets of these were bought, one for each of the 390 bands of the American Expeditionary Forces, with 10 sets as a reserve. The music was bought from some 27 music publishers, the largest suppliers being Carl Fischer, the Waterson, Berlin & Snyder Co., the Leo Feist Co., the Jerome H. Remick Co., and G. Schirmer (Inc.), all of New York, and the Oliver Ditson Co., of Boston.

Each complete set was packed in a separate case so that each case upon arrival in France could be sent immediately to a band of the American Expeditionary Forces without being disturbed. The sorting and packing of this consignment of sheet music was handled by Sergt. Berlin and a staff of technical musical assistants, who, at his request, contributed their services.

The supply of music was but one of hundreds of enterprises required to make the Army efficient, comfortable, and happy, quite aside from the more obvious ones of supplying guns and ammunition, artillery, aerial observation, and food and clothing. And these scattered undertakings in military supplies accounted for the expenditure of hundreds of millions of dollars. Nearly all of them were quartermaster enterprises. But before we lift the curtain on this, one of the most interesting branches of our military preparation, involving, as it did, the scientific solution of problems ranging from the production of supergasoline for the fighting airplanes to the proper and most economical method of cutting up the carcass of a steer, let us continue the musical overture by observing how the Army secured its band instruments.

There was a special branch of the Quartermaster Corps which concerned itself exclusively with the musical requirements of the Army. This branch bought in all approximately 143,000 musical instruments. These were secured at a saving of about $500,000 under the prices which the Government had been paying for such instruments prior to the war. Without going into the details of how this economy was effected, one typical instance may be cited. For years it had been the custom of manufacturers of musical instruments to embellish the trumpets and brass horns of bandsmen with engraving, chasings, and other markings. These were decorative only and had nothing to do with the quality of tone produced. By eliminating all such markings from the specifications, a substantial saving in cost was attained.

The principal suppliers of musical instruments were the Wm. Frank Co., of Chicago; J. M. York & Son, of Grand Rapids, Mich.; and the H. M. White Co., of Cleveland, Ohio. C. S. Conn & Co., of Elkhart, Ind.; the Eugene Geisler Co., of Chicago; and the Rudolph Wurlitzer Co., of Cincinnati, also supplied several thousand musical instruments.

FUEL, OIL, AND PAINTS.

During the months of hostilities the American public was constantly informed in advertising literature that fuel would win the war, and indeed fuel would win it, and did win it, in the sense that without fuel or with any grave shortage of fuel we could not have won. In this sense there was no commodity contributing to success in the great drama more important than coal. Coal not only furnished the power that transported the khaki-clad millions to France, but it furnished the manufacturing power in the United States and supplied the coke, which is essential to the manufacture of steel, thus entering into every rifle and piece of artillery.

MINING COAL FOR THE GOVERNMENT'S NEEDS WITH ELECTRICALLY OPERATED MACHINES.

6-TON ELECTRIC LOCOMOTIVE PULLING A STRING OF COAL CARS.

PICKING TABLES AND LOADING BOOMS IN A WEST VIRGINIA COAL MINE.

TWENTY-TON ELECTRIC LOCOMOTIVE BRINGING COAL OUT OF MINE.

America began keeping the records of coal mining in the year 1807. Woodrow Wilson was inaugurated President of the United States in 1913. In the 106 years between 1807 and 1913, and including those years, American mines produced a total of 9,844,159,937 tons of coal. In the succeeding five years of President Wilson's administration American mines turned out 2,960,938,597 tons of coal, almost one-third as much as was mined in the entire 1807–1913 period, and almost one-fourth of all the coal mined in the United States since records have been kept.

The American coal miners in 1918 met the war emergency by producing 150,000,000 tons of coal more than they had dug in 1914. The shortage of coal in the winter of 1917–18 was due not to the inability of the mines to produce the required tonnage but to inadequate railroad transportation facilities and severe weather conditions.

The war-coal project was in the hands of the United States Fuel Administration, but the office of the Quartermaster General assisted in the effort. Army officers were stationed at the offices of the various district representatives of the Fuel Administration throughout the country. These officers kept in constant touch with the factories making war supplies and saw to it that coal was diverted from less essential enterprises to the munitions factories. This service operated with such excellent effect that few manufacturers working on Government contracts were compelled to suspend operations because of the lack of fuel, and those who did have to suspend were able to resume again within a few days.

During the summer of 1918 the usual seasonal slack in the demand for fuel was taken up by the action of the fuel branch in absorbing practically all of the excess coal in the United States and storing it at Army posts, camps, and stations. This action kept the mines working at maximum capacity during a period when there is normally a curtailment in output. Of course, at the time there was no realization that the fighting was to end so soon, and this policy was adopted in preparation for unchecked industrial activity during the winter of 1918–19.

The Army itself was a heavy user of fuel, requiring it not only at its various manufacturing establishments but also at the great camps for heating purposes. The following table shows the Army purchases of fuel for the calendar year 1918.

	Amount.	Cost.
Anthracite coal .... tons..	710,304	$4,362,237
Bituminous coal .... do....	2,706,737	11,711,335
Total	3,417,041	16,073,572
Coke .... do....	9,576	80,643
Wood .... cords..	786,177	5,178,161
Total fuel	....	21,332,376

The Army was an enormous consumer of oil, the total oil purchases, both in the United States and in France, in the nine-month period from April 1 to December 31, 1918, amounting to $30,522,837. There were 49 items in the oil-purchasing schedule for the troops in the United States alone, including lubricating oils, fuel oils, oils for paints and varnishes, gasoline for motor trucks and airplanes, axle grease, floor oil, tempering oil, oil for the preservation and waterproofing of shoes, harness, and other leather equipment, and numerous other varieties of oils. The gasoline purchases were heaviest of all, Army motor trucks and cars in the United States requiring 484,282 barrels of it, worth $5,448,570, in the nine months between April 1 and December 31, 1918. The American Army motor trucks and cars with the American Expeditionary Forces were supplied with 703,104 barrels, worth $10,104,437, in the same period. For Army airplanes in the United States during the same months there were purchased 306,082 barrels of special aviation gasoline, at a cost of $3,906,650, and for the planes in France 146,780 barrels, worth $2,748,839.

To give the American aviator the hottest, most instantaneously explosive, and surest-fire gasoline ever produced, the American refiners turned out a naphtha along specifications drawn by the Government that was the highest refinement of gasoline ever produced in large quantities. This was done by taking the best gasoline that had ever been produced in commercial quantities and giving it another run through the distilling retorts. Thus it was literally the cream of the cream, containing only the most combustible elements of liquid fuel and nothing else.

This refinement became known as "257° fighting naphtha," and the Army confined its use to the service planes actually at the front. It was not supplied to the aviation training camps, either in this country or in France. In order to distinguish this naphtha as the finest engine fuel available and to mark it so that it would not be wasted by accident in any use other than that of service at the front, it was colored red with aniline dyes. The Army did not even trust 257° fighting naphtha to bulk transportation on tank ships, but stored it in steel drums and freighted it across the ocean in this form in cargo boats.

America has always been the largest producer of gasoline, and the experience and development in this country has resulted in many grades of the fuel. The ordinary commercial gasoline comes in five grades, the best grade being known as "straight-run" gasoline and the other grades, in the order of their cost and purity, as "casinghead," "blended," "pressure-still," and "cracked." For motor fuel for the Army the quartermaster specifications would accept nothing but "straight-run" gasoline, unblended and without dangerous additions which have a damaging effect upon motor cylinders. This gasoline, the best that could be bought by the civilian users, is known as "428° gasoline;" and it was the fuel used universally in our motor trucks and motor cars.

Above that were the three grades of gasoline, or rather, naphtha, produced specially for the American Army airplanes. The lowest grade of these was called domestic aviation gasoline, and it was the best commercial gasoline refined until its boiling point had been brought down to 347° F. This fuel was used by our aviators in this country and was known as "347° domestic aviation naphtha." A still greater refinement was the splendid "302° export aviation naphtha," which was used by planes in France, other than those at the front. The fighting naphtha was obtained by taking the cream of export aviation naphtha. Although purchased in enormous quantities, it cost the Government more than 41 cents a gallon. The Government paid slightly less than 22 cents a gallon for its motor gasoline.

Another new development in the oil industry brought about by the Government's war needs was known as "Liberty aero oil." This was an airplane lubricating oil of pure mineral origin, a refined lubricant of excellent viscosity and a low cold test, an oil which proved itself to be capable and reliable under the ever-changing atmospheric and pressure conditions of mechanical flight at the front. Liberty aero oil was a success. Most of it which was shipped overseas was made from paraffin base oils, although in this country we used successfully many aero oils of asphaltum base.

The Ordnance Department submitted a requisition for a three-months' supply of pure neat's-foot oil, which was in quantity almost twice the total American production of neat's-foot oil in the preceding year. The Government oil experts worked out a satisfactory substitute by combining animal and mineral oils. This was not only equal to neat's-foot oil under tests, but it was considerably cheaper.

The American Expeditionary Forces submitted a rush order for 6,000,000 pounds of dark axle grease. The specifications called for containers made of tin. But it was almost impossible to secure the tin for such a shipment. Experiments were conducted with all possible haste, and the result was a container made of black iron sheets treated with a special varnish to prevent the moisture in the grease from rusting the iron. This container proved to be satisfactory.

BRUSHES.

Offhand, one would scarcely say that brushes play any part of vast importance in the life of an individual; yet to buy the brushes for the Army required a special organization, competent to spend money by the millions of dollars and get value received for it.

Indeed it was quite surprising how many brushes in variety the Army required. The tooth brush, the shaving brush, the hair brush, the clothes brush, the shoe brush, and the paint brush might occur to anybody as necessities; but the Army used all these and in addition, artists' brushes, bottle brushes, chimney brushes, whitewashing brushes, gun-cleaning brushes, floor brushes, roofing brushes, stove brushes, horse brushes, and dozens of other kinds. In all, the Government bought 9,224,210 brushes, at a cost of $3,039,000. It required 59 factories in the United States to manufacture these brushes. The most numerous class of all were the tooth brushes, more than 1,500,000 of these being ordered from one company alone.

Brushes are made from many different materials, such as bristle, horsehair, fiber of various kinds, imitation bristle, split quills, and the like, but the most important is bristle. Only a little bristle is produced in the United States in comparison to the demand for it, the bulk of the supply coming from China, India, Siberia, and Russia. The procurement of bristle was no small part of the problem of supplying brushes for the Army.

Not one in every 10 tooth brushes used in the United States was of American manufacture before the war, the rest coming from Japan, France, England, Germany, and Austria. When the European supply was cut off, Japan became the principal source of supply. The problem of tooth brushes was further complicated by an embargo on bristles coming into this country and another on the exporting of bone to Japan.

The Army bought no shaving brushes made of horsehair, even in part, since horsehair is known to be the carrier of the much dreaded anthrax germ. The Government specified a shaving brush with an abbreviated handle, making it more convenient to carry. A handleless hairbrush was also specified. Paint brushes were largely standardized, but it was impossible to standardize toilet brushes because there were not enough facilities in the country to turn out sufficient quantities, if machinery had to be remodeled to meet Government specifications.

ROLLING KITCHENS.

Those in charge of general quartermaster purchases designed and produced the liberty rolling field kitchen, an equipment which could cook for 200 men. Rolling field kitchens were not new to our Army or the trade, there being about six types of commercial kitchens manufactured at the time we entered the war. Most of these were being produced on foreign war orders. In order, however, to secure a standardized kitchen with interchangeable parts, thus insuring a constant supply of spare parts, the division designed the liberty kitchen. There were two types of it—the horse-drawn type and the motor-drawn or trailmobile type.

Each kitchen consisted of a stove and a limber. The stove unit contained a bake oven and three kettles. The limber contained four bread boxes, which were also used as water containers, one cook's chest, four fireless cookers, and four kettles. In July, 1918, contracts were awarded for 15,000 complete kitchens, including the necessary cooking and camp utensils. Deliveries of these kitchens eventually reached a rate of over 200 per day.

Two factories adopted and installed track conveyor equipment on which the assembling process was carried forward from operation to operation until the finished kitchen, painted and boxed, was delivered to the car for shipment to the port of embarkation. The kitchens were packed each in a single crate, ready to be delivered to the front after arriving in France.

Before this kitchen was designed the Army had been paying from $700 to $1,050 apiece for rolling kitchens. The average price of the liberty kitchen was $502. Subsequent orders brought the total projected purchases of mobile kitchens to 25,000, of which 10,000 were of the animal-drawn type.

Substantial shipments of these kitchens had been received overseas before hostilities ceased, and in November deliveries were expanding at a rate which would have exceeded several times the 3,000 liberty kitchens required by the American Expeditionary Forces by January 1, 1919. About 7,000 rolling kitchens of all types were shipped to France.

TOOLS AND TOOL CHESTS.

Another important result accomplished in the purchase of general supplies was the standardization of tool chests. At one time the Army was buying and using approximately 100 different kinds of quartermaster tool chests. A committee to standardize tools and tool chests was appointed, and this committee reduced the number of types of tool chests to seven standardized ones—the carpenter's chest, the blacksmith's, the farrier's, the saddler's, the electrician's, the plumber's, and the horseshoer's emergency chest.

The committee also standardized the tools. Many varieties of such things as drawknives and handsaws had been purchased previously. This committee adopted a standard type of draw knife and a standard handsaw, and also standardized many other tools. Standardization of tool chests effected a large saving in transportation space by keeping the dimensions to a minimum. The standardized carpenter's chest occupied 3½ cubic feet less space than the older type wooden chest.

Since at the time the armistice was signed the Army was in the market for approximately 135,000 tool chests of the seven standardized types, the saving in shipping space would have been no slight achievement. But there was also in sight an enormous saving of money, not to speak of the fact that standardization would greatly increase the rate of manufacturing the chests.

HARDWARE.

The general supplies division of the quartermaster organization operated much of the Army's hardware store. In this work the division not only standardized Army tools, but also standardized the proportions in which the various tools were bought. This was not only an intensely interesting development, but it was of utmost importance to the American people, since it saved large sums of money and great quantities of shipping space.

The supply officers of the American Expeditionary Forces early began making up their estimates of the materials that must be produced in the United States and shipped to France, to maintain the efficiency of an indefinitely growing Army over a protracted period of time. In the matter of hardware these estimates came originally from the company units. Each repair unit, for instance, would look over the future, and its officers would estimate kinds and quantities of tools required for such and such a period. These little estimates came together in larger groups, and so on, the consolidation of figures continuing until eventually in the case of a certain tool there would be one figure on file at headquarters. Then one day one of those long daily cablegrams from France, signed "Pershing," came to Washington, bringing the future requirements for tools and other hardware.

Theoretically it might be assumed that the proportioning of items in these requisitions would be correct and that the American Expeditionary Forces might be expected to need tools in the proportions named. Of course, Sergt. A, in a repair unit with the artillery, might estimate too many hammers and too few wrenches, but Machinist X, miles away in some base shop, might call for too many wrenches and too few hammers. These two estimates would thus balance correctly; and, following out this line of reasoning, it would seem that the entire American Expeditionary Forces' hardware requisitions, compiled as they were, would be properly proportioned.

Yet when these requisitions came to Washington and were found to call for the manufacture of such things as files and bolts by the tens of millions, the supply officers here would not accept the theory that the proportions of various sizes called for were correct, but turned the searchlight of science upon these estimates.

The method selected of checking these estimates was simplicity itself, yet unique in the history of American industry and almost majestic in the scope of its comprehensive vision. The officer in charge of the procurement of hardware, in the case of files, for instance, simply called together the entire file-manufacturing industry—

and that means that not a single manufacturer was overlooked—and asked that industry to assemble the results of its experience over a period of the last five or six years. Each manufacturer would show, for instance, how many flat files he had sold of each length and of each type of cutting surface—either bastard, second cut, or smooth—how many half-round files, how many hand files, how many round files, how many square files, how many warding files, how many knife files, how many taper files, etc., all by lengths and by cutting surfaces. Thus when all these experience figures were assembled, the officers in charge at Washington knew exactly in what proportions the whole American industrial world had used files of various types throughout a considerable period of time.

This procedure was followed with respect to many common articles in hardware. The Hardware Manufacturers' Organization for War Service was formed to give just such assistance, cooperating up to 100 per cent of the hardware industry. The consolidation of the experience figures in American hardware consumption resulted in a schedule of supplies known as the Army's hardware tariff, a schedule showing the proportions in which hardware might be expected to be consumed.

The hardware tariff disclosed some surprising errors in the estimates from the American Expeditionary Forces. The American Expeditionary Forces' requisitions, for instance, had called for a total of 127,180,387 bolts of various kinds. The experience of bolt consumption in the American industry was able to correct this to a total of 125,285,000 bolts, or a saving of nearly 2,000,000 in number of pieces. The requisitions had called for 39,945,458 large carriage bolts. The experience of American consumption showed that only 9,700,000 large carriage bolts would be required. The original specifications had called for 31,839,741 small carriage bolts. The experience in American consumption showed that 60,300,000 would be necessary. In other words, the off-hand estimates of the American Expeditionary Forces had called for 30,000,000 too many large carriage bolts and nearly 30,000,000 too few small carriage bolts.

The specifications from France called for 5,000,000 stove bolts of the five-eighths-inch dimension. Since this size was not used or was not made at all by stove-bolt manufacturers, the item was canceled, and 2,000,000 smaller-dimension bolts substituted.

All bolts were supplied in quantities and proportions determined by following the proportions of the scientific tariff. They were shipped to France in these proportions, whence reports from the American Expeditionary Forces showed that the quantities sent completely covered the needs of the troops in the field. The saving in the manufacture of bolts alone came to nearly $4,000,000, and this says nothing of the saving in railroad and ocean freight charges, or the still more important saving in ocean tonnage space, since the bolts supplied according to the scientific tariff occupied many hundreds of cubic feet less space than the bolts originally specified would have filled.

The same procedure was followed in the supplying of files. The hardware manufacturers consulted their records and on the basis of actual consumption in American industry discovered that a repair unit consisting of a machine shop, a horseshoeing shop, a blacksmith shop, and a woodworking shop, with 11 mechanics working in the unit, would consume 305 dozen files per year, the experience tables showing precisely the proportions of the various sizes of files in this consumption. Consequently, when the American Expeditionary Forces requested 439,200 dozen files, the quantities of each size, kind, and style as specified in the requisition from France were disregarded, and the so-called tariff proportions substituted. The files as supplied not only proved adequate in number in every style, but they cost $250,000 less than it would have cost to fill the original order. Moreover, by using tariff sizes the industry was able to make immediate shipments and to run at full production from the start, since it needed only to produce files in the proportions known in the regular trade.

What was done with bolts and files was done in many other lines of hardware. When the American Expeditionary Forces saw that its hardware was coming in correct quantities, its officers notified the hardware supply organization to ship all tools and hardware materials in accordance with the so-called tariffs. The executive committee of the Hardware Manufacturers' Association for War Service, which made possible this achievement in commercial science, consisted of Messrs. Murray Sargent, Alexander Stanley, Charles W. Asbury, Fayette R. Plumb, and Isaac Black.

The standardization of proportions in the hardware supply succeeded in cutting an original requisition of the American Expeditionary Forces for 8,750 tinners' machines to 860, and an original requisition for 21,600 tinners' assorted groovers to 240, and still met every need of the Army's tin shops in France.

The Army hardware office was also called upon to supply such small hardware as fasteners for gas-mask knapsacks and pistol holsters, and some metallic parts for cartridge belts and similar goods. Less than two months before the armistice was signed orders were in sight for the manufacture of some 500,000,000 pieces of these small metallic devices. Most of them were to be made of brass. The uses of the Army in October, 1918, were calling for these articles in such quantities that it required approximately 250,000 pounds of brass per working day to meet the demand.

AMBULANCE BODY BOXED FOR SHIPMENT OVERSEAS.

PHOTO SHOWING METHOD OF CRATING CHASSIS FOR OVERSEAS SHIPMENT.

INTERIOR VIEW OF COAT FACTORY OPERATED BY THE PHILADELPHIA QUARTERMASTER DEPOT.

INTERIOR VIEW OF FLAG AND CHEVRON FACTORY OPERATED BY THE PHILADELPHIA QUARTERMASTER DEPOT.

At one time there came an order to procure 135,000,000 stud fasteners within approximately 90 days. The result was that one manufacturer, who had been producing 400,000 such fasteners in a day succeeded in raising his production to 1,000,000 per day, and this was only typical of the expansion elsewhere in the industry. The demands of the Army overtaxed the brass rolling-mill capacity of the land. As a result the hardware specialists investigated the possibility of substituting iron and steel for brass, and these substitutes were under consideration when the war came to an end.

Vast quantities of large sizes of rope were requested for overseas to replace steel hoisting cables, which could not be secured in sufficient quantities. Standard specifications drawn by the Government in cooperation with rope manufacturers insured the supply to the Army of rope only of the highest grades. Approximately 14,000,000 pounds of manila rope, 2,500,000 pounds of halter rope, and 2,000,000 pounds of cotton and jute twine were purchased at a cost of approximately $9,000,000.

Army hardware men bought 1,534,679 axes, at a cost of $1,838,979. They bought 1,256,994 shovels at a cost of $1,140,412, and 425,522 wrenches costing $395,776. They purchased 380,752 fire extinguishers at a cost of $1,761,711. They purchased 2,621,521 safety razors and 45,300,000 safety razor blades, the razors costing $3,171,806 and the blades $1,318,750. These items selected at random give some idea of the extent of the Army's hardware business.

QUARTERMASTER FACTORY ENTERPRISES.

It may not be generally known that the Quartermaster organization was an extensive manufacturer of war goods in Government shops. In another chapter has been described the method by which the Army was supplied with clothing. While many of the clothing contractors were private manufacturers, the Government itself manufactured more uniforms than it secured from any single outside source.

There were two Government uniform factories—one at the plant of the Philadelphia Quartermaster Depot and the other at the Jeffersonville (Ind.) Quartermaster Depot. The Philadelphia factory also manufactured chevrons, flags, and tents. The Jeffersonville depot produced army shirts in addition to outer clothing. The Jeffersonville depot expanded in size during the war until it became the largest shirt manufacturing establishment in the world. When the armistice was signed the Philadelphia uniform factory was rapidly gaining the eminence of being the largest clothing manufacturing plant in the United States.

The total value of the articles manufactured by the Philadelphia Quartermaster Depot during the war was $26,230,000. The garment factory at Philadelphia was started in June, 1918, and in five months it turned out 751,883 garments and 45,578 flags of various kinds. It was working toward an output of 12,000 pairs of trousers and 6,000 woolen coats per day. There were 3,000 employees in the shop and 2,000 outside seamstresses. The outside seamstresses made denim jumpers and trousers, white clothing, and olive drab shirts, the production of shirts alone reaching a total of 1,359,801 garments.

The Philadelphia factory attained an output of 5,000 pairs of chevrons per day, most of them embroidered by hand or by machinery. Before the war the Philadelphia factory had a maximum capacity of 68 pyramidal tents per day. This output was raised to 300 per day.

The Jeffersonville uniform factory was established in February, 1918. Jeffersonville is just a few minutes' ride from Louisville, Ky., which is a clothing center, and therefore there was little trouble in securing experienced workers. The factory was operated day and night with two shifts, each working eight hours. The plant reached a capacity of 750 woolen coats and 1,500 pairs of woolen trousers per day. The salaries of the women employees ranged from $50 to $80 per month. The Government established at Jeffersonville one of the most modern woolen cloth shrinking plants in the United States, costing approximately $50,000 and providing a capacity for sponging 10,000 yards of cloth per day. The Army supply officers pronounced the uniforms turned out at Jeffersonville to be the best and most honestly made clothing delivered to the Army during the war, yet the cost of manufacturing uniforms in this plant was at least 25 per cent under the average price paid to private contractors. The average cost of making a woolen service coat at Jeffersonville was $1.02, and the average cost of making a pair of woolen trousers was 54 cents.

The shirt factory at Jeffersonville was that depot's largest manufacturing enterprise. The Jeffersonville depot had been making army shirts since 1872. The shirt factory greatly expanded during the Spanish-American War, until it was employing nearly 2,000 operatives, mostly home workers. Thereafter the depot continued to make shirts at the rate of about 200,000 per year until the United States declared war against Germany, and in that time it had accumulated a roll of 2,000 sewing operatives who had worked for the factory at one time or other.

When the great demand for shirts came in the spring of 1917, the most expert of these seamstresses were hired outright by the month to act as instructors in the homes of new sewing women who had volunteered for the work. Advertisements were then sent out through the newspapers of that entire section for women workers, and presently the factory had a sewing force of 20,000 operatives from practically every town and village throughout southern Indiana

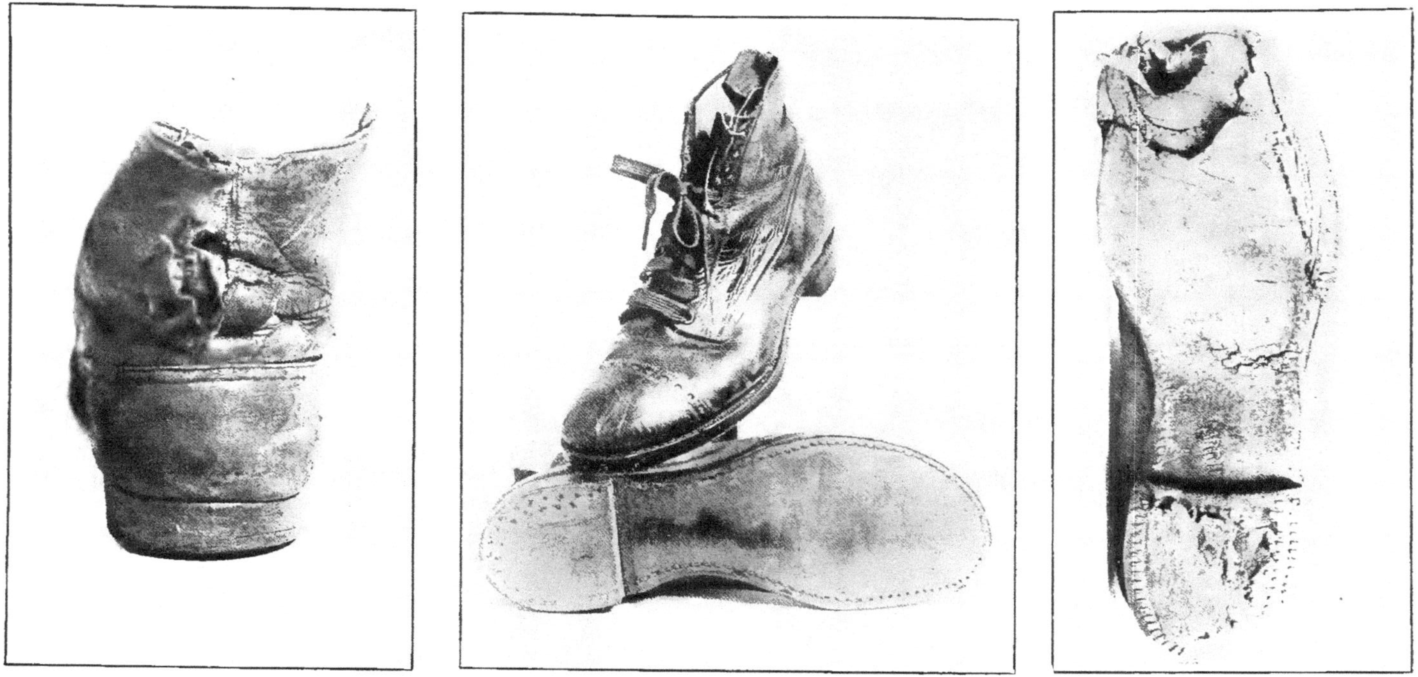

A PAIR OF ARMY SHOES[BEFORE AND AFTER BEING SALVAGED AT JEFFERSONVILLE QUARTERMASTER DEPOT.

TWO VIEWS OF THE RESCO SHOE-FITTING MACHINE.

SHOE-FITTING MACHINE WITH PLUNGER AND WINGS OPEN.

SHOE-FITTING SCHOOL, CAMP MEIGS, D. C.

and northwestern Kentucky. The output of shirts was increased from 600,000 per year to 8,500,000. Each home worker was supplied with one complete shirt to be used as a guide, and she secured from the factory as often as she needed it shirt material cut from the pattern and tied up in bundles of 10 sets. A large corps of sanitary inspectors was employed to visit the thousands of homes and see to it that the shirts were made under proper conditions. All shirts accepted from the home workers were thoroughly fumigated before being issued from the depot.

SHOE FITTING.

The Quartermaster Department, along with its other activities, was a school-teacher on a large scale. Without going into a general description of the quartermaster schools and the branches they taught, we will here consider some of the most interesting educational enterprises such as the shoe-fitting schools, the schools for butchers, and the school of goods packing.

Elsewhere in this volume the mechanical system of shoe measuring, perfected and adopted by the War Department, was described. Studies made at the camps at various times during 1917 and 1918, studies which examined nearly 59,000 men, showed that a little more than 70 per cent were wearing shoes too short, more than 9 per cent were wearing shoes too long, while less than 19 per cent were correctly fitted. It is probable that these proportions ran clear through the Army before shoe fitting was scientifically taken up, and there is no reason to believe that in civil life the averages of correct shoe fitting are any better.

After the so-called Resco system of shoe fitting was adopted, schools for shoe measuring were held at Camp Meigs, D. C., and at Jefferson Barracks, Mo. Each camp and cantonment in the country sent two officers to one or the other of these schools. The course of instruction lasted five days and consisted of lectures by experts and demonstrations of the various appliances. In this way the science of correct shoe fitting was scattered throughout the Army.

MEAT CUTTING.

It is no easy trick to teach a man to cut meat properly; butchering is a skilled trade. As soon as it was apparent that the American Expeditionary Forces in France were to be greatly expanded in size, our officers overseas sent requests that several trained and experienced butchery companies be sent over to cut meats properly for the organizations abroad. In order to comply with this request there was added to the curriculum of the quartermaster training camp in Florida a butchery course in the cutting, boning, rolling, and tying of fresh and frozen beef.

In this course there was developed an entirely new method of cutting beef known as the "natural guide" method; and by it men who had never cut meat before were developed into practical meat cutters in less than eight weeks of instruction and practice. The natural guide method, which was found to be far superior for Army use to any other meat-cutting system which had been known, was exactly what it was named, as it was essentially a separating rather than a cutting process. The beef quarters were boned and divided into their major parts by following the natural separations between muscles, tissues, and bones.

This method, which is not at all like that in commercial use, proved to be more economical than any meat-cutting system known, because it utilized every ounce of meat and produced a greater proportion of choice cuts suitable for pot roasts and other roasts than the older Army Cooks' Manual method of meat cutting. The Cooks' Manual method was similar to the method used by the retail butcher, in that it cut meat along artificial indetermined lines. The natural guide method actually produced 3 per cent more edible meat than the other method, since even the most expert meat cutters can not remove all meat from the bones by the Cooks' Manual method. Moreover, by the natural guide method all cuts are uniform, and the fats, suets, and bones are separated as clean, sweet, edible products.

Butchery companies were trained by the natural guide method and sent overseas in numbers sufficient for the requirements of the American Expeditionary Forces.

After the discovery of this method and the fact that it produced at least 3 per cent more meat than even the expert cutters could secure by the artificial cutting system, it was evident that further research work along this line would be profitable. Even expert butchers, in spite of all their skill and care, wasted meat. What must be the conditions in the mess kitchens of the Army where the cooks, with no expert knowledge of butchery, cut the meats? It was evident that numerous edible by-products of meat, such as fats and marrow, were going into the kitchen garbage pails and thence to the rendering plants.

The result of the investigation was a project to establish central meat-cutting and rendering plants for all large concentrations of troops, where all meats would be cut, boned, rolled, and tied, by experts, and delivered direct to the company kitchens ready for roasting or cooking in any manner. The fat and suet at such plants would not be soiled or made unsound by handling, and so it could be rendered and its food value retained. The oil could be cooked from the bones as a valuable by-product, the bones could be dried and sold commercially, and the plant could also have machinery for making sausage and hamburg steak. A plant of this character was put in operation during the summer and autumn of 1918 at Camp Johnston, the quartermaster training camp, and it proved to be a complete success. When the armistice was signed, the General Staff was considering the proposition of establishing these centralized meat plants at all the larger camps.

The meat experts also effected notable economies in ship space by developing what was known as shankless beef. Shankless beef was beef quarters with the four shanks removed. Quarters thus prepared occupied 14 per cent less freezer, cargo, and shipping space than quarters with their shanks.

A still further economy in shipping space was projected in the plan to bone all beef at the packing plants and ship it boxed or frozen in molds and wrapped in burlap. This method saved about 50 per cent of cargo space, and it began to be extensively used during the winter of 1918–19. One set of packages included tenderloins, sirloins, butts, loin steaks, top rounds, and shoulder steaks. Another set of packages contained roasts, including prime ribs, rumps, bottom rounds, and bottom chucks. A third set was for stews, including flanks, plates, blades, necks, shanks, and trimmings.

PACKING.

American exporters generally for many years have had the reputation of packing goods improperly for overseas shipment. Time and again travelers and investigators in foreign lands have pointed out that if America expected to compete successfully with other manufacturing nations in foreign trade, she must learn to pack goods so that the packages will not break en route and damage the contents. When we sent an Army of over 2,000,000 men to France, it was evident that unless we learned quickly how to put up our supplies properly for overseas shipment, our lack of knowledge would be costly to us.

Accordingly the packing service branch of the Quartermaster Department was established. One of its first acts was to set up a school of baling, packing, and crating, this school being located at the Forest Products Laboratory at Madison, Wis., where studies of packing were being made by scientists. The school started in July, 1918, and before the armistice came it had graduated 400 students from its six-weeks' course.

Now, while it was important that Army supplies reached the other side in good condition, it was soon seen that of even greater importance would be the economy that might be effected in shipping space by the scientific packing of goods. This obscure and little known packing service branch was really one of the most important agencies in the whole war organization, since the results which it accomplished in the saving of ship space were nothing short of astonishing. These economies came at a time when the German submarines were still highly destructive to American and allied shipping, and the shortage of ocean tonnage was one of the most disturbing factors in the whole war situation. The American packing service, in saving thousands of tons of shipping space, in reality offset the operations of the U-boats over a considerable period of time.

These space economies resulted usually from specifications drawn by the packing experts reducing the sizes of packing cases that were too large for the goods contained, and also by packing articles more compactly. For instance, these experts studied the rolling kitchen and determined the most compact assembly of its parts in a crate. The crate was then carefully designed to occupy a minimum amount of space. Some 18,000 rolling kitchens were packed ready for shipment to France. Had all of these been floated, a total of 22,500 cubic tons of ship space would have been saved, or the equivalent of five or six whole shiploads. As it was, room aboard ship could be found for only 6,940 rolling kitchens, which by being scientifically packed occupied 8,700 cubic tons less cargo space, or about two whole shiploads, than they would have occupied otherwise.

Wherever possible, entire units of such heavy articles as escort wagons and ambulances were packed in single crates. Wherever open spaces were inevitable in the crating, these vacancies were filled with various subsistence stores, such as dried peas or beans. Galvanized-iron cans, for instance, were packed with two sacks of flour inside each one.

The experts studied boxing to determine the best thickness of wood required by various commodities and the proper method of strapping or otherwise fastening the boxes. As a result there was a great improvement in the condition of goods arriving in France.

In no respect did the packing service effect greater space economy than in the packing of clothes for the American Expeditionary Forces. Formerly clothing had gone forward to troops packed loosely in wooden boxes. The packing service devised the system of baling all clothing, and a baling plant was set up at the Army supply base in Brooklyn. The service gave scientific attention to the proper folding of garments and eventually, after exhaustive experiments, developed a system of folding that allowed the maximum number of pieces which could go into a bale. It was found that these new methods saved two-thirds of the space that had been used formerly for the shipment of the same quantity of goods in boxes, to say nothing of the great saving both in labor and in boxing materials.

The baling plant at New York in the calendar year 1918 shipped to France approximately 1,000,000 bales of clothing and textile and other equipment that could be baled. The saving in money to the United States Government by this method of packing at this one plant in a single year amounted to approximately $55,000,000. The largest item in this economy was the matter of cargo space, which is estimated

A FIELD OF ROLLING KITCHENS AT NEVERS, FRANCE.

FIELD KITCHEN EQUIPMENT.

MULES AFTER MALLEINING.

Malleining is a serum test to determine if the animal is suffering from glanders.

at $49,080,000 saved to the Government. The complete statement of the financial saving in the shipment of these 1,000,000 bales is as follows:

Material	$1, 940, 000
Labor	583, 000
Tare weight	75, 000
Cargo space	49, 080, 000
Total	51, 678, 000
To these savings are to be added:	
Savings effected in rent and maintenance	700, 000
Freight to the shipping warehouses	490, 000
Rent of additional warehouse awaiting shipment	30, 000
Freight from warehouse to ship side	1, 250, 000
Recooperage	200, 000
Unloading at the forwarding stations	1, 000, 000
Total	3, 670, 000
Grand total	55, 348, 000

In addition to the financial saving there was a large saving in raw materials, which count for more than money to a nation engaged in a desperate war. This million bales of clothing saved 58,000,000 board feet of lumber, which would have been used in boxing had the old system of packing been followed. The lumber which might have gone into these boxes requires 30 years for its growth, but the burlap covering the bales was made of jute, which is raised in semiannual crops.

The size of the bale adopted was 30 by 15 by 14 inches and up to 19 inches. It is interesting to note that this size was determined upon because it was found that the burlap covering such bales of this size would cut into sandbags with a minimum amount of waste material. The Army abroad used great quantities of sandbags. Thus, by wrapping bales in burlap pieces of proper size, there was saved a considerable amount of cargo space previously occupied by baled burlap being shipped to France to be made into sandbags. It is also notable that baled clothing arrived in France in much better condition than clothing which had been packed in cases.

HORSES AND MULES.

The Quartermaster Corps was charged with the duty of providing horses and mules for the Army. This function is known technically as remount, and the buying of horses was in the hands of the remount division.

There were three permanent remount depots in the United States when the war began in April, 1917—one at Front Royal, Va., one at Fort Reno, Okla., and one at Keogh, Mont.,—an auxiliary remount depot at Fort Bliss, Tex., and a purchasing headquarters at Kansas City, Mo. When it became apparent that the Army would need a large number of horses, some of the most celebrated horsemen and riders in the country offered their services as buyers. Some 50 of them were commissioned as captains in the Quartermaster Reserve Corps and sent to the various purchasing headquarters for short training in the proper types of horses and animals required by the Army. These buyers purchased a large number of excellent animals.

In addition to the existing three remount depots there were established 33 additional auxiliary remount depots and two animal embarkation depots. The horses purchased were shipped to the various remount depots and there trained and conditioned for Army use.

It required a large number of officers and men to care for the remount establishment. Shortly before the armistice was signed there were approximately 400 officers and 19,000 enlisted men in the American remount service. The following statement shows the total numbers of horses and mules purchased for the American Army in the calendar years 1917–18, including those acquired by the remount service in France:

(a) *Number of horses and mules purchased from Jan. 1, 1917, to Jan. 1, 1919.*

Where purchased.	Horses.			Mules.		Grand total.
	Cavalry.	Light artillery.	Heavy artillery.	Draft.	Pack and riding.	
From French	21,450	61,944	42,973	2,181	7,160	135,708
From Spanish	1,400	423		13,329	3,295	18,447
From British	2,633	6,388	4,352	6,714	943	21,030
In United States	60,439	106,554	9,129	114,687	9,450	300,259
Private mounts	507	47				554
Young horses	4,474	1,045				5,519
Total purchased to Jan. 1, 1919	90,903	176,401	56,454	136,911	20,848	481,517

(b) *Number of horses and number of mules lost in America and abroad Apr. 6, 1917, to Jan. 1, 1919.*

	Horses.	Mules.	Total.
In America	24,144	6,040	30,184
Abroad	37,615	5,667	43,282
Total	61,759	11,707	73,466

(c) *Number of horses and mules on hand in the United States and insular possessions, Jan. 1, 1919.*

Horses:	
Cavalry	83,774
Draft	77,172
Mules:	
Draft	96,542
Pack and riding	13,950
Grand total	271,438

Thousands of American animals were shipped to the American Expeditionary Forces in France. Because of the lack of tonnage there were no animal shipments between March 26, 1918, and August

11. Between the declaration of war and March 26, 1918, a total of 30,329 animals were shipped abroad, and in the August 12–November 30 period 37,619 animals crossed the Atlantic, making a total of 67,948 American horses and mules sent to the American Expeditionary Forces.

The total expenditures of the Army both abroad and at home for horses and mules during the war period was $115,957,000, divided about half and half between the United States on the one hand and France, England, and Spain on the other.

The largest remount depot developed during the war is located at Camp Jackson, Columbia, S. C. This depot has a capacity of about 10,000 animals and its construction cost was about $300,000. Soon after the armistice was signed, when it became apparent that animals would no longer be needed, thousands of horses and mules at the different remount depots were sold at auction, these auction sales drawing large crowds of buyers.

STORAGE.

The problem of storing Army supplies became great only after hostilities had ceased. Before that time supplies were going through the warehouses and to the ships at the deep-water ports so rapidly that there was no backing up of the tide of them in the vast warehouse facilities that had been provided as a war measure. But as soon as the armistice was signed and the Army no longer grew in size but rapidly diminished as men were discharged, the manufacturing operations under way, necessarily continued for a time on a scale which had been developed in preparation for an Army nearly double the size of the one that existed on November 11, 1918, soon began filling up the warehouses.

The total storage capacity which the Army had on hand at the time of the fighting, exclusive of that at ports and that for the Department of Military Aeronautics, was as follows:

Zone.	Depot.	Warehouse.	Open.	Total.
		Square feet.	*Square feet.*	
1	Boston	1,295,460		1,295,460
2	New York	2,900,141		2,900,141
3	Philadelphia	2,333,124		2,333,124
4	Baltimore	1,468,572		1,468,572
5	Atlanta	1,499,848	14,300	1,514,148
6	Jeffersonville	2,000,000		2,000,000
7	Chicago	3,825,286	1,742,400	5,567,686
8	St. Louis	1,216,776		1,216,776
9	New Orleans	405,172	100,000	505,172
10	San Antonio	991,582		991,582
11	Omaha	130,472		130,472
12	El Paso	232,803	80,212	313,015
13	San Francisco	1,170,533		1,170,533
14	Newport News	234,879		234,879
15	Washington, D. C	815,603	342,100	1,157,703
	Grand total	20,520,251	2,279,012	22,799,263

	Warehouse.	Open.	Total.
	Square feet.	*Square feet.*	
Camps	5,326,590	5,104,901	10,431,491

The operation of one of the Quartermaster depot warehouses might be described at this point, and the general supply depot at Jeffersonville, Ind., is typical. During the war this depot procured for the Quartermaster Corps of the entire Army all horse-drawn vehicles and harness, and such items as barrack ranges, field ranges, and ovens, pack-train equipment, and other supplies.

The war deliveries began at Jeffersonville in the late summer of 1917. Receipts soon outgrew storage space. Adjoining lands were leased, and supplies, covered by paulins, were stored in the open. This early period of the war, prior to the spring of 1918, was a back-up period at all the warehouses, as supplies were produced faster than men were trained and transported to France. In the late spring of 1918 Jeffersonville began making heavy shipments of supplies overseas and from then on shipments exceeded receipts. For three months before the armistice was signed the Jeffersonville depot's shipments averaged 60 carloads a day and its receipts about 25 carloads.

After the armistice was signed, Jeffersonville was designated as the depot for the storage of all surplus horse-drawn vehicles and black harnesses therefor. Extensive temporary storage sheds were erected. Inbound shipments increased to about 80 cars a day. The depot has stored 4,000 rolling kitchens of the trailmobile type, these kitchens being packed in boxes, each package weighing about 4,300 pounds. The work of storing these kitchens is still in progress, and the pile of boxes will ultimately be 45 feet wide, 30 feet high, and 1,000 feet long. As the pile is made, corrugated-iron roofing is placed on the sides and top, thus forming a waterproof building.

Crated automobile trailers, weighing about 9,000 pounds per crate, are being handled in the same manner. Wagons are stored in galvanized-iron warehouses, each one capable of receiving 2,500 wagons, without wheels. Wagon wheels are stored in specially adapted sheds. About 2,000 automobile trucks have already been received for storage in specially constructed sheds. These trucks are mainly Nash Quads, four-wheel-drive trucks, and G. M. C. ambulance chassis. These chassis are stored on end, resting on the bumpers. The engines of all trucks are well oiled and the magnetos are covered with waterproof material.

As the supplies backed up into the warehouses, it became necessary for the Army to know where it stood in the matter of property; and a complete inventory was ordered, since there had been no time during the hurry and bustle of the war period to take stock. This inventory in itself was an enormous undertaking. To prepare for it the quartermaster training school at Camp Meigs, D. C., was completely transformed into a school for training experts for taking inventories. A standard scheme was worked out. The experts, after being trained in the standard method, were sent out into

STORAGE OF AUTOMOBILE CHASSIS, JEFFERSONVILLE, IND.

STORING ROLLING KITCHENS.

METHOD OF STORING WAGON WHEELS AT JEFFERSONVILLE QUARTERMASTER STORAGE DEPOT.

METHOD OF STORING HORSE COLLARS AT JEFFERSONVILLE DEPOT.

every zone in the country as instructors. In each zone they convened the so-called "town meetings." The town meeting was made up of Army storekeepers from each depot, post, camp, and station in the zone—any place where Army supplies were stored. These representatives were schooled in the inventory method and then sent back to their stations with instructions to start the inventory on December 31, 1918. The next operation was to organize an inventory factory in Washington as the consolidating point for all the inventories in the United States.

Some idea of the number of articles which Uncle Sam accumulated as a result of the war may be gained from the fact that the inventories received in Washington filled 40,600 sheets of paper, the size of an ordinary large letterhead, with typewriting single spaced. To take the inventory required a force in Washington of approximately 100 officers and 400 civilians, while there were probably over 10,000 officers and men engaged in the entire operation over the country. The inventory was undoubtedly the largest ever taken in the world.

Before the war the standard items of Army supplies had been 20,000. The inventory in the consolidation of its figures in Washington disclosed the fact that at the beginning of the year 1919 there were 120,000 standard items. and many of these stood for large quantities of individual pieces. As this report is written, a catalogue, or standard nomenclature list of supplies, comprising 120,000 items, is being prepared, to establish throughout the United States one language of supply for all items stored, distributed, and issued under the direction of the Director of Storage.

CHAPTER IV.

MOTOR AND HORSE-DRAWN VEHICLES.

The punitive expedition into Mexico in pursuit of Francisco Villa marked the real beginning of the use of motor transportation for the Army, although for many years the motor truck had received some attention for military purposes.

In 1904 a few progressive officers at West Point made preliminary tests of 1½-ton trucks, but these tests, while demonstrating that the truck would doubtless be of value to the Army in the future, were not sufficiently successful to create any particular interest. A few trucks were in use in the Army in 1907, but no systematic tests were made until 1912. At that time officers were seriously studying the motor transportation needs and problems of the Army.

In 1914 the Society of Automobile Engineers, having learned from the experience of European nations then at war that motor transportation is one of the most vital factors in the success of any army, offered its services to our War Department for the purpose of making a complete survey of the automotive industry, in the hope that the interests of the industry and of the Army could be coordinated so that in an extreme emergency the industry might be able to provide the necessary motor equipment for the Army, and that the Army might be able to use such equipment in the most efficient manner.

Pursuant to this offer, on April 28, 1916, the War Department asked the society's cooperation in issuing revised specifications for the purchase of 1½-ton and 3-ton Army trucks. In May of the same year, a committee consisting of the engineers from five companies manufacturing trucks, from five companies assembling trucks, and an engineer from a truck company not making the types of trucks under consideration, was appointed to cooperate with Army officers in making plans to provide our troops with motor vehicles suitable to their needs. On this committee were representatives of the Locomobile Co. of America, the Packard Motor Car Co., the Peerless Motor Car Co., the Pierce-Arrow Motor Car Co., the Kelly-Springfield Motor Truck Co., the Selden Motor Vehicle Co., the Commercial Truck Co. of America, the White Co., and the General Motors Co. This committee went over the Government specifications for the 1½-ton and 3-ton trucks, which had been proposed by the Army, and after a few changes had been made, the specifications were drawn up for what then seemed to be the ideal trucks for Army use in these two sizes.

Trucks at this time were urgently needed for our forces along the Mexican border and for the punitive expedition entering Mexico. Consequently rush orders for the first large quantities of trucks ever purchased or used by our Army were placed with the White Co., the Packard Motor Car Co., The Garford Motor Truck Co., the Kelly-Springfield Motor Truck Co., the Four Wheel Drive Co., and the Jeffery (Nash) Quad. It was with the trucks of these concerns that our Army officers obtained their first real experience on a big scale with motor transportation. The trucks themselves also received the most severe tests while in service on the border and in Mexico.

Not only did the Army officers secure valuable experience in operating this motor equipment but the manufacturer also took this exceptional opportunity to study motor vehicles in actual operation under Army conditions, and early in 1917 revised specifications for Army trucks were issued as a result of the numerous conferences that had been held between representatives of the War Department and the automobile industry.

In May standard specifications for the so-called class A (1½-ton to 2-ton) and the class B (3-ton to 5-ton) motor trucks were established, showing that the fundamental requirements of motor trucks for the Army were as follows: low-gear reduction, larger engines, 4-speed transmission (with very low first speed), maximum ground clearance, demountable tires of standardized size and specifications, large gasoline tank, electric lighting system, 3-point engine suspension, locking differential, extra quality alloy steel springs, and larger radiators.

After deciding on the requisites of an Army truck, the matter of standardization began to receive definite attention, it being the belief of many of the Army officers that it would be entirely possible and practicable so to standardize Army vehicles that but one type of truck would be sufficient for each size, and it became quite evident if this ideal could be worked out, the maintenance of Army vehicles would be a simple matter. Without some standardization, the providing of the proper stock of spare parts became a problem of extreme difficulty.

In the early summer of 1917 an appropriation of $175,000 was set aside by the Quartermaster Department for the purpose of financing the cost of designing and drawing up specifications for a complete new vehicle, which would become a standardized truck for our military forces. On August 1, 1917, there were assembled in Washington 50 automotive engineers who had been in touch with the truck needs of the Army; and these men, with the help of Army officers, began the task of designing a sample standardized truck, first centering their efforts on the 3-ton size, as this was at that time most urgently needed by the Army. On October 10 of this same year

109287°—19——32

the engineers had finished designing the new type of truck and had completed the first two sample trucks of this type, afterwards known as the "Standardized B." These two sample trucks were driven to Washington on October 19, were formally presented to the War Department, tested, and pronounced wholly successful.

Orders for 10,000 of these class B trucks were placed within the next few weeks. Five additional trucks were rushed through the plants as a check on tools and were completed January 10, 1918. In April, actual production having begun on the first 10,000, the purchase of an additional 8,000 was authorized, and orders for them were placed in May. In September, 1918, additional orders were placed for 25,000, but on account of the signing of the armistice no trucks were delivered under this last order.

Production of these standardized class B trucks was directed by the following men, who were called to Washington: Christian Girl, head of the Standard Parts Co., of Cleveland; James F. Bourquin, Continental Motor Co., Louisville, Ky.; Percy W. Tracy, of the Premier Motor Co., Indianapolis; Walter S. Quinlan, of the Maynard H. Murch Co., Cleveland; Guy Morgan, of the Mitchell Motors Corporation, Racine, Wis.; J. G. Utz, of the Standard Parts Co., Cleveland; G. W. Randels, of the Foote-Burt Co., Cleveland; and A. G. Drefs, of the Miller-Franklin Co.

All materials for the building of a standardized truck were mobilized through officials at Washington. In general, it was the idea to have at least three or four sources of supply for each part that went into the standardized truck, and as a result 150 parts manufacturers were given contracts.

During the time in which the Quartermaster Department was attempting to standardize all Army cargo-carrying vehicles, and up to May 15, 1918, the other branches of the Army were buying commercial trucks of different makes for their special uses. The Ordnance Department had concentrated on the Nash and F. W. D. trucks for ammunition and other ordnance work, and had ordered approximately 30,000 of these two types. The Signal Corps had specialized in the light and heavy aviation trucks, these being assembled from known and tried units, such as motors, axles, transmission, etc., and equipped with special apparatus for the Signal Corps. Approximately 4,000 of the light aviation and 4,600 of the heavy aviation trucks were ordered. The Engineer Corps had adopted the Mack 5½-ton truck and had ordered approximately 3,600. The Medical Corps had gone in for the G. M. C. model 16 for ambulances, of which approximately 5,800 had been ordered, and they had also purchased approximately 2,600 Ford ambulances.

These five branches of the Army had purchased trucks of other makes as well, and during the winter of 1917 and 1918 it became evident that the buying of so many makes and from so many different manufacturers was not a logical solution of the motor transportation problem. Each corps had its own ideas as to the type of truck required, and the sum of these ideas resulted in a decided lack of standardization for the Army as a whole, and no complete standardization for any corps as a unit.

During the first year of operations in France the American Expeditionary Forces had purchased various types of vehicles abroad in order to fill their immediate requirements, and the result was that over 200 different makes of motor vehicles were actually in use by the American Expeditionary Forces. This diversity in types was to some extent caused by lack of shipping space in which to transport motor equipment abroad. Not being able to secure sufficient trucks from the United States, due to shortage of ocean tonnage, the American Expeditionary Forces were compelled to purchase a miscellaneous assortment of foreign-made vehicles, thus complicating the maintenance problem beyond the possibility of a satisfactory solution.

The buying of motor equipment by so many different agencies of the Government was not only confusing to the manufacturer, who was selling to five different corps, but it also precluded any possibility of real standardization; and with a view of eliminating these two evils, Special Order 91, W. D. 1918, and General Order 38, W. D. 1918, were issued. The first created a standardization board and the second consolidated the procurement of all motor vehicles in the Motor Transport Service, which service operated under the direction of the Quartermaster General.

Under the special orders the standardization board was charged with selecting and approving the proper types for the use of the Army, the board being composed of representatives from each of the various corps. In this manner the various ideas of the different corps were coordinated through the discussion of the board, and the final result was that the following chassis were standardized for use:

Passenger cars: light, Ford and Dodge; heavy, Cadillac.

Ambulances: G. M. C. and Ford (with longer wheel base).

Trucks: one-half-ton to $\frac{3}{4}$-ton, Ford and Dodge (same chassis as in passenger cars); $\frac{3}{4}$-ton to 1-ton, G. M. C. Model 16; $1\frac{1}{2}$-ton to 2-ton, White; 3-ton to 5-ton, Quartermaster standardized "B."

The 4-wheel-drive TT type, called the "Militor," was also standardized, this being a special truck tractor designed by the Ordnance Department. These latter vehicles were never furnished to the Army, as production had not progressed far enough at the time of the signing of the armistice.

On this limited number of chassis could be mounted any bodies required by the Army. For instance the White $\frac{3}{4}$-ton to 1-ton chassis could be used as a passenger car when equipped with pneumatic tires and with the ordnance staff observation or reconnaissance body. The Ford and Dodge chassis could be used either with the regular passenger-car bodies or with light-delivery or light-repair bodies. The G. M. C chassis could be used either as an ambulance or could be equipped with a light-cargo body, thus providing a ¾-ton to 1-ton truck. The standardized "B" chassis could be used with special machine-shop bodies, special Signal Corps apparatus, or the regular "B" cargo body, etc. The work of the board was painstaking and thorough, and decisions were rendered only after experts had been consulted and exhaustive tests made.

While the board was standardizing on the types of vehicles to be purchased in the future for the Army, the Motor Transport Service was being formed, and by June 1, 1918, the consolidation of procurement, inspection, production, maintenance, etc., was well under way.

The needs of the American Expeditionary Forces for motor equipment were increasing by leaps and bounds, and the Motor Transport Service found that it was impossible to purchase the trucks standardized by the motorization board in sufficient quantities to meet the overseas requirements. It was therefore decided, after the consent of the board had been received, that certain other types of vehicles should be procured to fill the requirements of the Army until such time as production of the standardized truck could be increased. Therefore, the makes of trucks which were already in use in large quantities with the American Expeditionary Forces were temporarily made standard to meet the immediate needs of the Army. It being extremely difficult to purchase sufficient trucks, even of these additional makes, to meet the needs overseas, it was decided that still other makes of trucks would be procured for use in the United States, thus allowing all the makes standardized for overseas use to be shipped to France.

The Motor Transport Service operated from May 15 until August 15, 1918, when the Motor Transport Corps was organized under General Order No. 75. This order created a separate corps under the operations division of the General Staff for the operation as well as the procurement of all passenger and cargo carrying motor equipment for the Army. A few weeks later, however, Supply Circular 87, P., S. & T., was issued, placing the procurement of the above under the Director of Purchase, Storage and Traffic (Motors and Vehicles Division), but the operation and maintenance of vehicles was left with the Motor Transport Corps. From September on these organizations remained unchanged up to and after the signing of the armistice.

The table appended shows the status of the procurement and production of motor-vehicle orders as of November 1, 1918, 11 days prior to the signing of the armistice. It will be noted that there had been ordered for delivery before July 1, 1919, a total of 185,000 trucks, 23,053 ambulances, and 38,462 passenger cars.

The Army in April, 1917, possessed 3,039 trucks, 437 automobiles, 670 motor cycles, and 12 tractors. One and one-half years later it owned approximately 85,000 trucks, and had the war continued until July 1, 1919, there would have been approximately 185,000 trucks provided for its use by American industry. In addition, this same industry would have provided 30,000 ambulances, 40,000 passenger cars, 70,000 motor cycles, 70,000 bicycles, making a grand total approaching 400,000 vehicles, costing (with spare parts) over $700,000,000.

From the very beginning the Government received the hearty cooperation of the entire industry. The need was urgent, the demand tremendous, and many manufacturers were called upon to sacrifice their own product in order to meet the needs of the Army, and many were on a 100-per-cent war-work basis.

Motor truck production.

VEHICLES STANDARDIZED FOR OVERSEAS SHIPMENT.

Name.	Class.	Capacity.	Manufacturer.	Address.	Total ordered to Nov. 1, 1918.	Total completed to Nov. 1, 1918.	Floated overseas.
G. M. C. standard	AA	1-ton	Commerce Motor Car Co	Detroit, Mich	450		
Do	AA	do	Velie Motors Corp	Moline, Ill	1,000		
Do	AA	do	Dort Motor Co	Flint, Mich	1,000		
Do	AA	do	Elgin Motor Car Corp	Elgin, Ill	500		
Do	AA	do	Lexington Motor Co	Connersville, Ind	500		
Do	AA	do	Auburn Auto Co	Auburn, Ind	500		
Do	AA	do	Hupp Motor Car Corp	Detroit, Mich	500		
Do	AA	do	General Motor Truck Co	Pontiac, Mich	1,000		
Do	AA	do	Saxon Motor Car Co	Detroit, Mich	500		
Do	AA	do	Campbell Motor Car Co	Kingston, N. Y	200		
Do	AA	do	Columbia Motor Co	Detroit, Mich	300		
Do	AA	do	Moon Motor Car Co	St. Louis, Mo	300		
Do	AA	do	Liberty Motor Co	Detroit, Mich	450		
G. M. C. ambulance	AA	do	General Motors Truck Co	Pontiac, Mich	5,811	5,553	4,001
White	AA	do	The White Co	Cleveland, Ohio	2,695	2,196	809
Total					15,706	7,749	4,810
Dodge light delivery	AA	½-ton	Dodge Bros. Co	Detroit, Mich	9,352	2,644	1,802
Dodge light repair	AA	¾-ton	do	do	1,012	1,012	
Ford light delivery	AA	do	Ford Motor Co	do	12,002	5,492	7,206
Ford ambulances	AA	do	do	do	10,042	5,340	4,362
Total					32,408	14,488	13,370
White standard	A	1½-ton	The White Co	Cleveland, Ohio	9,201	1,813	1,532
Do	A	do	Peerless Motor Co	do	3,000		
Total					12,201	1,813	1,532
Light aviation	A	1½ to 2 ton	General Motor Truck Co	Pontiac, Mich	2,400	1,888	1,829
Do	A	do	Paige Motor Car Co	Detroit, Mich	500	480	
Do	A	do	Republic Motor Truck Co	Alma, Mich	500	354	
Do	A	do	Denby Motor Truck Co	Detroit, Mich	500	488	
Garford	A	1½-ton	Garford Motor Truck Co	Lima, Ohio	5,010	1,010	499
Packard	A	do	Packard Motor Car Co	Detroit, Mich	831	636	526
Pierce-Arrow	A	do	Pierce-Arrow Motor Car Co	Buffalo, N. Y	4,023	2,423	534
Total					13,764	7,279	3,388

Standardized B	B	3-ton	Packard Motor Car Co	Detroit, Mich	5	5	
Do	B	do	Gramm-Bernstein Motor Truck Co	Lima, Ohio	3,750	1,000	
Do	B	do	Kelly-Springfield Motor Truck Co	Springfield, Ohio	1,045	391	
Do	B	do	Indiana Truck Corporation	Marion, Ind	2,545	475	
Do	B	do	Service Motor Truck Co	Wabash, Ind	1,795	337	
Do	B	do	Republic Motor Truck Co	Alma, Mich	3,750	967	
Do	B	do	Pierce-Arrow Motor Car Co	Buffalo, N. Y	1,000	975	
Do	B	do	Selden Truck Sales Co	Rochester, N. Y	3,750	1,000	
Do	B	do	Bethlehem Motors Corporation	Allentown, Pa	2,725	675	
Do	B	do	Diamond T Motor Co	Chicago, Ill	3,203	638	
Do	B	do	U. S. Motor Truck Co	Cincinnati, Ohio	2,545	490	
Do	B	do	Brockway Motor Truck Co	Cortland, N. Y	2,202	587	
Do	B	do	Velio Motors Corporation	Moline, Ill	3,045	455	
Do	B	do	Sterling Motor Truck Corporation	Milwaukee, Wis	1,795	479	
Do	B	do	Garford Motor Truck Co	Lima, Ohio	1,750	978	
Do	B	do	Hurlburt Motor Truck Co	New York, N. Y	100		
Do	B	do	Midland Motor Truck Co	Oklahoma City, Okla	500		
Do	B	do	Atterbury Motor Car Co	Buffalo, N. Y	750		
Do	B	do	Standard Motor Truck Co	Detroit, Mich	750		
Do	B	do	Maccar Co	Scranton, Pa	500		
Do	B	do	Clyde Cars Co	Clyde, Ohio	500		
Do	B	do	Rowe Motor Manufacturing Co	Lancaster, Pa	500		
Do	B	do	J. C. Wilson Co	Detroit, Mich	500		
Do	B	do	Lewis Hall Iron Works	do	500		
Do	B	do	Denby Motor Truck Co	do	1,500		
Do	B	do	Winther Motor Truck Co	Winthrop Harbor, Ill	500		
Do	B	do	Vim Motor Truck Co	Detroit, Mich	500		
Do	B	do	Signal Motor Truck Co	do	500		
Do	B	do	United Motors Co	Grand Rapids, Mich	500		
Total					43,005	9,452	7,655
Packard	B	3-ton	Packard Motor Car Co	Detroit, Mich	10,106	4,856	3,479
Heavy aviation	B	do	United Motors Co	Grand Rapids, Mich	200	188	
Do	B	do	Federal Motor Truck Co	Detroit, Mich	1,000	1,000	
Do	B	do	Kelly-Springfield Motor Truck Co	Springfield, Ohio	2,225	1,725	2,110
Do	B	do	Standard Motor Truck Co	Detroit, Mich	250	186	
Do	B	do	Velie Motors Corporation	Moline, Ill	1,000	700	
Riker	B	4-ton	Locomobile Co. of America	Bridgeport, Conn	3,690	1,690	1,351
Pierce-Arrow	B	5-ton	Pierce-Arrow Motor Car Co	Buffalo, N. Y	3,170	1,970	1,660
Mack	B	5½-ton	International Motors Co	New York, N. Y	5,575	2,563	1,365
Total					27,216	14,878	9,965

Motor truck production—Continued.

VEHICLES STANDARDIZED FOR OVERSEAS SHIPMENT—Continued.

Name.	Class.	Capacity.	Manufacturer.	Address.	Total ordered to Nov. 1, 1918.	Total completed to Nov. 1, 1918.	Floated overseas.
Fuor-wheel drive	TT	2-ton	Nash Motors Co	Kenosha, Wis	14,684	8,139	
Do	TT	do	Hudson Motor Car Co	Detroit, Mich	3,000	219	
Do	TT	do	National Motor Car & Vehicle	Indianapolis, Ind	3,000	95	7,034
Do	TT	do	Paige Motor Car Co	Detroit, Mich	3,000	145	
Do	Winch	do	Nash Motors Co	Kenosha, Wis	100	100	
Do	Winch	3-ton	Four-Wheel Drive Auto Co	Clintonville, Wis	800	364	
Do	TT	do	do	do	7,150	4,233	
Do	TT	do	Mitchell Motors Corporation	Racine, Wis	5,023	855	4,748
Do	TT	do	Premier Motor Co	Indianapolis, Ind	4,500	1,307	
Do	TT	do	Kissel Motor Car Co	Hartford, Wis	3,500	997	
Total					44,757	16,454	11,782
Cadillac, limousine	Pass		Cadillac Motor Car Co	Detroit, Mich	1,043	222	1,503
Cadillac, open	Pass		do	do	2,970	1,734	
Dodge, open	Pass		Dodge Bros. Co	do	8,559	5,065	3,390
Dodge, winter top	Pass		do	do	3,201	2,126	
Ford, touring	Pass		Ford Motor Co	do	20,201	5,379	2,344
Total					35,974	14,526	7,237

MOTOR VEHICLES FOR DOMESTIC DISTRIBUTION.

Name.	Class.	Capacity.	Manufacturer.	Address.	Ordered to Nov. 1, 1918.	Completed to Nov. 1, 1918.	Completed to Dec. 1, 1918.	Total floated overseas.
Commerce	AA	1-ton	Commerce Motor Car Co.	Detroit, Mich.	1,548	1,548	1,548	272
Ford trucks		¾ to 1 ton	Ford Motor Co.	do	2,494	2,474	2,474	1,772
Total					4,042	4,022	4,022	2,044
White Unstandardized	A	1½-ton	The White Co.	Cleveland, Ohio	394	394	394	327
Kelly-Springfield	A	do	Kelly-Springfield Motor Co.	Springfield, Ohio	356	16	16	
Denby	A	do	Denby Motor Truck Co.	Detroit, Mich	500		182	
International	A	do	International Harvester Co.	Akron, Ohio	1,125	125	485	
Miscellaneous American	A	1¼ to 2 ton			949	945	949	220
Wilson	A	1½ to 2 ton	J. C. Wilson Co.	Detroit, Mich.	200		75	
Moreland	A	do	Moreland Motor Truck Co.	Los Angeles, Cal	85	43	76	
Miscellaneous American ambulance	A	do			78	78	78	
Total					3,687	1,601	2,255	547
White	B	3-ton	The White Co.	Cleveland, Ohio	306	306	306	280
Peerless	B	do	Peerless Motor Car	do	385	385	697	385
Mack	B	3½-ton	International Motors Co.	New York, N. Y	368	368	368	278
Velie	B	do	Velie Motors Corporation	Moline, Ill	125		28	
Gramm-Bernstein	B	do	Gramm-Bernstein Motor Truck Co.	Lima, Ohio	100		66	
Federal	B	do	Federal Motor Truck Co.	Detroit, Mich	500		85	
Standard	B	do	Standard Motor Truck Co.	do	287	35	89	
Selden	B	do	Selden Motor Truck Co.	Rochester, N. Y	171	71	121	5
Republic	B	do	Republic Motor Truck Co.	Alma, Mich	250			
Moreland	B	4-ton	Moreland Motor Truck Co.	Los Angeles, Calif	60		40	
White	B	5-ton	The White Co.	Cleveland, Ohio	48	48	48	34
Packard	B	do	Packard Motor Car Co.	Detroit, Mich	65	60	60	17
Hurlburt	B	do	Hurlburt Motor Truck Co.	New York, N. Y	200			
Federal	B	do	Federal Motor Truck Co.	Detroit, Mich	300		53	
Miscellaneous American	B	3 to 5 ton			278	278	250	209
Total					4,043	1,551	2,211	1,208
Dodge Sedan	Passenger		Dodge Bros	Detroit, Mich	10			
Dodge Roadster	do		do	do	550	22	175	
Ford Roadster	do		Ford Motor Co.	do	435	185	186	
Ford Closed	do		do	do	16	16	16	
Miscellaneous American	do				958	958	958	
Total					1,969	1,181	1,335	

MOTORCYCLES, SIDE CARS, AND BICYCLES.

The need of the Army for motorcycles, side cars, and bicycles was so tremendous that for many months during the war practically the entire output of these vehicles of the kinds selected as being most suitable for Army use was taken by the Government.

It was found that the Indian and Harley-Davidson motorcycles were best adapted to meet the necessities of the Expeditionary Forces in France, and these types were standardized for overseas shipment. Orders for a total of 39,070 Indian motorcycles were placed with the manufacturers at Springfield, Mass., and before the end of 1918, 18,081 of these had been delivered. From the Harley-Davidson manufacturers at Milwaukee, Wis., the Government received 14,666 machines of the total of 26,487 ordered before the end of 1918. In addition to the Harley-Davidson and Indian machines, 1,526 Cleveland motorcycles, made in Cleveland, Ohio, were contracted for, and 1,476 delivered previous to 1919.

Side-car equipment for the Indian and Harley-Davidson machines was bought in almost as great quantities as the motorcycles themselves. In fact, the demand for motorcycles and side cars from these two concerns was so great that they were working at 100-per-cent capacity for the Government before the summer of 1918.

The needs of the Army for machines increased so steadily and the requirements were so vast that both the Indian and Harley-Davidson concerns had made large additions to their plants for meeting the Government needs at the time the armistice was signed.

A standard military type of bicycle was turned out for the Army by the Westfield Manufacturing Co., at Westfield, Mass., and other bicycles were ordered from the Great Western Manufacturing Co., at Laporte, Ind., and the Davis Sewing Machine Co., at Dayton, Ohio.

Motorcycles, bicycles, and side cars.

MOTORCYCLES.

Name.	Manufacturer.	Address.	Ordered to Nov. 1, 1918.	Completed to—		Shipped overseas.
				Nov. 1, 1918.	Dec. 31, 1918.	
Cleveland	Cleveland Motors Mfg. Co.	Cleveland, Ohio	1,526	780	1,476	1,003
Harley-Davidson	Harley-Davidson Motorcycle Co.	Milwaukee, Wis.	26,485	12,376	14,666	20,007 (Harley-Davidson and Indian)
Indian	Hendee Mfg. Co.	Springfield, Mass.	39,870	14,054	18,018	
Total			67,881	27,210	34,160	21,010

SIDE CARS.

Name.	Manufacturer.	Address.	Ordered to Nov. 1, 1918.	Completed to Nov. 1, 1918.	Completed to Dec. 31, 1918.	Shipped overseas.
Harley-Davidson	Harley-Davidson Motorcycle Co.	Milwaukee, Wis.	26,099	11,934	14,332	19,160 (Harley-Davidson and Indian)
Indian	Hendee Mfg. Co.	Springfield, Mass.	39,124	13,863	16,804	
Total			65,223	25,797	31,136	19,160

Motorcycles, bicycles and side cars.—Continued.

BICYCLES.

Military Standard..	Westfield Mfg. Co.....	Westfield, Mass...	36,002	19,164	22,502	26,407
Do............	Great Western Mfg.Co.	Laporte, Ind......	15,750	....	3,750	
Do............	Davis Sewing Machine Co.	Dayton, Ohio.....	15,750	1,992	3,252	
Total.......	....	....	67,502	21,156	29,504,	26,407

HORSE AND HAND DRAWN VEHICLES.

It was early realized by Army officers upon our entry in the war that procurement of horse-drawn vehicles for the Army would require mobilization of practically the entire wagon-making industry of the Nation. Consequently one of the first steps taken to provide the Army with the necessary vehicles of this type was to call into conference representatives of the four largest manufacturing companies in the industry.

R. V. Board, of the Kentucky Wagon Co.; A. B. Thielens, of the Studebaker Corporation of America; E. E. Parsonage, of the John Deere Wagon Co.; and R. W. Lea, of the Moline Plow Co., were named members of an advisory committee to assist the Army in the procurement of vehicles.

Our first requisition called for the manufacture of 34,000 escort wagons. This order, with the necessary spare parts for these vehicles meant the building of the equivalent of about 50,000 wagons.

At the beginning of the war the manufacture of vehicles from kiln-dried lumber was almost unknown, as there had always been a sufficient amount of air-dried lumber on hand to meet every demand for farm-wagon construction. Our first order, however, practically used up all of the air-dried lumber then in existence in the country. In order that dry lumber could be obtained in sufficient quantities to keep up with the demands for Army vehicles, the War Department entered into an arrangement by which dry-kilns were built by contractors with the Government defraying half of the cost, the wagon manufacturer being reimbursed at the rate of $10 for each wagon produced, or on a basis of $10 for each $185 worth of spare parts fabricated.

Despite the fact that ordinarily six months were required even with kiln-drying before a log was ready for fabrication into a vehicle, all orders for the War Department were filled on time and in accordance with the plans outlined. To make this possible every manufacturer of the industry capable of turning out the class of vehicles desired did his part and did it so well that up to the signing of the armistice approximately 110,000 horse-drawn or hand-drawn vehicles had been delivered, of the total of 185,727 for which contracts had been placed.

Escort wagons formed the large bulk of the requirements at the start, but as the war progressed, the necessity was created for different designs of vehicles. So from time to time there were designed drinking-water carts and wagons, medical and ration carts, combat wagons, veterinary ambulances, sprinkling wagons, and various other types to meet special needs.

In the early spring of 1918 it was found that the wagon industry had about reached its limit so far as output was concerned, and that, if the war continued another year, new sources of supply would have to be developed. Then it was that the furniture industry was called upon to produce spare parts for vehicles. Under the presidency of P. B. Schravesande, president of the Grand Rapids School Equipment Co., the Furniture and Fixture and Light-wood Industry War Service Committee was organized to cooperate in arranging to have furniture makers enter the field of manufacture of spare parts.

It was arranged that the furniture manufacturers were to produce 75 per cent of the spare parts then requisitioned, totaling in value about $8,000,000. While the furniture industry was preparing its plants for the manufacture of these parts, the wagon industry continued to manufacture 25 per cent of the required parts in order to keep up a satisfactory flow.

When the armistice was signed practically all the furniture manufacturers had prepared to fill the spare-parts orders, but none of them had reached quantity production.

Automobile-wheel manufacturers were induced to turn out the immense quantity of wheels needed for escort wagons.

There were in all about 250 manufacturers of wagons, wagon parts, and wheels. Among the prominent wagon companies engaged in this work were: Bain Wagon Co., Oshkosh, Wis.; Columbia Wagon Co., Columbia, Pa.; Deere & Co., Moline, Ill.; Emerson-Brantingham Co., Rockford, Ill.; Florence Wagon Co., Florence, Ala.; Hackney Wagon Co., Wilson, N. C.; International Harvester Co., Memphis, Tenn.; Moline Plow Co., Moline, Ill.; Mogul Wagon Co., Hoskinsville, Ky.; Owensboro Wagon Co., Owensboro, Ky.; Pekin Wagon Co., Pekin, Ill.; Peter Schuttler Co., Chicago, Ill.; Springfield Wagon Co., Springfield, Mo.; Stoughton Wagon Co., Stoughton, Wis.; A. Streich & Bros. Co., Oshkosh, Wis.; Thornhill Wagon Co., Lynchburg, Va.; Tiffin Wagon Co., Tiffin, Ohio; Eagle Wagon Works, Auburn, N. Y.; A. A. Cooper Wagon & Buggy Co., Dubuque, Iowa; Winona Wagon Co., Winona, Minn.; White Hickory Wagon Co., Atlanta, Ga.; Kentucky Wagon Co., Louisville, Ky.; Studebaker Corporation, South Bend, Ind.; American Car & Foundry Co., Jeffersonville, Ind.

Among the leading automobile-wheel manufacturers who were given contracts for escort-wagon wheels were the following: Mutual Wheel Co., Moline, Ill.; Roger Wheel Co., Aurora, Ind.; Crane & McMahon (Inc.), St. Marys, Ohio; Hayes Wheel Co., Jackson, Mich.; Imperial Wheel Co., Flint, Mich.; Kelsey Wheel Co., Detroit, Mich.; Binnel Spoke & Auto Wheel Co., Portland, Ind.; Archibald Wheel

Co., Lawrence, Mass.; Chattanooga Wagon Co., Chattanooga, Tenn.; Hoopes Bros. & Dalington (Inc.), Westchester, Pa.; Prudden Wheel Co., Lansing, Mich.; Standard Wheel Co., Terre Haute, Ind.; Avoca Wheel Co., Avoca, N. Y.; New Wapakoneta Wheel Co., Wapakoneta, Ohio; Piedmont Wagon Co., Hickory, N. C.

Furniture factories that assumed contracts for making spare parts for horse-drawn vehicles included about 30 furniture manufacturers of Rockford, Ill., as well as the following: Grand Rapids School Equipment Co., Grand Rapids, Mich.; Sherman Bros. Co., Jamestown, N. Y.; Ramsey-Alton Manufacturing Co., Portland, Mich.; Connersville Furniture Co., Connersville, Ind.; P. Derby & Co., Gardner, Mass.; Ebert Furniture Co., Red Lion, Pa.; S. Karpen & Bros., Chicago, Ill.; Chas. P. Limbert & Co., Holland, Mich.; The Macey Co., Grand Rapids, Mich.; Thos. Madden Son & Co., Indianapolis, Ind.; Sidney Manufacturing Co., Sidney, Ohio; Basic Furniture Co., Waynesboro, Va.; Brecht Co., St. Louis, Mo.

Production data for horse-drawn vehicles.

	Number ordered between Apr. 6, 1917, and Nov. 11, 1918.	Value of orders.	Quantity delivered up to Nov. 11, 1918.	Quantity shipped overseas.	Value of quantity shipped overseas.
Commercial vehicles	181,077	$41,247,911	89,024	28,918	$7,247,522
Spare parts	....	39,690,255	....	....	2,551,642
Total	181,077	80,938,166	89,024	28,918	9,799,164

The following table shows the principal items listed under commercial vehicles, with the total number ordered, value of the order, and the total number delivered up to November 11, 1918:

Item.	Total ordered.	Value of number ordered.	Total delivered.	Value of number delivered.
Wagons.				
Ambulance	3,339	$1,168,650	3,319	$1,161,650
Escort, model A	1,000	242,000	1,000	242,000
Escort, J-118	102,078	20,415,600	37,613	7,522,600
Combat	15,500	7,750,000	7,099	3,549,500
Drinking water	2,687	1,262,890	2,687	1,262,890
Mountain, 3-seat	1,000	300,000	1,000	300,000
Sprinkling	1,056	496,320	1,056	496,320
Bottom, dump	436	71,940	255	42,075
Milk	4	1,400	4	1,400
Buckboard	1,859	232,375	1,155	144,375
Total	128,959	31,941,175	55,188	14,722,810
Carts.				
Drinking water	22,000	6,050,000	14,729	4,050,475
Ration	15,000	1,875,000	10,185	1,273,125
Medical	5,500	819,500	2,350.	350,150
Hand	7,309	211,961	4,607	133,603
Dump	1,183	118,300	1,037	103,700
Sanitary	1,009	201,800	811	162,200
Veterinary ambulance	80	20,000	80	20,000
Disinfecting spray	37	10,175	37	10,175
Total	52,118	9,306,736	33,836	6,103,428
Grand total	....	41,247,911	....	20,826,238

This table shows the total number of horse-drawn vehicles sent overseas during the war; also the value of these shipments and the unit prices.

Horse-drawn vehicles.

Article.	Unit.	Number.	Items shipped.	
			Unit value.	Total value.
Ambulances	Each	507	$350	$177,450
Escort wagons	...do	15,979	230	3,675,170
Combat wagons	...do	2,672	500	1,336,000
Spring wagons	...do	147	235	34,545
Water carts	...do	5,314	275	1,461,350
Ration carts	...do	3,231	125	403,875
Medical carts	...do	1,068	149	159,132
Total				7,247,522

CHAPTER V.

MEDICAL AND DENTAL SUPPLIES.

Lest it be thought that the American Army was dependent in any way for its hospital facilities and surgical supplies upon private contributions, it may be said that the Government during the period between April 6, 1917, and November 11, 1918, placed contracts for medical supplies amounting to $424,761,031. Contract cancellations after the armistice was signed amounted to $56,000,000. The remaining $370,000,000 approximately represents the cost to the United States of medicine, surgical instruments and dressings, ambulances, hospital furniture, equipment and supplies, and dental and veterinary supplies for the war.

This was considerably more money than was contributed by the American people to the American Red Cross, a great part of whose funds went to the relief of civilian populations in Europe, or to any other war charity. Thus it will be seen that the Government with billions of dollars to spend could well afford the few hundreds of millions necessary to give the American soldiers who needed it the best possible hospital attention. It accepted the gifts of this sort, ranging from gauze bandages to fully equipped motor ambulances, as the offerings of the people whose hearts overflowed with love and gratitude to the American soldiers and took this means of showing their concern; but the Government in no sense was dependent upon these donations.

Before 1914 four-fifths of all surgical instruments used in the United States were imported from Germany. This country, too, was practically dependent upon Germany for many of its most important medicines, including the potassium salts and such drugs as digitalin, salvarsan, atropin, etc. While in a way we had been developing substitute sources of supply in the United States for these indispensable commodities in the months between the outbreak of the great war and the date of our participation in it, the raising of a vast army and the project to send this army to the bloody battle fields of France created an American demand for medicines and surgical instruments beyond anything ever known in the United States. Yet, through the cooperation of manufacturers and the officers of the Medical Department's general purchasing office, which was on November 15, 1918, incorporated in the office of the Director of Purchase and Storage, sufficient supplies were developed, not only of medicine but of surgical instruments.

The development of the production of medicines for the use of troops in the field was particularly notable. The important drug, salvarsan, used in the treatment of syphilis, was a patented formula and had been furnished formerly by a single German manufacturer. In this country we produced arsphenamine as a substitute, gradually increasing the supply and constantly improving the drug until at length its toxicity had been so reduced that it equaled or even excelled the German product.

The facilities of the American drug and tablet manufacturers were taxed to the utmost to supply the Army. For example, during the year 1918 a total of 46,000,000 quinine tablets was produced, while 172,000,000 aspirin tablets were manufactured during the same period, and 835,000 pounds of calomel ointment, 45,000,000 iodine swabs, 10,250,000 tins of foot powder, and 300,000,000 tubes of iodin-potassium. All other items of medicines, antiseptics, and disinfectants, required by the Medical Department, were increased in proportion. This production not only strained the facilities of the manufacturers of chemicals and drugs but also called upon the glassware manufacturers for the necessary bottles and tubes in which to pack these medicines satisfactorily. Here again was an effort that required close cooperation between the trade and the Medical Department in order to meet the demand.

When it became evident that a declaration of war against Germany was imminent, the Medical Department proceeded to analyze the country's resources of medical supplies. These resources were to a large extent limited. The Allied nations had been making heavy and constant demands for these materials, so much so that even the mobilization of such a relatively small number of troops as were centered along the Mexican Border put a severe burden upon the medical supply facilities of the country.

The Council of National Defense took up the medical supplies problem at the outset. The various manufacturers sent their representatives to consult with the Surgeon General, and committees on surgical instruments, surgical dressings, medicines, and other important supplies were formed. These committees allocated among the various manufacturers the first emergency orders for these materials. The result was that the base hospitals at the 32 mobilization camps in 1917 were equipped in an amazingly short time. The New York Medical Supply Depot, which was then the largest purchasing agent, was called upon to supply 500 hospital beds each to 22 of the camps. This work was handled so rapidly that in some cases the shipments had to be held back to wait for the completion of the hospital buildings.

Perhaps the most difficult task was to determine what quantities of medical supplies would be needed for a given period. It is a comparatively simple matter to estimate the quantity of clothing necessary for a certain number of troops, or to figure what food they will require; but it is not possible to forecast the number of men who will be sick at a given camp at a specified time, nor is it possible to foretell the nature of diseases or injuries. An epidemic of measles or mumps requires different treatment than an epidemic of influenza, and makes necessary the use of a different variety of medical supplies. Experience sheets of supplies actually used in the past formed the basis of our requirements schedules.

Eventually there was worked out a system of supply based on the initial requirements of the unit of 25,000 men in the Expeditionary Forces and the automatic supply of replenishment of this equipment. In this system use was made of the knowledge and experience obtained by the British and French medical forces during their nearly three years of warfare before America went in.

The following statement of estimated expenditures for the fiscal year 1920 illustrates the difference in the medical requirements of an army of 500,000 men under peace conditions and an army of 5,000,000 men in a war such as the recent one:

	Peace—500,000 men.	War—5,000,000 men.
Surgical dressings	$3,059,525	$121,230,924
Textiles, hospital supplies	880,124	70,682,673
Miscellaneous hospital supplies	230,477	12,626,848
Medicines, etc	1,969,901	18,431,614
Hospital furniture and equipment	500,000	16,600,184
Surgical instruments	200,000	34,727,863
Dental instruments, equipment, and supplies	150,000	6,256,482
X-ray, equipment, and supplies	200,000	5,004,900
Field supplies	300,000	3,604,695
Veterinary	701,692	6,656,894
Laboratory	852,673	7,858,004
Stationery	159,183	2,228,634
Total	9,203,577	305,909,719

Civilian experts in various lines of medical supplies were brought into the organization to supply the wide range of specialized knowledge required in such a universal buying program as the Medical Department was about to conduct. Before the war the Army's purchases of instruments for oral and brain surgery, orthopedic supplies, Dakin outfits, and other special apparatus were practically negligible. During the war period these purchases amounted to millions of dollars. It may be seen readily that the purchasing office had to possess more than a superficial understanding of the materials to be bought.

Orders customarily went to the lowest bidders, with a careful review in Washington of all prices named in contracts. The inspection of material was an important phase of the work. This inspection was handled through the New York Medical Supply Depot, which

109287°—19——33.

called in as assistants the United States Board of Customs Appraisers at New York City. That corps of men had had long years of experience in inspecting and determining the value of surgical supplies, as most of these supplies in the past had come through the customhouse from foreign countries. The inspection of drugs was handled by the Medical Department's laboratories, the Army Medical School, and by the Bureau of Standards, which rendered valuable assistance in examining and testing samples. In addition the Medical Department maintained a corps of inspectors to travel from one factory to another, keeping in close touch with the progress and assisting in procuring raw materials and expediting deliveries.

The medical supplies were divided under the following classifications:

(*a*) Hospital equipment, such as beds, bedside tables, enamelware, etc.
(*b*) Surgical dressings.
(*c*) Surgical instruments.
(*d*) Medicines, antiseptics, and disinfectants.
(*e*) Field supplies (chests and units for extended field service).
(*f*) Dental supplies.
(*g*) Veterinary supplies.
(*h*) Laboratory supplies.
(*i*) Motor ambulance supplies.
(*j*) X-ray supplies.

The New York depot was intrusted with the purchase of miscellaneous hospital equipment and dental and X-ray supplies. The St. Louis depot purchased the veterinary supplies, and the field medical supply depot at Washington purchased the laboratory and field supplies. The motor ambulance supply depot, established at Louisville, Ky., purchased ambulances and ambulance spare parts. Appreciating the necessity for a certain amount of cooperation where the purchase of conflicting articles by the various depots was concerned, the general purchasing office of the Medical Department was organized at Washington. This purchasing office bought all surgical dressings, surgical instruments, and medicines and such items as were used in the field, post, veterinary, and dental stations.

In connection with the production of surgical instruments in this country it was necessary for the Medical Department to educate in the manufacture of these instruments certain concerns which had been engaged in the production of similar devices. Men skilled in the manufacture of instruments, with long years of experience, were sent to these factories to work out with the forces there satisfactory processes. It was necessary to recruit toolmakers, jewelers, and cutlery manufacturers in order to build up a sufficient supply of forged and finished instruments.

Surgical needles, for instance, had never been made in this country, but had all been obtained in England. As a war measure the British

Government placed this item on its list of restricted exports. After long and continued effort the general purchasing office developed American sources of supply of needles with remarkable success.

In one month we shipped 65 tons of surgical instruments to France. A few of the principal instruments, quantities purchased, and the prices paid were as follows:

	Average cost, each.
1,301,476 hæmostatic forceps	$1.04
284,600 tissue forceps	.59
348,500 minor operating knives	.57
225,000 probes	$.047\frac{7}{9}$
309,548 surgical scissors	.741
2,102 general operating cases	159.55
3,400 small operating cases	45.30
10,000 instrument cases for officers' belts	5.28
300,000 enlisted men's belt cases	1.35

Each general operating case contains more than 50 instruments and the small operating case more than 30 instruments, and in these two items alone are more than 207,000 forgings, practically all handwork.

The quantity of surgical dressings used in peace times was relatively small, so that the sources for supplying this material had to be increased enormously. To do this the Government went out into the cotton goods industry and induced such concerns as curtain makers and manufacturers of waists and white goods to make bandages for surgical uses. The Government obtained the raw material, gray gauze, and turned it over to the various manufacturers for bleaching, cutting, sterilizing, and packing in the necessary cartons.

Among other items during the last year of the war a total of 12,000,000 individual dressing packets were purchased and 795,000 boxes of gauze bandages, 574,400,000 yards of bandage, 10,000,000 first-aid packets, and 108,000,000 yards of gauze. During the same period a total quantity of 3,814,000 pounds of absorbent cotton was also bought.

Among the miscellaneous items obtained were approximately 1,600,000 blankets, 258,000 litters, and over 1,000,000 clinical thermometers. The rate of output of clinical thermometers was not all that the Medical Department thought it should be, and as a result a large quantity of thermometers was obtained on mandatory orders.

The heaviest buying period during the war was between July 1 and November 30, 1918. The supplies purchased or ordered in that period were the following, with their costs:

Medicines, antiseptics, and disinfectants	$19,728,715
Hospital furniture and equipment	8,220,297
Hospital supplies, textiles	69,321,787
Hospital supplies, miscellaneous	1,808,465

Surgical instruments	$6,576,238
Surgical dressings	75,762,383
X-ray equipment and supplies	2,466,089
Dental equipment and supplies	4,932,178
Laboratory equipment and supplies	2,301,683
Veterinary equipment and supplies	3,258,119
Motor ambulances and supplies	25,625,000

It is interesting to note that the purchases made in France for the Medical Department consisted mostly of large and bulky items, mainly hospital furniture and equipment, which, if transported from the United States, would necessitate the use of considerable valuable cargo space. Foreign purchases were made primarily to save ship space and not because of any shortage or failure to function in this country.

Although America is famous throughout the world for her dentists and dentistry, yet the participation of this country in the war created a demand for dental supplies that the American manufacturing facilities in existence in 1917 were unable to fill. For that reason it was necessary to extend the production capacity. The manufacturers in the trade rose to the occasion, and as a result the Government was able to supply to the A. E. F. from the United States all dental materials required, the only purchases made in France being of exceedingly bulky apparatus.

The total amount allotted for dental supplies for an army of 5,000,000 men in 1919 was $6,256,482. During the five months between July 1 and November 30, 1918, the dental purchases amounted to approximately $5,000,000.

The six leading dental items purchased by the Medical Department and the quantity and cost of each were as follows:

Items.	Quantity.	Cost, each.
Tooth-extracting forceps	47,319	$2.86
Dental chairs:		
Hospital equipment	1,112	167.06
Portable	3,200	49.00
Lathes:		
Unique	110	10.00
Electric	70	43.96
Fountain cuspidors	1,253	32.51
Burs	3,836,776	.081
Engines:		
Portable	1,790	63.00
Electric	814	122.80

CHAPTER VI.

SALVAGE.

Tables of statistics are apt to be tiresome affairs; but in the annals of the War Department, as part of the record of the American Army in the great war, there is a table of statistics that is replete with human interest. This is the table which depicts the activities of the salvage operations of the Army, both at home and abroad.

Until the war came to America and brought to us the necessity of being provident, thrift and economy could not be called characteristic American qualities. As virtues in the individual we were apt to despise them. Paris can live on what New York throws away, runs the old saying. For the prudent man we invented opprobrious names. Such names and phrases were but the surface outcroppings of a national tendency to be wasteful.

But the war came along to put a stop to waste and to raise thrift high in the esteem of America. Partly because of the mounting prices of food and clothing and partly because of well-organized and well-conducted propaganda on the part of various agencies of the Government, chief among them being the United States Food Administration and the Liberty Bond and War Saving Stamp organizations of the Treasury Department, America began to practice economy in the use of materials.

How much of the credit for this change can be claimed by the Government itself we may never know; but this may be said—in urging the people to save materials in their own homes, the Government did not, as it had done in previous wars, allow the traditional wastes of military campaigns. The Government practiced what it preached. It cleaned up its own back yard and utilized every scrap of useful material. It mended the shoes and clothing of the Army; it darned the socks; it tinkered the tin cans; it starved the garbage pails by economy in the mess kitchens and recovered the valuable components of garbage at rendering plants; it collected the junk; it swept the stables and put the manure on the land, and then produced crops from the increased fertility. All of these adventures in conservation and reclamation were known to the Army simply as Salvage; which after all was but the scientific attention which the Army paid to the "p's" and "q's" of military housekeeping—it was household economy on the scale of a family of 3,500,000 members.

The figures of the Army's thrift are most impressive. The figures of our war salvage are as follows:

Financial summary of the activities of the salvage service in France for 1918.

Depots and shops.		Kitchen economics branch.	
Month.	Value of output.	Month.	Recoveries.
January	No record.	May	$1,350.65
February	No record.	June	17,881.03
March	$850,000.00	July	74,167.31
April	900,000.00	August	23,581.20
May	1,500,000.00	September	35,677.02
June	2,000,000.00	October	109,013.84
July	3,500,000.00	November	120,158.63
August	5,500,000.00	December	92,685.43
September	7,251,512.40	Total	474,515.12
October	8,007,980.39		
November	8,072,042.08		
December	9,436,839.14		
Total	47,018,374.01		

Battle-field recoveries.

October	$8,000,000.00
November	4,000,000.00
December	3,100,000.00
Total	15,100,000.00

Capitulation of salvage activities in France for year 1918.

Value of output, depots and shops	$47,018,374.01
Battle-field recoveries	15,100,000.00
Kitchen economies	474,515.12
Waste sales	39,680.23
Rubber salvaged, 1,591,565 pounds, at estimated value of 10 cents per pound	159,156.50
Wool cloth shipped to British, 359,920 pounds, at estimated value of 20 cents per pound	71,984.00
Lumber salvaged, 1,737,940 board feet	69,025.20
Total	62,932,735.06

Financial summary of the activities of the salvage service in the United States from Apr. 1 to Dec. 31, 1918.

Activity.	Total articles.	Estimated cost of service.	Estimated value reclaimed articles.	Estimated net saving.
Clothing, shoe, and hat repair	12,635,458	$7,103,940.00	$37,632,158.05	$30,528,218.05
Cot repair	486,892	779,027.20	1,752,811.20	973,784.00
Canvas repair	122,480	611,900.00	3,023,418.79	2,411,518.79
Total				33,913,520.84
Dry-cleaning operations	4,686,415	2,247,292.52		[1] 166,233.29
Government laundry operations	72,263,964	1,888,823.93	[2] 3,115,847.31	1,227,023.38
Grand total				35,306,777.51
DISPOSITION OF WASTE MATERIAL.				
Estimated value of material turned over to organizations				941,709.00
Cash sales:				
Scrap material from June 1 to Dec. 31			$1,194,395.01	
Garbage, May 1 to Dec. 31			414,832.60	
Dead animals, May 1 to Dec. 31			7,043.91	
Manure, May 1 to Dec. 31			225,227.29	
Condemned hay, straw, etc., May 1 to Dec. 31			50,158.92	
Total				1,891,657.73
Farm products invoiced to quartermaster, May 1 to Dec. 31				107,271.79
Estimated net saving to Government				38,247,416.03

[1] On articles dry cleaned in Government shops. [2] Receipts and operating credits.

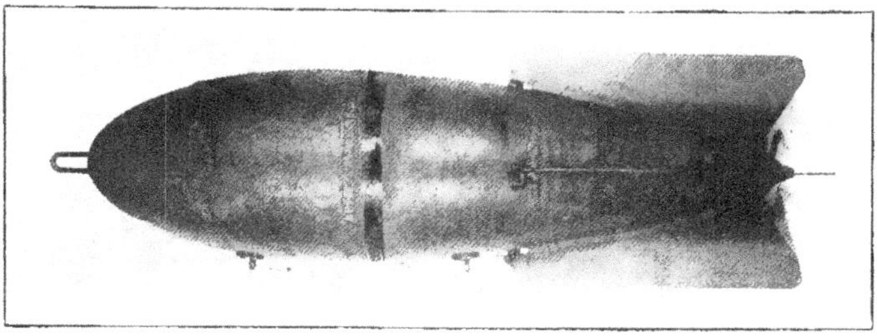

A 250-POUND DEMOLITION BOMB CARRYING 125 POUNDS OF EXPLOSIVE AND HAVING HEAVY CAST-STEEL NOSE AND PRESSED SHEET STEEL REAR BODY.

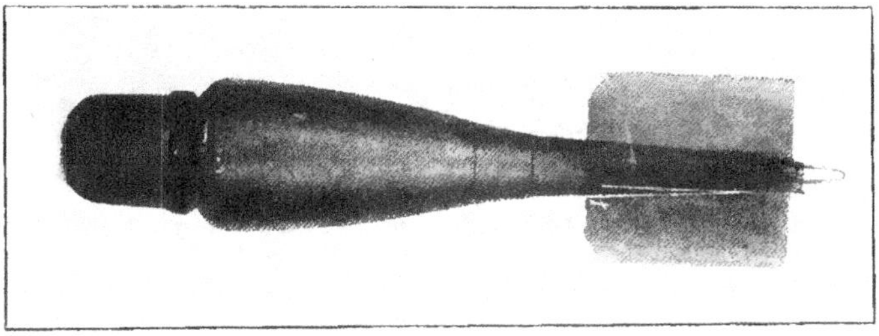

A 25-POUND FRAGMENTATION BOMB CARRYING 3 POUNDS OF EXPLOSIVES DESIGNED FOR USE AGAINST TROOPS

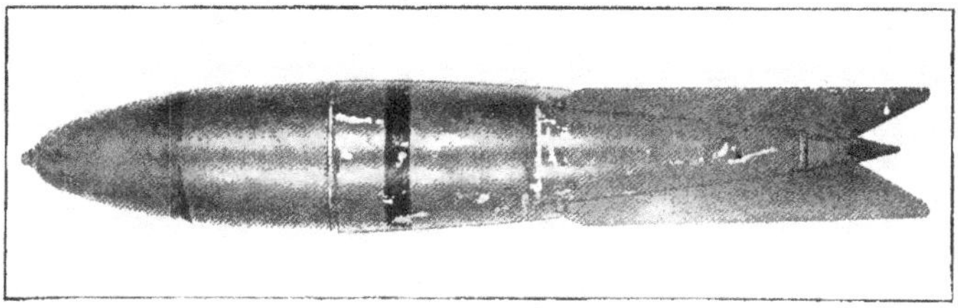

A 40-POUND INCENDIARY BOMB OF THE INTENSIVE TYPE, WITH STEEL NOSE AND FUSIBLE ZINC REAR CASING.

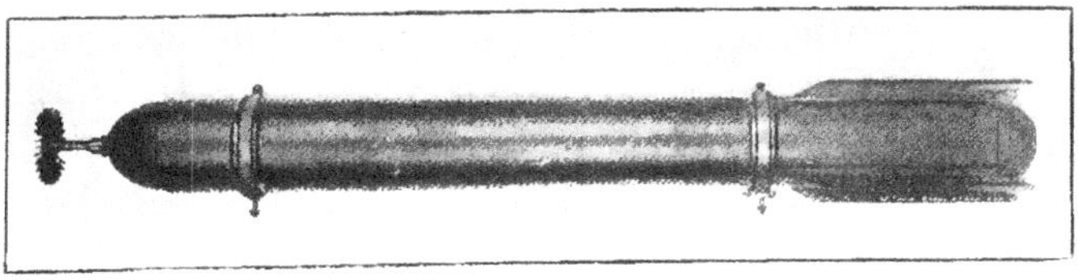

AIRPLANE FLARE.

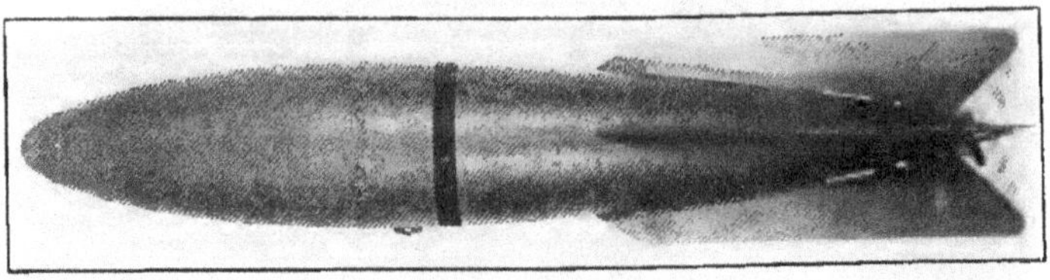

MARK I, HIGH CAPACITY DROP BOMB. A 105-POUND DEMOLITION BOMB, CARRYING 55 POUNDS OF EXPLOSIVE.

MARK II, HIGH CAPACITY DROP BOMB, NOW OBSOLETE, HAVING BEEN FOUND TOO SMALL FOR DEMOLITION PURPOSES.

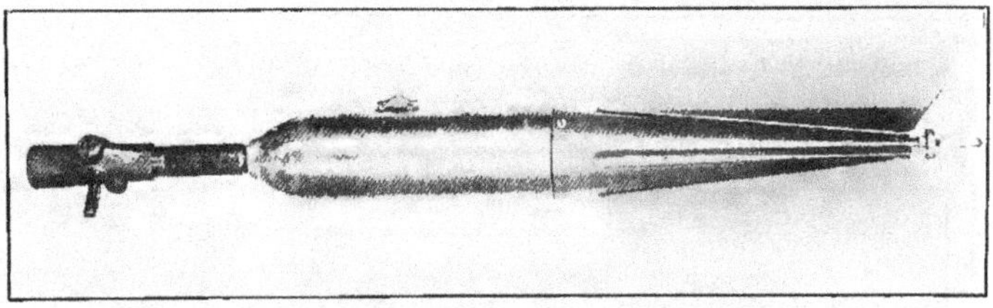

MARK II—A FRAGMENTATION DROP BOMB.

A 20-pound fragmentation bomb, made from a converted 3-inch artillery shell, carries 1½ pounds of explosives to be used against troops. Projection at nose causes burst to take place above ground.

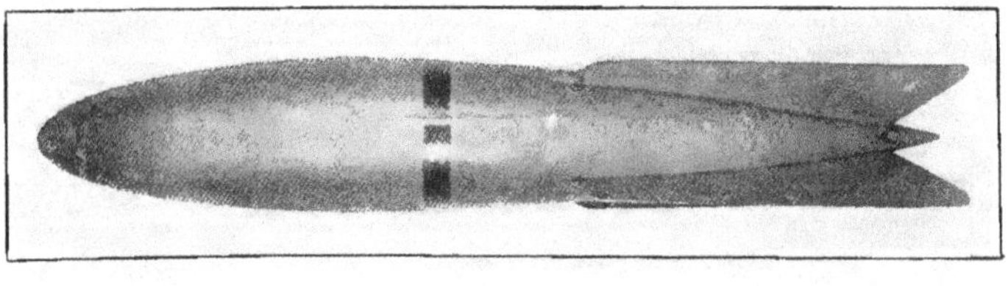

DROP BOMB, MARK III.

A consolidation of these figures shows that the total amount returned to the Government in money value by the savings of the salvage service of the Army for the single calendar year of 1918 was $101,-180,151. With this figure some interesting comparisons may be made.

In 1912, to meet every expense of the American standing Army, Congress appropriated $99,676,767.43; in 1913 the appropriations were $100,292,855.04. Salvage, reclaiming the materials once wasted and using them over again, saved enough in 1918 to have maintained the entire Military Establishment in 1912 or 1913.

But there is even a more striking comparison. During the fiscal year of 1898—the Spanish-American War year—the entire appropriations for the support of the Army amounted to $70,394,739.96. Salvage in 1918 saved $30,000,000 more than was appropriated to fight the Spanish-American War up to July 1, 1898, at which date the fighting was nearly over.

Take the cost of clothing the Army raised to fight against Spain, and add to it the appropriations for clothing the Army and equipping it with shoes, leather and rubber goods, and textile equipment for the years 1913, 1914, 1915, 1916, and 1917, and you have a total Government expenditure of $100,050,271.65. The savings of salvage in 1918 could pay this entire cost with $1,129,880 to spare.

It cost $20,280,000 for the clothing and equipage of the Army for the year ending June 30, 1917, at which date the war with Germany had begun. Salvage in the United States alone in 1918 saved to the Government $17,967,416 more than this appropriation.

Salvage undertakings touched intimately every soldier in the Army. This service which taught economy in the use of materials could with equal facility operate a laundry or dry-cleaning plant, or run a farm, or drive a good bargain in the sale of junk, or return goods that did not meet specifications and be reimbursed for them.

Wherever the experts of the service saw a leak through which the Government's money might flow out, they plugged it. An innovation in warfare as we knew it, it had to fight its way against prejudice at the start, but it developed what formerly was waste and a liability into a tremendous asset. Yet when the armistice came salvage had only commenced to show its possibilities. It had merely scratched the surface, but it had opened up unlimited fields for utilizing worn-out or unserviceable products or by-products of war. It had saved thousands of tons of shipping space in the transportation of supplies to the American Expeditionary Forces by using over again in France the things that otherwise would have had to be replaced by new; it saved this tonnage at the time when every ton saved counted heavily. In this and in the saving of materials at a time when all the raw materials of the earth would scarcely meet the insatiable demands of Moloch, the value of salvage can scarcely be measured by the money figures of its record.

SALVAGE IN THE UNITED STATES.

War salvage in the United States started on October 5, 1917, when the conservation branch was created in the Quartermaster Department. It started with an executive force of two commissioned officers and one stenogapher. When the armistice was signed about 13 months later, the salvage service in the United States alone had a force of approximately 500 commissioned officers, 20,000 enlisted men, and 2,000 civilian employees.

In this period the method of clothing and feeding the American soldier had been revolutionized. The old way was to issue a uniform to a soldier and hold him responsible for the repair and cleaning of it. He owned his uniform and had to keep it in good condition at his own expense out of the $15 per month the Government paid him. The new way was for the Government itself to retain ownership of the uniform and to repair and clean it at public expense. The soldier was required to pay only for his laundry work at a uniform charge of $1 per month, much under what he would have had to pay at commercial rates.

Formerly the soldier had to repair his own shoes. The soldier prefers repaired shoes to new ones for campaign service because the former are broken in and are comfortable. After the salvage service was established the Government retained ownership of Army shoes and repaired them at Government expense.

Once the Army seldom conducted sales of the boxes and crates in which supplies were packed. Salvage undertook such sales, thereby bringing considerable revenue to the Army.

But in these and similar economies, it was not so much the saving of money that was important as it was the saving of materials at a time when the supply of all materials was scarcely adequate to the war demands. When our troops first reached France the officers were surprised at the emphasis placed upon salvage operations by the British and French armies. They were soon to learn that salvage was stressed because it supplied materials which were scarce. Glycerine, a component of high explosives, had become so short in supply that the British Ministry of Munitions paid as much as $1,250 a ton for it. The British army distilled its garbage and procured from the operation glycerine at a cost of $250 per ton. This was a financial saving of $1,000 a ton; but, more important, it supplied glycerine at the time when money did not count. The British Ministry of Munitions got the glycerine, which meant explosives for use against the Germans, which was the main thing.

The British appreciated the importance of salvage so much that one of the officers sent with the British mission to the United States early in the war was a salvage expert, included in the mission so that we might early have the benefit of the British experience in this work.

Although the salvage service of America was authorized in the autumn of 1917, it was not until winter was declining into the spring of 1918 that the service became a working organization fully clothed with authority. Consequently its record was accomplished within a period of 9 or 10 months. The purpose and ideals of the service were embodied in its code, known as special regulations No. 77, promulgating rules and regulations for the conservation and reclamation of Army supplies and materials. The principal provisions of these regulations were as follows:

(*a*) The repair of clothing and equipage.

(*b*) Laundering and dry cleaning of clothing and equipage.

(*c*) Supervision of contracts for the renovation of clothing and equipage.

(*d*) The development of agricultural, mineral, and forest lands for the benefit of the Army.

(*e*) The organization, discipline, and training of men of special units, companies, battalions, and regiments for salvage work.

These regulations likewise—

(*a*) Created salvage companies of 7 officers and 588 enlisted men each, based on a camp strength of 27,000 troops where all salvage utilities exist.

(*b*) Provided methods of receiving and disposing of clothing and equipage turned in for renovation.

(*c*) Provided methods for the conservation of food and the reduction of waste through a systematic check on the disposal of garbage.

(*d*) Fixed the responsibility of organization in connection with salvage work.

(*e*) Provided definite rules and instructions to be followed by organization commanders and their commands in the care of clothing and equipage in order that the cost of maintenance might be reduced to a minimum and that usefulness might be conserved to the utmost, consistent with the health and well being of the troops.

(*f*) Provided a basis for the operation and financing of camp laundries and dry-cleaning plants operated in connection with base salvage plants.

(*g*) Provided a uniform monthly laundry charge of $1 per enlisted man at each camp and station where there was a Government-operated laundry, as well as a uniform charge for service rendered to officers, civilians, hospitals, and other camp agencies.

Thus it may be seen that special regulations No. 77 were not only a charter for the salvage service but a code of conduct in economy and thrift for the soldier of the American Army. Although the regulations did not become official until midsummer of 1918, they had a profound effect in the few months before the fighting in Europe came to an end.

Prior to July 1, 1918, all reports of garbage collection, etc., in the military camps in this country indicated that the American soldier in training wasted on the average of 2 pounds of food per day. This was not excessive, judged by civilian standards, since our large cities, a great part of whose population are fed not nearly so well as soldiers were fed in the camps, show a food waste nearly as great.

But the camp waste of food was regarded as excessive by the salvage officers. Special regulations No. 77 contained minute directions for conserving food in the camp kitchens. The result of these regulations was that in the four-month period beginning July 1, 1918, the average mess waste per man in the camps fell to 0.3 of a pound per day. Since there was an average strength of 1,500,000 men in training during these four months, the reduction of waste amounted to many thousands of tons of food.

These regulations also set up a salvage equipment for the use of the Army. As a rule each camp had a shoe-repair shop large enough to fix 400 to 500 pairs of shoes per day; a clothing-repair shop large enough to take care of the everyday mending of 30,000 troops; a hat-repair shop sufficient in size to restore the headgear of 30,000 men; and other miscellaneous shops.

But at the change of seasons there could be expected an exceptionally large turn in of worn-out clothing, and to handle these periodical floods of garments large base salvage plants were established at Fort Sam Houston, Tex.; Washington, D. C.; Atlanta, Ga.; New York City; Philadelphia; El Paso; and Newport News, Va.; with a base salvage plant for rejuvenating shoes at Jeffersonville, Ind. Smaller base plants were established at Chicago, New Orleans, San Francisco, and at the United States Disciplinary Barracks at Fort Leavenworth, Kans., and at Alcatraz Island, Calif. Other base plants to receive and classify and dispose of waste materials were established at New York, Philadelphia, Baltimore, Chicago, St. Louis, Fort Sam Houston, and Atlanta.

The shoe-salvage base plant at Jeffersonville Depot was more than a repair shop in the accepted sense of the term, for it became one of the most complete shoe factories to be found anywhere in the country. When this shop was being projected as a plant to take care of the overflow of worn shoes from the camps and depots, the United Shoe Machinery Co. agreed to furnish machinery sufficient to repair 2,000 pairs of shoes a day, supplying this equipment for a period of six months without any expense to the Government, except upkeep and the cost of supplies.

At the Jeffersonville shop shoes went through the mill from department to department much as machines are assembled in the familiar quantity-production manner. Shoes arriving were first counted, and then sorted and graded as follows:

Class 1. Unrepairable.

Class 2. Inner soles required.

Class 3. Regular run of repairs (half soles or full soles, heels, patching of uppers, and other minor repairs).

Class 4. Civilian.

The shoes arriving at this plant were in a condition that would have resulted in their being discarded altogether in the old days. The experience at Jeffersonville showed that 65 of each 100 pairs arriving at the factory could be repaired, and repaired cheaply. In January, 1919, of 132,112 pairs of shoes sorted, 45,000 were in irreparable condition and had to be thrown away. There were 11,475 pairs of class 2 shoes, 74,362 pairs of class 3, and 1,175 pairs of class 4.

From the sorting room the shoes went to the wash room, where they were disinfected and cleaned in a bath containing a solution of 40 ounces of formaldehide and 1 pound of castile soap to each 10 gallons of water. After being washed the shoes were placed on rolling racks, each rack holding 24 pairs of the same size and width. The loaded racks were wheeled to the lasting section where lasts were inserted according to sizes.

Next, machines cut off the worn portions of the old heels, after which the shoes went to the stripping bench, where the old soles were removed and the shank pieces skived to prepare a smooth joint for the new half sole. The next process was welting. The welts were prepared, and tarred felt was glued to the old inner sole to fill out uneven parts and prevent squeaking. The next operation was to lay on the half sole in a setting of rubber cement. Another machine rough rounded the soles to conform with the shape of the shoes.

Then the shoes reached the stitching machines, where the soles were sewed on, and then the leveling machines, which smoothed out the wrinkles of the inner soles. The next step brought them to the heeling machines, where the complete heels were attached in one motion. Next, machinery for nailing soles and heels, and then the trimming machinery for smoothing off the work. The final mechanical operation was on the scouring and finishing machines. Meanwhile, if the shoes needed patching on the uppers, this work was done by women operating sewing machines.

The final process was to give the field shoe a thorough coat of waterproof dubbin. A good polish was put on the russet shoes. A split leather insole was inserted in each shoe to insure perfect smoothness of the bottom. A pair of laces was tied to each pair of shoes, and then the shoes were packed in boxes of 24 pairs each and turned in to the Army stores.

The Jeffersonville shop repaired 222,135 pairs of shoes in seven months of operation. Thousands of pairs of shoes were discovered to have been fitted too short. This was shown by the fact that many of the shoes were worn out entirely in the toes. A shoe that is too long will turn up at the toes, while one that is too short will stub with nearly every step taken.

On August 8, 1918, the Secretary of War authorized the expenditure of $5,287,852 for the construction of laundries to serve from

20,000 to 40,000 men in each of 20 camps and posts. About this same time repair shops were authorized at each of the training camps and special dry-cleaning plants at Atlanta, Fort Sam Houston, El Paso, and Alcatraz Island. Before the armistice was signed many of these plants were in operation. In addition to these the salvage service eventually operated printing plants, wagon repair shops, and carpentry shops, so that at the time the armistice was signed there was hardly anything of quartermaster issue not subject to rehabilitation by the salvage division.

Each shoe-repair shop at the training camps had equipment sufficient to repair 500 pairs of shoes per day, utilizing the services of 40 to 50 men. When the shops were officially authorized, an inventory of the Army's old shoes showed there were approximately 1,500,000 pairs on hand in need of rehabilitation. In order to assist the camp shops in the work, the salvage service brought between 50 and 55 shoe factories into the reclamation effort, these private factories repairing about 500,000 pairs until the camp shops were able to catch up.

Because of the shortage of linen thread it was decided to use nails for attaching half soles, particularly in the repair shops in France. More than 2,500 nailing machines were bought and shipped to the American Expeditionary Forces. The American Expeditionary Forces adopted the English system of company cobblers and regimental repair shops. Upward of 11,000 cobblers' kits were shipped to France. In July, 1918, the American Expeditionary Forces requested machinery for a base shoe-repair plant in France. This machinery was shipped considerably before hostilities ceased.

The service maintained a corps of civilian instructors, who traveled from camp to camp and improved the efficiency of the Army cobbling. The accumulation of worn shoes at the embarkation camps was sent to various contractors for repairs. By November, 1918, the shoe-repair facilities of the Army had reached full operation, 500,000 pairs of shoes being repaired that month, a figure representing all the repairing required by 1,500,000 men. All shoe-repairing activities were under the direction of Philip H. Fraher, who was assisted by Joseph Caunt, of Pasadena, Calif., a retired shoe manufacturer with a wide experience both in this country and in England.

In its clothing-repair activities the salvage service dry-cleaned uniforms and woolen equipment, repaired and renovated hats, and reclaimed outer clothing and underclothing.

For the first time in dry-cleaning history, a method was worked out to destroy all living organisms and a considerable amount of bacteria, a process which is likely to have a lasting effect in the dry-cleaning industry. The specifications of this process were the result of cooperative laboratory research by the Bureau of Standards, the Public Health Service, and the salvage division. In addition to destroying germs and bacteria the process also thoroughly cleansed the garments. Experts from the salvage division were on hand to see to it that the various contractors lived up to the standard specifications. The authorized Government-owned dry-cleaning plants, which were to be the last word in what such establishments should be, were not completed, due to the signing of the armistice. Dr. Harry E. Mechling, a graduate physician and president of the Swiss Cleaners & Dyers, of Louisville, Ky., was in charge of the Army's dry-cleaning activities.

In the repair of clothing the service received much assistance from the Red Cross. Local Red Cross units in the vicinity of camps worked in conjunction with the officers of the salvage service in the reclamation of such garments as woolen shirts, underclothing, sweaters, helmets, socks, and gloves.

The base salvage plants at Atlanta and Fort Sam Houston reached a high state of efficiency in the repair of clothing. Shipment after shipment was made from such congested centers as Newport News and Hoboken to these plants, and within a comparatively short time the property was ready for reshipment and reissue. Capt. Harvey A. Rosenthal, a graduate of the first officers' training school, and in civil life in the clothing business, was in charge of clothing repairs.

All of the camps had shops for renovating and repairing hats. The average cost for repairing a hat was 35 cents, whereas the lowest contract price was 65 cents, and the quality of work at the Government shops was far better than that obtained from private contractors.

The following table gives an idea of the approximate saving to the Government in hat-repair operations:

Place.	Quantity.	Cost of repair of each.	Estimated total cost.
Government shop	181,764	$0.35	$63,617.00
Contract	539,495	.65	350,671.00
	721,259	....	414,288.00

Estimated value of hats made serviceable $1,000,746
Estimated net saving 586,458
Cost of 721,259 new hats 1,334,329

Mr. E. LeRoy Cummings, of the John B. Stetson Co., of Philadelphia, was in charge of hat-repair activities.

Extensive repairs to canvas materials were confined to the base plants at Philadelphia, El Paso, Fort Sam Houston, and Atlanta, and, on a smaller scale, at Jeffersonville. Minor repairs were conducted at camp shops, some of which were only in the course of construction when the armistice was signed. Tents were generally reconditioned while standing. Patches to tents were attached by means of a nitrocellulose cement, the best cement for the purpose which the salvage service found, being called vanite. Experiments at the Bureau of Chemistry resulted in the adoption of three waterproofing compounds named Preservol, Candeline, and Truscon. These compounds were applied to both standing tents and tents which had been taken down, with complete and effective results.

Laundering was not a new activity for the War Department, since when the war was declared the Government already owned 14 small steam laundries. Later the Government went into the laundry business on the scale demanded by the great chain of training camps, building cantonment laundries at a cost of approximately $300,000 each. Experienced laundrymen were placed in charge of camp laundries. Through the cooperation with the Government's insect experts of the Bureau of Entomology, laundering processes were worked out successfully to disinfect all clothing while washing it and to free it from vermin without shrinking fabrics or causing other damage. Government laundries during the war operated 24 hours per day with three labor shifts and cleaned an average of 10,909,850 pieces of clothing monthly, with gross receipts of over $500,000 per month, approximately half of which was profit.

One of the most interesting features of laundry activity was the development of mobile laundry units for overseas use near the front. The men to operate these units were trained in a special school at Camp Meigs, D. C. Each mobile unit required a crew of 37 men. The men of the Army nicknamed these special troops the "Fighting Chinamen."

The need of the American Expeditionary Forces for wash-up and delousing stations at the front, so that even troops engaged in battle might have clean clothes, called the mobile laundry units into being. The first experimental equipment was designed and constructed early in 1918. After that the salvage service produced 50 others, 32 of which were shipped to France.

Each unit consisted of a large steam tractor and four trailers, an outfit which on the road made up a train over 100 feet long. The trailers could be placed together in the field to form a building 30 feet long and 28 feet wide, the tractor acting as the power plant. On the trailers were washing machines, wringers, drying machines, tanks for water and soap, a pump, and a dynamo to supply electric lights. One of these plants working 24 hours per day could do the washing of 10,000 men. This unit was designed by officers of the salvage division.

Army laundry activity was in charge of three New York laundrymen: J. E. Dann, president of the Pilgrim Laundry, of Brooklyn,

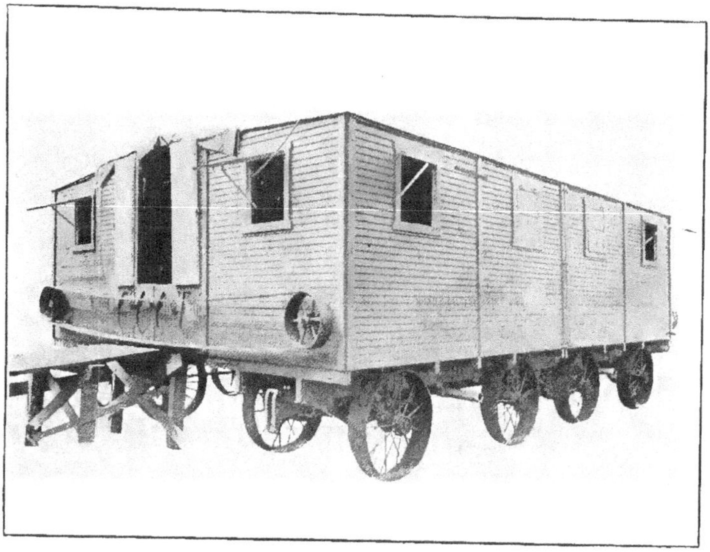

ONE OF THE ARMY'S MOBILE LAUNDRIES.

INTERIOR VIEW OF MOBILE LAUNDRY.

SALVAGING ARMY HATS IN FRANCE.

This photo was taken at Tours, France, in February, 1918.

SALVAGING ARMY SHOES IN FRANCE.

Picture taken at Tours in March, 1918, showing a few shoes worn out by the soldiers overseas.

and his assistants, William Longfelder, of H. Kohnstamn & Co., and E. D. Tribbett, of the American Laundry Machinery Co.

Wherever possible waste materials were reclaimed for use by the Army instead of being sold as junk. This was particularly true of bags and burlap. Hundreds of thousands of bags and great quantities of burlap furnished by the salvage division were utilized for army purposes. Without salvage, all of this would have been thrown away or sold at junk prices. When the fighting ended the base salvage plant at Chicago was being equipped to repair about 15,000 bags daily.

The purpose of garbage separation was threefold—the reduction of mess waste, the increase in revenue to the Government, and the recovery of glycerine contents for military purposes. Before the war with Germany the Army regulations required the burning of garbage at camps. When the great training camps were established, the Government adopted generally the policy of selling garbage to contractors, except at Camps Fremont, Hancock, McClellan, Sevier, and Shelby, where it was incinerated. Originally contracts had been let on a per-man basis, the contracts extending for several months, comparable to municipal contracts for the disposal of garbage. Later on, however, the policy was adopted of letting contracts for periods of one month only, since the number of men at the camps was continually increasing, and the garbage grew correspondingly bulkier. It is estimated that this change in contracting saved the Government considerably over $400,000. Contract sale prices ranged from 1 cent to 9 cents per man per month, the latter price in most cases including the manure from the stables.

With a view to obtaining glycerine, the War Department authorized the construction of 16 garbage-rendering plants at the larger concentration centers, but only one, that at Camp Lee, was actually constructed, since it was determined later by an investigation of our national resources that the amount of glycerine to be obtained in this manner did not justify the expenditure of the money. Also the project of establishing piggeries at camps was abandoned, after investigation by the salvage service, since it would have required 18 months to clear the investment, and in the meantime the Government would have been deprived of revenue from the sale of garbage.

The disposition of waste materials was under the direction of Louis Birkenstein, of Chicago, assisted by R. D. Cunningham, of Troy, N. Y.

In response to an insistent demand that farms be operated at various camps, the salvage division, on May 15, 1918, secured $60,000 from the vocational fund for the training of soldiers and allotted sums to 15 camps.

On November 4, 1918, Congress appropriated $250,000 for the same purpose, but little of this money was expended. The total acreage under cultivation in 1918 was 3,483, and the equivalent revenue derived from the camp farms amounted to $108,000. The farm work was under the direction of Capt. Henry G. Parsons, a practical and scientific farmer.

Salvage activities, in general, in the United States were under the direction of Philip W. Wrenn, who was chief of the salvage division during the most active period of its existence.

SALVAGE IN FRANCE.

Salvage in France was under the direction of the chief quartermaster with the American Expeditionary Forces. At first, it was undertaken in a smattering way, but as the American Army grew in size, salvage increased, until the salvage service became one of the features in the field, with thousands of men and women working in salvage activities, with salvage plants, branches, and depots, large and small, saving, repairing, conserving, and putting into shape, ready for reissue, materials of all sorts and descriptions. The word "salvage" became the watchword and pride of many an organization at the front.

Each field army had its chief salvage officer. Each division of troops had its salvage organization under a salvage officer. Each organization had its salvage dump, in which it took a just pride, and there was a spirit of friendly rivalry between different organizations as to which could save the most for the Government.

In the flood times of battle, when waste materials piled up on the fields, the regular salvage specialists were assisted in various ways. In some divisions the regimental bands were designated to act as emergency salvage companies. Sometimes after an engagement whole battalions and regiments were enlisted to clean up an area, and there is one instance on record where a wise general of the American Expeditionary Forces turned out his entire field army to clean up for salvage the area which it had just won from the enemy. The salvage service in France handled not only the recovery of quartermaster supplies, but it also collected and disposed of all materials captured from the enemy, including ordnance materials, and also all materials abandoned by our own troops and found on the battle fields. When troops moved into combat they took with them only such equipment as they could carry on their backs or on the meager transportation facilities allowed. Thus they frequently left behind them an enormous quantity of their possessions, including personal baggage. The salvage units went through such areas, visited every billet, and collected all Government and personal property and cared for it. As an indication of the magnitude of this work, there was one salvage dump in France 40 acres in area piled as high as goods could be thrown from trucks.

The salvage operations in France were conducted over an area of 4,000 square miles, and there were approximately 4,000 men in the salvage service field force. The various salvage depots and shops in France occupied a floor space of 736,000 square feet and had a personnel of 11,632 on December 31, 1918. Even before the war, the Quartermaster Corps of the Army was a good-sized organization, yet there were more French women and girls mending clothing for the American Army in France at one time than there were commissioned officers and enlisted men in the whole Quartermaster Service before the war.

Clothing generally for the American troops in France was repaired at special shops and at the homes of seamstresses in the small towns and communes. Each town had a forewoman who distributed the damaged clothing, after it had been disinfected and laundered, and kept all counts. There were 880 of these home workers, nearly all of them from needy families. The best record for darning socks was made by an old French grandmére, aged more than 80.

Numerous soldiers were discovered in the American Expeditionary Forces who were unfit physically for the hard service on the front line. These were permitted to go into the various salvage depots and shops, where they learned to be shoemakers, harness makers, saddle makers, wood workers, painters, metal workers, tailors, laundrymen, electricians, mechanics, checkers, warehousemen, etc., occupations in which many of them expected to engage after their separation from the military service.

The salvage troops in France were in five classes—the salvage headquarter detachments, depot battalions, field salvage battalions, laundry units, and the clothing and bathing units. One of the last named was attached to each division to handle field bathhouses and delousing and disinfecting plants, to receive old clothes, and to issue new or reclaimed serviceable garments.

The ordnance property salvaged in the field in the period between January 1, to October 31, 1918, included 5,000,000 rounds of small-arms ammunition, 71,909 shell of the 75-millimeter size, and 16,195 of the 155-millimeter size, more than 32,000 rifles, and 21,000 machine guns and automatic rifles. The unexploded or "dud" shell is a menace to life, and the duty of destroying these in immense quantities fell to the salvage service.

Some of the salvage squads in the field were composed of men who through lack of education or lack of knowledge of the English language were unable to do front-line service. They were largely composed of American troops of alien birth.

109287°—19——34

The divisional salvage squads sorted the materials at the railhead dumps for shipment to the various depots. When trucks brought up supplies to the front and unloaded, the salvage detachments there filled them up again with all sorts of materials which had been picked up, and the trucks carried their loads back to the railheads, the railroad stations of the division. This queer conglomeration of trash consisted of everything from a hairbrush to a 77-millimeter enemy gun. To show the sorts of articles that are picked up in an area over which an army has fought, there is given here the following list of items selected at random from the check of a salvage shipment from the railhead of the Twenty-sixth Division on August 12, 1918:

1,100 pairs of leggins.
21 pairs of shoes.
30 leather gun cases.
21 bags of harness.
350 mess kits.
750 condiment cans.
750 bacon cans.
150 first-aid packets.
50 feed bags.
300 pistol holsters.
1 wagon bed.
275 German rifles.
3 boxes tent poles.
7 boxes gun repairs.
150 rifle grenade throwers.
4 German machine guns.
200 German canteens.
6,000 gas masks.
50 saddlebags.
1,400 canteens.
200 caps.
900 helmets.
1,025 pack carriers.
750 canteen covers.
1 wagon.
76 wagon parts.
1 ammunition cart.
4 ration carts.
17 wagon wheels.
4 boxes artillery material (telephones, etc.).
1,400 American canteens.
400 American rifles.
47 German automatic guns.
75 gun bolts.
100 respirators.

The kitchen economics branch of the salvage branch of the American Expeditionary Forces in the recovery of fats and glycerine and other kitchen by-products during the month of September, 1918, saved $57,404.19 to the Government. The value was increased in October to $109,013.84, and in November to $120,158.63. In addition to this saving, kitchen salvage in October produced over 25,000 pounds of grease and over 14,000 pounds of dubbin for waterproofing shoes. This branch of the service also had the disposition of unserviceable food supplies, entailing the salvage of large quantities of flour, sugar, rice, and beans damaged in transportation or injured by exposure to weather so as to become unfitted for troops. Such vegetables as peas and canned corn, unsatisfactory for use, were dried and ground and sold for chicken feed or hog feed, bringing in a considerable revenue.

The question of laundering for the field hospitals, particularly after hard fighting, was a vital one. During the month of December, 1918, a total of 7,811,566 pieces of laundry was handled by the laundry branch of the salvage service. This included 3,700,000

OUTPOST AND FIELD WIRE SALVAGED FROM BATTLE AREA, GIEVRES, FRANCE.

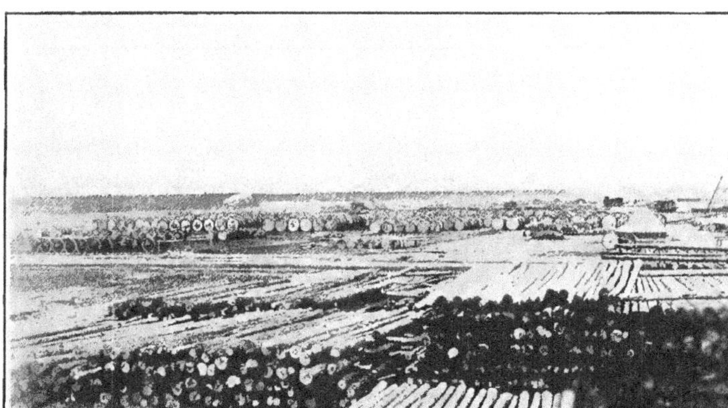

PORTION OF OPEN STORAGE YARD. SAPPINETS FOR TEMPORARY LINES IN FOREGROUND, CABLE REELS IN BACKGROUND.

GERMAN PRISONERS WORKING FARM LAND.

100 SOLDIERS HARVESTING SNAP BEANS AT CAMP GORDON FARM NEAR ATLANTA, GA.

pieces for the hospitals alone. The American Expeditionary Forces were required to establish three large shops for mending clothing sent to the laundries.

The salvage service in France rendered a peculiar service in being the repository for lost articles. The baggage branch of the salvage service worked in close cooperation with the Army transportation service, railroad transportation service, the central records office, the graves registration service, the effects depot, the French railroad officials, and other agencies which might assist them to recover and handle all lost baggage for the members of the American Expeditionary Forces or for their heirs in the United States.

The garden service of the American Expeditionary Forces was operated as a separate branch of the Quartermaster Corps, but a word about its work may not be amiss here. In addition to gardens at the camps and hospitals in France, there was a large central farm at Versailles, near Paris, where American officers and men were assembled to learn intensive farming before being sent to the various stations to assume charge of garden work. This service was composed entirely of men who had been wounded or gassed, or were otherwise physically unfit for service at the front. The garden operations provided welcome additions of fresh vegetables to the American Expeditionary Forces' diet and also gave many Americans an insight of the French methods of intensive farming.

The 85,000 German helmets used in advertising the American Government's fifth war loan—the Victory loan of April, 1919—were all collected and shipped to the United States by the salvage service of the American Expeditionary Forces. In fact the immense quantities of dunnage and junk collected by the service are expected to be of untold historic value as time goes on. Various historical societies and museums are taking steps to secure collections of this war material.

Civilians in Europe are now wearing shoes built originally for American troops, later worn out by them, and still later reconditioned by the salvage service in France. A large number of these shoes recently sold for approximately $4.30 a pair. Since the average total cost to repair shoes was $1.05 a pair, the Government realized a net gain of $3.25 from every pair of these shoes.

In connection with the conservation of waste materials the salvage service conducted a considerable manufacturing enterprise in France. It turned waste into a large number of small articles, such as metal markers for graves or effects of deceased soldiers, sheet tin (this from discarded tin containers) for lining the stables at the remount depots, large shipping bags, cement sacks, collar ornaments, divisional insignia, brassards, overseas caps, guidons, curtains for engine cabs, and many other things. The service took discarded campaign hats and old uniform and overcoat cloth and made hospital slippers with cloth tops and felt soles.

Such things as waste cotton scrap, waste paper, shredded rope, tin cans, and woolen rags collected in France were saved and sold, but nothing was sold that could be utilized for repairing or manufacturing purposes. Leather scrap was used to make leather straps or shoe laces, and the worst of the leather scrap was burned at the power plants of the salvage depots as a fair substitute for coal. Old harness, books, small scraps, leather washers, and the like, canvas and burlap scrap, went to the camouflage screen makers. Woolen rags were shredded and used over again for making cloth. Cotton rags too poor for other purposes went to the paper mills. Rubber scrap became new rubber material. Nothing which had a value was allowed to go to waste.

The salvage depot at Tours, France, alone in the period from March to November, 1918, inclusive, produced goods to the value of $19,383,353.58, at a total expense of $268,955.37, giving the Government a net profit of $19,114,398.21.

The value of all this work went far beyond the value of the figures in dollars and cents, which is the only concrete way in which it can be expressed. The saving in raw material alone which it effected was an important factor in the war; yet of even greater service was the salvage production of materials, particularly ordnance materials, which took much time to manufacture at home and after that required a long haul to get them to the American Expeditionary Forces. Some of the materials recovered on the battle field were scarce and hard to get, and every pound of them recovered added that much to the power of the American Army in France.

BOOK VI.

THE CONSTRUCTION DIVISION.

CHAPTER I.

CANTONMENTS AND CAMPS.

As soon as America had arranged to raise an army by selective conscription, the Government proceeded to provide living quarters for the soldiers to be mobilized for training.

This was a job magnificent in its proportions, carried out with a speed that was little short of magical. At 16 points, widely scattered over the country, the construction expert and the civil engineer struck the earth with their potent wands; workmen swarmed forth; the staccato of myriads of hammers and the whine of saws merged into a rolling chorus of industry; and 16 new cities arose—almost overnight, it seemed—built of wood to be sure, raw and unpainted it is true, but snug and taut and equipped with every necessary convenience known to the dwellers of modern American cities.

The United States had been wont to measure other public works by that of the Panama Canal, which had been the largest construction operation ever undertaken by America, or any other nation, prior to the great war. The construction cost of the Panama Canal was approximately $375,000,000 and the operation continued over a period of 10 years. The 16 cantonments for the National Army and the 16 camps for the National Guard cost about seven-tenths as much as the Panama Canal, but they were completed in shorter time than it takes to build an ordinary suburban dwelling house.

The science of warfare had made mighty strides since America's last great war, that of 1861 to 1865, but in no respect more than in those matters relating to the individual soldier's comfort and bodily welfare. The soldier of 1863 lived in a tent, or in the chance shelter of a billet. When the weather was cold he might alternately toast and congeal at his camp fire, and at night he rolled himself in his blanket and reposed on a pallet of straw.

His grandson warrior who went to the training camp in 1917 found life comfortable in a substantial barrack, warmed with steam heat or stoves. A good mattress on a hygienic metal bed wooed his slumbers after a hard day of training.

The soldier of 1861 bathed where he could and when he could. He of 1917 kept clean daily under the shower bath. The soldier of 1861 slaked his thirst at neighboring wells or streams; and water-borne diseases, such as typhoid fever, reaped a harvest of lives. His successor drank water which was tested and filtered, sterilized when necessary, and the once fatal epidemics of armies were kept away from his cantonment. This water, moreover, came to him in a pipe under a pressure sufficient to throw a stream from a nozzle clear over his barrack, an efficient safeguard against the fire that might destroy his wooden city.

The soldier in the Civil War washed out his own clothing on the infrequent occasions when he possessed both water and leisure. The National Army recruit received his khaki immaculate from a modern laundry equipped with the latest types of labor-saving machinery. The latter's grandfather suffered from scurvy because of the limitations of his diet. The soldier of 1917 ate tender beef and green vegetables kept fresh in ammonia-cooled refrigerators. The fighter of 1861 relished the hoecake baked in the ashes; his successor partook of white bread fresh from camp ovens.

The camps of 1861 were arranged in haphazard fashion; those of 1917 were laid out by expert city planners. In the spring the soldier of 1861 waded and toiled through mud; the soldier of 1917 walked dry-shod upon walks or drove his autotruck upon macadamized or concrete or brick camp roads. The illumination of the camps of 1861 was the light of the stars and the bivouac fires; the 1917 barracks were built along streets radiant with electrical incandescence. For amusement the soldiers of 1861 had their campfire choruses and rough military sports, but the private in the National Army had the theater, the motion picture, a library of good reading matter, the Y. M. C. A. and similar clubhouses, the gymnasium, the post exchange where he could buy periodicals, candy, fruit, and other small luxuries.

And so the contrast might be carried along. The marvel of the cantonments of 1917 was not that a grateful Republic gave to its conscripted soldiers the conveniences of existence enjoyed by all urban communities, but that it provided them in such short time. Ninety days after the first spade struck into the ground the cantonments were ready to receive two-thirds of their men, while one or two of the largest of all were complete in every essential respect.

The houses of the National Army and the National Guard went up at the rate of $2,000,000 a day. It is almost impossible to visualize this speed or to comprehend the feat of nailing fifteen hundred and some odd million board feet of lumber into place in about three months, or that of stringing wire in that time in length sufficient to reach from New York to San Francisco and back and westward again to Cleveland, or that of tacking enough rain-tight roofing paper to make a canopy for the island of Manhattan, another for Atlantic City, with nearly a square mile of it left over.

The cantonment job took so many nails and spikes that it created an actual shortage in that industry. All of the factories in the United States that make metal pipes could not turn out enough to supply the needs of the water, sanitary, and heating systems of the cantonments, and so wooden pipes made from staves were used in most of the camps for piping water.

In the matter of lumber alone it has been computed that the total amount ordered and mobilized by the Washington office for the 32 camps would build a board walk 12 inches wide and 1 inch thick to the moon and half way back again. In addition to this vast quantity of lumber there were used in the camps and cantonments about 100,000,000 square feet of wall board—wall board being universally used instead of lath and plaster to line and ceil rooms—12,000,000 square feet of window glass, and 100,000,000 square feet of roofing. The 2,000,000,000 eight-penny nails used, if placed end to end, would girdle the earth three and one-half times. The heating systems would make a single steam radiator 100 miles long, while the heating boilers were equivalent to one boiler 6 feet in diameter and 3 miles long.

The rate of the flow of materials to the army of 200,000 workmen on the cantonment jobs gives an inkling of the speed with which they put lumber and nails together to make barracks. It took 12 heavy freight trains a day, 50 loaded cars to the train, to keep wood and metal supplies at the builders' elbows. These builders erected the camps at the rate of 30,000 tons of material a day. America had never seen construction progress to equal that.

We liken swift building to the mushroom's growth, but almost always this figure of speech is used as hyperbole. Some of the feats of the cantonment builders, however, equaled the growth of the mushroom in fact. In more than one instance at spots where the sod sparkled with dew at dawn, when the builders put away their tools at sunset there stood structures roofed and inclosed, needing only a few interior touches each to be ready to shelter 200 men.

Efficiency in war is a matter of teamwork. Every vital branch of the military organization must do its part well if the whole effort is not to fail. The weak link in the chain might well have been the construction organization of the Army, for here was an emergency job on a scale beyond anything in the experience of our greatest builders. America in her first 18 months of war was able to send to France across 3,000 miles of dangerous water more men than the United Kingdom in a similar period could send to the front over the netted and laned 40 miles of English Channel. No slight share of the credit for this achievement must go to the cantonment builders, who despite great difficulties had the housing ready for the new armies when they were called forth for training.

As soon as the United States entered the war the Government sent out the call for technical experts of all sorts. For a quarter of a century the United States had been specializing in technical training for its young men, and now in the hour of need the ability existed to conduct the war, which in its first year was to be largely a matter of construction and manufacture of equipment. There was a wonderful outpouring of these men of action, the technicians who had been building the bridges and skyscrapers of the Nation, developing its mines, providing its water systems, designing its machinery, organizing and commanding its trained and untrained workers, engaging in public and private works of every description and magnitude.

The Army prior to the 6th of April, 1917, consisted of a relatively insignificant force of men. For this Army the construction of barracks and other quarters had been in the hands of the Quartermaster General.

The officer in charge of the Construction and Repair Division of the Quartermaster General's Office was relieved of his former duties and placed by the Secretary of War in charge of a new and almost entirely independent division, reporting directly to the Secretary, called the Cantonment Division, and charged with providing the necessary construction and camp facilities for the National Army and the National Guard. This was in May, 1917, at which time the commissioned personnel of the division consisted of only three officers. This step was recommended by the General Staff, acting in accordance with the advice of civilian construction experts on the Council of National Defense.

One year later the personnel of this division had grown to 263 officers and 1,100 civilians in Washington, the best constructors, engineers, draftsmen, managers, purchasing agents, and other specialists obtainable by the Government; there were hundreds of other officers and civilian experts in the field for this organization; it had an enlisted personnel of some 16,000 men and employed over 200,000 laborers and craftsmen; it had jobs on hand, complete and incomplete, aggregating $600,000,000, or nearly twice the cost of the canal at Panama, while future works then being planned and later actually undertaken came to another $600,000,000; it had now become the Construction Division of the Army, attached directly to the office of the Secretary of War, charged with all the army construction within the United States. Such was the expansion of one branch of the Army to meet the emergency. Construction operations for the Army overseas, conducted principally with troop labor, was in charge of the Corps of Engineers.

Congress passed the selective service bill on May 18, 1917. Before the end of May the military authorities had decided to call the first levies of the National Army on September 1. The little Cantonment Division, which had in the week after its birth grown to a personnel of 30 officers and numerous civilian experts, received orders to have the camps—16 complete cities to accommodate 40,000 inhabitants each and 16 tent camps, with many incidental buildings and public utilities—ready in 90 days.

Actually the time allowed for construction was much shorter than that, for the last site was not approved until July 6. About 60 days later, on September 4, the National Army cantonments were ready for 430,000 men, two-thirds of the first draft. Although some construction, subsequently authorized, was not entirely complete until later, the cantonments nevertheless were at all times prepared to receive the conscripted soldiers faster than the Army could assimilate them.

However irksome to the impatient construction officers the interval between the time when the cantonments were ordered and the day when the last sites were approved, it was not time wasted by any means. There was much preliminary work to be done. The magnitude of the task ahead was appalling. Yet the Cantonment Division, with scarely anything to start with, with not even the ground selected for a single camp site, must design and adopt types for buildings, mobilize materials, standardize everything possible, adopt an emergency contract that should protect the Government from the grafter and the profiteer, locate stores of materials, commandeering them if necessary, and also discover manufacturing plants capable of turning out supplies as rapidly as they were needed, build up an organization to handle the work in every detail, and be ready to start hammering in the nails on the day the materials arrived on the jobs. Actually these officers had something less than 20 days in which to accomplish this feat.

There had been, however, a measure of pioneering in several of these directions. The Council of National Defense had an organization of civilian experts in many lines gathered together in Washington to give advice to the military authorities. Through its committees the council prepared a form of contract upon what came to be known as the "cost-plus with sliding scale and fixed maximum fee" plan, which limited the cantonment contractor in each case to a maximum fee of not more than $250,000, the Army itself retaining control of the cost of materials and the wages paid to labor.

Since the cantonments cost anywhere from $8,000,000 to more than $12,000,000 each, the average fee to the contractor was slightly less than 2½ per cent, out of which the contractor had to pay overhead expenses, such as his main office expenses and the like; so it will be seen that the United States drove a close bargain with its cantonment builders in spite of the breathless haste to get the work done.

It was not until the 1st of June that the war authorities decided upon wood construction for the 16 National Army cantonments and canvas tentage for the 16 camps of the National Guard. According to the original plan, so far as could be foreseen, the cantonments were to be permanent camps to receive fresh contingents of selectives as long as the war should last, whereas after receiving its training the National Guard would go to France and leave its American camps deserted. The wood construction was much more expensive than tentage—amounting to $215 per man of the first draft, as it proved—but it was permanent; once installed it made no further demands for materials, and in convenience and comfort, especially in winter, it was far superior to tentage.

Meanwhile the Cantonment Division had designed a model barrack building, 43 feet wide and 140 long, to house 150 men, or one company, as the company was in the spring of 1917. Here, in the adoption of this model and general camp plans, there might easily have occurred in Washington a fatal indecision. Both the British and the French armies had found by experience that a company of 250 men was a more convenient size for trench warfare than a smaller one. There was some question whether the American Army would be guided by this experience. Gen. Pershing, was to decide this matter, but he did not reach Europe until June 15. A weak executive control in Washington might have justified itself in waiting for this decision before starting in at full speed to build the cantonments. Those in charge of the program took upon themselves the responsibility of building the 150-man barrack, trusting to their own ability to adjust the buildings later to changed conditions. As a matter of fact, when the company unit was enlarged to 250 men, it was readily possible to house two companies in three barracks, leaving space in two of them for the kitchen and mess room. Still later the Construction Division built smaller barracks for 66 men each, providing four such barracks to the company.

Before a single site was selected the experts in Washington had designed the buildings and mapped out the future cities. America, leaving behind her the decorative atrocities of the old Victorian days, had been seeking beauty; and this yearning had produced a new profession, that of town planning. Town planners in the Construction Division grouped the 1,500 buildings required by each cantonment into two typical arrangements, known as the straight-line and the U-shape layouts. Later at each cantonment there was a town planner who adapted one or the other of these plans to his local camp topography.

The selection of camp sites was in the hands of boards of officers designated by the commanding officers of the six Army departments. Early in May these boards set forth on their quest. Then ensued a lively bidding on the part of American cities to secure cantonments for their own neighborhoods. The Government took the utmost advantage of such inducements as were offered. The city of Tacoma, Wash., sold its municipal bonds to the amount of $2,000,000

Start at 7 a. m.

1 p. m.

10.30 a. m.

5 p. m.

A DAY'S WORK IN BARRACK CONSTRUCTION.

CAMP GRANT AT ROCKFORD, ILL.

This photograph, taken from kites at an elevation of 1,000 feet, shows a typical cantonment, of which sixteen were built to train the National Army in 1917.

and with the proceeds bought 61,000 acres at American Lake and presented the land to the Army. This tract became the site of Camp Lewis, most beautiful of all the cantonments.

The site for Camp Upton, at Yaphank, Long Island, 15,198 acres, was provided for the Government at an annual rental of $1 per acre. San Diego, Calif., gave 8,000 acres at Linda Vista to the Government rent free for five years. This became the location of Camp Kearney for the National Guard. Camp Fremont, another National Guard training center, was pitched on 7,203 acres of ground donated rent free for one year by the city of San Francisco. Louisville gave the site of Camp Zachary Taylor for the National Army rent free for two years. And there were numerous other similar inducements.

In all the National Army cantonments occupied 167,741 acres which the Government obtained at an average annual rental of $3.93 per acre after the second year of occupancy. The National Guard camps covered 78,639 acres, at a rental of $112,042 the second year of occupancy and thereafter.

The clearing of the sites was no mean part of the cantonment job. The site of Camp Upton at Yaphank, Long Island, proved to be covered with underbrush, and when this was cleared off it was discovered that the remains of an old forest were still there, and stumps were thickly scattered over the entire site. These had to be blasted or pulled out before any building operations could proceed. The sites in character ranged from the sandy loam of Camp Devens, in Massachusetts, to the red clay of Virginia and South Carolina; from the farm lands of Michigan to the prairie on which was built Camp Travis, Tex. Some were flat; some rolling; all were different in shape and extent; and the layout of the camp and the arrangement of the buildings in each case had to be adapted to the local conditions by the constructing quartermaster on the job.

To give a picture of a typical cantonment, let us take Camp Grant, at Rockford, Ill., as an illustration. It cost approximately $11,000,000; it could accommodate 45,000 men and 12,000 horses; its buildings numbered 1,600. Water was supplied from six wells drilled 175 feet deep. There were 38 miles of water main, while the reservoir tanks could hold 550,000 gallons. Its electric lighting system entailed the use of 1,450 miles of copper wire, 1,200 poles, and 35,000 incandescent lamps.

During the construction period 50 carloads of building material were unloaded at Camp Grant every day, and an average of 500,000 board feet of lumber was put up every day over a period of weeks. In the Camp Grant schedule we find such items as 50,000,000 feet of lumber, 700 tons of nails, 4,000,000 square feet of roofing, and 3,000,000 square feet of wall board.

Only a builder of secure financial standing could handle such a construction contract. Frequently his pay roll and current bills for supplies amounted to half a million dollars in a week. Let the Government delay a few days in its payments to him and he might find himself obliged to raise a million dollars in cash on the instant to meet his immediate obligations.

To avoid such embarrassments the Construction Division adopted the policy of paying its bills the day and sometimes the hour they were incurred. At each cantonment job it stationed a disbursing officer with a checkbook. This officer reported by telegraph each night, and the next morning there was deposited to his credit in the Treasury a sum of money sufficient for his immediate needs. On numerous occasions this officer paid for materials the instant they were unloaded from the cars and checked. The Government maintained an auditing organization at each job. This organization checked and inspected all material as it was received, comparing the delivery in each case with the original order, and counted the workmen at least twice a day.

An intense rivalry sprang up in the construction of the 16 cantonments. Sixteen teams, with an average of 10,000 men on each team, began racing for the goal, which was to be 80 per cent completion by September 5, the date when the first contingents of selective soldiers were to be received by the cantonments. It was more exciting than any campaign in any professional baseball league, because the time was shorter and the stake vastly greater. The Cantonment Division kept alive this spirit of rivalry by posting each day in each cantonment the figures showing the rates of construction at all of them. The team, from the superintendents down to the humblest unskilled laborers, discussed these ratings as fans talk about the baseball averages.

The race was a close one, being won by Camp Taylor at Louisville, which was 79.4 per cent complete on the day the contest ended, lacking only six-tenths of 1 per cent of coming up to the maximum 80 per cent of completion regarded as possible in the time available. But other construction gangs were pressing that at Camp Taylor closely: Camp Travis, with 78.6 per cent of completion at the end of the contest; Camp Lee, 78.5 complete; Camp Devens, 74; Camp Lewis, 72; and Camp Sherman, 70. The Camp Lewis percentage was taken on August 31.

All construction work, including numerous additions not contemplated in the original plans, was virtually complete by November 30. These additional structures included cantonment base hospitals, on which the Government spent $10,000,000 for the National Army and $7,500,000 at the National Guard camps. With one or two exceptions these hospitals each had facilities for 1,000 patients at once, being the largest in the United States, at that time.

At several of the cantonments the water problem was simplified by the near presence of water mains of the systems of adjacent cities. At the other camps, however, it was necessary to construct independent water systems sufficient to furnish 55 gallons of water daily to each of 45,000 men. This is nearly twice as much water as was then being furnished to the average European army camp, and it meant in each case a system of centrifugal pumps and gravity tanks with a capacity of 2,250,000 gallons daily.

The local water supply was secured either from running streams or from dug wells. If the purity of the water was in doubt it was sterilized by the chlorine process and sometimes filtered in addition. The purity of the water and the care exercised in screening kitchens, mess halls, and later the dormitories themselves, from flies is seen in the hospital record for the first year of the war. Of all the soldiers who sought hospital treatment for sickness during that first year, only one patient in 5,000 was suffering from a water-borne disease.

In each cantonment there were about 1,500 wooden buildings, presenting a constant fire menace. As a protection from such disasters each cantonment organized its own fire department with modern motor equipment stationed in three or more fire houses. The men of the fire companies were usually those who had had previous training as members of city fire departments. There was not a single serious fire at the cantonments during the war period.

Besides keeping soldiers well and clean, the camp facilities provided them with opportunity for moral and healthy amusement. Various organizations combined to supply the camps with amusement facilities. There were the library buildings, the Red Cross buildings and halls, the Y. M. C. A. buildings, the Knights of Columbus buildings, the Salvation Army buildings, Y. W. C. A. buildings, and the Jewish Welfare Board buildings.

Although the American soldier bought liberally of Liberty bonds, took out War Risk insurance, and sent to his family the greater portion of his monthly pay, still he had a small amount of money to spend for little luxuries or necessities. These included small supplies such as candy and fruits, and they were on sale at the usual post exchange or company store. This was a small building, usually with a broad covered porch or shelter around three sides, so that the men in bad weather could be dry as they waited in line for their turn at the windows. The Y. M. C. A., K. of C., Red Cross, Y. W. C. A., Jewish Welfare Board, and Salvation Army buildings offered reading and writing rooms and general gathering places; yet these were insufficient for a total camp population averaging 40,000. Hence the Commission on Training Camp Activities built through the agency of the Construction Division a Liberty Theater at each camp.

The Liberty Theaters were of temporary construction, but in size compared favorably with the largest theaters in our modern cities. To provide these amusement facilities the Construction Division put up approximately 5,000,000 board feet of lumber, 9,000,000 square feet of wall board, and 40,000 square feet of roofing.

The average large city laundry was insufficient in capacity to handle the laundry work of an average of 6,000 people per day, which was the requirement to keep some 40,000 men in clean clothing. Consequently the camps and cantonments were provided with their own laundries built by the Construction Division. This put such a demand upon the manufacturers of laundry machinery that it created a shortage, and later there was a shortage of soap and powder. The 30 laundries built used up 13,000,000 board feet of lumber and 300,000 square feet of wall board.

The householder may throw his old shoes into the trash box and sell his old suit to the ragman, but the Army threw nothing away. Consequently the Construction Division was called upon to build reclamation plants at many stations. Usually one large plant was built at a center convenient to several camps, and to this center were sent the worn-out uniforms, shoes, leggins, and all other equipment.

Every camp of considerable size in the United States was provided with model bakeries. The total capacity of all the baking equipment installed by the Construction Division would turn out 1,000 tons of bread per day. This is a total of 2,000,000 loaves of 1 pound each. Each camp bakery oven would take care of 4,500 men per day in two 8-hour shifts, or it could bake 5,000 loaves in 24 hours.

There were required at all the camps and cantonments numerous storehouses for materials to be used immediately by the troops. The Construction Division built 789 of these small storehouses at the National Army cantonments alone.

The question of cold-storage facilities for the camps offered a knotty problem. Certain camps generally relied upon cold-storage space obtained near by, or else upon refrigerator cars iced in the vicinity, but in the other camps it was necessary to build special refrigeration plants. These had an ice-making capacity ranging from 6 to 35 tons daily. The ice consumption of the American soldiers in the United States proved to be 2¾ pounds per man per day.

The kitchens in the camps and hospitals would be paradise to any woman who had drudged with old-fashioned methods and equipment in cooking. As far as possible the Army's housekeeping was done by machinery. The bread slicer in common use would cut 200 slices of bread per minute and stack the slices automatically, the loaves feeding automatically into the slicer. The meat choppers would cut up 20 pounds of meat or vegetables in five minutes, and the electrically operated potato peeler would peel 40 pounds of potatoes in three minutes. The meat slicers would cut meat at the rate of 40 slices to the minute. Vegetables were cooked and meats roasted by high-pressure steam. The vegetable cooker could turn out 35 gallons of cooked product in 15 minutes. The dish-washing machines could wash, dry, and sterilize 10,000 dishes per hour. At the hospitals the food was taken from the central kitchen to the outlying wards in mobile fireless cookers, designed to keep the food hot until served.

To prepare food for 45,000 men, 350 kitchens were required by each cantonment. The National Army in training used 9,000 hotel ranges.

In most of the cantonments, particularly those in the South, the heating of quarters in winter was accomplished by means of room heaters and cannon stoves. The constructors installed 75,000 of these. The officers' quarters everywhere and four entire cantonments in the North—Devens in Massachusetts, Grant in Illinois, Custer in Michigan, and Dodge in Iowa—were heated by steam either from central heating plants or by means of ordinary boilers such as are used in residences. The total surface of the steam radiators installed would make five gigantic stoves 300 feet square and as high as the Woolworth Building in New York.

Besides the camps and cantonments used by the line troops, the Construction Division also built various special camps required by the mobilization, training, and transportation of the Army. These included the quartermaster training camp at Jacksonville, Fla., accommodating approximately 35,000 men; and the camps for the Engineering Corps, camps for heavy and light artillery training schools, and for other special units.

Camp Joseph F. Johnston at Blackpoint, the quartermaster camp near Jacksonville, is a good example of one of these special training camps. This consisted of quarters for 150 officers, 32 barracks to house 200 men each, together with barracks for wagon companies, pack companies, truck companies, and a bakery company, as well as stables for 1,200 animals and 50 riding horses, together with storeroom buildings and truck and auto garages.

Camp Holabird, near Baltimore, used for teaching men to repair and crate motor trucks and vehicles, had accommodations for about 7,500 men. Another special camp was Camp Humphreys, for the training of men in the Corps of Engineers, located a few miles down the Potomac River from Washington. This camp could accommodate 33,000 men in 1,350 buildings located on a camp area of 2,500 acres. Every foot of this site had to be cleared of timber and underbrush

100287°—19——35

during one of the severest winters of recent years. All material had to be hauled in trucks over fearful roads pending the construction of a 5-mile spur of railroad track, yet the job was completed approximately on time.

Other special camps included Camp Bragg for training Field Artillery, located at Fayetteville, N. C., with quarters for 11,000 men; Camp Knox at Stithon, Ky., for 30,000 men, having an area of nearly 60,000 acres for training troops in the use of Field Artillery; and Camp Franklin, located on part of the Camp Meade cantonment reservation, a special camp for Signal Corps instruction, with accommodations for 11,000 men.

Then there was the Coast Artillery training cantonment, Camp Eustis at Lee Hall, Va., which had accommodations for 17,000 men; Camp Meigs at Washington, D. C., a quartermaster camp, providing accommodations for 4,000 men; and Camp Benning, at Columbus, Ga., an infantry school of arms, to accommodate 5,040 men, on a camd area of 98,000 acres. At Camp Raritan, at Raritan River, N. J., the Ordnance Department established a training school for 6,250 men.

Camps and cantonments built by Construction Division.

Name.	Location.	Maximum capacity.	Approximate cost.	Contractors.
Camp Beauregard	Alexandria, La	29,121	$4,300,000	Stewart McGehee, Hudson Construction Co., J. W. Snyder.
Camp Bowie	Fort Worth, Tex	41,879	3,400,000	J. W. Thompson, H. G. Bush.
Camp Bragg	Fayetteville, N. C	11,831	11,000,000	James Stewart Co.
Camp Cody	Deming, N. Mex	44,959	3,800,000	J. W. Thompson Co., H. G. Bush.
Camp Custer	Battle Creek, Mich	49,014	13,000,000	Porter Bros., R. G. Phelps, W. E. Wood Co.
Camp Devens	Ayer, Mass	36,832	11,800,000	Fred T. Ley Co., Coleman Bros.
Camp Dix	Wrightstown, N. J	42,806	12,300,000	Irwin & Leighton Co., J. S. Rogers Co., J. W. Ferguson & Co.
Camp Dodge	Des Moines, Iowa	49,229	10,800,000	Charles Weitz Sons.
Camp Doniphan	Lawton, Okla	46,183	2,706,000	Selden-Breck Construction Co., Trope & Carney.
Camp Abraham Eustis	Lee Hall, Va	16,759	11,700,000	Winston & Co.
Camp Fremont	Palo Alto, Calif	30,000	2,556,000	Lindgren Co., E. A. Hettinger.
Camp Funston	Fort Riley, Kans	42,806	10,500,000	George A. Fuller Construction Co., Henry Bennet & Son, Gray Construction Co.
Camp Gordon	Atlanta, Ga	46,612	11,100,000	Arthur Tufts Co., Mackle Construction Co., Southern-Ferro Construction Co.
Camp Meade	Odenton, Md	52,575	16,200,000	Claiborne Johnson Co., Smith, Houser & McIsaacs.
Camp Meigs	Washington, D. C	3,774	655,000	Philip F. Gormley, Frank L. Wagner.
Camp Merritt	Dumont, N. J	39,079	14,500,000	McArthur Bros. Co., W. H. Fissell & Co.
Camp Mills	Garden City, L. I	25,000	13,000,000	Clough-Bourne Co.
Camp Hill	Newport News, Va	5,852	16,125,000	Westinghouse-Church-Kerr Co., Hampton Roads Engineering Construction Co., Boyle-Robertson Construction Co.
Camp Stuart	do	24,234		

Camps and cantonments built by Construction Division—Continued.

Name.	Location.	Maximum capacity.	Approximate cost.	Contractors.
Camp Oglethorpe, Fort Greenleaf Forrest		24,457	$5,600,000	Park-Grimes Co.
Camp Pike	Little Rock, Ark	55,010	12,700,000	James Stewart & Co. (Inc.), Stewart McGehee Co.
Camp Polk	Raleigh, N. C	(1)		Holliday-Krouse Co.
Camp Sevier	Greenville, S. C	41,693	6,508,000	Gallivan Building Co.
Camp Shelby	Hattiesburg, Miss	36,010	5,400,000	T. S. Moudy Co., Richard McCarthy Co.
Camp Sheridan	Montgomery, Ala	41,593	3,500,000	Algernon Blair, Paschen Bros.
Camp Sherman	Chillicotho, Ohio	49,112	12,900,000	Thomas A. Bently & Sons, D. W. McGrath & Sons.
Camp Taylor	Louisville, Ky	45,424	8,000,000	Mason & Hanger, Alfred Struck Co.
Camp Travis	San Antonio, Tex	42,809	8,200,000	Stone & Webster, McKenzie Construction Co.
Camp Upton	Yaphank, L. I	43,567	13,500,000	Thompson-Starrett Co., Mark C. Tredennick Co., C. H. & R. C. Peckworth (Inc.).
Camp Grant	Rockford, Ill	62,675	14,400,000	Bates & Rogers Construction Co., Ross T. Beckstrom Co., Henry Erickson Co.
Camp Green	Charlotte, N. C	48,305	4,300,000	Consolidated Engineering Co., J. A. Jones Co.
Camp Hancock	Augusta, Ga	45,099	6,000,000	T. O. Brown Co., William Crawford.
Camp Humphreys	Belvoir, Va	32,434	12,745,000	Phillip F. Gormley Co.
Camp Jackson	Columbia, S. C	44,009	10,000,000	Hardaway Construction Co., Columbia Lumber & Manufacturing Co., H. B. Hann.
Camp Jackson, No	do	(1)		Hardaway Construction Co.
Camp Johnston	Jacksonville, Fla	18,265		T. A. Bentley & Sons., J. Y. Wilson.
Camp Kearney	San Diego, Calif	32,066	5,838,000	W. E. Hampton Co., John Roberts Co.
Camp Knox	Stithton, Ky	27,805	18,733,184	John W. Griffith & Sons.
Camp Las Casas	San Juan, Porto Rico	13,265	2,500,000	Purdy & Henderson Co.
Camp Lee	Petersburg, Va	60,335	16,500,000	Rhinehart & Dennis (Inc.), Harrison Construction Co., John T. Wilson & Co. (Inc.).
Camp Lewis	American Lake (Tacoma), Wash.	46,232	8,400,000	Hurley Mason Co., The Construction Co.
Camp Logan	Houston, Tex	44,899	3,300,000	American Construction Co., Horton & Horton.
Camp McArthur	Waco, Tex	45,074	4,000,000	Fred A. Jones Construction Co., Blome & Sinek Co., Edgar H. Bruyere.
Camp McClellan	Anniston, Ala	57,746	9,800,000	John O. Chisholm & Co., Labarre & Erwin, A. W. Stoolman.
Camp Wadsworth	Spartanburg, S. C	56,249	4,000,000	Fiske Carter Co.
Camp Wheeler	Macon, Ga	43,011	3,200,000	W. Z. Williams.

1 Abandoned.

CHAPTER II.

MISCELLANEOUS CONSTRUCTION.

Great as was the job of building the army camps and cantonments, it was nevertheless only a part of the work which fell to the Construction Division, and much the smaller part at that.

On November 11, 1918, the Construction Division was conducting 535 building operations in 442 localities in the United States. These involved an aggregate expenditure of more than $1,000,000,000. Including the various camps and cantonments, these activities were being conducted or had been conducted in every State of the Union but one. An average of more than 200,000 workmen, principally of the building trades, had been engaged continuously for months.

In the executive administration of the work the organization required 1,487 officers and 12,355 civilian Government employees, of whom 2,555 were located at the offices of the division in Washington. Merely for the maintenance and the operation of the various completed projects a force of 16,359 enlisted men was required. In a little more than a year the organization had grown from a handful of clerks and executives to one of this size. The brigadier general who headed the Construction Division had been a Captain when war was declared.

In this period the organization had housed a population equal to that of the city of Philadelphia in 40 large camps, each in number of inhabitants comparing in size to such cities as Racine, Wis., Cedar Rapids, Iowa, or Wheeling, W. Va. It had constructed storage depots and warehouses that would cover 890 acres. It had built hospitals with beds for 128,378 patients. It had purchased and nailed up 2,647,605,426 board feet of lumber, enough to stretch around the Equator twenty times in boards 12 inches wide and 1 inch thick. Loaded on freight cars to their capacity this lumber would require a train reaching from Washington, D. C., to Kansas City. It had used enough brick to build an 18-foot road from Kansas City to Chicago. It had constructed 645 miles of railroads and made 1,081 miles of wagon roads, mostly of concrete. These are only a few of the high points in this building record.

There are few undertakings of mankind in all history which can be compared with this enterprise. The price paid for the Panama Canal and the Canal of Suez, the cost of damming the Nile and tunneling the Alps, and the money spent on building the Government railway into the heart of Alaska might be lumped together and still the aggregate would not equal the cost of providing the buildings, exclusive of those of the training camps, which the American Army had to have in the United States after it went to war.

We can gain a picture of the size of this construction by considering the building records of the United States. In this country there are about 150 cities large enough and ambitious enough to keep annual building statistics as the indices of their prosperity. In these cities, whose populations range in size from that of New York down to those of communities of 20,000 or 25,000 inhabitants, dwell nearly a quarter of all the Americans. They are metropolitans, the people who demand most of the builder for their comfort and luxury. Yet in no one year had the building construction in these 150 largest American cities combined approached in amount within $250,000,000 of the cost of our military construction undertaken during the war.

The Government became not only the greatest of customers for the building industry but almost the sole customer. This whole great industry, one of the largest in the country, which had been busy in its interminable task of providing the mansions of peace, was suddenly converted under military direction into a machine for building a titantic war plant. Before the Nation could mobilize its material resources or train its human ones for war it must have buildings—headquarters for its executives, barracks for its men, structures for its various arsenals for the manufacture of explosives and chemicals, warehouses for the storage of reserves of material, terminals for the transfer of overseas shipments, schools, laboratories, proving grounds for testing its weapons, hospitals, embarkation depots, and a vast number of structures for less conspicuous activities.

It was the work of the Construction Division to provide these facilities. Exclusive of the cantonments themselves, this work fell into projects ranging in size from small building groups costing a few thousand dollars to enormous powder plants, huge terminal docks, vast warehouses and other great undertakings costing $10,000,000, $16,000,000, $25,000,000, $40,000,000, and as high as $70,000,000 for a single project.

ORDANCE CONSTRUCTION.

Perhaps the most striking of these undertakings were the various construction jobs called for by the ordnance program. There were more than 60 of these, and they ranged in cost from $100,000 up to $70,000,000.

One of the larger of these projects was that of the Aberdeen proving grounds on Chesapeake Bay, not far distant from Baltimore. This reservation, with its area of 35,000 acres and its magnificient testing and observation ranges, 75 miles in length, over the waters of the bay, will undoubtedly be retained permanently by the Government.

As the plant exists to-day it has a capacity of testing 5,000 shell between daylight and dark.

At Aberdeen the Construction Division built barracks to house 8,000 men, quarters for 230 officers, and all the accessories of convenience and amusement which a community of that size should have.

Guns came to the proving grounds unassembled, so that it was necessary to build an assembly plant. This building is 165 feet wide and 500 feet long and cost $1,000,000. As an adjunct to the assembly plant there is a machine shop which is one of the largest in the United States.

Mention should be made of the 25 miles of standard railway trackage which the Construction Division put down at Aberdeen. This was exclusive of the spur tracks for the heavy guns mounted on railway carriages. These tracks approached the firing range on apparently outlandish curves and at every variety of angle. The guns were fired at these various angles to determine if the recoil would push the carriages from the track or would spread the rails.

The development of aerial bombing and the necessity for testing our own aerial bombs required the construction at Aberdeen of hangars and quarters for an aviation squadron.

In addition to these facilities, the project involved the construction of powder magazines, shell-loading plants, and warehouses. There were built 15 miles of concrete roads and 30 miles of roads of other types. Garage accommodations were provided for 100 trucks and automobiles. The firing ranges required observation towers of various sorts. The observation dugouts had to be of special strength, because certain of the tests at Aberdeen involved the actual bursting of gun barrels, making necessary specially heavy protection for the observers.

Aberdeen is equipped with a complete waterworks system and with a hospital for 250 patients. An interesting laboratory constructed on the grounds is that in which the so-called "dud" shell, or those which fail to explode, are analyzed for their defects. The Aberdeen project was started in December, 1917, and first tests were made within a month. The entire project cost over $30,000,000.

One section of the Aberdeen reservation, about 4,000 acres of it, was set apart for the uses of the gas-warfare organization of the Army and was later known as the Edgewood Arsenal. The progress at Edgewood is indicative of the manner in which chemical warfare increased in importance during our period of belligerency. It was originally estimated that $250,000 would provide a plant at Edgewood sufficient for our chemical-warfare needs. The actual cost of the Edgewood Arsenal at the date of the armistice, so great had been the expansion of chemical warfare, was about $43,000,000. At that time there had been constructed or were in process of construction filling plants that could turn out 120,000 loaded gas shell per day. The equipment at Edgewood includes a cantonment for 10,000 men, some of it of permanent construction. There were built 10 miles of macadam road at Edgewood and 15 miles of railway, in addition to large warehouses and a dock where loaded shell could be freighted upon lighters to deep water.

Another project made necessary by the expansion of chemical warfare was the gas proving grounds at Lakehurst, N. J., the entire project costing $1,500,000. The site of 5,000 acres provided space for two target ranges, each 4 miles long. Extensive laboratories were built at Lakehurst, and the proving ground was operated by a force of 1,500 men. In addition to this there was located at Lakehurst a camp for 3,400 troops in training. All buildings for these facilities were provided by the Construction Division.

In addition to Aberdeen and Lakehurst the Construction Division built a proving ground at Clear Springs, Md., used for testing out 37-millimeter guns; another such institution at Port Clinton, Ohio, for testing 155-millimeter and 240-millimeter howitzers; and others at Scituate, Mass., and Savanna, Ill. The combined cost of these last four projects was $6,507,520.

One important undertaking of the Construction Division was that of providing warehouse depots for ordnance materials. These supplies differ from ordinary Army supplies in the important particular that they must be treated gently and handled with care. A quartermaster storehouse can be of emergency construction type, that is, more or less built of wood, but an ordnance storehouse, since it usually contains high explosives, must be strictly fireproof. In undertaking the creation in record time of a number of ordnance depots the Construction Division faced not only the problem of the type of building required but also the location of these buildings. It was necessary to locate them at deep water in order to avoid frequent handling of the high explosives, yet no depot could be situated in any thickly settled center because of the danger to the civilian population. At most deep-water points on the Atlantic coast which had railway connections the available sites were already occupied. The result was that the ordnance depots had to be built on marshes and meadows, on land which for construction purposes had always been regarded as impossible. Yet they had to be completed in as much of a rush as any buildings which the Army demanded.

There are now five of these great ordnance depots on the Atlantic coast built by the Construction Division: at Metuchen, N. J.; Curtis Bay, Md.; Pig Point, Va.; Charleston, S. C.; and Pedricktown, Del. The largest of these is the one at Metuchen, known as the Raritan Arsenal. The Raritan site contains about 2,200 acres of salt marsh. High tides used to submerge the whole area almost completely. Before any building could be started the Construction Division had to build a dike 9 miles long around the whole reservation. The entire project was perched on piles, and these piles, by the tens of thousands, were driven into the frozen ground during the severe winter of 1917–18. Labor was hard to get and hard to keep. After the laborers' quarters had been built and a few powder magazines had been erected, it became almost impossible to keep men on the job because of the danger of working in a powder arsenal.

Most of the Raritan buildings are of terra-cotta construction. There are 85 completed magazines, each 51 feet wide and 218 feet long, for the storage of shell, black powder, and miscellaneous items, this number not including 12 magazines of sheet-metal construction, each 26 by 42 feet, for the storage of high explosives. When the armistice was signed the Construction Division was building 60 similar magazines intended for the storage of smokeless powder.

At Raritan was also located a school of instruction for ordnance troops, with a cantonment to accommodate 10,000 men. A 150-bed hospital was part of the equipment, as was also an assembly shop and a motor-instruction school.

Along the river a dock was built 2,000 feet long. On the dock were constructed several huge warehouses for the storage of material. Fifty miles of railway were constructed. The project on the armistice date was probably the best equipped ordnance depot in the world. It cost about $14,000,000.

The next largest ordnance depot is that at Curtis Bay near Baltimore. It is half the size of the Raritan project and cost about $7,000,000.

The Pig Point ordnance depot is located at Hampton Roads, about 12 miles from Portsmouth, Va. In order to obtain berthing facilities for trans-Atlantic ships it was necessary to build a dock more than a mile long out to deep water. The dock is said to be the longest wharf in the United States south of Philadelphia. The Pig Point job cost about $3,500,000.

The Charleston Arsenal cost $5,000,000; while $7,000,000 was the amount provided for the arsenal at Pedricktown. The Pedricktown job, however, was started late, and not over $2,000,000 had been spent when the armistice was signed.

In addition to these five terminal depots the Construction Division provided two other ordnance warehouses for the storage of miscellaneous supplies—one at Middletown, Pa., costing $1,250,000, and the other at Augusta, Ga., costing $250,000.

The description of the powder bag-loading plants, which were built by the Construction Division, is contained in the chapter of this report relating to the production of powder and other explosives. There were three of these plants, one located at Woodbury, N. J., another at Tullytown, Pa., and the third on the historic battle ground at Seven Pines, Va. Since these plants were perforce located at isolated places it was necessary in each case to provide housing accommodations for the workers, many of whom were women. The Construction Division built the houses at Tullytown and Woodbury while those at Seven Pines were provided by the United States Industrial Housing Corporation.

These bag-loading plants cost from $4,500,000 to $6,000,000 apiece, and they were erected in a remarkably short time. Work was started on the Woodbury project on March 19, 1918, and the plant was ready for operation on May 28, although the plant did not actually start operating until June 15. The spade was first struck into the ground at Tullytown on March 6, 1918, and on July 17 the 250 buildings of the project were ready. The work at Seven Pines began April 24, and the plant was ready for operation on August 24, 1918.

AVIATION SCHOOLS AND TESTING FIELDS.

The Signal Corps, needing a special type of construction, undertook the work itself at the start of war, but by October, 1917, the Construction Division had proved itself so efficient that all Army building work in the United States, including that of the Signal Corps, was turned over to it. The work for the Signal Corps entailed the construction of the necessary buildings for flying fields, testing fields, aerial photography, and gunnery schools, balloon observation schools, repair and testing shops, and tremendous storage depots that had to be of special fireproof construction because of the inflammable nature of the oil and other materials used by the Signal Corps.

At the aviation fields a special type of portable hangar built of steel, 65 by 140 feet, was adopted. For the big bombing planes larger hangars were required but of the same type of construction. At each aviation field were barracks for a large number of men, together with water and sanitary conveniences. There were 31 of these fields located principally in the West or Southwest.

In addition to these there were four testing fields for the aircraft service, located in the eastern half of the United States where the flying machines and the engines were being produced. One of these was at Dayton, another at Buffalo, a third at Detroit, and a fourth at Elizabeth, N. J. The aerial gunnery school at Miami, Fla., was one of the largest of the aviation-construction projects. This plant included buildings, target ranges, steel hangars, photographic laboratories and other equipment—all built at a cost of $1,500,000.

The balloon school at Lee Hall, Va., cost $1,000,000, and that at Arcadia, Calif., $500,000. At each of these schools there were barracks for the men, quarters for the laborers, and experimental buildings, not to speak of the huge sheds, 200 or more feet in length, in which the balloons were housed.

QUARTERMASTER BASES AND WAREHOUSES.

Construction for the Quartermaster Department involved the building of warehouses on a scale hitherto unknown in the United States. The warehouse plan was carefully worked out as part of the strategy of conducting the war, the Council of National Defense making the first investigations of the proper locations of supply depots, and these early findings being later amended by the General Staff. Several important considerations determined the locations of these depots and warehouses. In the first place we would require great storage and shipping facilities at tidewater; yet, if these were all to be located in one spot or in one general region, there was a possibility that a submarine blockade off the Atlantic coast could stop the shipment of supplies to the American Expeditionary Forces. Thus the first project was to locate the great supply bases at New York, Boston, Philadelphia, and Norfolk. But it was evident that a relatively small number of enemy submarines operating in a comparatively restricted area could block shipment from these four ports. Therefore Charleston, S. C., and New Orleans were added to the supply-base project.

There was another thing to be taken into consideration, namely, the providing of sufficient warehouse space, so that, if there should be any blockade of ocean transportation, the manufacture of supplies could continue at its war rate and still find places to which to ship its important products. Yet if the storage were all provided at the tidewater bases, there would be danger of railroad congestion at the ports. Consequently, as auxiliary to the terminal warehouses, there was provided a system of enormous warehouses built in the interior of the United States.

The system eventually worked out to give seven expeditionary supply bases located, six of them, at the cities just named, in addition to one that had been built at Port Newark, N. J., during the winter of 1917, and nine interior depots located respectively at Baltimore, Chicago, Columbus, Ohio, Jeffersonville, Ind., New Cumberland, Pa., Philadelphia, Pittsburgh, Schenectady, N. Y., and St. Louis. These latter were central in various producing districts.

The terminal projects alone involved construction on a scale that was without precedent. The interior depots and the huge terminal bases provided 690 acres of storage space, all inclosed in reinforced concrete of the most modern type. They were all built in a little over 12 months. Into them went construction enough to build a concrete building 70 feet wide from New York to Philadelphia and a wharf nearly 8 miles long, with berthing accommodations for 65 ships at once. The facilities included 650 miles of railroad and 1,000 miles of concrete roadway.

NEW ORLEANS ARMY SUPPLY BASE.

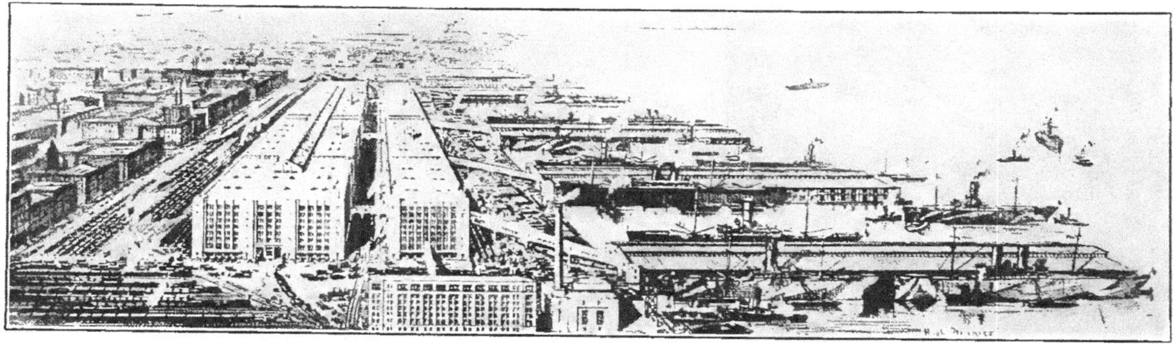

BROOKLYN ARMY SUPPLY BASE.

BROOKLYN ARMY SUPPLY BASE.

BOSTON ARMY SUPPLY BASE.

The Army supply bases at Brooklyn and at Boston are examples of the immensity of the expeditionary depots built along the Atlantic seacoast. The base at Brooklyn has approximately 4,000,000 square feet of storage space in its two huge 8-story reinforced concrete warehouses. One of these warehouses is 980 feet long by 200 feet wide and the other is 980 feet long by 300 feet wide. In addition to these, the installation at the base consists of three double-deck piers, each 150 feet wide and 1,300 feet long, and one open pier 60 feet wide and 1,300 feet long. In its railroad yards there is storage space for 1,300 cars at one time. The capacity of the base is 700,000 tons of supplies, or the equivalent of about 100 shiploads. Twelve ships of 8,000 tons, dead weight each, can be loaded at one time, and the loading of these vessels can be completed within 24 hours, so vast and complete are the facilities at this project.

Construction started at the site on May 15, 1918. More than 7,000 workmen were engaged on this job at one time, and the entire project was to be completed before July 1, 1919, while it was to be in partial operation by January 1. When the armistice was signed, 4,387,360 square feet of floor space had been completed and 187,173 cubic yards of concrete had been poured.

The base at Boston is 8 stories high, built of concrete, and gives 2,750,000 feet of storage room. Its wharves are 4,000 feet long. The work on it was started May 14, 1918, and ended October 3. In that time 200,000 cubic yards of concrete had been poured, 22,000 tons of reinforcing and structural steel put into position, 3,000,000 brick laid, 30,000 piles driven, 1,500,000 cubic yards dredged, and 30 miles of track laid. In all, 7,000 carloads of material were handled in this building.

The Norfolk base is located at Bush Bluff, 4 miles from the city. The chief feature of this project is a group of eight 1-story concrete buildings providing 2,000,000 square feet of storage space. The pier sheds are built of concrete blown by compressed air upon steel lath. The docks total a mile in length. The base can handle 600 cars of supplies in a day. In addition to the storage and shipping buildings themselves, the Construction Division provided quarters for a regiment of stevedores and a battalion of guards. A 120-bed hospital was erected at the project. The wharfage front was made of concrete piles weighing 12 tons apiece, and 217 acres of land were made by dredging outside of the piles and filling in behind them. The work was started in May and was nearly finished when the armistice was signed.

The Norfolk and Hampton Roads district has the distinction of being the center of more war construction than was conducted at any other point in the United States. There were located here the Navy arsenal, the Navy yard, the Navy training station, and the great Norfolk naval base. The largest construction project of all at Norfolk was the quartermaster terminal which the Construction Division built there. But in addition to these there was the Pig Point ordnance depot, described above, Camp Stuart and Camp Hill, both embarkation camps, the Artillery school at Fortress Monroe, Camp Eustis, and the Langley aviation field of the Army.

With so many construction undertakings going on at once, the labor problem proved to be an early embarrassment at the Norfolk quartermaster terminal job. However, good quarters and good food for the construction gangs at the terminal largely solved this problem. At one time in the development on the shores of this part of Chesapeake Bay, the street car and electric lighting system of Norfolk broke down under the strain. The Government thereupon took the power house and operated it thereafter for the duration of the emergency.

The interior storage depots of the Quartermaster Department provide 12,000,000 square feet of storage. They are all of permanent construction. They range in size from the one at Pittsburgh, with 184,000 square feet, to that at Schenectady with 2,500,000 square feet.

The depot at Chicago was built by the Central Manufacturing District as contractor on a site sold by its trustees to the Government. This structure, costing $3,000,000 and giving 29 acres of storage, was built completely in the period between March 4 and September 15, 1918.

MOTOR TRANSPORT CONSTRUCTION.

There was extensive construction for the Motor Transport Corps. Few civilians perhaps realize the size attained by this branch of the Army, with its more than 3,000 officers and 100,000 men in the United States and abroad. The Construction Division designed a standardized repair shop to be used in this country or to be transported overseas as desired. There were three centers of repair, shipment, and the training of men for the Motor Transport Service, the largest being Camp Holabird at Baltimore and the others, Camp Jesup at Atlanta, and Camp Normoyle at San Antonio, Tex.

When the armistice was signed the Army had in this country thousands of motor trucks, motorcycles, and ambulances. Perhaps 80 per cent of these were located in the districts subsidiary to Camp Holabird. Consequently, there were constructed at Holabird an enormous repair shop and a shop for the taking down and shipping of motor vehicles. The machine shops at these camps were of permanent construction. Large storage facilities were also furnished.

The experiences of the Army in Mexico taught that to be effective motor transport must have adequate facilities for repair. The

QUARTERMASTER STORAGE DEPOT BUILT BY CONSTRUCTION DIVISION AT ST. LOUIS.

CHICAGO PERMANENT QUARTERMASTER DEPOT WAREHOUSES.

standardized army repair shop is of glass, steel, and concrete construction. It is operated by 55 officers and 1,400 men.

One of the most interesting buildings provided for the Motor Transport Corps was the crating shop at Camp Holabird. The first motor trucks sent overseas were shipped completely assembled. In addition to taking up unnecessarily much-needed cargo space on the transports, they frequently arrived in poor condition, due to the effects of the salt air upon their metal parts.

Consequently, it was decided to ship trucks disassembled in crates. One of these huge vehicles could be taken apart and, except for its truck body, wrapped up in a parcel 20 feet long, 40 inches wide and 40 inches deep. The truck body could be packed in a crate 12 feet long and 6 feet wide and 1 foot thick. These crates were moisture proof.

The crating of trucks saved 75 per cent of the ship space formerly required. The crating crews became so facile that they could take down and pack in a single day from a mile and a half to two miles of trucks. This unique shop cost $500,000.

The construction at Camp Holabird started February 4, 1918. The camp now occupies 144 acres and has a cantonment for 7,000 men. The 22 buildings of an abandoned distillery on the ground were remodeled to serve as permanent storehouses for the millions of dollars' worth of tools and motor-vehicle parts which the Government acquired in the war.

ARMY HOSPITALS.

For the Surgeon General's Department the Construction Division constructed hospitals in this country providing accommodations for a total of 121,000 patients, 12,000 nurses, 4,000 doctors, and 34,000 hospital operation and maintenance troops. There were 294 of these hospitals in all, built at a total cost of $127,725,000 and divided into three types: base hospitals located at the various training camps, departmental hospitals located at various other Army posts, and general hospitals for the reception of sick and wounded men returning from France. The construction of general hospitals did not cease with the signing of the armistice, and at a recent date it had provided 97,000 beds for patients.

The builders adopted a standardized type of hospital construction. The unit in this type was a single-story ward building of frame construction lined in the interior with gypsum board or some similar material. An open porch along the entire length of one side of the ward building provided opportunity for convalescents to be wheeled outdoors. Each ward had room for 34 beds and also had a diet kitchen, a nurses' room, a doctors' room, lavatories, and an inclosed solarium at the end. These buildings were connected with each other by inclosed corridors running through the clear. At the Fox Hills Hospital, Staten Island, N. Y., there was a mile and a quarter of this corridor construction. The corridors in each case fed in toward the central administration group of buildings in which were located the operating rooms and the various laboratories.

As crews developed their teamwork some marvelous instances of speed were shown in putting up the buildings. One crew of 566 men at Fox Hills erected a complete hospital wing in one working day. At 7 o'clock in the morning the ground of the site was untouched. That night at 6 o'clock a 40-bed ward was standing finished on the site—painted, equipped with heating and ventilating apparatus, all plumbing installed, the last electric bulb screwed in, and in every respect ready for occupancy. It was like magic; yet soon thereafter the Construction Division had set the period of 10 hours as the standard time for building one of these wings.

General Hospital No. 3 located at Otisville, N. Y., has a capacity of 579 beds and is a complete military hospital plant designed for the treatment of tubercular cases. A short summary of the work done upon it will give an idea of the general nature of the construction problems incident to the building of the military hospitals during the winter and spring of 1918.

On February 2, 1918, the Constructing Quartermaster with a few officers and clerks arrived at the site of the hospital, about 37 acres of land, near the village of Otisville, Orange County, N. Y., on the southern slope of Shawangunk Mountain. The contractor and his organization came on the same day. They found the site covered with snow and with no accommodations even for the office force of the Constructing Quartermaster or the contractor, except an old creamery building, which was promptly rented and into which the two organizations moved the next day.

Actual work was started February 5 and continued through every severity of weather until the project was completed in the early part of July. The work was interrupted, hindered, and hampered by snow and mud, by transportation congestion preventing the delivery of materials on time, by the absence of local labor and the necessity for importing labor from near-by markets wherever it could be procured, and by the consequent necessity for running special trains to transport laborers to and from the job. No local facilities existed for housing any of the workmen, and temporary accommodations had to be built to accommodate 200 Italian laborers both in the matter of shelter and food.

The average height above sea level of the site of this hospital is a little over 1,000 feet. It was found after construction began that the site was full of springs, which caused further trouble and difficulty in developing the building operations.

The cost of the project was $1,681,000. About 300 carloads of material were used, including approximately 3,000,000 feet of lumber. The largest number of laborers employed at any one time was 1,795, with a working day, as a rule, of nine hours, union and nonunion labor being employed without discrimination. Water supply, sewer construction, roads, 330 feet of railway siding, sewage disposal plant, electrical installation, and boiler houses, all had to be built. The work was all completed early in July, 1918. In addition to the bed capacity of the hospital, accommodations were provided for a hospital personnel of 224.

One of the largest hospitals involving entirely new construction is General Hospital No. 21, Denver, Colo. This is also a hospital for the treatment of tuberculosis cases. The construction is permanent, hollow tile and stucco, and the hospital facilities are sufficient to accommodate 2,000 beds for recuperative and curative work. The plant consists of hospital wards, tuberculosis wards, officers' quarters, nurses' quarters, mess halls, storehouses, laundry, schoolhouse, power house, water and sewer installation, and all of the necessary utilities for the adequate operation of a completely self-contained unit.

The original authorization covered only 1,000 beds. Subsequently there was authorized an addition equally as large, and the entire project was completed on March 1, 1919. The actual capacity of the hospital is in excess of the capacity estimated when the work was begun, and it is estimated that 2,486 patients can be cared for, the cost per bed running less than $1,350. In view of the nature and character of the permanent construction involved, and the fact that this is a military hospital with the usual numerous construction details not found in civilian hospitals, the construction at this figure is regarded as exceptional.

INCIDENTS OF THE WORK.

When the Government undertook the whole enormous military construction program, it was found that there were few builders in the United States who had equipment enough to handle the bigger jobs. Consequently the Construction Division adopted the policy of acquiring equipment of various sorts, usually paying rent for it. Such equipment included locomotive cranes, concrete mixers, locomotives for trench machines, road machinery, and other heavy apparatus. This equipment was rented under an agreement that whenever the rent paid had aggregated the cost of the article rented, the latter should become the property of the Government. In this way the Government has acquired property of this sort worth about $3,000,000.

The Construction Division at all times procured raw materials for the contractors engaged upon the projects. During the summer of 1918 the division was procuring material at an average rate of nearly $1,000,000 per day.

There were many interesting incidents in connection with this activity. In the summer of 1917, when the cantonments were going up, it became necessary to provide some 60,000 stoves and heaters, yet there were not that many stoves for sale in the country including all existing stocks, nor was the capacity of the various stove works sufficient to make up the number in the three months' period before the soldiers would be going to the camps. Accordingly, officers of the division were sent out to make addresses to the workmen at the stove factories, and as a result of such efforts the companies speeded up their output until they were able to supply all of the camps with heating facilities by October 1, 1917. In this effort the Government went into the market and procured the pig iron, coke, and other supplies for the stove foundries.

The Construction Division was also able to obtain 15,000 Army kitchen ranges in three months, although that number is a normal year's output of the entire manufacturing facilities of the country.

When the project for the expeditionary supply base at Port Newark was taken up—in the late fall of 1917—the Construction Division set about it to get 63,377 piles for the foundations of this construction. There were 64 pile drivers on the job, driving in a total of 1,566 piles in a day; and to supply these the woodsmen of Maryland, Virginia, North Carolina, and New Jersey were called upon for their best efforts. Due to the unforeseen severity of the winter, the rivers were frozen and the railroads choked with freight. Near by was the Hog Island shipbuilding project at Philadelphia, needing more piles than the railroads could deliver. The trees in the woods were frozen and often broke to pieces when they fell. In the southern logging districts the negro woodsmen refused to stay on the job because of the cold. The Construction Division then took hold, sent soldiers into the woods, felled the trees, and then put guards on the cars of piling to see that they were not lost in transit. As a result, the piles for the Port Newark job were delivered on time.

In addition to procuring materials for its own contractors, the Construction Division also procured building materials for the Shipping Board and for the Bureau of Industrial Housing. The peak load of labor on the Army construction jobs came in the summer of 1918, when 230,000 men were on the pay roll, drawing $7,626,800 a week in pay, and still the jobs were short 150,000 unskilled men. In general, the union scale of wages and hours of labor were adopted, but the open shop was maintained. Labor troubles were infrequent and not serious. To prevent strikes the Government formed the Cantonment Adjustment Commission, consisting of three members, the Army representative being Col. J. H. Alexander, of the Construction Division. Of all the strikes that hampered our war activities, less than 1 per cent were strikes of the building trades.

When the labor shortage of 1918 was most acute the Construction Division turned to Porto Rico and the Bahamas for unskilled labor, importing 2,600 Bahamans and 13,000 Porto Ricans. This imported labor was exclusively used on southern building projects and was sent back home when the armistice was signed.

The Construction Division had charge of the operation and maintenance of the utilities of the various training camps, a work requiring a force of 452 officers and 16,559 men. In all there were 54,808 buildings at these camps to be kept in repair. This was done at the cost of $8.10 a year for each man housed. The Government supplied electricity to the camps at an average cost of $0.02½ per kilowatt hour. In a single year of operation, the 32 camps burned about 2,000,000 tons of coal for heating. This was at a cost of approximately $10 per man.

The utilities of the camps were under the management of men who could qualify to be city managers. They had the operation of water systems, fire departments, and other common conveniences of cities. Water was supplied at the rate of 55 gallons daily per man. The purity of the water and the adequacy of the sanitation may be gauged from the fact that in July and August, 1918, the annual death rate at the camps was 2.8 per thousand. In our Mexican War the annual death rate of American troops from disease was 110 per thousand; in the Civil War it was 65 per thousand; in the Spanish War 26 per thousand; and among Japanese troops in the Russo-Japanese War it was 25 per thousand. The death rate in civil life for men of the draft age is 6.7 per thousand.

Each camp and cantonment was adequately protected by fire companies equipped with the most modern apparatus, nearly all of it motorized. Each camp fire company had 60 men. A low annual fire loss in civil life is $2 per capita. In 1917, 20 American cities of about 31,000 population each showed an annual fire loss of $2.15 per capita. The average for the United States is $2.42 per capita, At the training camps, in spite of their inflammable construction, the average annual loss from fire per capita was only 46 cents.

109287°—19——36

TABLE 1.—*Army supply bases and interior depots built by Construction Division between April, 1917, and January, 1919.*

Project.	Square feet of floor area.	Approximate cost.	General contractor.
Port Newark terminal	1,652,400 [1] 367,600	$10,260,090	Mason & Hanger-McArthur Bros.
Norfolk Army supply base	2,016,000 [2] 884,500	25,975,770	Porter Bros.
Philadelphia Army supply base	845,570	15,510,112	Snare & Triest Co.
Charleston quartermaster terminal	1,152,000 [1] 379,200 [2] 269,000	12,675,000	Mason & Hanger.
Boston Army supply base	1,651,104 882,000	28,040,000	W. F. Kearns Co.
Brooklyn Army supply base	3,936,000 [2] 1,143,000	32,500,000	Turner Construction Co.
New Orleans Army supply base	1,792,000 [2] 280,000	11,660,875	George W. Fuller Co.
Interior storage depots.			
Baltimore interior depot	663,800	2,143,676	Sanford & Brooks.
Boston temporary warehouses	327,600	1,265,079	W. F. Kearns Co.
Chicago interior depot	580,400	809,300	Central Manufacturing District.
Chicago permanent warehouse	1,230,400	3,691,800	Do.
Columbus interior storage depot	2,657,600 [1] 246,400	6,128,022	Hunkin-Conkey Co.
Jeffersonville quartermaster interior depot.	320,000	1,282,563	Caldwell & Marshall Co.
Hoboken quartermaster expeditionary depot.	162,540	201,000	Barney-Hooks-Ahlers.
New Cumberland interior storage depot	1,568,000 [1] 400,000	4,700,270	Bates-Rogers Construction Co.
Philadelphia interior storage depot	917,800	3,470,940	William Steele & Sons.
Philadelphia quartermaster expeditionary depot.	203,900 [1] 51,400	1,091,050	Do.
Pittsburgh interior storage depot	178,600	630,900	The Austin Co.
Schenectady interior storage depot	2,080,800 [1] 537,600	6,051,550	Feeney & Sheehan.
St. Louis interior depot	384,300	1,368,540	Westly Construction Co.

[1] Sheds. [2] Pier sheds.

TABLE 2.—*Hospitals built by Construction Division between June, 1917, and January, 1919.*

Name.	Location.	Capacity (in beds).	Approximate cost.	Contractor.
Azalea T. B.	Azalea, N. C.	1,500	$2,715,000	Gude & Co.
Boston City, west department.	West Roxbury, Mass.	[1]	47,000	Chas. Logue Building Co.
Cape May General Hospital No. 11.	Cape May, N. J.	700	46,150	Cauldwell-Wingate Co.
Carlisle General Hospital No. 31.	Carlisle, Pa.	120	180,000	Warren Moore & Co.
Colonia General Hospital No. 3.	Colonia, N. J.	1,650	2,367,884	Cauldwell-Wingate Co.
Cooper-Monitah No. 32.	Chicago, Ill.	500–625	158,368	Chas. Logue Building Co.
Denver General Hospital No. 21.	Denver, Colo.	2,266	3,100,133	C. S. Lambie Co.
Eastfield General Hospital No. 38.	Westchester, N. Y.	1,550	237,000	Frank H. Goble.
Elks and Robt. Bringham General Hospital No. 10.	Boston, Mass.	700	8,000	Chas. Logue Building Co.
Ford Hospital, General Hospital No. 36.	Detroit, Mich.	2,000	107,750	Albert A. Albreacht.
Fort Bayard General Hospital.	Fort Bayard, N. Mex.	1,046	650,000	V. E. Vare.
Fort Des Moines Hospital	Des Moines, Iowa	1,266	244,148	Chas. Weitz Sons.
Fort Douglas Hospital	Fort Douglas, Utah	227	690,000	Lynch-Cannon Eng. Co.
Fort McHenry General Hospital No. 2.	Baltimore, Md.	2,800	2,327,975	J. Henry Miller Co.
Fort McPherson Hospital	Fort McPherson, Ga.	2,000	680,000	Gude-Krebs & Co.
Fort Oglethorpe General Hospital No. 14.	Fort Oglethorpe, Ga.	1,300	967,377	Park-Grimes Co.
Fort Ontario Hospital No. 5.	Fort Ontario, N. Y.	660	557,000	J. J. Turner & Sons.
Fort Root General Hospital No. 33.	Little Rock, Ark.	500	16,455	
Fort Sheridan General Hospital No. 28.	Fort Sheridan, Ill.	5,000	2,515,786	Summer-Sollitt Co.

[1] Part of General Hospital No. 10, Boston, Mass.

TABLE 2.—*Hospitals built by Construction Division between June, 1917, and January 1919*—Continued.

Name.	Location.	Capacity (in beds).	Approximate cost.	Contractor.
Fort Snelling Hospital	Fort Snelling, Minn.	1,780	$489,600	H. N. Leighton Co.
Fox Hills Debarkation Hospital No. 2.	Fox Hills, Staten Island.	1,808	2,929,556	Thompson-Starrett Co.
General Hospital No. 16	New Haven, Conn.	700	360,670	Sperry Engineering Co.
Grand Central Palace Debarkation Hospital No. 5.	New York City	3,500	215,000	Day labor.
Greenhut Building Debarkation Hospital No. 3.	New York City	3,130	420,000	Do.
Infirmary	St. Louis, Mo.	1,000	65,000	March Construction Co.
Lakewood General Hospital No. 9.	Lakewood, N. J.	1,020	232,735	Geo. D. Morrow.
Letterman General Hospital.	Presidio, Calif.	1,250	112,716	
Madison Barracks Hospital.	Madison, Wis.	273	125,535	J. J. Turner & Sons.
Markelton General Hospital No. 17.	Markelton, Pa.	500	93,580	Dawson Construction Co.
Nassau General Hospital No. 39.	Long Beach, Long Island, N. Y.	1,500	25,000	Day labor.
Norfolk State General Hospital No. 34.	Norfolk, Mass.	700	120,000	Chas. Logue Building Co.
Otisville General Hospital No. 8.	Otisville, N. Y.	1,020	1,685,766	R. H. Howes Construction Co.
Park View General Hospital No. 24.	Pittsburgh, Pa.	1,200	208,940	Day labor.
Plattsburg General Hospital.	Plattsburg, N. Y.	2,000	231,948	D. Callahan.
Richmond College Debarkation Hospital No. 51.	Richmond, Va.	960	51,750	John T. Wilson Co.
Roland Park General Hospital No. 7.	Baltimore, Md.	500	329,617	J. Henry Miller Co.
General Hospital No. 35.	West Baden, Ind.	1,500	125,000	Chas. W. Gindle Co.
United States Army General Hospital No. 12.	Biltmore, N. C.	551	37,050	Gude & Co.
United States Army General Hospital No. 18.	Waynesville, N. C.	400	73,700	Gude-Krebs & Co.
United States Army General Hospital No. 23.	Hot Springs, N. C.	750	103,200	Do.
Walter Reed General Hospital.	Takoma Park, Washington, D. C.	2,093	1,675,151	Skinker & Garrett.
Whipple Barracks General Hospital No. 20.	Prescott, Ariz.	1,000	1,629,683	James E. Morgan.

TABLE 3.—*Approximate quantities of materials purchased by materials branch, Construction Division, June 15, 1917, to Dec. 15, 1918.*

Item.	Quantity.	Cost.	Remarks.
Brick	325,783,400	$2,631,799	Would build a road 18 feet wide approximately 380 miles long, or equal to the distance from Indianapolis to Pittsburgh; loaded in cars having a capacity of 15,000 brick, would require 21,720 cars to load.
Lumber	[1] 2,647,605,426	69,773,825	If considered 1-inch pieces 12 inches wide and placed end to end, would reach 500,000 miles, or 20 times the circumference of the earth. It would take 175,000 cars to haul this material, figuring 15,000 feet board measure material to each car. Figuring each car 35 feet long, would make a train 1,160 miles long, or the distance from Washington, D. C., to Kansas City. Total weight, 5,000,000 tons.
Wood stave pipe	[1] 2,900,000	2,325,000	If placed in one single line, would extend 550 miles. This is approximately the same miles of pipe as all of the water and gas lines in the city of New Orleans, La., which supplies a population between 400,000 and 500,000 people.
Structural steel	[2] 64,000	6,311,092	This would require 1,600 cars having a capacity of 80,000 pounds each. It would require structural-steel shops employing 100 men fabricating 2,000 tons per month three years to complete the fabrication. This is sufficient material to erect twenty 20-story buildings 75 feet square.

[1] Feet. [2] Tons.

TABLE 3.—*Approximate quantities of materials purchased by materials branch, Construction Division, June 15, 1917, to Dec. 15, 1918*—Continued.

Item.	Quantity.	Cost.	RemarSs.
Roofing	[1] 1,400,000	$3,650,000	Sufficient to roof 5 square miles, or would roof 100,000 ordinary houses of 14 squares to each house.
Cable	[2] 1,103,560	257,233	Approximately 210 miles.
Copper wire	[2] 33,283,000	1,409,600	6,303 miles.
Cement	[3] 6,181,194	13,723,153	36,148 car loads, approximately, of 171 barrels to car load. Loaded on cars 35 feet total length, would make one train 239 miles long, or greater than the distance from Boston to New York.

[1] Squares of 100 square feet. [2] Feet. [3] Barrels.

Procurement branch during period of 18 months, June 15, 1917, to December 15, 1918, purchased and allocated materials to the amount of $245,115,443.

BOOK VII.

THE SIGNAL CORPS.

CHAPTER I.

SIGNAL CORPS MATERIAL.

The spent runner who hurled himself through the gate of ancient Athens and with his dying breath gasped out the news of the brilliant success of the Athenian troops against the Persian at Marathon in the year 490 B. C. was the first famous soldier of a signal corps; but since then the exploits of the bearers of military tidings have filled the pages of legend and story. Just as other branches of military science have been brought to a high perfection in modern times, so in equal degree has the art of military signaling progressed in efficiency.

Where the ancient athlete once exhausted his strength in bearing military messages long distances in the field, the modern Mercury uses the wireless phone. In Civil War days the pony express rider brought from some desperate stand the story of the lack of ammunition; to-day the ammunition-supply organization is in constant touch with the front by means of telegraph or the long-distance telephone. In the Indian campaigns in our own West messages from beleagured parties were sometimes conveyed by signal smokes; the "lost battalion" in the Argonne sent news of its plight by carrier pigeon.

Modern warfare has indeed retained the old, but it has also developed the new, in transmitting military tidings. So important is this branch of fighting that it is put in the hands of a specialized organization, which in the American Army is known as the Signal Corps. The Signal Corps not only had charge of the operation of the various communicating devices in 1917 and 1918 in the field of operations (except latterly in the air), but it also had charge of the manufacture of the equipment for this work.

The production of signaling equipment was far greater than the uninformed person would imagine. As an instance, there was one special type of telephone wire, a form unknown to commercial use before the war, which, before November 11, 1918, was being produced at the rate of 20,000 miles a month, at a cost of $5,650,000 per month, requiring the complete capacity of the day and night operation of all fine wire machinery in the United States, except that which was working on Navy contracts. Many other production activities of the Signal Corps were carried through on a similar scale.

Until after the Civil War, the operation of large units of troops was greatly handicapped by the limitations of military signaling as then known. A force could not be effective in combat that could not be readily reached in all quarters by runners or riders or by visual signals. The development of the telegraph and telephone and the invention of radio changed all this, so that in the great war armies stretched out on fronts 100 miles or more in length with every part of them in immediate touch with every other part through the exact and complete systems of signaling on the field.

Military signals to-day include the telephone, the telegraph, radio telegraphy and telephony, the buzzer, the buzzerphone, panels, pyrotechnics, flags, smoke signals, pigeons, dogs, mounted orderlies, and runners. Each of these means of signaling is an adjunct to the others; when one fails, another is employed to get the message through. Some have special uses for branches of the service with peculiar requirements. The radiophone is especially suited for communicating from airplanes. Artillery fire is directed by wire and wireless communication. Trained pigeons are sometimes able to get messages through when all other means of communication have failed.

The Army did not have a great quantity of signaling equipment when it went to war with Germany, but what it did have was good. The American punitive expedition in Mexico, where long lines of communication over rugged country were required, had given opportunity for testing modern signal apparatus in the field. Many of the signaling devices used by the American Expeditionary Forces were, at least in type, in common use by the civilian population; yet the procurement of this equipment offered heavy difficulties. This was because the Army was much more exacting than was commercial demand as to the quality of material used. For instance, a telephone instrument for use in the field hardly can be compared with the telephone in a business man's private office. The field set demands stronger connections, better insulation against the dampness of outdoor work, and more rugged construction to withstand rough usage by an army on the march.

One of the larger tasks of the Signal Corps in France was that of providing facilities for communication for the service of supply. The first Signal Corps officers sent to France soon realized that the forthcoming American Army could not depend upon the French telegraph and telephone systems in the various zones of operation, because those systems were already overburdened by the uses of the French government. Consequently, it became necessary to set up our own telegraph and telephone systems, extending them from the ports of debarkation through the various bases and zones up to the battle regions. The magnitude of the system which finally was constructed is shown in the fact that on November 11, the date of the armistice, there were in France 96,000 miles of American telegraph and long-distance telephone circuits. This wire was all used by the service of supply and by the various Army bases behind the front.

Yet in the field of fighting the requirements for wire were even greater. At one time during the height of the operations it was evident that the time was not far distant when the Signal Corps would need 68,000 miles a month of what was known as outpost wire, for use simply in connecting up the telephone and telegraph systems carried along by the troops in their advances.

Outpost wire was entirely a development of the war against Germany. The original telephone system used at the front had been the single telephone wire grounded to complete the circuit. But all the armies in France perfected their listening instruments to such a degree that they could hear conversations conducted on the grounded telephone circuits, the sounds being detected in the earth itself. Consequently, it became necessary to carry forward with troops two-wire telephone circuits, thus doing away with ground connections. Even then care had to be taken that the insulation of this double wire was perfected, lest the impulses enter the ground through gaps in the insulation. Wireless for outpost communication was equally impracticable, since the enemy could easily listen in and hear radio phone messages.

Outpost wire insured secret communication at the front. Outpost wire was a twist of two wires, each single wire being made up of seven fine wires, four of them of bronze, and three of them of hard carbon steel. These were stranded together, coated, first with rubber and then with cotton yarn, and finally paraffined. The wire was produced in six colors—red, yellow, green, brown, black, and gray—for easy identification in the field, each unit employing its own color.

The wastage of outpost wire was enormous. In an advancing movement it was folly to undertake to pick up the wire. The abandoned miles of it had to be left in the field to be salvaged later by the clean-up parties.

The proposition of producing 68,000 miles of outpost wire every month staggered the wire manufacturers of the country. There were not enough braiding machines to complete such an order, and new ones had to be built before such a quantity of outpost wire could be attained.

In addition to the various means of communication, the Signal Corps was also called upon to supply in large quantities such other articles as wire reel carts, flag staffs, field glasses, photographic equipment, chests, tools, meteorological apparatus, and wrist watches.

In the production of its supplies, the Signal Corps was confronted with the same obstacles of inadequate industrial capacity, dearth of raw materials, and congestion of railroad transportation, that embarrassed almost every line of military production. To meet these difficulties the Signal Corps organized an elaborate inspection force which not only checked the work at the various factories for quality and rate of production, but was also constantly on hand to help the harassed manufacturer out of his difficulties as they arose. The Signal Corps never slept. At night and on holidays there was at least one officer on the job in Washington to receive telegrams or long-distance telephone messages and to be ready to act quickly in any emergency.

From the production standpoint, signal equipment was divided into several general classifications: (1) telephone and telegraph apparatus; (2) radio apparatus; (3) line-construction materials; (4) batteries; (5) wire and cables; (6) field glasses; (7) wire carts; (8) photographic supplies, pigeons, and pigeon supplies; and (9) chests, kits, and tools, mechanical signals, electric signals, meteorological apparatus, and wrist watches.

TELEPHONES AND TELEGRAPHS.

In the early days of the conflict the construction of signal materials in the United States was devoted to such basic supplies as wire, cable, tools, and the standard types of telephone equipment, such as telephone sets and switchboards. The first great task in France was to install the lines of communication for the service of supply, a system that required American equipment because it was planned to operate it with American-trained telegraph and telephone operators.

Now, there were numerous styles of commercial telephone equipment manufactured in the United States. The plan, therefore, was adopted of allowing the various manufacturing concerns to bid on a tentative production schedule, giving an exclusive contract to the lowest bidder in each type of apparatus. This exceptional policy was adopted in order to avoid multiplicity of types of equipment to be used abroad. If many makes were adopted in each type they would necessitate the procurement of many types of spare parts and replacement materials.

The concerns which produced the telephone equipment for the American Expeditionary Forces were the Western Electric Co., of Kansas City; the Kellogg Switchboard & Supply Co., of Chicago; the Stromberg-Carlson Telephone Manufacturing Co., of Rochester; the Frank Black Co., of Chicago; and the Reliable Electric Co., of Chicago.

At the signing of the armistice there were 282 American telephone exchanges in France, with 14,956 telephone lines reaching 8,959 stations. The 282 exchanges ranged from the small four-line monocord unit, such as may be seen in any business office, to the standard American multiple board of the city telephone exchange. Of these latter there were over 30 in use by the American Expeditionary Forces when the armistice was signed.

MULTIPLEX PRINTING TELEGRAPH. SCHOOL OF RADIO AND MULTIPLEX TELEGRAPHY.

SIGNAL CORPS SCHOOL FOR TELEPHONE ELECTRICIANS, UNIVERSITY OF MICHIGAN. INSTRUMENT AND SWITCHBOARD REPAIR CLASS.

FIELD WORK WITH UNDAMPED WAVE TRANSMITTER AND TRENCH ANTENNA

SIGNAL CORPS EQUIPMENT INSTALLED IN A DUGOUT.

Left to right—S. C. field telephone, type "EE-4;" field service buzzer; signal lamp (projector), type "EE-6"; S. C. radio set, SCR.-76 and S. C. radio set, SCR-54A.

The special telephones adopted for use in the field were different from any in commercial use in America. The Signal Corps had developed certain special instruments combining both telephone and telegraphic principles. The field telephone, model 1917, for instance, was a telephone which included a telegraph buzzer on its telephone circuit. This instrument was used when great secrecy in communication was required. The messages were sent in telegraphic code, the buzzers being heard by the receiver. Another instrument was known simply as the buzzer. This was an instrument which utilized the telephone receiver for telegraphic messages. It was a supreme development for use over defective lines. An instrument closely related to the buzzer, but which gave even greater secrecy, was known as the buzzerphone. The buzzerphone was put into production just before the close of hostilities.

The mobile switchboard in most general use by our troops at the front was developed originally by the French and was known as the monotype. It was designed in units and could be extended to accommodate up to 12 trunk lines leading away from the board. This apparatus was the "central" of the front-line dugouts. It could be put into operation in a few minutes and was easily carried by a soldier.

The switchboard of the dugouts was the only telephone equipment not of American design used by the American Expeditionary Forces. It was put into production in the autumn of 1917 in three American plants, under the general policy of the Signal Corps to contract with more than one factory for the production of any important device.

Another type of field switchboard when packed for transit resembled a salesman's trunk. It was used in the camps and provided for 40 lines. This board was being constantly redesigned as field needs developed. A new type of camp switchboard was coming into heavy production at the end of hostilities.

Still a third type of portable switchboard was built in units resembling the units of a sectional bookcase and was set up in the same way.

The telegraph apparatus of the lines of communication in the S. O. S. was designed along purely commercial lines. It included the latest type of printing telegraph equipment, the apparatus first adopted being the multiplex printing telegraph as used by the Western Union Telegraph Co. Later, the Morkrum printing telegraph was also adopted.

At the close of hostilities 133 complete telegraph stations with full equipment were in operation in the service of supply. The peak load of this service, just prior to the armistice, was 47,555 telegrams, averaging 60 words each, sent from these stations in a single day. The daily average in the final weeks of the fighting was 43,845 telegrams.

RADIO.

At the outbreak of the war, the field radio equipment in active use by the Army was limited to two sets, both of comparatively high power. On the other hand the allied forces had developed a complicated and extensive use of radio sets of small power, many of them operated from airplanes, and the Signal Corps found itself confronted with the task of developing an entire new line of complicated electrical apparatus, and putting it into large quantity production in the shortest possible time. The progress made is indicated by the fact that at the signing of the armistice the number of types of complete sets on which development work had been carried out was 75. Of these approximately 25 were in quantity production. When it is remembered that each of these sets consisted of hundreds of parts, many of which required careful study and experimentation as well as design, the magnitude of the problem is appreciated.

The initial step in the reorganization of this branch of the Signal Corps' work consisted in the establishment of a radio section in Washington and a corresponding section in France. The former was charged with the design of apparatus and the preparation of manufacturing drawings and specifications, while the latter served as the first hand observer of actual service requirements and approved all equipment before it was used in the field. An important auxiliary of the development organization in Washington was the radio laboratories established at Camp Alfred Vail, where all necessary technical facilities, such as model shops, drafting rooms, research laboratories, a completely equipped flying field, etc., were maintained. With this engineering organization and the production organization which handled all Signal Corps equipment, the work here detailed was carried out.

Shortly after the declaration of war, the French government sent to this country a distinguished commission which included the foremost radio experts, who were thoroughly familiar with the latest military developments. Technical information and samples of radio apparatus were also obtained from British sources. With this beginning, the engineering work naturally divided itself into two general problems—first, to duplicate the approved foreign designs, and then to create designs for new types of apparatus which would be superior to any in service. Work on these two groups of problems was prosecuted simultaneously with the result that there were soon in production American equivalents of a number of French and British sets, together with improved original types of American radio apparatus.

Probably the most noteworthy technical development during the war, in so far as radio communication is concerned, was the extensive use made of vacuum tubes. These "bottles," which make practical

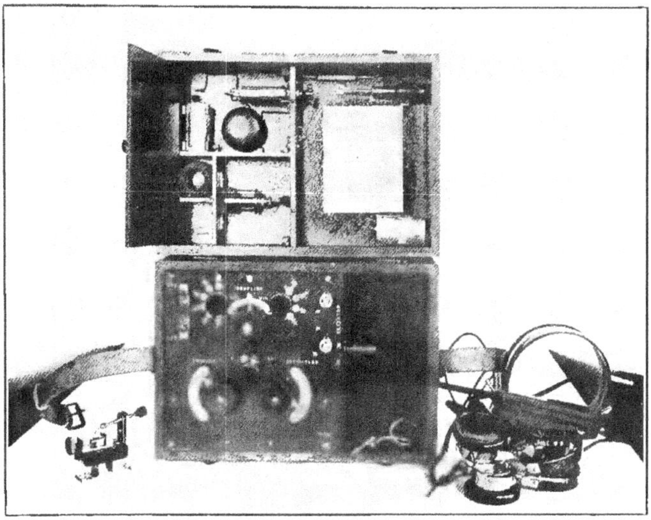

RADIO RECEIVING SET, SCR-54, USING CRYSTAL DETECTOR.

REAR PANEL VIEW, RADIO TELEPHONE GROUND SET. TYPE SCR-67.

RADIO TELEGRAPH TRANSMITTING SET IN CASE.

RADIO TELEGRAPH TRANSMITTING SET WITH CASE REMOVED.

use of the electrons of the new physics and which are sometimes called audions or pliotrons, are literally marvels in the realms of engineering, and their applications are as yet hardly realized. One form was used for the reception of signals prior to the war; but the military developments, particularly in France, had so progressed that when this country entered the war they were used both for receiving and transmitting signals, and most of the more important sets depended on them. To meet this demand the services of the three foremost vacuum-tube engineering organizations of the country were enlisted, and under the direction of the Signal Corps radio engineers the progress toward satisfactory design and construction of the required types was rapid. Within less than six months standardized tubes were turned out by the quantity production method at rates sufficient to insure the requisite supply. Work was continued, however, on the development of still better types of tubes. The improvements that have been made from time to time have been incorporated in the tubes being produced on a large scale, so that tubes of recent manufacture are a great improvement over those made a year ago.

As indicative of the extent and variety of the radio development work which was carried on, there is given below a partial list of the types of sets which were completely developed and placed in production during the war period:

LAND RADIO EQUIPMENT.

Spark sets, sending and receiving (three types).
Continuous-wave Army radio telegraph sets (three types).
Radio telephone set (one type).
Tank radio telegraph set (one type).
T. P. S. (earth telegraphy) (four types).
Wavemeters.
Battery-charging sets.
Radio operating and repair trucks.
Miscellaneous special equipment.

AIRPLANE RADIO EQUIPMENT.

Interphone sets (for use of two to five persons).
Radio telephone sets (three types).
Radio telegraph sets (three types).
Direction-finding radio-receiving set (one type).

The magnitude of the production of special items involved may be gathered from such figures of expenditures as the following:

For vacuum tubes	$1,650,600
For storage batteries	5,315,350
For dry batteries	602,470
For battery charging sets	1,524,400

These are, of course, only some of the items. The total production authorized was valued at approximately $45,000,000.

The remarkable development and improvement of military radio equipment which has taken place under the direction of the Signal Corps during the last 18 months will undoubtedly materially change the system of Army communications and even the tactical use of military equipment and personnel. A typical example of this development is the airplane radio telephone, described elsewhere, the use of which has made possible the "voice-commanded air squadron." The military value of an air squadron has been enormously increased by virtue of this device, which enables the squadron commander to direct the movements of the individual airplanes in any manner which circumstances may require.

Certain other radio devices recently perfected, the nature of which can not be revealed, will undoubtedly affect the tactical use of troops to such an extent as to make certain kinds of radio equipment as indispensable to the operations of military units as the rifle or the machine gun.

LINE EQUIPMENT.

The first requisition for line equipment for France called for the construction of 500 miles of telephone and telegraph main pole lines, carrying 10 copper telephone and telegraph wires. It was found that ship space could not be spared for poles in such quantity. Consequently a forestry unit was sent to France to get these poles from the French woodlands. All of the other materials for the 500 miles of line, together with materials for approximately 600 miles of extensions, were procured in the United States and shipped to France within six months after the requisition was received. This material was secured in such short time only by the cooperation of the large commercial companies in the United States, who literally stripped their warehouses bare of their supplies.

In the late summer of 1918 the American Government began anticipating the advance of the allied forces into Germany, and the Signal Corps put into production a reserve equipment for long distance line approximating 500 miles. Soon there was received from France a cablegram asking for the shipment of this material, and it was all floated before the armistice. However, as it turned out, this equipment was never required, since the terms of the armistice gave the American forces the German telephone and telegraph lines in the occupied territory.

This line equipment was all of a type standard in the United States. For the fighting zone special line equipment was required. Before the war with Germany American signal troops had set up their emergency telephone and telegraph lines on the standard "lance poles." These poles served admirably in open warfare, but proved to be impracticable for the static conditions of fighting in France. After a considerable supply of lance poles had been shipped abroad their production was curtailed. Thereafter the trench telephone and telegraph lines were supported on short stakes with special cross arms, in appearance the conventional telegraph poles in miniature. The enormous mileage of trench lines called for a great quantity of insulators and cross arms. The wastage of these fittings, due to their being exposed to artillery fire, became increasingly greater in the closing months of the war.

In wire itself, the American production was enormous. This production included the commercial type of copper line wire and the drop wire for connecting up individual telephones to the pole lines. Much commercial cable for connecting congested centers with branch switchboards was also required. Yet all of this wire used in the system within the Service of Supply was but a small quantity compared with the requirements in the fighting zone.

The production of double-conductor wire, or the so-called outpost wire of No Man's Land, which had relegated to the scrap heap the standard field wire of open warfare, necessitated an extraordinary effort. The wire had to be light enough for easy transportation and laying, strong enough to withstand the abrasions from traffic crossing it as it lay on the ground, and exceedingly well insulated. The first estimate was that an American Army in the field might use 1,000 miles per month of outpost wire. When the first American force actually went into action, in the spring of 1918, a reserve supply of 20,000 miles of outpost wire was in the American warehouses in France, with a vast quantity of cable in reserve. Cable, at first used in large quantities at the front, was invariably buried several feet underground and abandoned at every change of headquarters.

As the fighting grew more intense and covered a wider and wider area, the wastage of outpost wire became enormous. The demand of our forces for cable dropped to a negligible quantity, but wire requirements rose. Outpost wire became the main dependence of ourselves and the allies for all communication in the active sectors. A higher quality of wire was specified. So great was the destruction of wire that by July, 1918, the original estimate of 1,000 miles per month to be supplied by American factories had jumped to 20,000 miles.

As a substitute for outpost wire to fill the immediate needs the familiar twisted drop wire, with which the ordinary telephone is connected with the main circuit, was adopted. Our field officers liked drop wire, its only objectionable feature being its relative bulk. All available drop wire in the United States was shipped across, and its manufacture was pushed until the new type of outpost wire could be produced.

The Signal Corps supplied the mounting needs of the American Expeditionary Forces through August and September, 1918, with the available drop wire plus the growing production of the new outpost wire. In early August all the wire makers in America were summoned to a conference, in which the Signal Corps made known the necessity of pushing production. The result was an expansion which reached a total production of 40,000 miles of outpost wire in November.

Just before the armistice was signed, the American Expeditionary Forces indicated that they would require 50,000 miles of outpost wire every month, beginning in January, 1919. This requirement had already been fully anticipated, since the American manufacturers had set for themselves a maximum production of 68,000 miles per month by August, 1919.

To secure this production every wire mill in the United States worked 24 hours per day. When the production was at its height, inquiries came from the allied governments, indicating that they would call on American wire makers for a quantity of wire equal to what the latter were already producing for the American Expeditionary Forces. In other words, this proposition called for the doubling of a production which had already attained great size. Yet, had the fighting continued, there is every reason to believe that the industry would have risen to the demand.

The production of outpost wire was an intricate operation. To fill the demand for 50,000 miles of outpost wire a month called for 300,000 miles of steel strand and 400,000 miles of bronze strand every month. The steel strand had to be given repeated heat treatments before it had acquired the necessary tensile strength.

ELECTRIC BATTERIES.

The American Expeditionary Forces consumed great quantities of electric batteries, the familiar dry battery of commerce being most used. Toward the end of the fighting arrangements were being made to establish in France a plant at which dry batteries would be assembled by French labor, utilizing parts made in America. The necessary apparatus and materials for the first operation had reached France prior to the armistice, but the plant was not in production at that time.

Storage-battery requirements of the American Expeditionary Forces were heavy and exacting. The storage battery was the only practicable source of electrical energy for the operation of small portable radio outfits. Field conditions required a storage battery that would not spill its contents, with a jar not easily broken, the whole equipment being as light as possible. A rubber composition jar was finally adopted.

The chief reliance of the American Expeditionary Forces was in storage batteries of European manufacture, which were to be used until American production got underway. When by the summer of 1918 America had perfected her own designs of radio equipment, the Signal Corps took up the matter of storage batteries for radio and decided upon types. This was in July, 1918. A conference of battery manufacturers was called and the orders were allocated among practically all the storage-battery plants in the United States that were in a position to undertake quantity production. The end of hostilities stopped this production on the eve of heavy deliveries.

FIELD GLASSES.

When the war began, the Signal Corps had the duty of providing field glasses for all branches of the Army, issuing them to noncommissioned officers and selling them at cost to commissioned officers engaged in combat. The first estimates showed that these glasses would be needed by the tens of thousands, whereas the manufacturing facilities in the United States had turned them out merely by the hundreds.

The optical-glass industry had never been developed in America, our field glasses being supplied with lenses of European glass, and principally German glass. In 1914 the imports of optical glass were $641,000 in value. The following year they were almost nothing. The advance of the German army toward Paris encompassed the glass plants of Belgium and many of those of France. England needed the entire output of her own glass factories.

In the autumn of 1914 the American optical-instrument makers began to develop an optical-glass industry, largely stimulated by the possibility of obtaining heavy orders at high prices from the British, French, and Russian governments. The most important work was done by the Bausch & Lomb Optical Co., of Rochester, N. Y.; the Spencer Lens Co., of Buffalo, and the Pittsburgh Plate Glass Co., of Pittsburgh. They were aided by the United States Bureau of Standards and by the geophysical laboratory of the Carnegie Institution. The Bureau of Standards established a laboratory at Pittsburgh where experiments were conducted with 30-pound pots of glass.

Optical glass differs greatly from ordinary glass. It must be clear, without striae, and there must be no strains in it, resulting from the final stirring and cooling. It must give a high transmission of light.

About the time of America's declaration of war the American experiments had produced glass suitable for optical instruments. This glass, however, was being turned out in quantities quite insufficient to meet the demand during the first few months.

109287°—19——37

In addition to the difficulties surrounding the glass supply, there was only a limited number of establishments capable of manufacturing field glasses after the glass was procured. These concerns were located principally in Rochester, N. Y., where they had been manufacturing a wide variety of optical instruments, including opera glasses, camera lenses, scientific and educational apparatus, battery commanders' telescopes, marine glasses, microscopes, and gun sights. In order to meet the war requirements of America for field glasses, these factories had to install large quantities of new equipment and to run day and night. The equipment consisted of lens-grinding apparatus, lathes, dies, and automatic screw machinery.

In addition to the Rochester factories there was a concern in Denver, Colo., the Weiss Electrical Instruments Co., which in a smaller way had been manufacturing surveyors' levels and other engineering apparatus. The Talbot Reel & Manufacturing Co., of Kansas City, had been making fishing reels in a small plant about 30 feet square. This factory was purchased in 1917 by Mr. L. Harris, who, after finishing a contract for gun sights for the Ordnance Department, built a factory especially for the production of Army field glasses and reached the quantity manufacture of these instruments before the armistice came. The chief center of supply, however, continued to be Rochester, where the plants of Bausch & Lomb, the Gundlach-Manhattan Optical Co., and the Crown Optical Co. are located. These factories expanded many times, and the output of field glasses went beyond what the executives at the outset of the enterprise imagined could be possible.

The Bausch & Lomb Co. was started in Rochester about 50 years ago by J. J. Bausch, who was born in Germany. The plant developed gradually, making a full line of spectacle lenses and optical instruments. The Carl Zeiss Works, of Jena, Germany, had a financial interest in the plant, and Bausch & Lomb had a financial interest in the Zeiss plant. This connection, however, was dissolved in 1915, when Bausch & Lomb took on contracts for the manufacture of field glasses for the British, French, and Russian governments.

Before 1914 this concern had never manufactured more than 1,800 pairs of field glasses in a year. The output was speeded up until in November, 1918, a total of 3,500 pairs was being produced each week, while the development was aiming toward an output of 5,500 pairs of glasses per week beginning in January, 1919. At the date of the armistice the Bausch & Lomb factory had a floor space of 32 acres and employed 6,000 men and women.

The Gundlach-Manhattan Co., which had made camera lenses chiefly, was eventually able to produce 600 pairs of field glasses a week. The Crown Optical Co. was not so rapid in its expansion; and in late 1917 the Navy Department commandeered it and thereafter operated it in charge of Lieut. Commander L. C. Scheibla. Under naval management the output of this factory increased so that the Signal Corps was able to obtain from it about 1,200 pairs of high quality field glasses each week, the plant continuing also to supply the needs of the Navy.

Out of a situation that seemed impossible at the outset the Signal Corps built up an industry within a comparatively few months which provided all the field glasses that were necessary in the operations of the American Expeditionary Forces. Often to keep the optical factories equipped with sufficient workmen the Signal Corps obtained the furlough of drafted men with experience in this line so that they might go to work making field glasses.

All Army organizations except Artillery were supplied with a six-power glass having an angular field that took in a view 150 yards wide at a distance of 1,000 yards. The glasses were of the prismatic type with individual focus for each eye. Each glass was provided with a leather carrying case and shoulder strap. On the top of the case a compass was mounted.

The Artillery organizations were supplied with eight-power field glasses, all of which were purchased in France.

The total requirements of the American Expeditionary Forces for field glasses of the six-power type during the period of hostilities were approximately 100,000 pairs. The total shipments from America were approximately 106,000 pairs.

MISCELLANEOUS SUPPLIES.

The Signal Corps took up with three concerns—the Hampden Watch Co., the Illinois Watch Co., and the Elgin Watch Co.—the matter of providing wrist watches for the Army. A 7-jewel movement was adopted as standard for issue to troops and a 15-jewel movement for sale to officers. A waterproof case was adopted, bearing the serial number of the movement on the outside, the case being so constructed as to require a special tool to gain access to the movement.

The production of wire carts for the Signal Corps did not exceed 25 per year prior to 1917. The demand for these carts, which were hard to build, increased at such a rate that during the autumn of 1918 the matter of procuring them was one of the most serious production problems faced by the Signal Corps.

The Holmes Automobile Co., of Canton, Ohio, abandoned the production of automobiles and in September, 1918, turned over its entire plant to the production of wire carts. Other manufacturers were the George B. Marx Co., of Brooklyn; the J. G. Brill Co., of Philadelphia; the American Instrument & Tool Co., of New York; and the Wesel

Manufacturing Co., of Brooklyn. In all, 721 wire carts were manufactured and 327 shipped overseas.

A total of 2,402 tool chests for the Signal Corps was produced during the war period. The plan eventually adopted was to split up the orders for tools among the various manufacturers and to give the manufacture of the empty chests to prison labor at Fort Leavenworth, where the tools were to be shipped and packed in the chests. This plan, however, required the construction of a special building at Fort Leavenworth, and in the meantime the assembling of tool chests was conducted at the Signal Corps supply depot at Philadelphia and at the port of embarkation. The armistice stopped the construction of the assembling factory at Fort Leavenworth.

The Signal Corps produced a suitable number of gas alarm signals known as strombos horns. This equipment consisted of an alarm horn operated by air pressure acting against a diaphragm and thereby producing a loud and distinct chatter. Compressed air was supplied in small steel cylinders connected to each horn by hose. The air tanks were charged behind the lines from a portable air compressor which could pump into several cylinders at once. The horns were manufactured by the Klaxon Co., of Newark, N. J., the cylinders by the Harrisburg Pipe & Pipe Bending Co., Harrisburg, Pa., and the air compressors by the Ingersoll-Rand Co., of New York.

Flag kits were not used to any great extent by the American Expeditionary Forces, although large quantities of these were produced in this country.

The Signal Corps originally had jurisdiction over all war photography, either of land or air, except for a small amount conducted by the engineers in connection with their own operations; but later aerial photography became a branch of the Military Aeronautics and Aircraft Bureaus. After that the Signal Corps was charged with taking all photographs of historical nature or other interest.

In connection with this work two types of cameras were necessary—still cameras and motion-picture cameras. Later in the war there was being developed a new motion-picture camera which was expected to be the ideal type for use in the field.

It was with great difficulty that a sufficient number of photographic lenses was obtained for the use of military cameras, since the large lens factories of America were tied up with other war orders. A campaign was conducted by the leading newspapers and magazines of the country which resulted in the Government's securing from amateur photographers a large number of high-grade lenses, mostly of foreign manufacture.

The Signal Corps scattered its camera operators broadcast over the country, photographing cantonments and other war activities to the most minute details. These photographs and films were then made public in newspapers, periodicals, and motion-picture theaters throughout the United States, with the result that the people saw with their own eyes how their soldiers were preparing themselves for the defense of the Nation.

An interesting development of war photography was the production of motion pictures showing the training of soldiers. Many pictures were taken to show graphically on the screen the different chapters of the Army drill regulations. These pictures will have a future use to the Government in training soldiers efficiently in the shortest possible time.

The Signal Corps photographers also developed a new kind of history of the war, a history written entirely in pictures for future generations to scan.

PIGEONS.

Although nearly every European army for 40 years has trained the carrier pigeon to be a field messenger, the American Army never adopted the bird until 1917. In a single year the Signal Corps established hundreds of pigeon lofts in this country and overseas and bought and trained more than 15,000 pigeons for service in France. In actual use on the field the pigeons delivered more than 95 per cent of the messages intrusted to them, flying safely through the heaviest shell and gas barrages.

The standard pigeon loft adopted by the Signal Corps had a unique trap arrangement which permitted the entry but not the exit of returning pigeons, and an electrical alarm which automatically notified the attendants of an arrival. Such lofts, however, were of the stationary type and not practicable for use in France. For the American Expeditionary Forces the Signal Corps purchased mobile lofts. It was found that pigeons would come home as well to mobile lofts, which were constantly changing position, as they would to stationary lofts. The first mobile lofts built in the United States were top-heavy, but this defect was overcome by increasing their width and adding heavier wheels. They were all built by the Trailmobile Co., of Cincinnati, Ohio.

Civilian pigeon fanciers were appealed to and urged to breed young birds to stock the Government lofts. The Signal Corps distributed small aluminum bands to be put on the legs of squeakers, as the newly-hatched pigeons are called, which were intended for sale to the Government. The uniform price of $2 per bird was paid, and over 10,000 youngsters were bought for stocking purposes.

Tons of pigeon feed were purchased and shipped to Europe. Some of this grain, such as millet, Argentine corn, pop corn, hemp seed, and Canada peas, was hard to obtain; but nevertheless the supply was well maintained. It was shipped in hermetically sealed containers to prevent it from becoming mildewed.

The American Army copied the French and English models of willow and reed baskets to hold the birds. One type of basket was carried on the back of the soldier and contained small corselets in which the pigeons were securely fastened. Corselets were suspended from the sides of the basket by elastic contrivances permitting considerable joggling without injury to the birds. All of these baskets were made by the A. L. Randell Co., of Chicago.

Message books were manufactured in accordance with a French model. After the message had been written, it was placed in an aluminum capsule which was fitted in a holder of aluminum. This holder was attached to the pigeon's leg by means of aluminum bands. These bands were found to break easily, and pure copper bands were later substituted. The message holders were manufactured by Thomas A. Gey, of Norristown, Pa.

Thousands of items of supply were included in the supply schedules of the Signal Corps. In the following lists some of the more important items are shown, the production indicated in each case being that between April 6, 1917, and November 11, 1918.

Telephone equipment.

	Produced.	Floated overseas.
Batteries, dry	970,171	396,427
Bells:		
Extension	865	470
Vibrating	13,756	12,934
Blocks, connecting	6,500	6,500
Cabinets, wire chief, testing	225	225
Coils:		
Induction	255	50
Repeating	801	801
Condensers	10,205	6,788
Cords:		
For telephones	5,000	5,000
For switchboards	23,539	11,890
Fuses, for monocord switchboard	670,000	341,000
Receiver diaphragms	2,700	2,050
Receivers, telephone	12,950	9,354
Repeaters, telephone	362	362
Staples, insulated	912,300	809,800
Switchboards:		
Camp, 40-line	111	68
Commercial types	304	304
Switchboard, telephone, monotype	14,462	13,264
Telephones:		
Artillery type (W. E. 1375)	66,544	46,123
Camp	38,456	32,668
Commercial types	2,669	1,514
Telephone offices, truck	1	1

Telegraph equipment.

	Produced.	Floated overseas.
Buzzers, service	3,983	3,478
Connectors, stud	8,027	8,027
Disks, cipher	6,157	6,157
Keys	1,830	1,830
Relays	1,672	1,147
Sounders	1,998	1,998
Switchboard, telegraph	1,321	550
Telegraph office, truck	1	1
Typewriters	020	880
Vibroplex, transmitter	470	420

Radio equipment.

	Total produced.		Total produced.
Airplane interphone set	4,263	Ground radio telephone set	527
Airplane radio telegraph receiving set	7,029	Storage batteries	227,139
Airplane radio telegraph transmitting set	3,971	T. P. S. receiving sets	2,510
Airplane radio telephone set	3,186	T. P. S. transmitting sets	1,995
Amplifiers	1,250	T. P. S. two-way set	2,010
Battery charging sets	455	Vacuum tubes	446,818
Ground radio telegraph receiving set	8,052	Wavemeters	8,042
Ground radio telegraph transmitting set	2,637		

Photographic equipment.

	Production.	Shipped overseas.
Chemicals ... pounds	50,723	41,881
Cameras:		
4 by 5 speed graphic	541	283
4 by 5 RB graflex	310	237
6½ by 8½ cyclo graphic	280	249
Miscellaneous still	40	40
Motion-picture	470	145
Paper, photo, all sizes ... gross	21,364	16,364
Do ... rolls	5,186	3,686
Lenses	2,797	696
Tripods, M. P	558	147
Plates ... dozen	68,873	48,873
Holders, plate	28,293	18,293
Film, M. P ... feet	7,500,000	4,000,000
Film, still ... rolls	48,814	28,814

Line construction material.

	Produced.	Floated overseas.
Anchors, guy	17,360	17,120
Belts, lineman's	6,733	5,332
Bolts, lag screws, etc	1,139,648	1,137,928
Braces, cross-arm	123,162	98,440
Brackets:		
Oak	237,000	237,000
Pole, galvanized	13,929	13,929
Transposition	26,905	26,280
Cable:		
Telephone, circular loom ... feet	662,978	396,250
Telephone, lead covered ... miles	80,202	1,696
Cable compound ... pounds	6,160	3,760
Carts, reel	737	327
Clamps, guy	39,250	26,000
Climbers, with straps ... pairs	11,828	10,619
Cross arms	38,500	38,500
Gloves, lineman's ... pairs	849	748
Hangers, cable	51,000	51,000
Insulators, for lance poles	291,124	26,700
Insulators, glass	1,158,836	1,158,836
Insulators, porcelain	2,411,670	1,793,220
Knives, electrician's	264,754	141,920
Knobs, wooden	953,540	400,000
Marlin ... pounds	4,527	1,427
Muslin, cable splicers ... yards	1,251	1,251
Nails, for insulators ... pounds	36,000	36,000
Pikes, wire	1,992	1,728
Pins, cross-arm	807,653	477,400
Pliers, lineman's	193,533	84,642
Poles, lance	209,000	23,685
Reels, breast	6,109	6,108
Sleeves, copper, splicing	146,934	51,934
Solder ... pounds	21,806	14,606
Tape:		
Friction ... do	106,042	90,318
Rubber ... do	20,523	16,713
Terminals, cable	1,290	1,290
Wire:		
Stranded, messenger ... feet	2,470,577	2,470,357
Telephone—		
Copper, bare ... miles	126,654	78,880
G. I., bare ... do	14,411	7,970
Single, insulated ... do	23,950	14,011
Twisted pair, insulated ... do	75,022	45,457
Wire carts	721	327

CONCLUSION.

The reader who has come to this point has before him the picture of the Nation's industry at war—the whole teeming effort in its main outlines, its myriad ramifications, its boundless activity, its ten thousand enterprises, its infinite toil, its hosts of workers, its wonders of scientific achievement, its attainments, even its failures—in short that humming complex of work, planning, ambition, disappointment, triumph, shortcomings, ability, and driving force which was a mighty people concentrated with all of its powers upon a single objective.

It remains now to describe the place occupied by this effort in the whole strategic plan of the war against Germany. We did not go into the struggle as if we expected to fight a single-handed war. Whatever we did either with military personnel or with munitions we did with reference to what the nations associated with us were doing or could do in the same respects. The whole plan was coordinated more or less perfectly, and these international understandings and agreements touched and influenced even the most trivial of our enterprises.

The reader who has in mind the record set down on the preceding pages is now prepared to comprehend the force and extent of the international cooperation in the war and to judge how well America played her part in the general scheme. Let us go back, therefore, and review the history of these agreements.

For many months before America came into the struggle, England, France, and Italy had been engaged in grappling with the scientifically organized forces of German military autocracy. The world war had become a conflict of materials, almost as much as of men. All participants had mobilized their industrial resources in a manner and to an extent undreamed of in times of peace.

The allies had marshaled all available raw materials and factory production in their own lands, and still faced colossal deficiencies in supplies for their military programs. They had been forced to reach out into the markets of the world to meet these deficiencies. They had come to America and placed huge orders for raw materials and finished products. The normal capacity of America's peace-time production had been insufficient to meet their overwhelming needs.

In August, 1914, the total factory capacity in the United States for the manufacture of powder was 6,000,000 pounds a year. In April, 1917, under the stimulation of orders placed by the allies, the capacity had been increased more than sixtyfold. England, France, and Italy were taking this entire production and asking for more.

They had absorbed our entire output. A huge stream of materials, supplies, and ammunition was flowing steadily from America to the front line trenches in France. The allied governments had moulded their military programs in reliance upon the continuation of this source of supply. Their troops were on the front and in contact with the enemy. Failure of supply meant disaster.

The flow of materials from America to the armies in France could not, under any circumstances, be interfered with or curtailed. This fact was promptly recognized by the United States, and the allied governments were assured that America's military program would be formulated and performed without interference with the allied programs of supply from this country.

America's industrial contribution to the war, as a nation, was to be over and above the industrial contribution to the allies then being made by our individual producers. This fundamental plank in the interallied platform of cooperation was laid down at the very commencement of America's preparation, and it was strictly adhered to until the end of hostilities.

A comprehensive cooperative plan for America's industrial participation in the war remained to be worked out. A survey had to be conducted of the new partner's strength and weakness in supply. A determination had to be made of what the allies could give to the new partner, and what they must receive from her. This was done by the Interallied Munitions Council sitting in Paris, by the foreign missions in Washington in conference with the War Department, and by the allied war ministries and Gen. Pershing abroad.

An analysis of the facts of the situation disclosed that:

A. The world over—

(1) There was a critical shortage of ocean tonnage which promised to become more critical as time passed on account of the success of German submarine operations.

B. In France and England—

(1) The output of factories was being seriously curtailed and limited by lack of raw materials and semifinished products.

(2) If an adequate supply of raw materials and semifinished products could be made available, the factories had a substantial surplus manufacturing capacity which could be placed at the disposal of the United States.

C. In the United States—

(1) A surplus of raw materials and semifinished products for transport to France and England could quickly be made available.

(2) It would be impossible, within less than a year, to build up additional manufacturing capacity in the United States sufficient to supply a large army.

The lack of ocean tonnage was recognized by all as the vitals of the problem. France, Italy, and the United States had comparatively little merchant tonnage. England's vast tonnage was suffering rapid depletion by submarine losses and was totally inadequate to meet allied needs. Ships were the biggest single deficiency in the interallied program.

The cooperative industrial program of the Allies and the United States had to be geared into the shipping problem. To do this the determination of what materials should be shipped from the United States had to be decided first on the basis of what economies could be effected in shipping space. If raw materials for aircraft occupied less cargo space than the finished product, the maximum utilization of available tonnage demanded the shipment to France of these raw materials to be made into the finished product there. If, on the other hand, finished nitrocellulose powder for artillery shell propellants, or finished picric acid for artillery shell explosives, occupied less cargo space than the raw component materials used in their production, the shipping shortage demanded manufacture of these explosives and propellants in the United States. Not a single ship could be freighted with an extra pound or cubic foot of cargo which by any effort could be saved.

The French Mission in the United States early recognized this fact and urged the manufacture in the United States of picric acid to be used as explosive in 75-millimeter and 155-millimeter shell, pointing out that the finished product occupied but one-nineteenth as much cargo space as the raw materials.

Gen. Pershing recognized the point, and in August, 1917, cabled as follows:

> A joint French-American commission has examined the question of the production in France of powders and explosives and reports as follows: France must import by December 4 the greater part of the raw materials used in the manufacture of powders and explosives. The weight of raw materials required is 10 to 20 times the weight of the finished product. The shipping situation is such that by December the output of France will be limited by the amount of raw material produced in France or easily obtainable. * * * The present outlook is that in December the French output will not be more than half of the present output. To avoid calamity the United States must not only furnish powder and explosives for all of its own forces but must supply about half of the French requirements. It is therefore recommended: (A) that the United States Government furnish all powders and explosives needed for present contracts with French Government; (B) that the United States Government prepare to furnish by December 300 tons per month of explosives and 200 tons per month of powder for French consumption; (C) that study be immediately commenced for the purpose of adapting American powders to French cannon of different types, this study to be made both in the United States and in France by competent experts; (D) that the French Government put at the disposition of the American Government competent experts both in the manufacture and use of these powders in the guns * * *.

Subsequent computations made on this side of the ocean indicated that in the case of picric acid and other explosives this ratio between raw materials and finished product in bulk was too great, yet in principle these computations did not affect the desirability of shipping the finished product rather than the raw materials.

Again, Gen. Pershing cabled to the Chief of Staff in the United States urging the purchase of completed artillery, artillery ammunition, and airplanes abroad, in order that "saving of tonnage" might be effected, and pointing out the saving of cargo space resulting from the shipment to France of raw materials instead of finished products, saying:

Following is comparison in tonnage of the principal manufactured articles of ordnance obtained in Europe and the replacements in raw materials contracted for the same. All tonnage ratios shown are in favor of raw materials:

Field Artillery guns	1 to $7\frac{1}{2}$
155 millimeter howitzers and ammunition	1 to $1\frac{1}{4}$
8 millimeter ammunition	1 to $3\frac{6}{10}$
Trench mortars	1 to $12\frac{1}{2}$
Grenades	1 to 4
In airplane production:	
Packed airplanes, in weight	1 to 2
Packed airplanes, in cubic capacity	1 to $2\frac{1}{2}$
Packed airplanes in area covered by boxes on board ship	1 to 9

In the above comparison in the ammunition item, finished explosives are regarded as raw materials.

The Interallied Munitions Council, sitting in Paris and containing among its membership the best military and industrial brains at the command of the allied cause, including Gen. Pershing, Gen. Robertson, chief of the imperial general staff of Great Britain, and Gen. Foch, then chief of the general staff of the French Army, came to the same conclusion, and Gen. Bliss transmitted its findings in a memorable cable, a part of which was reproduced in the preface to this report.

Every mind was in accord. Tonnage must be saved. It could be saved and in vast amounts by calling upon the United States to supply the raw and semifinished materials, and upon the French and British war factories to utilize these raw and semifinished materials in the manufacture of the finished products.

But could this solution of the vital shipping question be dovetailed into the industrial situations of the various nations concerned? Could the United States supply the essential raw and semifinished materials in quantities equivalent to the amounts consumed in the manufacture of the finished product? Did the French and British factories, with these materials laid down in their yards, have available a sufficient manufacturing surplus to supply the needs of their own armies and also to produce in part for the armies of America?

The foreign missions were in Washington. They knew intimately the economic and industrial situations in their respective countries; they knew the military plans of their general staffs; they knew in what respects their programs of supply for their armies in the field needed assistance, and in what respects these programs could be met or exceeded. With this information available, they were prepared to furnish the answer as to the manufacturing capacities of allied Europe.

The British War Mission in Washington communicated to the War Department a cable from the British minister of armament, setting out the position of the British Government on reciprocal supply:

The British Government is willing as far as possible in matters of urgency to manufacture for use of the Americans any products necessary to the more speedy equipment of the Americans that the Americans consider they can obtain in England more promptly or better than in the United States. Furthermore, the situation as to manufacture of steel products is better than it has been. The British Government will help to its utmost ability without making actual and immediate replacement of raw material an indispensable condition when any order is given. On the other hand the general principle of replacements of raw materials as soon as possible should be observed. It has become more a question of furnishing supplies promptly to the allies than a mere question of replacing what has been furnished American troops; in other words, the needs of the allies should be considered as one, and England should manufacture for the allies anything that is necessary or best got that way, and America should in the general interest of the allies furnish as soon as convenient raw material to replace that used. * * *

Writing to Maj. Gen. Crozier, Chief of Ordnance, the French high commission urged the placement in France of orders for artillery and artillery ammunition and pointed out the existence of surplus factory capacity available for their production. The commission summarized the industrial situation in France in the following language:

Even in such remarkable technical conditions as yours, it takes time to realize such a program, to organize manufactures and to have men to direct them. You will take less time than we did in France, where the output of big guns was not adequate to our needs before the end of 1916. But time—more or less—had to be an essential factor, so that after careful consideration, it has been found that the only plan to be carried out in order to supply the first American divisions with material on their landing in France was to avail ourselves of the surplus capacity of production of the French factories, which had been since the beginning of the war very powerfully equipped and were able to turn out greater quantities than those corresponding to our supply of raw material.

The allies could deliver the artillery, artillery ammunition, and airplanes if America could deliver the raw and semifinished materials. America answered that she could and would produce and transport to Europe raw materials and semifinished products in amounts equivalent to the amounts consumed by allied factories in manufacturing the completed guns, shell, and airplanes.

The details remained to be worked out. The French high commission submitted statements showing the amounts of each component material consumed in French factories in the production of guns and ammunition of the various calibers. There were to be supplied by America 6 tons of steel for each 75-millimeter gun, 40 tons of steel for each 155-millimeter howitzer, and 60 tons of steel for each 155-millimeter gun, and proper proportionate amounts of necessary materials used in the manufacture of artillery ammunition.

The program of industrial and economic cooperation between the United States and the allies thus took form. It used in the most efficient manner every nook and cranny of every available ship. It utilized to the utmost the surplus manufacturing capacity of France and England. It brought into the war at the earliest moment the resources of America in raw and semifinished materials. It spanned the period during which America could go forward with her gigantic mobilization of manufacturing power and later convince the Central Empires of the futility of further struggle.

With the program mapped out, reciprocal agreements for supply remained to be made. Orders were promptly placed.

The United States ordered from France a total of 5,854 pieces of field and trench artillery of various calibers, of which 3,834 were delivered to the American Expeditionary Forces prior to the armistice.

By August, 1917, more artillery ammunition was on order with the French Government than was fired by the American Expeditionary Forces from January 18, 1918, when the first complete American division entered the line, until November 11, 1918, when the end of hostilities was announced to the world. Of the amount ordered 10,000,000 rounds were delivered before firing ceased.

In aircraft equipment, the French factories also had a surplus capacity and delivered to Gen. Pershing up to November 11, 1918, a total of 4,881 finished airplanes.

By the terms of our agreement with the French Government, America obligated herself to supply the raw materials and component parts of the finished products delivered to our forces in France. This agreement America performed twice over. For every ton of raw materials and semifinished products America agreed to furnish to France, she furnished two tons. According to French statements, our replacement obligation in raw materials was 350,000 tons. America furnished over 800,000 tons.

In exchange for the artillery and artillery ammunition of French manufacture fired by Pershing, America supplied to France in metals alone over 700,000 tons of steel, 30,000 tons of pig iron, 5,000 tons of brass and spelter, and 50,000 tons of copper.

In addition, and for use in the artillery ammunition received from French factories, America manufactured and supplied to France in a finished state all the principal materials used in loading all shell delivered to the American Army. These materials consisted of smokeless powder, used as a propellant to drive the shell from the guns, and of picric acid, used as a high-powered detonative to burst over the enemy lines. The French used 12,000 tons of smokeless powder in our shell. America delivered an equivalent amount of finished powder. The French consumed 18,000 tons of picric acid in loading shell for American use. America supplied 18,500 tons.

In exchange for the finished airplanes, again America supplied the raw materials and component parts. For the framework of the French planes driven by American aviators, America furnished 34,500,000 feet of spruce, fir, and cedar, enough to manufacture over 16,000 finished planes; for the propellors, America furnished 7,000,000 feet of mahogany and walnut, enough for 40,000 propellors; 4,000 tons of aluminum, enough for thousands of planes; and dopes for painting airplane wings, and miscellaneous aircraft materials and supplies far in excess of the number of finished planes delivered to Gen. Pershing. Under special contract made in August, 1917, and in addition to the above, America furnished to France all materials for 5,000 finished planes and all parts for 8,500 finished airplane engines, which were to be assembled in France for the American Expeditionary Forces. The engine parts were in forgings and needed only to be machined. For the use of the French Government in machining these engine parts, America built and delivered the necessary equipment and machinery.

Thousands of additional smaller items of all kinds were supplied by the various governments to each other from day to day. No deficiency in the military programs of any of them was permitted to exist, if it could be made good by any of the others.

All of America's vast contribution to the allied program of supply was not only produced in America, but it was taken to France in army transports. From August, 1917, to November 11, 1918, an average of 2,000 tons of American materials for French factories left American ports every day aboard American army transports. Through a submarine-infested ocean, in which the Germans sank over 21,000,000 tons of dead-weight shipping, these materials were carried in army transports manned by American crews, and laid down at the doors of French factories.

By February, 1918, Gen. Pershing estimated that 2,000,000 tons of cargo space had been saved by the adoption of this program of international and reciprocal supply, a saving of more tonnage than was then available for the use of the American Expeditionary Forces.

The Franco-American commission on explosives estimated a reduction of 75 per cent in cargo space in the shipment of explosives alone.

So the silent drama of international cooperation was carried out. The story of British and American mutual aid during the war is the same story in substance as that of Franco-American cooperation, with changes only in the figures. Economy of shipping was effected. British and French factory capacity was utilized. The vast reservoir of American raw materials and explosives was thrown against the enemy. International cooperation on a scale and in a spirit of cordial, mutual helpfulness, such as the world had never dreamed of, helped to equip 2,000,000 American soldiers in France.

And it was done, all of it, without curtailment of the huge stream of material which was flowing from America to the allies when the United States entered the war. France and England received ever-increasing quantities to the last day. The more than 800,000 tons of replacement materials for artillery, artillery ammunition, and airplanes delivered to America was over and above the millions of tons secured by the allies for their own use directly from American producers.

It was partly by reason of the adoption of this program and its complete performance that Gen. Pershing, after the armistice, could say:

> During active operations extending from January, 1918, when our first division entered the line, until the close of hostilities on November 11, our troops were supplied with the equipment and ammunition necessary to carry their work to a successful conclusion.

Beyond all this, our Government, as part of the interallied program, created vast facilities for the manufacture of supplies which England, France, and Italy still required for their own needs and which a comprehensive consideration of the entire program, with particular reference to shipping, showed could be best produced in this country. Factories for the production of immense additional quantities of picric acid, powder, and other materials were built by our War Department to fill the deficiencies in the military programs of our associates in the war.

And beyond and behind all this America went forward with her own gigantic preparations for the conquest of the dark forces which threatened world civilization. It was this mobilization of her might almost as much as the effect of her immediate force which helped to convince the German general staff of the futility of further resistance and assisted to bring the war to an early end.

○

Made in the USA
Monee, IL
05 November 2026

67886783R10449